ASTRONOMY FROM LARGE DATABASES

SCIENTIFIC OBJECTIVES AND METHODOLOGICAL APPROACHES

Garching, 12–14 October 1987

PROCEEDINGS

Edited by F. Murtagh and A. Heck

January 1988

ESO Conference and Workshop Proceedings No. 28

EUROPEAN SOUTHERN OBSERVATORY
Karl-Schwarzschild-Str. 2, D-8046 Garching bei München
Federal Republic of Germany

ISBN 3-923524-28-5

Foreword

The rôle of databases is increasing daily in astronomy. So too is the need for the modern astronomer to be up to date on applicable methods to use on large collections of data. The need for focussed statistical analysis and the need for experimentation in knowledge based approaches are both indirectly related to the greatly growing quantities of astronomical data. The latter areas — the *scientific objectives and methodological approaches* part of the title — were the initial aim of this conference. But it is clear also that there is an urgent need to coordinate technical database issues. In both of these aspects, the papers in this Proceedings Volume will be found to provide valuable introductory and reference material.

Approximately 150 were in attendance at the Conference held on the premises of the Physics Department of the Technical University of Munich. The extensive surroundings and infrastructure kindly provided by the Technical University allowed the type of discussion and debate which is vital for the fields of database research and of related data analysis in astronomy.

In the three days of the conference, 74 communications were presented. The written versions of these presentations are presented here. The promptness and timely professionalism of the authors is appreciated.

Among those who worked hard for the success of this event was Hans–Jürgen Kraus. He ensured the essential logistics of the 3–day link–up to the nearby European Southern Observatory headquarters. Christina Stoffer was on the Reception Desk, as was Chantal Bruneau, and Ulrike Kurz who provided valuable travel assistance when necessary. Britt Sjöberg bore more than a key part in Conference arrangements over a period of many months: in processing the successive mailings, handling accommodation, arranging hotel/conference bus details, and throughout the conference itself, she was indispensible. The success of the Conference certainly owes much to her.

The technical preparation of this Volume was in the expert hands of Kurt Kjär. We thank him for his helpfulness and care.

Fionn Murtagh (Garching),
André Heck (Strasbourg).

INTRODUCTION

Ladies and Gentlemen,

When we started considering the organisation of this Conference, the launch of the Space Telescope, a major producer of astronomical databases and a central subject in our original Conference programme, was about two years ahead and we were thinking that it would have been appropriate to meet and discuss the issue of large archives at the time ST just started collecting data.

Now that we meet for real at the Conference, because of well known unfortunate events, ST is still about two years ahead. Nevertheless we decided to hold the Conference as initially planned with the consideration that its subject is indeed ripe for discussion; and it is so, in my opinion, because of two main reasons. On one side more and more astronomical space experiments are designed and operated as proper observatories: they produce a vast amount of data which are characterized by a high degree of uniformity because the instruments are well defined and the observing modes are limited in number, they are out of reach of astronomers wishing to "improve" their performance, and their efficiency is well monitored. All this leads to the build-up of a controlled archive of calibrated data which, contrary to what is still the case in ground based astronomical archives, can indeed be re-analysed by different scientists pursuing different scientific aims. Concurrently, on the other side, recent developments in both computer hardware and software are transforming these otherwise static sets of data into a live, flexible and readily accesssible tool for everyone's research: mass-storage technology, particularly optical disks technology, is becoming affordable and allows almost real-time random access to large archives, computer networks allow the concept of distributed databases to become a reality and finally advanced software packages, e.g. Database Management Systems and Expert System Shells, show promising results in finding a friendly optimum path through a rather intricate "forest" of different tools and data.

It seems therefore that it is the appropriate time to openly discuss in this forum how and to what extent Astronomy can benefit from all of this. And I do hope that in few years we can reassess the situation, at that time putting also the ST gigabytes into context.

However, before introducing the first invited speaker of this morning, let me conclude my welcome message with a word of warning. We are here to discuss what kind of Astronomy and Astrophysics can be extracted from large databases and related tools: but if I scan through the abstracts' booklet I still find too many "technical" contributions and too little science. Possibly it is correctly so, after all we are just starting using these new techniques and I may be expecting too much; however I know by experience that in this game it is just too easy to be carried away by the "system" and its fancy tools/toys and sometimes one has the impression that the work is finished when the puzzle is completed and the "user interface" responds in as friendly a manner as it was defined in the "requirements".

Let's try not to forget that our task is the advancement of scientific knowledge and if at the end of the day we have not learnt something new in science, all technological effort will have been wasted.

P. Benvenuti

TABLE OF CONTENTS

SCIENTIFIC COMMITTEE

P. Benvenuti	(ST-ECF, Munich)
P. Grosbøl	(ESO, Munich)
A. Heck	(CDS, Strasbourg)
C. Jaschek	(CDS, Strasbourg)
M. Johnston	(STScI, Baltimore)
J.L. Mead	(GSFC, Greenbelt)
F. Murtagh	(ST-ECF, Munich)
B.G. Taylor	(ESA, Noordwijk)
U. Zimmermann	(MPE, Munich)

DEMONSTRATIONS

Intelligent Text Management System (Information Access Systems, Inc.)

KEEconnection (IntelliCorp Inc.)

STARCAT and SIMBAD (ST-ECF and CDS)

LITERATURE DISPLAYS

European Space Agency

European Southern Observatory

D. Reidel Publishing Co. (Kluwer Academic Publishers)

Springer Verlag

BUSINESS MEETINGS

Working Group for Modern Astronomical Methodology

Technical Committee 13 (Astronomy and Astrophysics), International Association for Pattern Recognition

IAU Commission 5, Working Group on Astronomical Data

Astrophysics from Large Databases

A MULTI-SPECTRAL ANALYSIS SYSTEM USING LARGE DATABASES

Edward W. Brugel, Gitta Domik and Stephen A. Voels
Center for Astrophysics and Space Astronomy
University of Colorado
Boulder, CO 80309 (USA)

ABSTRACT.

The Center for Astrophysics and Space Astronomy (CASA) at the University of Colorado is in the process of designing, testing, and implementing a multi-spectral data analysis system that incorporates several large astronomical datasets (in particular *IUE*, *IRAS*, and *HEAO2*). The system provides interactive scientific analysis and intercomparison of multi-mission/multi-spectral data from satellite and groundbased observatories. The design of this system is based upon the experience gained from serving as one of two Regional Data Analysis Facilities for the International Ultraviolet Explorer (*IUE*-RDAF); and on the experience gained by several local scientists carrying out multi-spectral research programs.

1 MULTI-SPECTRAL ANALYSIS CENTER CONCEPT

It is well accepted that broad scientific benefit can be derived from studying celestial objects at a wide range of wavelengths. Often, it is the only recourse to develop a full physical model of an astronomical source. The list of astrophysical problems that are best treated with multi-spectral analyses is nearly as long as the list of all astronomical research objectives today. The availability of large archives of astronomical data, coupled with the research interests of local astronomers, are the major motivations for us to develop a multi-spectral analysis capability.

A guiding principle in the design of our multi-spectral analysis center is to make it possible for a user to sit at a single terminal; locate appropriate files in separate archives; access those files; incorporate the user's own groundbased observations; carry out reduction and data correction programs; convert the data to a mutually compatible format; and then merge the data into a coherent system appropriate to the scientific task. In this manner, an astronomer could locate and analyze all the relevant data on a particular object or a particular area on the sky, and effectively and efficiently carry out a study that otherwise might require independent data analysis efforts at several geographically separated centers.

The analysis of archived satellite data produced by individual missions implies a dependence upon these missions for general data management, and for supplying calibrated data corrected for major instrumental signatures. We recognize and encourage the efforts of the separate satellite projects and their associated centers. We are not attempting to re-do the "first" level data calibration and data management efforts; our facility is primarily concerned with "second" level data reduction and scientific analysis. A key ingredient in this process is the *integration* of data from different observatories (satellite or groundbased). Much of the overall system has been designed and several individual components have been implemented and tested.

Each of four fundamental components to our multi-spectral analysis facility is discussed in the following sections: (1) the available archived datasets; (2) the computer hardware; (3) the analysis software system; and (4) the database management system used to facilitate access to the data.

2 AVAILABLE ARCHIVED DATASETS

The analysis of both one-dimensional spectral data and two-dimensional images from the *IUE*, *IRAS*, and *HEAO2* archives is the initial goal of our facility. Incorporating user supplied groundbased optical and radio data is also part of the initial design. Access and intercomparison of the following large datasets are currently available: *IUE* - Condensed Archives; *IRAS* - HCONs, LRS, AO; Groundbased optical - user data; and Radio - user data.

Both of the satellite datasets reside locally on either 6250 or 1600 bpi tapes. Eventually the datasets will be transferred to optical disks for on-line access. If access to these and future astronomical data archives become easily available via computer networks, then it will not be necessary to have the data stored locally. At present, the satellite observatories are far superior to groundbased facilities in providing access to their archived databases. Although the vast majority of data acquired at groundbased telescopes is digital, and thus easily archived, these facilities have generally ignored the possibility of establishing an accessible data archive. Hopefully this oversight will be remedied soon.

Although the *HEAO2* data were acquired over seven years ago, and we had requested copies of the full archives more than a year ago, we have not as yet received the HRI or IPC data. We have been assured the complete dataset would be available in the near future. Nonetheless, we do have a small sample of the *HEAO2* image data which we have successfully accessed and analyzed. Data from future missions (such as HST, ROSAT, Lyman, etc.) will be included in the system as they become part of the public domain.

3 HARDWARE CONFIGURATION

The requirements for a multi-spectral analysis facility dictate the general computer hardware configuration. These requirements are: (1) a multiple user environment; (2) an image processing capability; (3) a large data storage capacity; (4) and interactive (not batch) operations. Thus, the computer system must supply relatively rapid processing of the data (fast CPU) and quick data transfer from storage devices to display screens (fast I/O). The size of the datasets involved also dictate the need for several large storage devices. A number of distinct computer systems would meet these computational requirements. However, we have added requirements for (1) compatibility with past software develop efforts (the *IUE*-RDAF spectral analysis package represents significant programming effort); (2) availability of existing hardware; (3) the capacity for future expansion; (4) and the cost of the system must fit within the limited budget of a moderate size university department. Based upon these requirements, we have acquired the computer system listed in table 1.

Table 1 - Hardware Configuration

CPUs	Data Storage Devices	Terminals/ Image Display	Other Peripherals
2 MicroVAX IIs	456 MB Disk Drive (x2)	AED 1024 Color Display	TK50 Tape Drive (x2)
16 MB memory	159 MB Disk Drive	I^2S IVAS Image Display	Kennedy 9100 (x2)
13 MB memory	71 MB Disk Drive (x2)	Matrix Instruments Camera	Kennedy 9610
	2.2 GB Optical Disk	Graphic Terminals (x27)	Laser Printers (x2)

One MicroVAX II is primarily an image analysis computer and operates as a single user system for image processing. The second MicroVAX II is a multi-user system, whose main task is the analysis of vector data. The system of two CPUs operate as a Local Area Vax Cluster. In addition to our primary hardware system, we are also connected via a local area ethernet to an Alliant F/X. The latter machine is used for AIPS processing of VLA radio data. The reduced data, in the form of two-dimensional maps, are transferred to our system for further analysis and comparison with images from other datasets.

4 ANALYSIS SOFTWARE SYSTEM

In order to give the individual researcher control over the data reduction process, we have designed the analysis software system for interactive processing. This is achieved through the continuous graphic display of the data during each processing step; and the requirement that the user make specific decisions on how the processing should proceed. Following our experience with the *IUE*-RDAF, we have chosen the *Interactive Data Language* (IDL) for most of our software development. In addition, we have several "standard" astronomical data analysis packages available for our users (e.g. IRAF, AIPS, VISTA, FIGERO).

The data analysis software system involves three major areas of programming: (1) Spectroscopic analysis system; (2) Image analysis system; and (3) the combination and correlation of separate databases (or Input/Output system). It is the correlation and combination of separate datasets that make this system distinct from a single instrument (or wavelength region) reduction package.

4.1 Spectral Analysis Subsystem

The spectral analysis subsystem is concerned with the handling of vector data representing spectral information. This is the simpliest of the three subsystems and the one most nearly completed. Essentially we have converted the *IUE*-RDAF spectral analysis system into a wavelength independent spectral data reduction package. The only required inputs are calibrated wavelength and flux vectors. Determining these vectors is a process often unique to a given instrument and is done outside the spectral analysis subsystem. Separate reduction packages for several instruments (such as *IUE*, CCD long slit spectra, etc.) already exist. The spectral subsystem includes all of the general math and fitting routines that are used on spectral data; for example, line identifications; equivalent width and integrated flux measurements; Guassian fits; continuum comparisons with Planck functions, Kurucz model atmospheres, exponentials and standard stars; cross correlations; and smoothing functions.

4.2 Image Analysis Subsystem

Since image data come in more "flavors" then spectral data, the image subsystem is considerably more complicated then the spectral analysis subsystem. The information contained within a multidimensional astronomical dataset may be an *image* (RA-DEC in the x-y direction and flux in the z-direction); or spectral data recorded on a two-dimensional detector that must be extracted and then sent to the spectral subsystem.

In the area of *image* data our programming effort has focused on general routines for data extraction; two dimensional display; and connecting the image data to the database system for object identification. Routines have been developed which apply to any image dataset and include: image display with a variety of color tables; applying image scaling and stretching in order to enhance contrast; provide area flux integration for various software controlled aperture sizes and shapes; overlay images with coordinate grids and/or contour plots; arthematic manipulations - co-add images, divide or subtract background images, image rotation, shifting images; smoothing techniques; cursor controlled interactions - identify objects, obtain RA-DEC and flux values from image positions. Other general purpose utilities under development are Fourier filtering techniques and maximum entropy image reconstruction.

We have also considered special techniques that must be applied to particular datasets. For example, correcting for the unique background in the released *IRAS* sky flux images or applying time series analysis to *HEAO*-2 IPC data.

For another special case, we are acquiring the IUESIPS software used for creating merged spectral data from the raw two-dimensional image.

4.3 Data Input/Output (IO) Subsystem

The main purpose of the data IO routines is to provide uniform data formats; a system for connecting image data to the catalogs in the database management system; and a means of communicating between various reduction packages (e.g. IRAF, SDAS, DAOPHOT, STARLINK). The following specific tasks are part of the IO subsystem. For example, in order to simultaneously display an *IRAS* image, an Einstein IPC image and a radio map, each dataset must be reduced to the same coordinate and flux system. In order to identify an object based upon a given cursor position (RA-DEC) on an image, it is necessary to connect the image coordinates to a large number of astronomical catalogs contained in the database management system. As an example of communication between reduction packages, an IDL program has been written to access and control the DAOPHOT analysis software.

5 DATABASE MANAGEMENT SYSTEM

There are two distinct types of astronomical databases which must be managed; (1) catalogs of information on astronomical objects (such as the SAO catalog); and (2) datasets of astronomical observations (such as the *IRAS* HCONs). In the latter case it is primarily important to ascertain what data is available and where is it stored. This information is obtained by accessing the catalog of observations for each particular dataset. The following requirements characterize a database management system (DBMS) used to retrieve data from astronomical catalogues: (1) expandability - a growing number of large data bases have to be managed; (2) capability of specifying various combinations of retrieval operators (e.g. all *IRAS* point sources with $F_{60\mu} - F_{100\mu} > x$); (3) fast access of the data.

Nearly all of the DBMS developed in recent years follow the relational model, in which data are stored in the form of tables. The tables are broken down into "tuples" that can be described as physical records. In our case each tuple would describe characteristics of one identified star. A tuple can be seen as a row in a table; attributes describe the columns of tables.

Relational algebra is a list of operators used to manipulate the search through tables. Thus relational operators extract a subset of all available rows, and a subset of all available columns to generate new tables. Operators for data retrieval appear as follows:

SELECT attribute(1) ,..., attribute(n) FROM table WHERE attribute(i) LO value(s),
where LO stands for "logical operator", e.g. $>, <$ or $=$.

Specification of such operators is application dependent; e.g. astronomers typically search within tolerance values of certain sky positions, therefore "tolerance" should be defined as a logical operator.

The *IRAS* point source catalogue was chosen as the first test case. The sheer number of entries (tuples) in that catalogue (app. 246,000) and the additional reference entries to each point source requires a fast solution to the retrieval problem. A prototype version of the DBMS was implemented in FORTRAN as a first step to anticipate difficulties and specify definitions for the overall system. The next step will be to test the performance of currently available relational database shells before a decision for the final implementation is made. The DBMS shells under consideration are: DEC VMS RDB, Oracle and Ingres.

The ultimate goal of the database system is to provide quick access to a large number of astronomical catalogs and dataset through a single search request. For example, a single request for an RA-DEC will produce all pertinent data from the SAO catalog; all *IUE*, *IRAS*, and *Einstein* observations; and additional catalogs as they become available. At present, only the *IUE* merged log of observations has a functioning data retrieval system.

6 FINAL COMMENTS

At present the above designed system is not fully integrated. There are separate analysis sections for *IUE* data, *IRAS* image data (HCONs), *IRAS* LRS data, *HEAO2* IPC data, *SMM* UVSP data, optical CCD images (including DAOPHOT) and radio data using the AIPS system. Combining these separate packages will be the primary effort during the next months. A fully implemented catalog access system is also not yet available; however the preliminary design has been tested and several catalogs can be accessed in a "user unfriendly" manner. We expect to have a major portion of the multi-spectral analysis system completed within the next year. Since the analysis system was designed as separate modular packages, productive research efforts are possible even during the final implementation stage. A large number of local scientist (≈ 40) are currently analyzing *IUE*, *IRAS*, *HEAO2*, groundbased optical and VLA data using parts of the system described herein.

SCIENTIFIC ANALYSIS OF THE IRAS DATA BASE AT IPAC

T. Chester and G. Helou
Infrared Processing and Analysis Center
California Institute of Technology

ABSTRACT

The original IRAS data base contains 50 Gbytes of detector data tagged with sky position. The processed data base includes 8 Gbytes of digital image data and 0.06 Gbytes of catalog data. Over the last several years at IPAC, we have developed a data system which allows rapid and easy analysis of this data base as well as other astronomical data bases. In this talk, we summarize the IRAS data base, describe the central features of our data system, and analyze how several major astronomical discoveries were made using the data base.

I. INTRODUCTION

The IRAS data base was the first "large" astronomical data base covering the entire sky produced in a uniform manner from one instrument in a relatively short period. The IRAS Point Source Catalog of 250,000 sources was matched only by the SAO catalog of 250,000 sources which was obtained from several different sources over a significant period of time. (The Durchmusterung catalogs have only been recently available as a data base as opposed to printed form.) The IRAS All-Sky Images are the infrared analog of the Palomar and ESO/SRC optical sky surveys, but their value is increased by their availability on digital tapes. Since the IRAS data base includes both the Point Source Catalog and the All-Sky Images, it is the prototype of the large astronomical database.

In this paper, we describe the IRAS data base in Section II. The analysis system for interrogating this data base is described in Section III. In Section IV we analyze the relationship between some of the scientific discoveries made using the IRAS data and the data base itself.

II. THE IRAS DATA BASE

The IRAS data base is summarized in Table I.

Table I. THE IRAS DATA BASE

Product	Description	Size (Mbytes)	Actual No. 6250 Tapes
Raw data	Detector data tagged with sky position	50,000	600
Explanatory Supplement	Detailed description of hardware and processing	2	-
Point Source Catalog (PSC)	245,889 point sources	44	1
LRS Spectra Catalog	Spectra of 5,426 PSC sources	5	1
Zodiacal History File	Time-ordered averaged data at 0.5° X 0.5° resolution	96	1
All-Sky Images	3 sets of 16.7° X 16.7° fields with 2' pixels	3,800	78
Additional Observations	13,853 images with 0.25-1.0' pixels, each covering ≈1 sq. deg.	3,800	66
Small Scale Structure Cat.	16,740 sources of size 1-8'	4	1
CPC data	1508 Chopped Photometric Channel images	30	1
Asteroid and Comet Data	IRAS and derived data on 1811 known asteroids and 25 comets	30	1
Serendipitous Survey Cat.	43,866 point sources derived from the Additional Observations	8	1

III. THE IPAC ANALYSIS SYSTEM

The Infrared Processing and Analysis Center (IPAC) at Caltech was designed as a facility for astronomers in the U.S. to fully utilize the IRAS data archive. The immediate predecessor of the IPAC was the IRAS Scientific Data and Analysis System (SDAS). The SDAS had been responsible for producing the primary mission data products of the IRAS mission. The IPAC was initially set up in 1985 by converting the people and software of SDAS to a user-oriented facility as opposed to a production-oriented facility. Since we had only 6 months to make the conversion, of necessity we began with a rudimentary analysis system that was not of one design.

Over the next two years, the experience that we gained gave us the insight to know how to design a more powerful and more friendly system. Thus we purchased a new computer and set ourselves the goal of designing and developing such a system. We took delivery of a CDC Cyber 180/850 in January, 1986 and proceeded to implement our design. As of November, 1987, the Phase I development of that system was finished and became operational.

This section describes the IPAC data system as it exists now and gives our plans for the future.

III.1 Design Goals

The requirements for our data system, in order of priority, are given below. Each of them is then discussed further. The requirements are:

1. Provide readily available astronomical and data system expertise to visitors.
2. Provide a responsive and powerful system for data analysis, including both catalog and image data. The basic system should be project and application independent, should provide syntax free dialog, allow remote accessibility, and provide data access with minimal operator involvement.
3. Provide compatability with analysis systems at other observatories to minimize visitors having to learn new systems.
4. Provide publication quality graphics.

The most important characteristic of a data system, requirement 1 above, is the human expertise that accompanies it. Without the expertise to answer sophisticated user questions, even the best data system is seriously compromised. This point is well-known, but always bears repeating.

The heart of the requirements is embodied in requirement 2 above. Each of the ingredients in a responsive and powerful system will now be discussed separately.

A powerful system must accomplish within a reasonable time period every reasonable task that it is given. Since the expectations of what constitutes a reasonable task changes with every new generation of computers, this implies a periodic upgrading of the basic computer system. The basic computer system itself generally needs to be a mainframe rather than a less powerful computer. However, as long as supercomputers are relatively rare, a supercomputer is not required.

A responsive system is one which never leaves the user hanging or which requires the user to perform contortions in order to get his tasks done. Thus the system must respond immediately to all commands by doing <u>something</u>, even if it only letting the user know that he must wait 30 seconds before the task at hand will be finished. Waits of several seconds to respond to commands must be eliminated.

Analysis tasks are typically composed of many separate pieces. For example, a catalog analysis task might consist of source selection by constraining one or more parameters within a catalog, a comparison to other catalogs, and then one or more plots involving any of the previous information.

A responsive system must make it easy to perform all the separate pieces of any analysis task. One good way to ensure this is to modularize the individual tasks (such as searching, comparing, and plotting) and set up a superstructure that makes it easy to pass data to and from the individual tasks. That is, all data analysis tasks should be well integrated.

To prevent redesigning the wheel every few years, it is important to make the basic data analysis system project and application independent. After all, most analysis tasks are not project specific. The examples above of searching, comparing, and plotting are general. Thus there is no reason to set up the data analysis system in such a way that it will only work with the data or task at hand. Of course, some application specific software will always be required, but it can be minimized by careful design.

The user interface to the data system should be made as simple and easy to use as possible. The user dialog with the system must avoid esoteric commands and should be syntax free. The data system must be molded to speak a human language instead of forcing humans to speak computerese.

In this age of telecommunications, any system for data analysis must not only allow remote access but must be designed so that all of its features are accessible to any remote user up to the limit of his particular terminal. This allows the data analysis system to be used by a much larger community, as well as permitting local users to access the system from the location of their choice, such as their office, home, or someone else's office.

Finally, a responsive and powerful system must provide data access with minimal operator involvement. After all, if you are accessing the system from home at an off-hour, you do not want to have to drive over to the computer to mount a tape. Thus the data system must offer a large amount of on-line storage, as well as the capability to automatically retrieve any other data that may be required.

Inside a responsive and powerful system will lurk tasks that are common to many different data analysis systems. It will obviously benefit any user who moves between several different systems if there is as much commonality as possible between them. Thus requirement 3 above demands compatability as far as possible. This must include major astronomical image processing systems such as IRAF and AIPS, and may eventually be expanded to include operating systems such as UNIX.

The end goal of any data analysis system is usually publication quality graphics. Thus requirement 4 above recognizes that, and demands that the data system be able to supply it.

III.2 Implementation

At IPAC, we have a staff of 6 Ph.D. astronomers who are assigned as the primary IPAC 'Friend' to every visitor. In addition, we have an engineering staff of 27 that is also available to help visitors. This staff includes all the people responsible for producing the IPAC data analysis system and most of the IRAS products given in Table I, and hence is a valuable reservoir of expertise on our products and analysis system. In addition, we have a Data Management Team of 10, a library staff of 3, a facility staff of 5, and a Management and Administration staff of 6.

Our host computer is a CDC Cyber 180/850, a mainframe computer. It offers the power necessary to accomplish the goals given above, which was the primary reason for its selection. Unfortunately, because the system software and hardware contained some deficiencies as delivered, it took almost two years to develop an acceptable data analysis system using the Cyber.

We set up the data system in a quite general way so that most tasks are independent of the formats or idiosyncracies of the IRAS data. For example, whereas previously we had a search program operating only on the IRAS Point Source Catalog, we now have a general catalog analysis facility which can search any catalog on any parameter. The key conceptual change was to use a DBMS system that operates using a catalog definition file. Once that file is defined for a given catalog, any parameter of that catalog is then accessible to the DBMS program.

In the same vein, the comparison program, which looks for positional matches between sources from different catalogs or source lists, was set up generally. We defined a pointer file with the ability to hold pointers to any catalog on the system. All results of any comparison are held in the pointer file. This allows subsequent analysis to operate on any parameter in any catalog.

Our plot programs operate similarly. Any quantity can be plotted versus any other quantity, making analysis quite simple.

Thus our analysis system is easily expandable to include new data without the need to rewrite much code. Once the catalog definition file is written, the full power of the data analysis system can operate on any new catalog.

The user interface to our data analysis system is the Transportable Applications Executive, which is used by almost all programs. It is an English language interface, requiring almost no previous training to be able to change parameters and run each program. New users can access program and help information through menus whereas expert users can bypass the menus and directly access any given program.

We are planning to implement IRAF as an available user interface in the future.

The entire data system has been designed for remote access. Information, including 'printouts', plots, and monitoring information can be routed to the user's terminal or to any hardcopy device. Any terminal can be supported to the maximum extent possible by defining the characteristics of that terminal to our system. We plan to hook up to all the major electronic networks in the near future.

Data access is handled by putting a large amount of data on-line. Data sets that cannot fit on-line will be placed on optical disks stored in an automatic 'juke-box' retrieval mechanism (not yet available). On-line storage for large temporary data sets is also available. Thus with minimal user intervention, almost any data on the system can be easily accessed.

An important aspect of our system is the large amount of help information present on-line. All data inventories, the IPAC User's Guide, and help for individual programs are all accessible.

IV. EXAMPLES OF SCIENTIFIC ANALYSIS OF THE IRAS DATABASE

The system we have described above resulted from the need to access and analyze the IRAS data during and after the mission. But how did the data analysis system contribute to the extraction of scientific results from the IRAS database? We now review some of the most important scientific discoveries or results derived from the IRAS mission, and assess the role of database analysis in the result. We have ordered the results beginning with those most likely to be discovered by analyzing a database, and progressing through those requiring more complex database functions, to those beyond the reach of present-day database technology.

1. The Vega phenomenon is the discovery of a large infrared excess around Vega, providing the first direct evidence outside of the solar system for the growth of large particles, and the possibility of planetary systems (Aumann et al. 1984). This is a discovery that could have quite naturally come out of an interrogation of a database: collect the infrared colors of main sequence stars, and set apart those that deviate from a blackbody. But this dicovery was made before the IRAS database was ready for such inquiries, because Vega was being used as a calibrator, and revealed its infrared excess at that stage.

2. The bulge of the Milky Way is delineated by IRAS sources of certain colors (corresponding to stars with massive dust shells) much better than by any other observable (Habing et al. 1985). This may be considered quintessential database work, requiring to sort through the IRAS Point Source Catalog, select sources with certain infrared color criteria, and map them in Galactic coordinates. The discovery was made using the IPAC database access and analysis system.

3. The IRAS colors of normal galaxies have revealed an unexpected anti-correlation, meaning that the cooler galaxies are in the 12-to-25 micron ratio, the warmer they are in the 60-to-100 micron ratio (Helou 1986). The color

sequence then turned out to be a star formation activity sequence which helps in the search for what determines the star formation rates. This is also simple database work, requiring a selection of IRAS sources associated with galaxies, and color-color plots. The work was done using the IPAC data analysis system.

4. IRAS-luminous galaxies, with Arp 220 as prototype, can be considered a new class of objects emitting more than 95% of their power in the mid and far infrared (Soifer et al. 1985). Their discovery depended on the ability to compare the infrared and optical data for all (or as many as possible) cataloged galaxies, looking for the most extreme ratio of infrared to blue emission. The work was also done using the IPAC system.

5. The IRAS signature of Seyfert galaxies is a well-defined color criterion that allows the discovery of new Seyfert galaxies based solely on the IRAS colors of sources (de Grijp et al. 1985). This is also a result that can be derived from database interrogation, requiring the examination of the IRAS colors of galaxies known to be Seyferts. The work was to a large extent done on the IPAC data analysis system.

6. The infrared-radio correlation in disks of galaxies is a remarkably tight relation suggesting a critical missing link in our models of galaxy-wide star formation activity (de Jong et al. 1985, Helou et al. 1985). Although it could have been discovered through database interrogation, it was not. One group based its conclusions on follow-up observations; the other group on a literature search followed by extensive re-evaluation of the radio data. This discovery cannot therefore be attributed to the interrogation of a database.

7. Infrared-loud quasars are a class of objects characterized by quasar-like luminosities and morphologies, with most of the power concentrated in the mid-infrared (Beichman et al. 1986). It is hard to imagine the discovery of such objects could have resulted from a database interrogation, because they are unlikely to have been classified and observed before IRAS. They were discovered through follow-up observations of IRAS sources with very faint optical counterparts.

8. The search for brown dwarfs using the IRAS database is still in progress (Low 1985, Chester et al. 1986), attempting to find such objects, or to place stringent limits on their potential contribution to the missing mass. This search is not possible without the IRAS database, but it is also impossible simply within it: it proceeds by IRAS source selection followed up by optical spectroscopy.

Two more important discoveries, namely the Zodiacal dust bands, and the infrared cirrus (Low et al. 1984) could not have been made by interrogation of the database analysis system at IPAC, since they depended on the examination of one- and two-dimensional image data.

In conclusion, it is clear that standard database access and analysis tools have been quite effective in the extraction of astronomy from the IRAS database. But it is also clear that some discoveries are simply outside the realm of present-day standard database techniques, especially those that depend on two-dimensional data analysis. We also conclude from this review of discoveries that the simplest database access and analysis tools are still the most valuable ones.

V. SUMMARY

As the volume of interesting astronomical data grows, it becomes increasingly impractical and senseless to set up analysis systems at the home institution of every researcher. Instead, a system of analysis centers, each offering a large data base, the expertise in its use, the tools to use it, and remote accessibility offers a vastly superior method for the individual astronomer to do research with large data bases. In spite of various limitations, standard database access and analysis tools are effective at extracting scientific results from astronomical databases.

ACKNOWLEDGEMENTS

We acknowledge the entire IPAC staff who have collectively designed, implemented and validated our new data system, and especially B. Poulson, who was responsible for many of the ideas embodied in requirement 2 of Section III.

REFERENCES

Aumann, H.H., et al. 1984, *Ap. J. (Letters)*, **278**, L23.

de Grijp, M.H.K. et al. 1985, *Nature*, **314**, 240.

de Jong, T. et al. 1985, *A. & A.*, **147**, L6.

Chester, T., et al. 1986, *B.A.A.S.*, **18**, 961.

Habing, H.J., et al. 1985, *A. & A. (Letters)*, **152**, L1.

Helou, G., et al. 1985, *Ap. J. (Letters)*, **298**, L7.

Helou, G. 1986, *Ap. J. (Letters)*, **311**, L33.

Low, F.J., et al. 1984, *Ap. J. (Letters)*, **278**, L19.

Low, F. 1985, in *Astrophysics of Brown Dwarfs*, ed. M.C. Kafatos et al. (Cambridge: Cambridge University Press), p. 66.

Soifer, B.T., et al. 1984, *Ap. J. (Letters)*, **283**, L1.

ASTROPHYSICS FROM LARGE DATABASES: THE EXAMPLE OF IUE

Sara R. Heap
Laboratory for Astronomy and Solar Physics
NASA/Goddard Space Flight Center

1. INTRODUCTION

The International Ultraviolet Explorer (or IUE) is an on-going satellite in geo-synchronous orbit devoted to obtaining ultraviolet spectra of astromomical objects. By any measure, it is an extremely successful satellite. The scientific accomplishments of the IUE have been summarized in a recent book, *Exploring the Universe with the IUE Satellite* (ed. Kondo 1987). The last chapter of the book, "Mining the IUE Archive," by David Giaretta, Jaylee Mead, and Piero Benvenuti is particularly relevant to the subject matter of this conference.

The title of this talk refers to "the example of IUE". Just what example has IUE archive set, and what impact on astronomy has it had? Also, what are the reasons for the effect it has had, and what has limited its impact? I will not try to address the first question on the scientific impact: *Exploring the Universe with IUE* does that more than adequately. Instead, we will "look behind the scenes" to find out how and why it has had such an effect. We will first need to know about the IUE data-base itself: much of its effect on astronomy could be predicted simply from its size, diversity, long-term coverage, and other characteristics. The next section briefly describes the IUE data-base.

A description of the IUE data-base is necessary but not sufficient explanation, because this fine archive would have had no effect whatsoever unless it were made available to the astronomical community and unless there were data-systems for acquiring and making use of the data. By international agreement (ESA, SERC, and NASA), an IUE observation is made available to the astronomical community world-wide six months after it is obtained.

Systems for making use of IUE data cannot be summarized so quickly, and the bulk of this talk will center on methodology: ways to access to the data-base, ways to process and analyze the data, and so on.

2. CHARACTERISTICS OF THE IUE DATA-BASE

By any definition of the term, be it number of observations, number of observed objects (about 4500) or sheer volume of data, the data-base of IUE observations is a large one. The number of observations in the public domain is

FIGURE 1: GALACTIC DISTRIBUTION OF OBJECTS OBSERVED BY IUE
UPPER FIGURE: BRIGHT ULTRAVIOLET OBJECTS ($m_{SWP} < 8^m$)
LOWER FIGURE: FAINT ULTRAVIOLET OBJECT ($m_{SWP} > 8^m$)

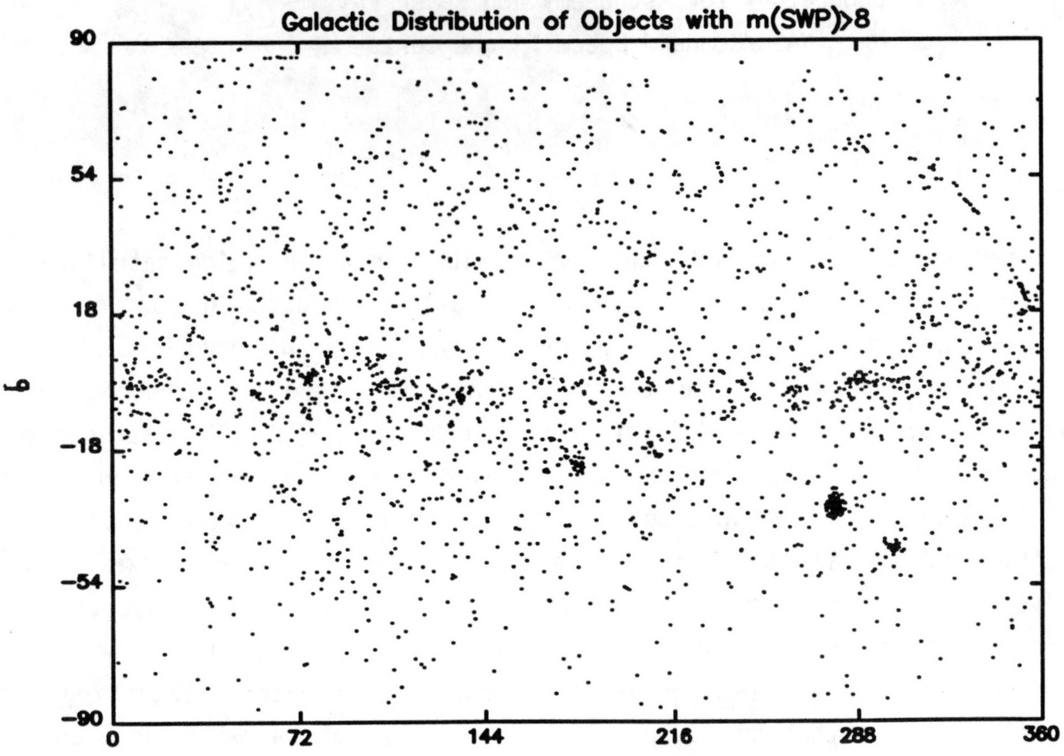

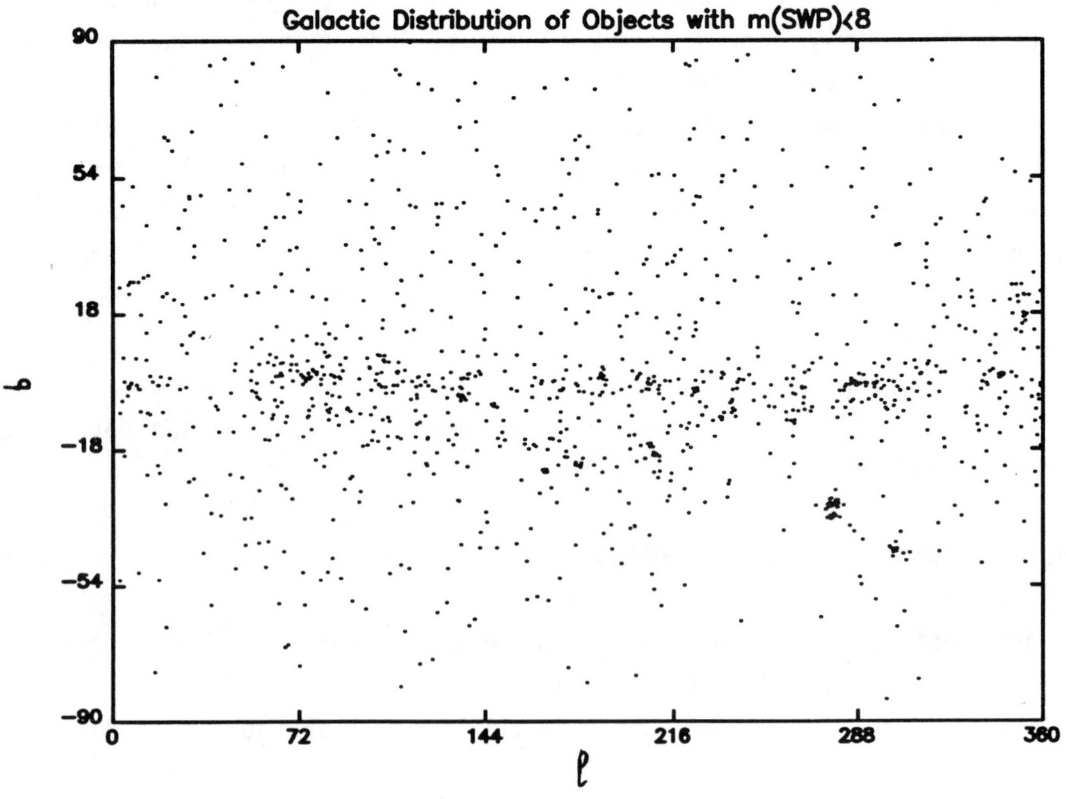

shown below, grouped by camera (spectral region) and spectral resolution:

TABLE 1

Camera	High Res.	Low Res.	Total
SWP	9,167	18,933	28,100
LWR	6,626	12,428	
LWP	2,725	5,402	27,181
---	-----	------	------
total	18,518	36,763	55,281

Size of Data-Set	(Mbytes)
raw image	0.6
corrected image	1.2
extracted spectrum	~0.7
-------------------	-----
total	2.5

Of course, the IUE is still operational. It produces ultraviolet spectra at the rate of about 6000 spectra per year.

Each raw observation consists of a 768x768 byte image. The processed and ancillary data for each observation add even more to the total volume. Multiply the 2.5 MBytes per observation by 60,000 observations and you get 150 Gbytes, which I think we would all agree is a large number.

The IUE data-base is broad and varied as well as large. Scientifically, these other two attributes are probably more important than total volume. The IUE has been used to observe practically every kind of object known to astronomers. Table 2 shows the distribution of observations grouped by type of object. Over the course of a year, the IUE can point to any region of the sky. Figure 1, in showing the galactic distribution of observed objects, demonstrates IUE's complete sky coverage. By utilizing an SEC Vidicon camera whose target has excellent storage characteristics, the IUE can have exposures ranging from less than a second to hours (Figure 2). This make for an exceedingly wide range in UV-brightness levels that can be accommodated. Figure 3 shows a plot of visual magnitude vs. broad-baseline color, where the ultraviolet magnitude is derived from the average flux in a low-dispersion SWP observation.

The IUE satellite will have its tenth birthday next January. A full decade's worth of observations, coupled with an enlightened observing policy, which has not sought to restrict repeated observations of the same object, has yielded a wealth of data on a number of astrophysically important objects. Sometimes, this has come about due to the sustained efforts of international campaigns, but at other times, this has just happened. Either way, there are some outstanding studies of Seyfert galaxies, x-ray binaries, Be-type stars etc., made possible by the IUE archive. Again, the possibilities of fruitful archival research increase with time as about 6000 observations per year are added to the IUE data-base.

TABLE 2: TYPE OF OBJECTS OBSERVED BY IUE

Object Class	No.	Description	Object Class	No.	Description
0	44	sun	50	188	R, N, S-type stars
1	27	earth	51	164	LP variables
2	112	moon	52	482	irr. variables
3	1509	planet	53	923	reg. variables
4	510	planetary sat.	54	1048	dwarf novae
5	126	minor planet	55	802	classical novae
6	805	comet	56	152	supernovae
7	598	IPM	57	1200	symbiotic stars
8	5	giant red spot	58	510	T Tauri stars
9	1		59	947	x-ray objects
10	514	WC star	60	265	shell stars
11	1095	WN	61	105	eta Carina
12	2156	O IV-V	62	6	pulsars
13	1083	O I-II	63	648	nova-like stars
14	209	Oe	64	81	other
15	353	Of	65	30	mid-identified targets
16	1939	sdO	66	553	interacting binaries
17	103	DO	67	1	
18	1		68	1	
19	110	other,bright	69	1	
20	3110	B0-2 V-IV	70	1404	Central Stars of PN
21	2584	B3-5 V-IV	71	338	Planetary nebula
22	1191	B6-b9.5 V-IV	72	719	HII region
23	2167	B0-2 III-I	73	80	reflection nebula
24	519	Be-B5 III-I	74	63	dark cloud
25	665	B6-B9.5 III-I	75	258	SNR
26	1293	Be	76	19	shock-ionized nebula
27	953	Bp	77	1	
28	172	sdB	78	1	
29	174	DB	79	1	
30	954	A0-3 V-IV	80	278	spiral galaxy
31	374	A4-9 V-IV	81	157	elliptical galaxy
32	261	A0-3 III-I	82	87	Irr galaxy
33	297	A4-9 III-I	83	461	globular cluster
34	167	Ae	84	1578	Seyfert galaxy
35	146	Am	85	720	quasar
36	734	Ap	86	134	radio galaxy
37	493	DA	87	315	BL Lac object
38	95	HB star	88	248	emission-line galaxy
39	1271	composite	89	1	
40	773	F0-2	90	7	intergalactic medium
41	1043	F3-9	91	1	
42	74	Fp	92	1	
43	51	late-type WD	93	1	
44	1742	G V-IV	94	1	
45	1130	G III-I	95	1	
46	1144	K V-IV	96	1	
47	1042	K III-I	97	1	
48	690	M V-IV	98	1190	internal Pt lamp
49	619	M III-I	99	3522	nulls and flat-fields

FIGURE 2

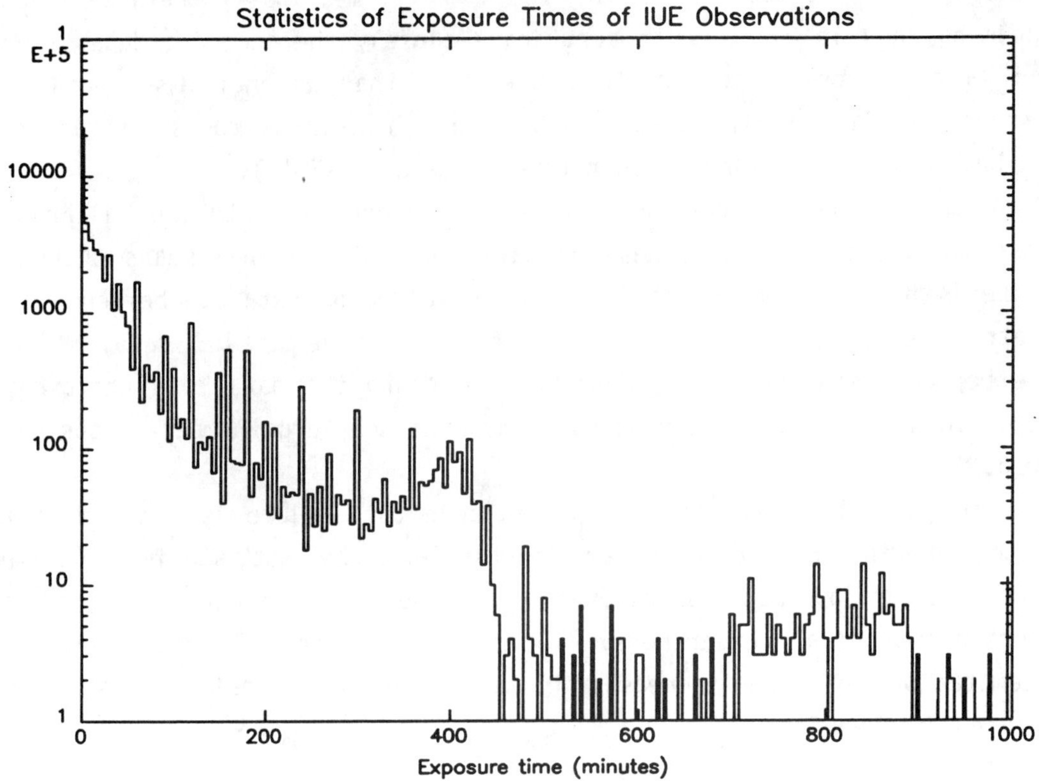

FIGURE 3

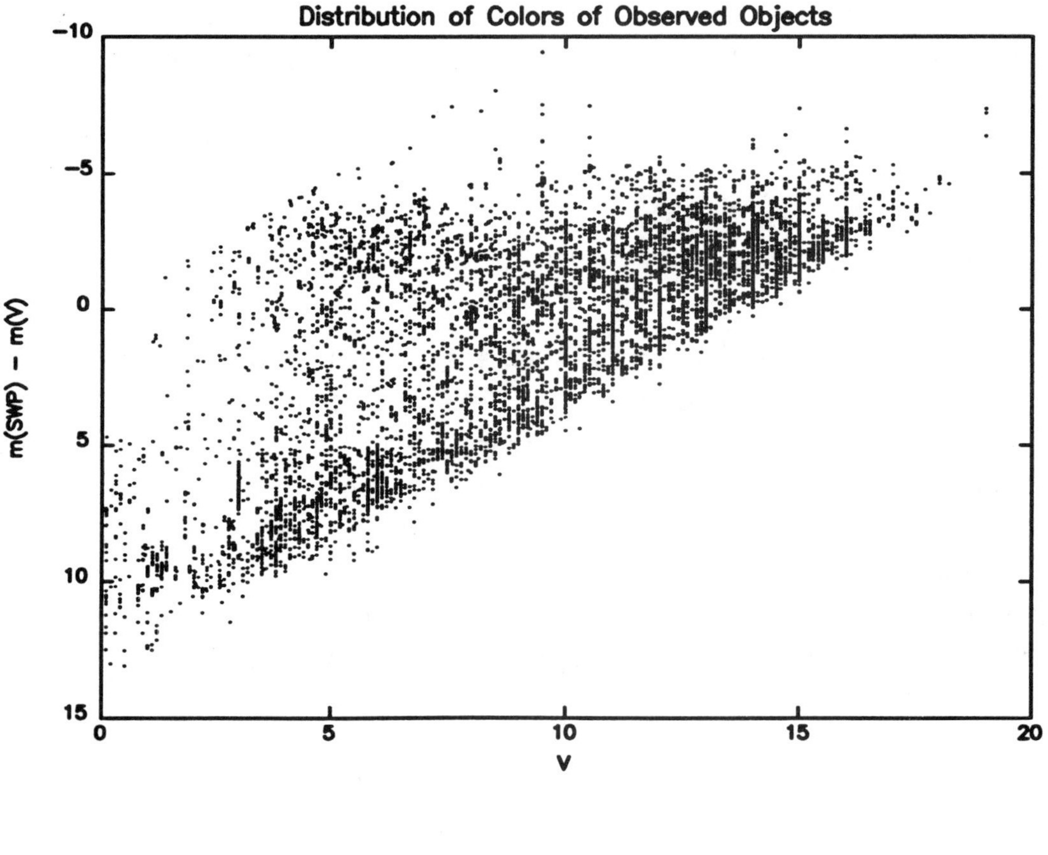

Breadth and volume do not come without disadvantages. The high data-rate, particularly at the beginning of IUE's lifetime, was sometimes close to overwhelming, and it was hard to scrutinize closely the incoming data. There was a major error in one of the calibration files that was not discovered until a year after launch, simply because there were inadequate tools for interactive image display and evaluation at that time (the late 1970's).

Over the course of time, improvements have been made to the calibration of the instrument and the procedures for data-processing. This means that the IUE data-base is not uniform in quality: the more recent data are better. The IUE Observatory is planning to construct a final archive of IUE data, which will involve reprocessing the entire data-base to bring the data to a uniformly high quality. This is a major undertaking because of the large number of observations involved.

Another potential disadvantage of a large data-volume is the temptation to limit the growth of a data-center, particularly by not storing intermediate data-products or by not distributing raw and intermediate data-products to observers and archival researchers. To its credit, the IUE Observatory resisted this temptation to limit increases in data-volume, and astronomers world-wide have reaped the benefits.

3. SYSTEMS FOR MAKING USE OF IUE DATA

As implied by the title, "The Example of IUE," the IUE in all its manifestations - involvement of the astronomical community world-wide, operations and data-systems -- has been a trend-setter. Let us go through some of these aspects one by one: scope of research, access to data, data-processing and analysis.

3.1 Scope of Research

In the past, astronomers in the U.S. were sponsored by the project offices of the particular satellite whose data they were using, and they had to confine their research accordingly. Now this year, NASA has a "Space Astrophysics Data-Analysis Program" (SADAP) which encourages meritorious archival research irrespective of origin (IUE, HEAO-2, IRAS, SMM, etc.) or color (wavelength). The statistics for this first year (*gratis* Y. Kondo) indicate that a quarter of the 154 programs selected by NASA involve the IUE archive either as the primary source of data (33 programs) or as a supplement (9 programs).

3.2 Selection of Appropriate Observational Data

Searching The IUE Catalogue. With 60,000 or so spectra to choose from, some means of selecting the appropriate observations is of utmost importance. Astronomers from SERC did a wonderful service by composing a list of object types (the list shown in Table 1) and persuading IUE observers to assign an object-class to each observation as it was taken. "Object-class" has turned out to be one of the most useful parameters in the observation catalogue. You can quibble with the reliability of the classification system -- one man's planetary nucleus (object class=70) may be someone else's Of star (object class=15) -- but, on the whole, the system has worked. It is now possible to acquire archival data defined by object class and other parameters in the IUE Observation Catalogue rather than listing the observations one by one, e.g. you can request all far-UV, low-dispersion spectra of quasars as a group.

Multi-Catalogue Searches. Because the IUE data-base is so extensive, the chances are good that IUE has been used to observe a particular object of interest that was observed by some other satellite, and it is often useful to combine IUE spectra with data from other astronomical satellites. We may wish, for example, to derive the complete flux distribution -- from x-ray wavelengths to the infra-red -- of a quasar or active galaxy, and so we need a DBMS that can tell us what satellites besides the IUE have data on a particular object of interest.

When IUE spectra are to be combined with data from other satellites or ground-based observatories then position becomes all-important, and in this respect, the IUE Observation Catalogue falls a little short. The 1950-epoch equatorial position of an object listed in the catalogue comes straight from the observer without cross-checks. Since the IUE has a 16'x16' target-acquisition field, the position need not be precise for the observation to be a success. But a precise position *is* essential for identifying the object in other catalogues, since this is the only attribute the catalogues have in common. I said earlier that the IUE had observed *about* 4500 astronomical objects. I was vague because the number of objects depends on the uncertainty in position. Given a tolerance of 2", 10", and 20", the totals become 4658, 4253, 3943 objects, respectively.

As preparation for multi-satellite research, we have installed the observation catalogues of the major astronomical satellites (IUE, EXOSAT, EINSTEIN, ANS, TD-1, and IRAS) and have developed a DBMS system within Interactive Data Language (IDL) to work with these catalogues. Table 3 lists the main data-base functions of this system with a brief explanation of each. The main point of the list is to show that they are independent of the specific catalogue(s) being worked on. Most astronomers need only learn the top six functions in the list. For those of us who find even that too hard, there is an english-language interpreter. Another important feature of this DBMS is that it is totally integrated within the general data-analysis system. The same system that helps to select the relevant observation data also helps to acquire and analyze the selected data. This point will become clearer as we go along.

Let us use AGN's and quasars as an example of a group of objects we wish to study using data from the IUE and other satellites. Out of the 56,908

Satellite	No. in Data Base	No. of AGN Observations	No. of AGN's	No. of Matches
Exosat	2,108	549	99	47
IUE	56,908	2,747	176	106
IRAS	245,889	6,943	6,943	

observations listed in the IUE catalogue, 2747 were of AGN or quasars. A few AGN's, such as NGC 4151, have been the object of long-term observing campaigns, so the number of observations outweighs the number of observed objects by a factor of 15X. Of the 176 AGN's and quasars observed by IUE, 47 have also been observed by Exosat, and 106 are also listed in the IRAS point-source catalogue. Table 5 shows the overlap.

A future version of this DBMS system will be able to assimilate data from several catalogues at once, taking the best parts of each to build a composite. For the quasars and active-galaxy example, the DBMS might extract the redshift from one catalogue, angular radius from another, etc. and combine these supplementary data with measurements that we make on the IUE spectra.

This work, although useful, does not diminish the need for a master catalogue of all objects observed by the major astronomical satellies. This is a job for the big data-centers, such as the Strasbourg Data Center or the National Space Science Data Center at Goddard.

3.3 Rapid Access to Data.

Knowing that the data are there only whets the appetite to make use of them. In the early years of IUE, getting IUE archival spectra involved: a user mailing in to the appropriate data-center a request for specific observations, a hapless data-technician at the data center locating which magnetic tapes contained the requested observations, physically mounting the tapes and carrying out the tape-copy, and finally(!) mailing the tape with the requested observations back

TABLE 3

MAJOR DB ROUTINES

USER ACCESS

```
DBOPEN   - open data base file
DBHELP   - information on data base files
DBFIND   - search for selected entries (observations) in file
DBPRINT  - print selected information on selected or all entries
DBSORT   - sort on up to nine items (fields)
DBEXT    - extract selected items for specified entries
DBVAL    - extract item for single entry
DB_OR    - merge two search lists made by DBFIND

db - pseudo-english query langauge interpreter
```

FILE MAINTANENCE

```
DBCREATE - create a new data base or modify specifications
               on an existing data base
DBTEXTIN - add entries to a data base from an ascii text file
DBINDEX  - index selected items
DBPUT    - place value for selected entry and item into data base
DBDELETE - delete selected entries from file
```

ADDITIONAL PROGRAMMER TOOLS

```
DBCLOSE     - close a data base
DBEXTIND    - extract values for indexed item
DBGETDESC   - get information of data base file
DBITEM      - get information on selected item(s)
DBRD        - read entry of data base
DBWRT       - write new or updated entry to data base
```

FUTURE ROUTINES

```
FT_TO_DB    - create or update a data base from a FITS table file
DB_TO_FT    - create a FITS table file from selected entries and items
                  in a data base file.
TAB_TO_DB   - create or update a data base from a IRAF/SDAS table file.
DB_TO_TAB   - create a SDAS table file from selected entries and items
                  in a data base file.
DBFITS      - add or update entry in a data base file using FITS image
                  header keyword values
```

to the user.

All that is changed now with the advent of massive on-line data-storage systems and networks. In the U.S., the first big change was in 1983 when Goddard got an IBM 3081 machine with a mass-storage device capable of storing 240 Gigabytes. Within a few months after it arrived, all the IUE processed data (*i.e.* extracted-spectral files) were put on the mass-store. Since then, over 50,000 IUE spectra have been retrieved from the mass-store via the IUE Regional Data-Analysis Facility at Goddard and the NSSDC. In the past year, we have also put the raw IUE images on the mass-store (or to be more precise, on the accounting system of the mass-store). The other big change was the installation of the GATEWAY, a means of connecting a VAX with the IBM machine. Astronomers doing interactive data-analysis on our VAX now routinely fetch IUE data (either the raw image or the extracted spectra) from the mass-store in about 5 minutes.

3.4 Data Processing.

In at least half the cases, the processed data are adequate for use in later data-analysis. In other cases, the data need to be reprocessed. Processing of IUE data from the raw image to extracted and calibrated spectrum is a major undertaking, until recently only performed by the two IUE observatories at Vilspa, Spain and at Goddard. In the past year, however, we have had running a system to process IUE observations on our VAX computer. The system seems to work well, is reasonably efficient, and most importantly, can be easily run by an astronomer.

With wide-area networks, such as SPAN, reprocessing can be performed by remote users as well as by Goddard scientists. In fact, graduate students at the University of Colorado were one of our most helpful test-users of the system. This means that it is no longer necessary for each astronomy group throughout the country to purchase a high-powered computer: via networks, an astronomer can access a larger computer that has all the necessary software and calibration files, carry out the appropriate processing, and then simply bring back the results to his or her workstation. This is *distributed computing* raised to its full potential.

Table 5 shows an actual session at a computer terminal in which a user acquires, processes, and displays an IUE spectrum of the central star of the planetary nebula, NGC 246, an exceedingly hot, carbon-rich star. The main IDL routines that are used in this session are:

(1) DBFIND, which searches the IUE observation catalogue for observations of interest;

TABLE 4

IUE Merged Log database				Exosat Observation Log				IRAS
object id	ra (1950)	dec (1950)	obj class	target name	right ascension (1950)	declinat'n (1950)	prop clas	source name
0014+81	00 14 4.8	+ 81 18 42	85	S50014+81	00 13 05	+ 81 18.7	AGN	
MRK 348	00 46 5.0	+ 31 41 00	84	MKN348	00 46 04	+ 31 41.0	AGN	00461+3141
0049+171	00 49 16.5	+ 17 09 41	85	PG0049+171	00 49 11	+ 17 7.1	AGN	
0052+252	00 52 11.1	+ 25 09 24	85	PG0052+251	00 52 07	+ 25 6.8	AGN	
MRK 352	00 57 9.1	+ 31 33 27	84	MKN352	00 57 08	+ 31 33.5	AGN	
MKN1152	01 11 21.8	- 15 06 37	84	MKN 1152	01 11 21	- 15 6.1	AGN	
0119-013	01 19 30.0	- 01 18 00	85	II ZW 1	01 19 30	- 01 18.0	AGN	01194-0118
C 526A	01 21 37.2	- 35 19 33	84	NGC 526 A	01 21 46	- 35 21.4	AGN	01216-3519
3C 66B	02 20 2.4	+ 42 45 58	86	3C 66A	02 19 25	+ 42 45.8	AGN	?02193+4247
931	02 25 18.0	+ 31 05 18	84	MKN 1040	02 25 21	+ 31 7.8	AGN	02252+3105
0235+164	02 35 53.0	+ 16 24 00	87	0235+164	02 35 53	+ 16 24.0	AGN	
3C 120	04 30 32.0	+ 05 15 00	84	3C 120	04 30 40	+ 05 16.8	AGN	04305+0514
AKN 120	05 13 39.7	- 00 12 16	84	AKN120	05 13 38	- 00 12.2	AGN	05136-0012
G+8-11-1	05 51 9.6	+ 46 25 50	84	MCG 8-11-11	05 51 20	+ 46 27.4	AGN	05511+4625
0754+395	07 54 38.0	+ 39 29 00	84	MKN382	07 52 03	+ 39 19.1	AGN	
0804+761	08 04 35.4	+ 76 11 32	85	PG0804+761	08 03 42	+ 76 10.8	AGN	
TON 951	08 44 33.9	+ 34 56 09	85	PG0844+349	08 44 22	+ 34 55.0	AGN	
0851+202	08 51 57.3	+ 20 17 59	87	OJ 287	08 52 08	+ 20 19.6	AGN	?08519+2017
NGC 3516	11 03 23.0	+ 72 50 00	84	NGC 3516	11 03 57	+ 72 50.2	AGN	11033+7250
MRK180	11 33 32.7	+ 70 26 00	87	MKN180	11 33 41	+ 70 28.5	AGN	
N 3998	11 55 20.0	+ 55 44 00	86	NGC 3998	11 55 06	+ 55 45.9	AGN	
4051	12 00 37.0	+ 44 48 42	84	NGC 4051	12 00 47	+ 44 49.3	AGN	12005+4448
C4156	12 08 18.0	+ 39 45 02	84	NGC 4151	12 08 26	+ 39 43.4	AGN	
1215+303	12 15 21.1	+ 30 23 40	87	NGC 4278	12 17 35	+ 29 33.6	AGN	
MKN 205	12 19 37.0	+ 75 35 00	85	MKN205	12 19 34	+ 75 35.2	AGN	12195+7535
1229+204	12 29 33.1	+ 20 26 03	85	PG1229+204	12 29 21	+ 20 25.8	AGN	
MKN 231	12 54 5.0	+ 57 09 00	84	MKN231	12 54 20	+ 57 9.9	AGN	12540+5708
1308+326	13 08 8.2	32 36 42	85	1308+326	13 08 14	+ 32 39.1	AGN	
MARK 279	13 51 53.6	+ 69 33 13	84	2A1348+700	13 51 52	+ 69 33.2	AGN	13519+6933
NGC 5548	14 15 44.0	+ 25 22 00	84	NGC 5548	14 15 55	+ 25 22.0	AGN	14156+2522
OQ 530	14 18 6.2	+ 54 36 57	87	1418+546	14 18 08	+ 54 37.1	AGN	
MKN 478	14 40 5.0	+ 35 39 00	84	MKN 478	14 40 13	+ 35 41.1	AGN	14400+3539
MK 841	15 01 36.3	+ 10 37 55	84	MKN 841	15 01 47	+ 10 39.9	AGN	15015+1037
MARK 290	15 34 44.8	+ 58 04 01	84	MKN 290	15 35 01	+ 58 6.0	AGN	
MK876	16 13 55.0	+ 65 50 20	84	MKN 876	16 13 48	+ 65 50.0	AGN	16136+6550
371	18 07 18.5	+ 69 48 58	87	3C 371	18 06 58	+ 69 46.8	AGN	18072+6949
3C382	18 33 21.1	+ 32 39 15	86	3C 382	18 33 12	+ 32 39.2	AGN	
3C390.3	18 45 38.5	+ 79 43 02	84	3C390.3	18 45 17	+ 79 40.6	AGN	18456+7943
4C73.18	19 28 49.4	+ 73 51 45	85	1928+73	19 28 49	+ 73 51.7	AGN	
3C 405	19 57 44.0	+ 40 36 00	86	CYGNUS-A	19 57 37	+ 40 33.0	AGN	19577+4035
II ZW136	21 30 1.3	+ 09 54 59	85	II ZW 136	21 29 58	+ 09 52.3	AGN	21299+0954
2141+175	21 41 13.7	+ 17 30 01	85	OX 169	21 41 13	+ 17 29.8	AGN	
2155-304	21 55 58.4	- 30 27 54	87	PKS2155-304	21 55 58	- 30 27.8	AGN	
BL LAC	22 00 39.4	+ 42 02 08	87	2200+420	22 00 39	+ 42 2.1	AGN	?22000+4208
3C 445	22 21 15.0	- 02 21 00	86	3C 445	22 21 15	- 02 21.2	AGN	
3C 446	22 23 11.1	- 05 12 17	87	PHL5200	22 25 54	- 05 34.2	AGN	22231-0512
NGC 7469	23 00 59.0	+ 08 32 00	84	NGC7469	23 00 44	+ 08 36.3	AGN	23007+0836

(2) IUEGWY, which uses the GATEWAY connecting the HRS VAX 8200 with an IBM 3081 machine at Goddard to fetch a raw IUE image from the mass-store;
(3) BSHELL (batch shell), which processes an IUE observation according to current standards;
(4) IUEPROCESS, which formats the spectral file into vectors of wavelength, flux, and data-quality flags (Figure 4).

These four routines represent the basic functions needed by an adequate data-system for astronomy: a DBMS system (DBFIND), a communications system (IUEGWY), a system for intensive (batch) processing, and a system for interactive display and measurement.

3.5 User Catalogues and Archives

Figure 5 shows the flux distribution for the central star of NGC 246 as given by *The Goddard Catalogue of UV Fluxes*, an on-line file containing the fluxes for all low-dispersion observations. This archive is an example of a "user-catalogue," that is, a subset of IUE data constructed by a user (or small group) that is small enough to be stored on-line. Not only is the *UV Flux Catalogue* useful not only for browsing the IUE archive, but it is also helpful for quickly finding out the flux of an object at some ultraviolet wavelength and computing exposure times on the Space Telescope, and it has been used to do statistical research on large groups of objects.

This brings up the general matter of user catalogues. Those who deal with a lot of IUE spectra or who make many measurements on a single spectrum have need to organize their material. What guidelines and tools are there to help an astronomer make new catalogues or working data-sets? Fortunately, the question of formats appears to be solved. By consensus, FITS-table format should be the standard format of any catalogue. The tools for making catalogues are already in place (Table 3). All that remains to be done is a little "astronomer-proofing" of the system.

4. TRENDS IN ASTRONOMICAL DATA SYSTEMS

In the U.S., NASA is currently holding a series of "Astrophysics Data-Systems Workshops" charged with defining the requirements for astronomical data-systems of the future (Squibb, this volume). The final recommendations of the participants have not been formulated as yet, nor have the implications of these recommendations been fully worked out, but certain trends are already apparent. As I list them below, I think you will be able to discern the hand of

TABLE 5: A SAMPLE SESSION

```
Username: HRSHEAP
Password:

$ xidl
VAX IDL, (C) 1983, RSI.
Site  171, Licensed for use by: HRS 8200

HRS>dbhelp                ; what catalogues are available?

GTO       IRASLRS     IRASLRSA    IRASLRSW    PLHRS       ANS      EINSTEIN
IRAS      AOLOG       HCON_DIR    IUE         PLHRSENG    EXOSAT

HRS>dbopen,'iue'          ; open up the IUE data-base

HRS>dbhelp                ; what does the IUE data-base contain?
  ENTRY        OBS_YR       OBS_DAY     OBS_TIME    CAM          IMAGE       APER
  DISP         PROGRAM      CATALOG     OBJECT      SPARES_R     SPARES_D    SPARES_VM
  SPEC_TYPE    SPARES_LC    OBJ_CLASS   EBV         LG_APER      EXP_TIME    STATION
  COMMENTS     GO_NAME      PROC_DATE   REL_DATE    NSSDC_INFO   SCF_INFO    RA
  DEC          RA_1950      DEC_1950    LUM_CLASS   CAM_NO       VIS_MAG     LII
  BII          MAG_1550     MAG_2250    MAG_2850    SPARES

---- Indexed Items ------
IMAGE   APER   DISP   OBJ_CLASS  RA  DEC   CAM_NO  VIS_MAG   LII   BII

iue contains       56908 entries

                                                                    ; get obs. of
HRS>ngc_246=dbfind('obj_class=70,00:44<ra<00:45,-13<dec<-12')       ; NPN of NGC 246
           15 entries found

HRS>ngc_246=dbsort(ngc_246,'disp,cam,image')                        ; sort them

HRS>dbprint, ngc_246,'object,cam,image,disp,aper,comments'          ; print them

        IUE Merged Log database
 ----------------------------------------------------------
   object     camera  image  disp  aper          guest
     id         id     no.   mode                observ.
                                                 comments
 ----------------------------------------------------------
   NGC 246     LWR     6807   H     L  509
       246     SWP     3353   H     L
   NGC 246     LWR     1502   L     S  UNDEREXP
   NGC 246     LWR     1503   L     S  UNDEREXP
   NGC 246     LWR     6806   L     L  502
       246     LWR    10910   L     L  C=290,B=32
       246     LWR    14706   L     L  C=140,B=27
   NGC 246     SWP     1556   L     S  VERY UNDEREXP
   NGC 246     SWP     1563   L     S  UNDEREXP
       246     SWP     3352   L     L
       246     SWP     3371   L     S
       246     SWP     3371   L     L
   NGC 246     SWP     7795   L     L  601
       246     SWP    14290   L     L  C=241,B=13
       246     SWP    18642   L     L  C=135,B=20

HRS>iuegwy,3,3353,'raw'                                        ; fetch SWP 3353 from
                                                               ; the IBM mass-store
    Ordering        camera         image          type
                    SWP            3353           raw

    Output will be placed into: HRS::  HRSHEAP

    Checking GATEWAY for access path to SCF...be patient

    Testing for IBM connection
    ..... Still testing.....
    Network is up - IBM JCL will be generated
    Job TESRH335 Submitted
```

TABLE 5 (CONTINUED)

```
HRS>bshell          ; set up for reprocessing SWP 3370 (raw)
  Enter input raw image file name...ie. S3370R for SWP 3370 RAW
  ? s3353r

  Here is the label...Please note: camera number, dispersion, aperture ***

                      0001000107680768   1 1 0130 3353              1  C
      *    *      *    *    *   6720*      *    *    *  *  *  *  *     *   2  C
SWP 3353,NGC 246,112 MIN,HI DISP,LG AP,HEAP,DAY 319                 3  C

SELECT CAMERA
(1) 1 [LWP]
(2) 2 [LWR]
(3) 3 [SWP]

SELECT CAMERA
3

SELECT DISPERSION
(1) HIGH
(2) LOW

SELECT DISPERSION
h

SELECT APERTURE
(1) LARGE
(2) SMALL
(3) TRAILED

SELECT APERTURE
1

SELECT PROCEDURE TERMINATION POINT
(1) S (LINE BY LINE SPECTRUM)
(2) L (MELO DATA SET)
(3) H (MEHI DATA SET)
(4) P (PHOTOMed IMAGE)

SELECT PROCEDURE TERMINATION POINT
h

 SEARCHING IUE LABEL FOR THDA INFORMATION
 OLD IMAGE BEING PROCESSED; NO THDA DATA
 Also checking SIPS History portion of label:

 CAMERA TEMPERATURES (TDC, THDA, TDAC) AT LAST EXPO START
        BEFORE READ OF S3353R.LAB
      LWP       2.7500        3.4319        27.479
      LWR       11.510        12.176        40.212
      SWP       7.8335        7.8335        37.258
      SWR       7.1606        6.8235        26.547

 Enter temperature...otherwise <cr>? 7.8

  INPUT                         = S3353R
  CAMERA         (1,2,or 3)     = 3
  DISPERSION     (h/l)          = H
  APERTURE       (l/s)          = L
  REGISTRATION   (a)            = A
  CALIB          (c)            = C
  OUTPUT FILE    (p,s,h,l)      = L
  TEMPERATURE    (#/ )          = 7.8

  Are these correct? (yes/no)? yes

Job BSHELL (queue HRS$SYS$BATCH, entry 774) started on HRS$SYS$BATCH

   ********** later *******

HRS> IUEPROCESS,'s3353h', header,wave,flux,eps      ; make a spectrum
HRS> LASER3, header,wave,flux,eps                   ; plot it on the
                                                    ; laser printer
```

FIGURE 4: HIGH-DISPERSION SPECTRUM OF NGC 246 (SWP 3353) IN "IUE FLUX NUMBERS"

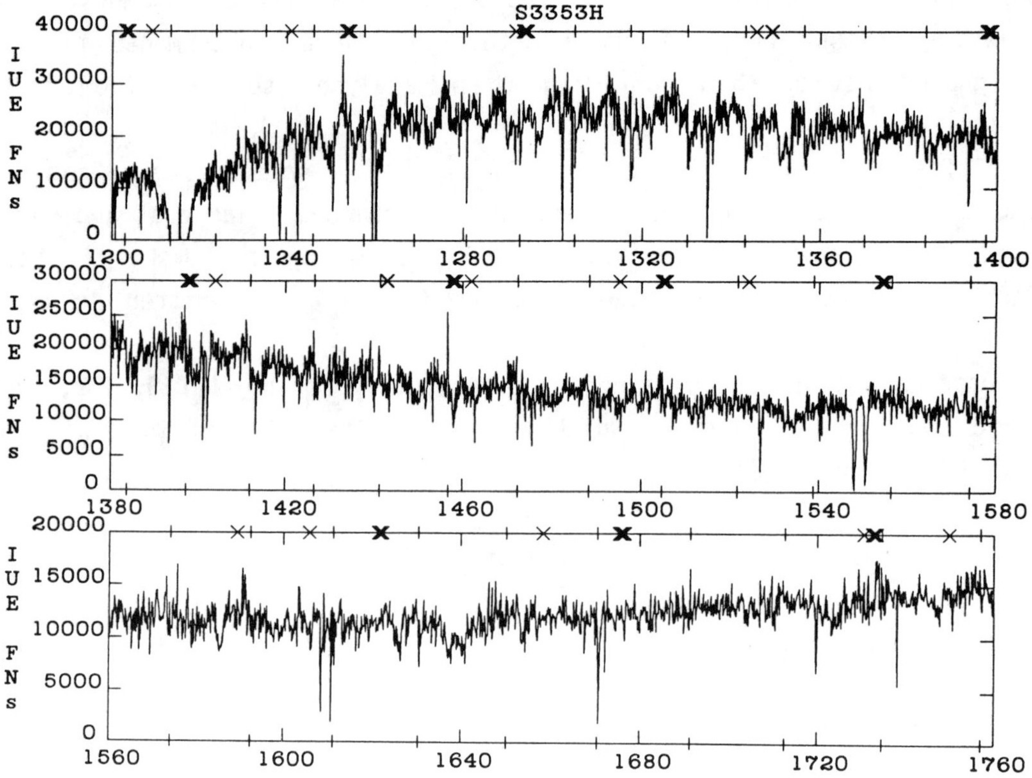

FIGURE 5: LOW-DISPERSION SPECTRUM OF NGC 246 (SWP 14290) IN ABSOLUTE FLUX UNITS (erg s^{-1} cm^{-2} A^{-1})

IDL>UUFLUX,3,14290

WAVE A	MAG	FLUX EXP -11	BP
1550	*6.52	0.90	700
1200	*5.72	1.87	100
1300	5.87	1.63	100
1400	6.26	1.14	100
1500	6.55	0.87	100
1600	6.85	0.66	100
1700	7.13	0.51	100
1800	*7.31	0.43	100
1900	*7.55	0.35	100
1180	*5.62	2.05	30
1216	5.96	1.50	30
1245	5.71	1.89	30
1300	5.88	1.61	30
1335	6.00	1.45	30
1360	6.11	1.31	30
1398	6.27	1.13	30
1489	6.49	0.92	30
1500	6.52	0.90	30
1550	6.67	0.78	30
1600	6.85	0.66	30
1640	7.06	0.54	30
1718	7.14	0.51	30
1750	7.23	0.47	30
1800	*7.31	0.43	30
1909	*7.64	0.32	30
1950	7.57	0.34	30

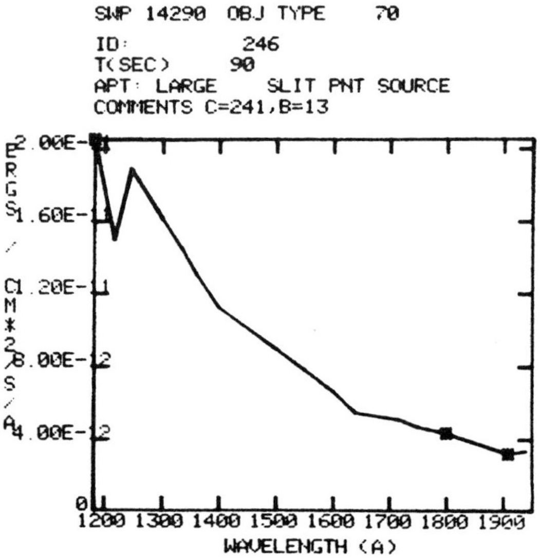

* = SATURATION OR RESEAUX WITHIN BANDPASS

IUE in starting these trends, or at least, in keeping pace.

Astronomical data-analysis will be defined in terms of a scientific problem rather than the satellite whose data are used to tackle that problem. There will no longer be a community of "IUE astronomers" or "IRAS astronomers," because astronomers will most likely be using data from several instruments in space or on the ground rather than confining their analyses to data from a single instrument.

More often than not, astronomers will obtain the relevant data from a major data center. The data may be transferred electronically via local and wide-area networks, or they may be shipped on magnetic tape or optical disk. If they are transferred by tape or disk, that tape or disk will be written in the FITS format.

Astronomers will often use a software package - probably MIDAS, IDL, IRAF, or AIPS - to analyze their data. Whichever one(s) they choose, they will insist that the software package:

(a) be an integrated system, including the functions of:
- o data-base management
- o communications
- o numerical manipulation of data
- o interactive graphics and image display
- o input-output to tape and disk;

(b) be callable either in batch mode or interactive mode, as appropriate;

(c) be independent of the specific instrument and capable of handling data obtained from all the major astronomical satellites.

Many astronomers will use workstations rather than a main-frame or mini-computer to analyze their data. The operating system on the workstation could be either the VMS operating system (on DEC machines) or UNIX. Thus, it will probably also become mandatory that an acceptable data-analysis sytem run on either system.

----- --- -----

Throughout this paper, I have referred to "we" without naming the people involved. It is high time to do so. David Stern (Research Systems, Inc.) is the author of Interactive Data Language (IDL), which is language used by the IUE Regional Data-Analysis Facilities in the U.S., by the Space Telescope's High Resolution Spectrograph team, and other space-astronomy groups. Don Lindler (Advanced Computer Concepts, Inc.) is the person responsible for developing software to process and analyze data that will come from the High Resolution Spectrograph. Lindler was also one of the major developers of the IUE Scientific Image-Processing System and the IDL-based analysis system at the IUE Regional Data Analysis Facilities. Dr. Dean Ahmad (Imad-ad-Dean, Inc.) and Don Lindler built the VAX-based system for processing IUE data, and Keith Feggans (ACC) played a major role in testing it out. Finally, Edward Sullivan (NASA/GSFC) should get the credit for getting all the IUE observational data onto the IBM 3081 mass-store.

X-ray Surveys from the EXOSAT Results Data Base

P. Giommi L. Angelini and G. Tagliaferri

The EXOSAT results Data Base is the collection of the results obtained from a standard scientific analysis performed on all EXOSAT data. It includes information on the X-ray emission from more than one thousand cosmic sources.

From this data set, using an Astronomical Data Base Management System developed at the EXOSAT Observatory, we have performed surveys of the high and low galactic latitude X-ray sky. In particular we have studied the space density, the time variability and the spectral shape of various classes of serendipitous X-ray sources.

THE ESO-LV PROJECT
automated parameter extraction for 16000 ESO/Uppsala galaxies

Andris Lauberts
ESO, Garching bei Munchen, W-Germany

Edwin A. Valentijn
Kapteyn Astronomical Institute, Groningen, the Netherlands

ABSTRACT: Since the completion of the ESO/Uppsala survey for 16000 galaxies in 1982 plans developed to extract all possible photometric and morphological parameters of these galaxies from the complete set of the ESO B and R survey plates. Clearly the great range in magnitudes (8 to 22 in B), angular sizes (10″ to 30′) and the enormous size of the input database (10 Gigabyte) would require a judicious combination of manual and automatic parameter extraction methods.

Essentially the new ESO-LV project contains the following parts: 1) Digitize 16000 galaxies with the PDS on 606 Blue and 606 Red original ESO survey plates. 2) Bring together existing photoelectric photometry + own measurements in a catalogue. 3) Calibrate the plates (allowing for several iterations), determine the properties of the galaxies almost fully automatically by running a newly developed set of programmes in the MIDAS environment and channelling all I/O through MIDAS tables. 4) Produce a catalogue on paper, magnetic tape and optical disk. 5) Investigate the data base scientifically. Each blue and red original Schmidt plate was calibrated by fitting a characteristic curve function to sensitometric marks and at least one standard galaxy with multiaperture photoelectric photometry. The catalogue is substantially complete down to a total blue magnitude of ~14.5 and for fainter magnitudes quickly becomes biased to relatively large galaxies of low surface brightness.

The accuracy has been judged from the residual in the photographic (*pg*) minus photoelectric (*pe*) versus surface brightness (*sb*). The smallest *pg-pe* scatter of the local *sb* in the range Blue *sb* 18.0 to 23.5 has a one σ of 0.14^m around the median, which in turn deviates only a few 0.01^m from zero. Central, possibly saturated regions, have been replaced by available photo-electric measurements. The accuracy of that technique is discussed. The error of the integrated magnitudes at half total light radius is found to be less than 0.10^m. At B *sb* fainter than 23.5 (larger apertures) the *pe* data become gradually less reliable due to increasing sky background contributions and superposed stars (always removed from the digitized *pg* images). Subsets of our database have been compared with other well known catalogues. In particular we have found a good agreement with an up to date version of RC2 by G. de Vaucouleurs *et al.*(1976). However, for the total B magnitudes of early type galaxies we find a disagreement with both de Vaucouleurs and Burstein et al. (1987), which is most likely due to errors resulting from the growth curve fitting technique they both have employed.

All numerical information is stored in MIDAS table files currently holding 170 parameters, including the 2D profiles, for each object. One version of the database is also implemented in the ESO intelligent database machine (IDM) allowing a very quick access to a selected subset of parameters, like identification, coordinates, magnitudes, dimensions, orientation and environmental parameters. The new ESO galaxy catalogue is expected to appear by summer 1988 together with an optical disk resident atlas comprising a total of 28000 B and R digitized images.

INTRODUCTION

A description of the new ESO catalogue containing an extensive set of 170 parameters of the ~13000 ESO/Uppsala galaxies (Lauberts, 1983) and ~1000 companions with Declination

$< -17.5^o$ and substantially complete for $|b| > 20^o$ is given in Lauberts and Valentijn (1983 and 1985, see also the abstract of this paper). In this paper we will give some information on the eventual products and the accuracy of the derived parameters. Of central importance is the way we have constructed the database by hierarchical blocks of increasing complexity with the built in possibility to reenter at any level. The fundamental pool of preprocessed data consists of a set of MIDAS table files, one for each survey field, which allows further refined analysis resulting in one single MIDAS table file, the catalogue. This catalogue has been designed to give the user a maximum freedom to use the 170 parameters and also facilitates the derivation of, for instance, additional photometric parameters.

COMPLETENESS

The original selection criterium for galaxies to appear in the original ESO-Uppsala catalogue was an apparent minimum angular diameter, 1′, on the original ESO blue Schmidt survey plate. These diameters were originally measured by eye and the limit of 1′ was chosen since it was considered the minimum angular size allowing a reasonably accurate morphological classification. For the final catalogue we will process $\sim 400 * 2$ survey plates out of the $606 * 2$ available plates, avoiding most of the plates with $|b| < 20^o$. The basic reason for avoiding the low galactic regions is that the extra work involved would be very disproportionate with the additional information that would result. ($\sim$ 10% more galaxies for at least 50% extra work). Moreover, the lack of extended galaxies on most of the low galactic latitude plates makes the photometric calibration of those plates extremely difficult and most likely much less reliable than the other plates.

With the newly determined parameters of the galaxies we can now review what the original ESO-Uppsala selection criterium means in terms of for instance the total B magnitude and the 25^m Blue isophotal diameter, D_{25} (figure 1). The dashed line in both panels of figure 1 indicates the expected distribution of these parameters for a uniform and isotropic distribution of galaxies. For $B_{tot} >\sim 14.5$ and a Blue $D_{25} <\sim 1.4'$ the catalogue becomes incomplete. The strictly complete samples contain 2861 galaxies within the 14.5^m limit and 2621 galaxies satisfying the 1.4′ diameter limit out of a total of, the presently processed, 11842 objects with no close companions. The catalogue is expected to be fairly complete down to $B_{tot} = 15.5$ (57% of the objects) and down to a minimum diameter $D_{25} = 45''$ (80% of the objects).

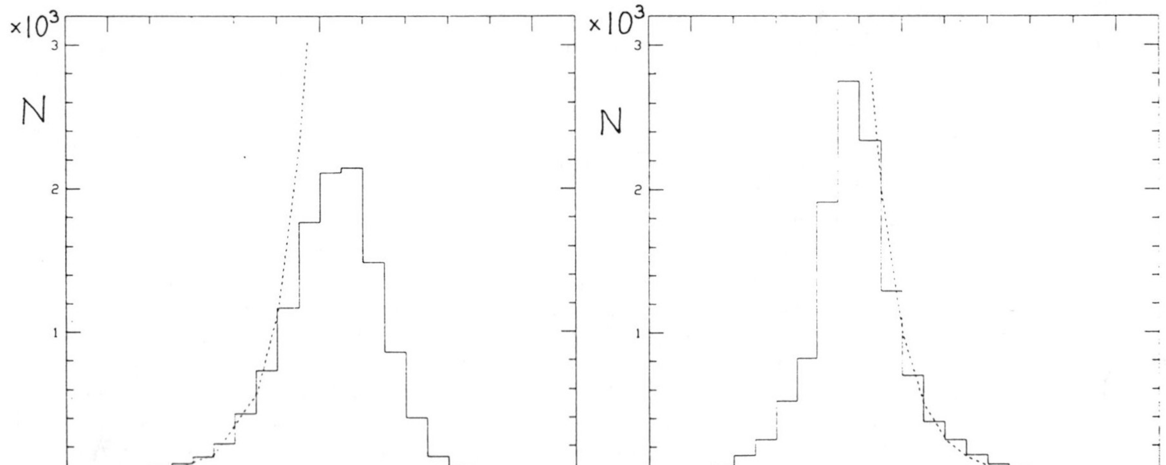

Figure 1: A histogram of total B magnitude (left side) and the Log D at the 25^m B isophote (right side) of the galaxies in the catalogue

ACCURACY OF MAGNITUDES

A comparison of our new diameters at the 25 Blue isophote with other standard photometric systems showed a good agreement (Paturel *et al.*,1987). We have performed two independent comparisons of our photographic data with photo electric measurements of others. In figure 2 the residuals of the photographic minus photoelectric ***surface brightness (sb)*** are quantified in median and one-sigma deviations, plotted as function of the B surface brightness ***sb*** for galactic latitudes $> 30^o$. The number of residuals in a particular ***sb*** bin depends on the galaxy's ***sb*** distribution and the availability of photoelectric observations. The composite picture has thus contributions from many individual galaxies. From this comparison we conclude that the one-sigma error in the local surface brightness does not exceed 0.14^m for both B and R in the B ***sb*** range 18 to 24. In the left panel of figure 2 we have also included the median for those fields which have a $|b| < 30^o$. In those low galactic latitude fields relatively more superposed stars are supposedly present , compared to the $|b| > 30^o$ fields. The effect of this inclusion of spurious stars on the photo electric measurements and hence on the residuals is demonstrated by the significantly higher median for the low galactic latitude fields in the B ***sb*** range 25-28.

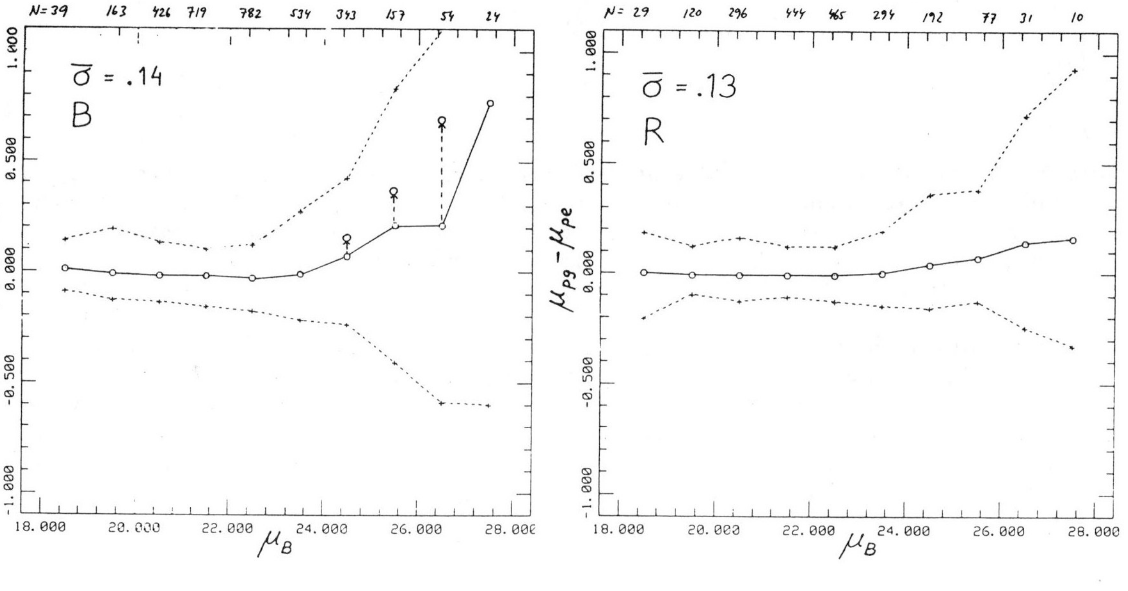

Figure 2: The residuals of the photographic surface brightness magnitudes minus photo-electric standard measurements as a function of the B sb. Left panel: the Blue sb, right panel: the Red sb. The number of objects used for each point are given at the top of each panel. The central line represents the median, while the two adjacent lines indicate the one σ deviation of that median. The average σ between B sb 18 and 24 is indicated at the upper left corner of each panel. The graphs represent data obtained at $|b| > 30^o$. In the left panel the arrows point to a new median representing fields with $|b| < 30^o$, showing the effect of faint stars on the photo electric photometry. In the photographic photometry the stars were removed.

In figure 3 the residuals of the *integrated* photographic B_e minus photoelectric magnitudes in circular apertures enclosing half of the total light are plotted as function of the B_e magnitude. About 50 percent of the objects with possibly saturated central regions, had their *pg* values replaced by available *pe* measurements. While the comparison in *sb* gave an indication of the errors in *sb* , the comparison in figure 3 shows how that error propagates in an r.m.s. error in the *integrated* half total light magnitudes. For the galaxies with a replaced central area the r.m.s. residuals are 0.07^m and 0.05^m for the B and the R, respectively. For the galaxies without a replaced central area the corresponding errors are 0.12^m and 0.08^m, respectively. The latter numbers should also serve as an indicative measure for the great majority of systems for which we did not have photo-electric photometry at our disposal. Note, that the errors presented in this paragraph are derived from residuals, using external aperture photometry. Since the aperture photometry technique involves some possible sources of errors, like aperture centering and faint stars, the internal error of the magnitudes in our catalogue will be less than those given above.

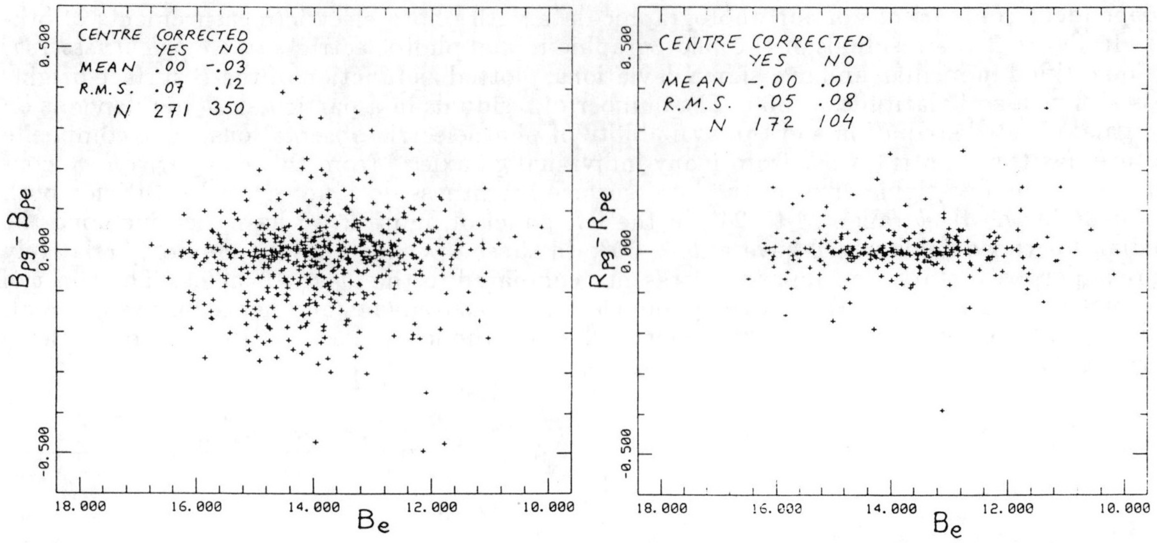

Figure 3 : The residuals of the integrated photographic (B_e) minus photoelectric magnitudes inside half total light circular apertures, as a function of the B_e magnitude. Left panel: Blue residuals, right panel: red residuals. The mean, the r.m.s. and the number of objects are printed at the upper left corner of each panel , both for a sample with the central region corrected and for a sample without a photoelectric correction.

By definition, the total magnitude of a galaxy means all integrated light out to an infinite radius. In practice, in order to derive B_{tot}, we used the radial surface brightness profiles and we have integrated these profiles at least out to B_{27} and stopped the integration between B_{27} and B_{28} at the radius where the integrated light started to fall off. By using the first drop off point of the integrated radial profile as the outer radius of the integration, we think we have optimized an empirical determination of the total magnitude.

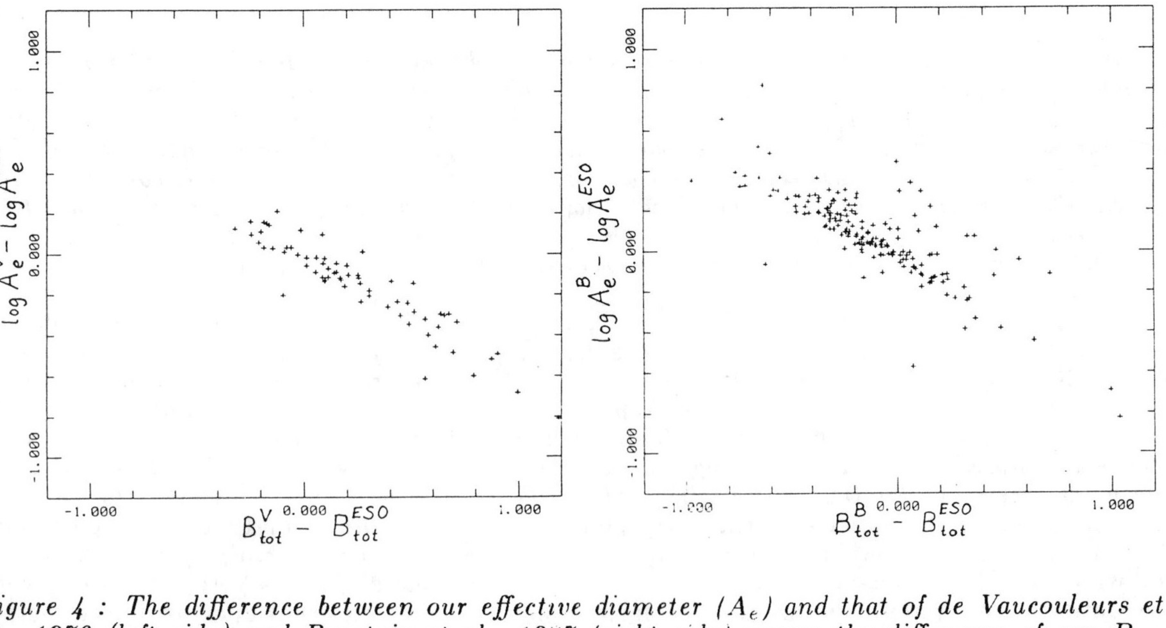

Figure 4 : The difference between our effective diameter (A_e) and that of de Vaucouleurs et al., 1976 (left side) and Burstein et al., 1987 (right side) versus the difference of our B_{tot} and that of the other authors for early type galaxies. The systematic errors demonstrate the problems with the growth curve fitting techniques.

We have also compared our B_{tot}, with those of de Vaucouleurs *et al.* (1976) and Burstein *et al.* (1987), in figure 4. The figure shows clearly that the difference in effective apertures is systematically correlated with the difference in the B_{tot}. Since such a correlation is very unlikely to emerge from our 2-dim photometry, we suggest that this is caused by problems with the growth curve technique used by the others. In this technique the B_{tot} is essentially derived from the effective aperture. If the effective aperture was estimated wrongly one would expect the B_{tot} to undergo a related error.

FINAL PRODUCTS

The catalogue, which is presently organized as a large MIDAS table file, contains a carefully selected number of parameters. In Table 1 we give a summary of the parameters obtained, with the main items being identifiers, coordinates, morphological types, magnitudes, colours, surface brightness profiles, structure descriptors, angular sizes, orientations, environmental statistics and available redshifts.

Table 1 : ABSTRACT OF PARAMETERS OBTAINED

	number of entries
ESO identifier	1
NGC or IC numbers	2
Morphological type, original and revised	2
RA and Dec, original(1950) and revised(2000)	4
Radial velocity and mean error	2
B and R magnitudes, total	2
B and R mag at B surface brightness 26 mag/sqarcsec	2
B and R mag at B surface brightness 25	2
B and R mag in 10",...,2512" diameter circle	26
B and R mag in 10",...,2512" diameter ellipse	26
B and R surface brightness at 10",...,2512" diameter ellipse	26
B-R surface brightness on ellipse enclosing half total light	1
B-R in 10" diameter circle	1
B-R total	1
Deviation (B-R) divided by 1-sigma (B-R) for morphological type	1
B and R difference photographic minus photoelectric	2
B major and minor radius in ellipse fit at B sb 20,...,26	26
R major and minor radius in ellipse fit at R sb 18,...,24	26
B and R position angle in ellipse fit to isophotes	26
B and R smallest error in ellipse fit to isophotes	26
B and R sky background magnitudes per square arcsec	2
Diameter ratio a/b E/U original, ellipse fit, adopted	3
Position angle E/U original, ellipse fit, adopted	3
Position angle of mean sb gradient at B sb 21,...,25	5
Major radius of ellipse fit at B sb 25, 26, 27	3
Major radius E/U orig, at 90, 80, 70, 50 percent of B total	5
Major radius at 0.5 mag contrast to tangent from B sb 25	1
Smallest log aperture with photoelectric data	1
Number of galaxies per square deg total, E, S0, Irr, Spirals	5
B and R sb on ellipse enclosing half total B	2
dBsb/dlog(Diameter) at B sb 25	1
Number of resolved blends using mixed exponential and Hubble law	1
Distance to 10th nearest neighbour	1
Identity of nearest and second nearest neighbour	2
Distance to nearest and second nearest neighbour	2

An example of the structural parameters obtained in this project is given in figure 5. Ellipses have been fitted to the images of early type galaxies at various surface brightness levels. Next, an overall twist angle was defined as the difference of position angle of the major axis between the *sb* 21 and 25 on the Blue plates and between *sb* 19.5 and 23.5 on the Red plates. Thus the Red and Blue twist angles have been determined entirely independently, and figure 5 gives an impression of the accuracy of the technique employed.

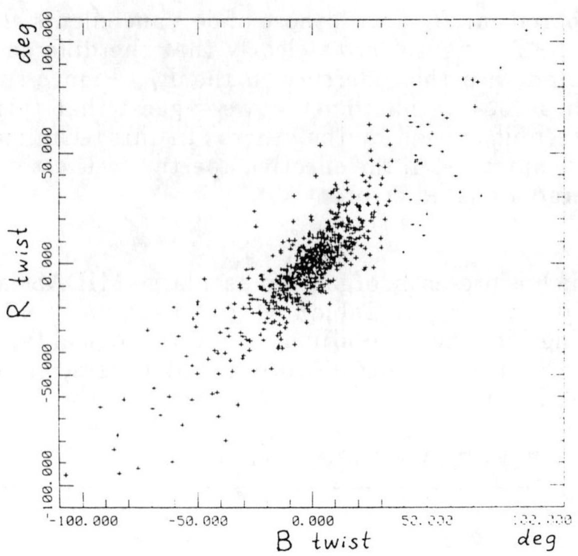

figure 5: Isophotal twists of ~1200 early type galaxies. The B twist versus the R twist.

A major scientific aim for analysing the catalogue is the study of parameters of galaxies as function of environmental conditions. For this reason several environmental entries have been created. Since the data has been sampled in a rather uniform way over most of the Southern sky, irrespective of the location of the galaxies in clusters, poor groups or isolated fields, we would expect that especially this feature allows a study, in a statistical sense, of galaxy properties (like bulge to disk ratio of SO's, twists, colour gradients) as function of the local galaxy density.

According to the present schedule the catalogue should appear in its final form by summer 1988, containing a total of 14000 ESO/Uppsala galaxies in 400 fields. The form of presentation will meet different demands on the volume and detail of information. A condensed version of the main catalogue together with full explanation of our procedure will be printed on paper (book). A version will be implemented in ESO's IDM database machine. The complete main catalogue + subsets (B and R photoelectric standards used for the calibration) will reside on magnetic tape, most likely in FITS format. The calibrated images will eventually be stored in surface brightness on optical disk.

REFERENCES

Lauberts, A. 1982: The ESO/Uppsala Survey of the ESO(B) Atlas, European Southern Observatory.

Lauberts, A., Valentijn, E.A. 1985: Proc. of the Eighth IAU European Regional Astronomy Meeting Toulouse: "New Aspects of Galaxy Photometry", Ed. J.L. Nieto, Publ. Springer Verlag , p 73

Lauberts, A., Valentijn E.A. 1983: The ESO Messenger **34**, 10

Burstein, D., Davies, R., Dressler, A., Faber, S.M., Stone, R.P.S., Lynden-Bell, D., Terlevich, R.J. and Wegner, G. 1987: Astrophys. J. Suppl. **64**, 601

Paturel,G., Fouque, P., Lauberts, A., Valentijn, E.A., Corwin, H.G., de Vaucouleurs, G., 1987: Astron. Astrophys. *in press*

Vaucouleurs, G. de, Vaucouleurs, A., Corwin, H.G. Jr, 1976: Second Reference Catalogue of Bright Galaxies, Texas University Press, Austin (RC2)

AN INFRARED-OPTICAL STUDY OF IRAS POINT SOURCES USING THE IRAS AND COSMOS DATABASES.

S.K. Leggett[1], R.G. Clowes[2], H.T. MacGillivray[2], Q.A. Parker[2], A. Savage[2], G. Wang[2] & R.D. Wolstencroft[2]

[1] Department of Astronomy, University of Edinburgh, Blackford Hill, Edinburgh, UK
[2] Royal Observatory, Blackford Hill, Edinburgh, UK

ABSTRACT

The Infrared Astronomical Satellite IRAS surveyed about 96% of the sky finding ~250,000 point sources. The full potential of this database can only be realised through a complete systematic optical identification programme, such as that in progress at Edinburgh. We have produced a unique optical-infrared database by combining data from the plate-measuring machine COSMOS with the IRAS catalogue. To achieve this we have developed our own software which is used in conjunction with the STARLINK database package SCAR. SCAR is also used for analysis of the data; a summary of the astrophysical information gained for IRAS stars and galaxies is presented.

1. INTRODUCTION

The Infrared Astronomical Satellite IRAS surveyed about 96% of the sky in bands centred at 12, 25, 60 and 100 μm (Beichman et al. 1985). Around 250,000 point sources were detected, of which about one-third have positional associations with objects in astronomical catalogues. The full potential of this database can only be realised through a systematic optical identification programme such as that in progress at Edinburgh, using high quality UK Schmidt IIIa-J survey plates, and the plate measuring machine COSMOS. The COSMOS system has recently been reviewed by MacGillivray & Stobie (1984), and is described by MacGillivray et al. in this conference. We have produced a unique infrared-optical database by combining the IRAS and COSMOS data.

The programme has so far covered the following sky areas : 200 square degrees at the Abell 1060, Coma, Fornax and Virgo galaxy clusters, 340 square degrees near the North Galactic Pole (NGP), and a large area of 2750 square degrees centred on the South Galactic Pole (SGP). These regions contain 5340 IRAS point sources; identifications for a small area of the SGP and the Virgo cluster area are given in Wolstencroft et al. (1986) and Leggett et al. (1987), respectively.

2. CREATING THE DATABASE

The first step in creating the infrared–optical database is scanning the central 28 square degrees of each UK Schmidt photographic plate in COSMOS, which takes about 6 hours. Typically around 60,000 sources are found above the limit we have set of $B \sim 21$. The COSMOS data set includes accurate positions and blue magnitudes for each object, as well as various image parameters (e.g. size and position angle).

The STARLINK Catalogue Access and Reporting package SCAR is used to select all IRAS point sources within the plate area. Typically there are about 30 IRAS sources per Schmidt plate. We then use our own software to extract all sources in the COSMOS data which have been measured to be within 1 arcminute of the IRAS catalogue position. On a VAX 11/780 this process takes about 3 minutes cpu.

Our own programs output lists of candidate optical identifications for each IRAS source, as well as plate overlays on which the IRAS sources are plotted. The candidates are inspected by eye using the overlays. Identifications with bright stars are unambiguous, and these give the initial IRAS positional errors for the field, which then allow identification of the other sources.

Finally, the formatted list of COSMOS data output by our programs is editted – flagging the correct identification and adding notes on the optical appearance or galaxy morphology. This formatted file, with a description file, constitutes a SCAR catalogue. The SCAR package is then used for joining the COSMOS and IRAS data for each object, and for analysis of these data. Furthermore, SCAR can be used to augment this data set by joining with other machine–readable catalogues (for example we have used the Binggeli, Sandage & Tammann (1985) Virgo galaxy catalogue and the Huchra catalogue of galaxy redshifts).

3. ASTROPHYSICS FROM THE DATABASE

The large sky area covered by this programme has provided not only a large sample of IRAS sources, but allowed us to study the effect of different environments on IRAS properties. Hence we can study many aspects of the infrared sky. At high galactic latitudes we find that 97% of IRAS point sources have an optical counterpart with $B<21$; in the field the sources are divided almost equally between stars and galaxies.

3.1 Stars

Studies of the infrared properties of stars gives information on the nature of circumstellar matter (e.g. proto-planetary disks, Gillett 1986) and mass loss from evolved stars (e.g. Olnon et al. 1984). We have found so far that our identifications (at high galactic latitudes) are dominated by ordinary K-type giants, and all such stars brighter than $V \sim 8$ are detected by IRAS. Our sample does contain some evolved stars that have undergone mass loss – Mira-type stars. Early type stars with infrared flux excesses (indicating circumstellar material) are relatively uncommon.

3.2 Galaxies

Studies of the optical and infrared properties of galaxies give information on star formation as well as on the general nature of infrared emission from galaxies (such as recent multi-temperature component models of galactic emission, e.g. Cox, Krugel & Mezger 1986). This programme has also enabled the study of the effects of both environment and morphology on the far-infrared properties of galaxies.

Leggett et al. (1987) show that IRAS galaxy sources are dominated by spirals; for galaxies with $B<16$, 4% of ellipticals and lenticulars are detected compared with 44% of Sc galaxies. 88% of Sc's brighter than $B=14$ are detected. The 'normal' IRAS galaxy is a spiral of type Sa-d with $B<14$, far infrared luminosity (L(IR))about equal to optical luminosity and $F_\nu(100\mu m)/F_\nu(60\mu m) \sim 3$. L(IR) is $\sim 10^9 L_\odot$, and for a given B, L(IR)/L(B) varies by about a factor of 10.

These properties show that the typical IRAS galaxy is indistinguishable from field disc galaxies with normal rates of star formation (Rieke & Lebofsky 1986). Furthermore, cluster environment has no effect on IR properties. The Virgo cluster galaxies are indistinguishable in the infrared from field galaxies at similar redshifts; this is true of even the very HI-deficient galaxies (Leggett et al. 1987, also Leggett, Brand & Mountain 1987).

Wolstencroft et al (1986), Leggett et al. (1987) and work in progress, show that far-infrared luminosity is correlated with the flux ratio $F_\nu(60\mu m)/F_\nu(100\mu m)$ and the luminosity ratio L(IR)/L(B). The optically faint background galaxies and the interacting galaxies are 'hotter' and more infrared luminous than the rest of the sample. Galaxies with $B>16$, $L(IR)/L(B)>4$, and $F_\nu(100\mu m)/F_\nu(60\mu m)<2$ may have $L(IR)>10^{11}L_\odot$. Redshifts obtained for faint field IRAS galaxies show that those with $L(IR)/L(B)>30$ may be ultraluminous with quasar-like luminosities of $L(IR)>10^{12}L_\odot$; this applies to about 1 in 20 IRAS field galaxies.

There is some evidence that the central regions of spiral galaxies become important at high infrared luminosities. For the effectively redshift-limited sample of the Virgo cluster (Leggett et al. 1987) infrared luminosities as high as $10^{10}L_{\odot}$ were only reached by later spiral types (Sc's); however the SGP sample contains a population of early type spirals with high values of L(IR)/L(B) (Wolstencroft et al 1986), implying values of L(IR)$\sim 10^{11}L_{\odot}$.

The COSMOS measurements of galaxy size have also allowed us to investigate whether there is a cosmologically significant anisotropy in the surface density of IRAS galaxies, as suggested by Rowan-Robinson et al. (1986). A comparison of our SGP and NGP samples does imply a disparity in surface densities, but our optical information shows that it is strongly concentrated in galaxies having optical, angular major axes greater than 130 arcseconds, for which the median distance is only $\sim 16h^{-1}$Mpc (Clowes et al. 1987). We conclude that the origin of the disparity is the local supercluster rather than a structure of much larger dimensions.

3.3 Empty fields

IRAS found Galactic emission in the far-infrared from interstellar grains heated by the interstellar radiation field, which is known as 'cirrus'. The 3% of IRAS sources that we find have no optical counterpart to B$\sim$21 are mostly due to the infrared cirrus.

However, about 1% of IRAS sources, which so far are all 60μm only sources, do appear to have an optical counterpart near or below the plate limit from CCD images. These sources seem to be extremely infrared luminous galaxies, which are probably interacting galaxies. The relationships between infrared luminosity and infrared to blue luminosity ratio and infrared flux ratio, mentioned in 3.2, indicate that such sources may have LIR$\sim 10^{13}L_{\odot}$.

REFERENCES

Beichman, C.A., Neugebauer, G., Habing, H.J., Clegg, P.E. & Chester, T.J. (eds), 1985. Explanatory Supplement to the IRAS Catalogues and Atlases, US Government Printing Office, Washington DC.

Binggeli, B., Sandage, A. & Tammann, G.A., 1985. Astr.J., 90, 1681.

Clowes, R.G., Savage, A., Wang, G., Leggett, S.K., MacGillivray, H.T. & Wolstencroft, R.D., 1987. Mon.Not.R.astr.Soc., in press.

Cox, P., Krugel, E. & Mezger, P.G., 1986. Astr.Astrophys, 155, 380.

Gillett, F.C., 1986. Proc. First International IRAS Conference, Light on Dark Matter, D. Reidel.

Leggett, S.K., Clowes, R.G., Kalafi, M., MacGillivray, H.T., Puxley, P.J., Savage, A. & Wolstencroft, R.D., 1987. Mon.Not.R.astr.Soc., 227, 563.

Leggett, S.K., Brand, P.W.J.L. & Mountain, C.M., 1987. Mon.Not.R.astr.Soc., 228, 11p.

MacGillivray, H.T. & Stobie, R.S., 1984. Vistas Astr., 27, 433.

Olnon, F.M., Baud, B., Habing, H.J., de Jong, T., Harris, S. & Pottasch, S.R., 1984. Astrophys.J., 278, L41.

Rieke, G.H. & Lebofsky, M.J., 1986. Astrophys.J., 304, 326.

Rowan-Robinson, M., Walker, D., Chester, T., Soifer, T. & Fairclough, J., 1986. Mon.Not.R.astr.Soc., 219, 273.

Wolstencroft, R.D., Savage, A., Clowes, R.G., MacGillivray, H.T., Leggett, S.K. & Kalafi, M., 1986. Mon.Not.R.astr.Soc., 223, 279.

ACKNOWLEDGEMENTS

The COSMOS group at the Royal Observatory Edinburgh supplied the measurements of the UK Schmidt survey plates; data were processed using the STARLINK facilities.

Mapping the Extragalactic Sky with IRAS PSC Data

E.J.A. Meurs[1], *H.-M. Adorf*[2], *R.T. Harmon*[3]

[1]European Southern Observatory
Karl-Schwarzschild-Str. 2
D-8046 Garching b. München, F.R. Germany

[2]Space Telescope – European Coordinating Facility
European Southern Observatory
Karl-Schwarzschild-Str. 2
D-8046 Garching b. München, F.R. Germany

[3]Institute of Astronomy, University of Cambridge
Madingley Road
Cambridge CB3 0HA, United Kingdom

Abstract: Extragalactic sources have been selected from the IRAS Point Source Catalog (PSC) on the basis of positional and flux criteria. A number of sources were labelled as likely being either extragalactic or galactic in nature. The resulting training sets were used for designing a maximum-likelihood classifier, which optimally discriminates between the various source categories. This classifier was applied to a relevant subset of the PSC eventually leading to sky distributions for each source category. The relative frequencies of each source category were used as 'odds' in a maximum-a-posteriori classification of the PSC subset. Further development of this work is indicated.

Introduction

The sky survey carried out by the Infra-Red Astronomical Satellite (IRAS) covered 96% of the sky. Data for ~ 250,000 point sources were obtained at 12, 25, 60 and 100 μm and are listed in the Point Source Catalog (PSC). The data are homogeneous all over the sky, except for a region close to the galactic plane. Identification programmes show the sources to be associated with a variety of objects. Most of the sources are stars, which are hot (bright at 12 and 25 μm) and concentrated towards the galactic plane and centre. Galaxies on the other hand are cool (bright at 60 and 100 μm) and are distributed all over the sky. There are still other galactic sources, generally cool objects, which are associated with Cirrus, HII regions, etc. and exhibit various scale heights in the Galaxy.

For many purposes it would be desirable to be able to extract the extragalactic content from the PSC. This has been attempted previously by Yahil et al. (1986) and Meiksin & Davis (1986), who had to exclude large areas of the sky where contamination by Cirrus seemed unavoidable. A recent investigation by Meurs & Harmon (1987) demonstrated a combination of infrared colours with the 100 μm flux to effectively separate extragalactic sources from galactic components in the PSC. Because of the evident interest in reliable IRAS galaxy samples (cf. Harmon et al. 1987), it will be important to assess the results of this selection using appropriate methods from multivariate statistics and decision theory. The large number of sources and the rather homogeneous way in which they have been obtained render the PSC indeed a promising case for statistical investigation. In this contribution we present the first stage of this work in which we generally follow the approach of Meurs & Harmon, aiming specifically at selecting a clean set of extragalactic sources from the PSC.

Construction of training sets

The results from source identification programmes and inspection of colour-colour diagrams for IRAS sources with good 60 μm fluxes indicate that galaxies occupy a fairly well-defined region in flux-space (cf. the colour-colour diagram in Figure 1). For a start, we generally omitted galactic stars by requiring $\log F_{25}/F_{60} < 0.2$. The remaining subset of the PSC comprised $\sim 32,000$ sources.

Meurs & Harmon (1987) describe in their 'handcrafted' approach two further galactic components, viz. "Cirrus" and "Thin Plane". The sky distributions of these two components and of the extragalactic sources are all different, and so we defined sky regions where the objects of one of the three categories are likely to dominate. Thus we arrived at labelled sources which — together with their four IRAS fluxes — constitute appropriate 'training sets' for *galaxies*, *Cirrus* and *Thin Plane* with ~ 7000, ~ 1000 and ~ 1000 objects, respectively.

Bayesian classification

The parameter distributions of the training sets were approximated by multivariate Gaussian distributions, which were used to compute class-conditional probabilities for all objects in the PSC subset. Based on the maximum-likelihood principle (Duda & Hart 1973; Hand 1981) quadratic discriminant functions were derived, which provide an optimal separation between the three source categories under consideration. The resulting classifier was applied to the PSC subset defined above and the class-specific parameter distributions are displayed in Figures 2a–c, in a colour-colour diagram as in Figure 1. The sky distributions corresponding to each of the three categories are shown in Figures 3a–c.

Relative frequencies of the three source categories taken from the maximum-likelihood classification step were subsequently used as 'odds' (bias) in a maximum-a-posteriori classifier. The resulting sky distributions look fairly similar to those derived with maximum-likelihood classification in Figure 3. Some differences are found in the galactic plane, where several sources change category from *Thin Plane* to *galaxy*.

Conclusions

Our classification results may be summarised as follows:

(1) The classification yields a sharp decision boundary between *galaxy* and *Cirrus* (Figure 2a and 2c).

(2) In the two-colour diagram the *Thin Plane* distribution and the distributions of the other two categories overlap with each other (Figure 1 and 2), which is alleviated by the 100 μm flux.

(3) The shape of the *Thin Plane* distribution is non-Gaussian and may be better described by a mixture of two Gaussians.

(4) From the mean 'spectra' (Figure 4) it can be read off that (a) *Thin Plane* objects are bright (as measured by F_{100}), *galaxies* are faint and *Cirrus* objects are in between; (b) discrimination between *galaxies* and *Cirrus* mainly depends on colour C_{34} and the flux F_{100}; colours C_{12} and C_{23} have little discriminative power for separating *galaxies* from *Cirrus*; (c) all fluxes contribute to discriminate between *Cirrus* and *Thin Plane*.

Future work

The prior probabilities used in maximum-a-posteriori classification may have to be related to positional information. It is intended to use fully identified source samples for the construction of training sets and Bayesian classifiers. Identified sources again are to be used to assess the hit rates and false-alarm rates of the statistical classifiers. Another possibility for evaluating the power of the classifiers would be to use Efron's bootstrap sampling method. Finally it is envisaged to use unsupervised classification techniques in order to study the clustering in the parameter space of IRAS point sources. A result of these investigations may be a further sub-division of source categories.

Acknowledgement: This work greatly benefitted from the availability of the MIDAS table system (Ponz & Murtagh 1987).

References

Duda, R.O., Hart, P.E.: 1973, *Pattern Classification and Scene Analysis*, John Wiley & Sons, New York

Hand, D.J.: 1981, *Discrimination and Classification*, John Wiley & Sons, Chichester

Harmon, R.T., Lahav, O., Meurs, E.J.A.: 1987, Mon. Not. R. Astron. Soc. **228**, 5P–10P

Meiksin, A., Davis, M.: 1986, Astron. J. **91**, 191–198

Meurs, E.J.A., Harmon, R.T.: 1987, *preprint*

Ponz, J.D., Murtagh, F.: 1987, *this volume*

Yahil, A., Walker, D., Rowan-Robinson, M.: 1986, Astrophys. J. **301**, L1–L5

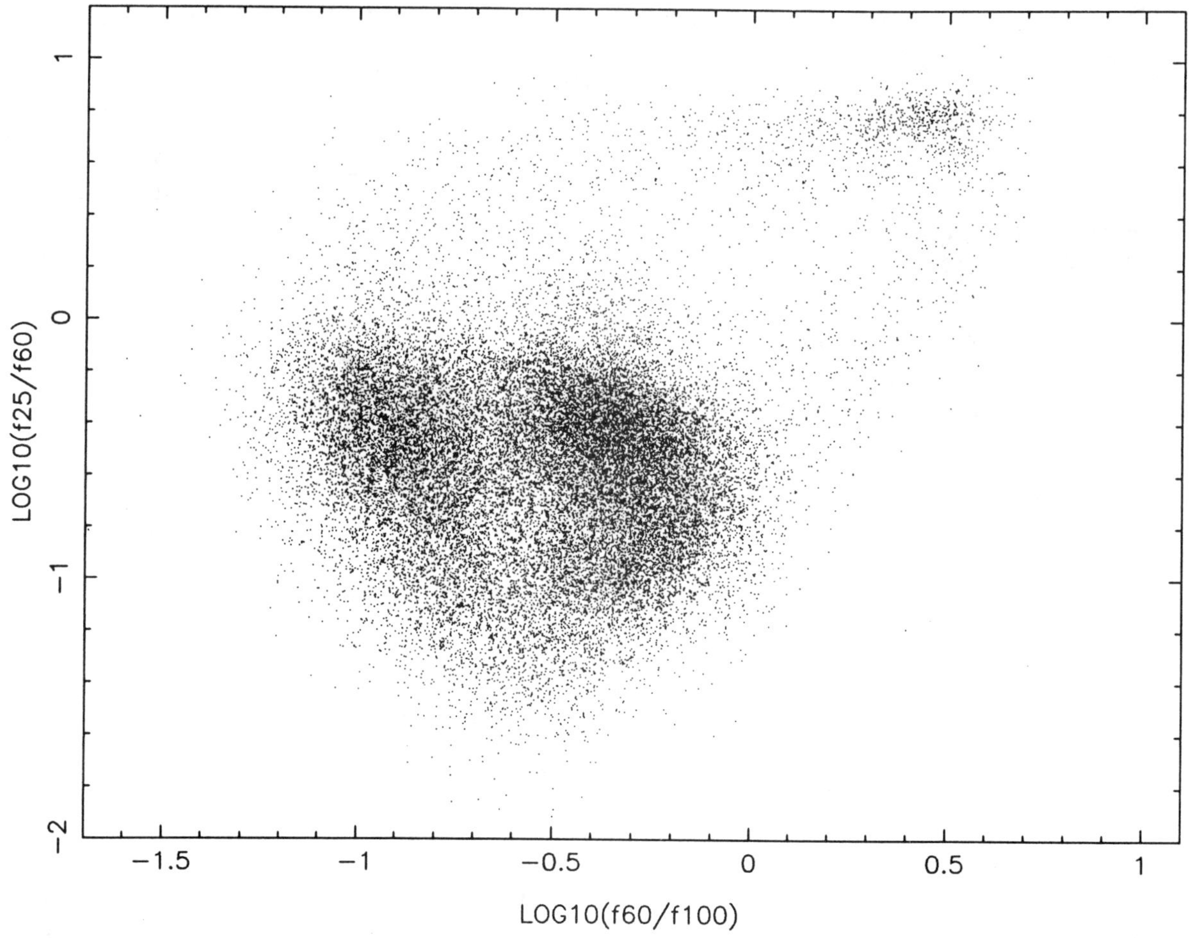

Fig. 1: The subset of IRAS PSC sources (see text) in the two-colour diagram of C_{23} versus C_{34}, where the colours are defined as $C_{23} := \log F_{25}/F_{60}$ and $C_{34} := \log F_{60}/F_{100}$.

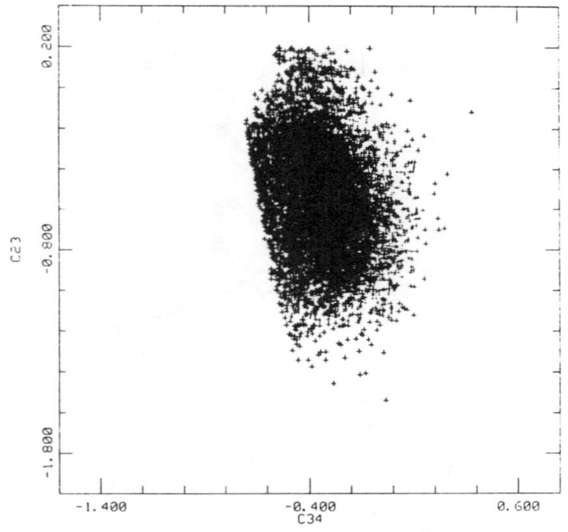

Fig. 2a

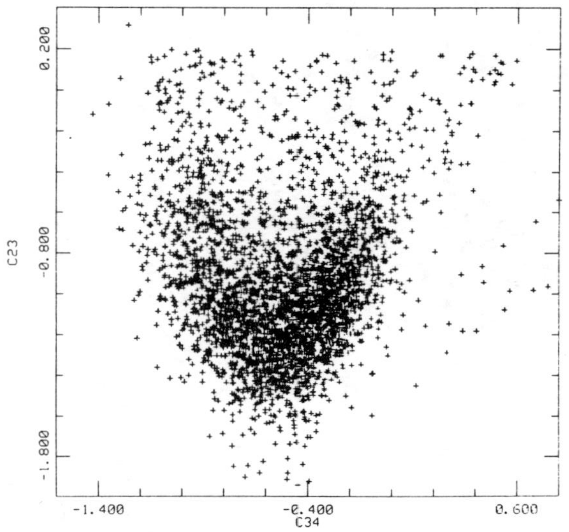

Fig. 2b

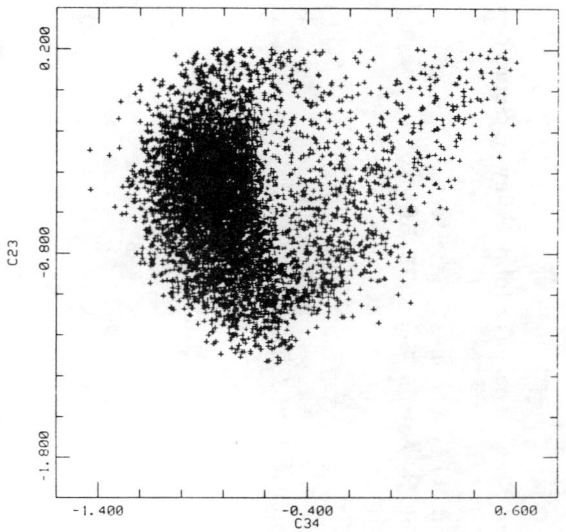

Fig. 2c

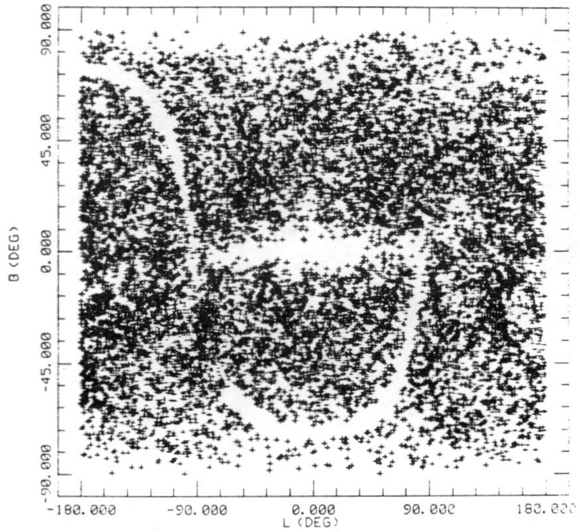

Fig. 3a

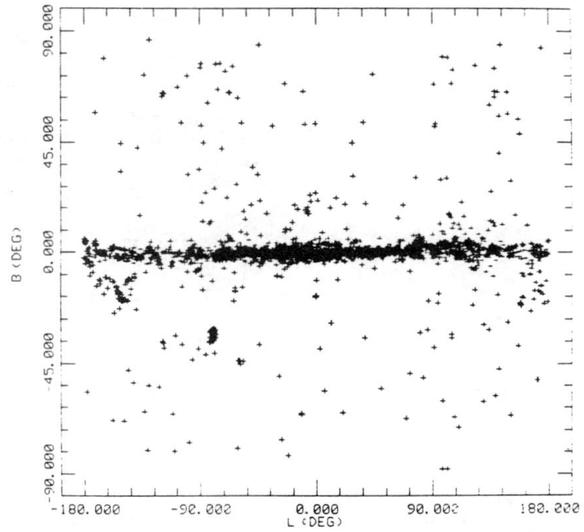

Fig. 3b

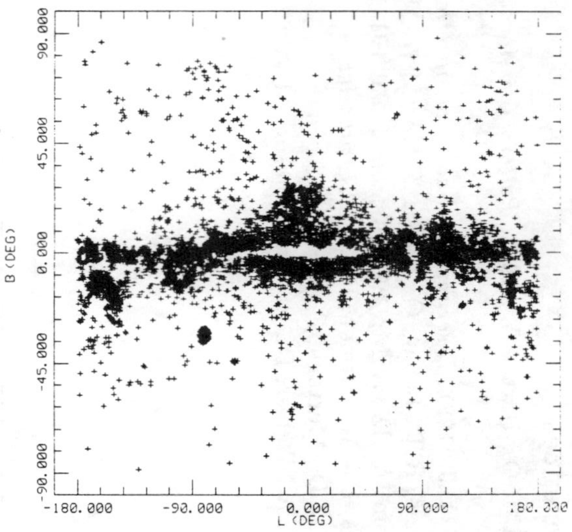

Fig. 3c

(see previous page)

Fig. 2: The two-colour distributions of the sources selected with the maximum-likelihood classifier: (a) *galaxies*, (b) *Thin Plane* and (c) *Cirrus*.

Fig. 3: The sky distributions of the same categories as in Figure 2. The empty region in the middle of (a) and (c) is caused by source confusion. (The curved blank strips in (a) represent the 4% of the sky absent from the PSC.)

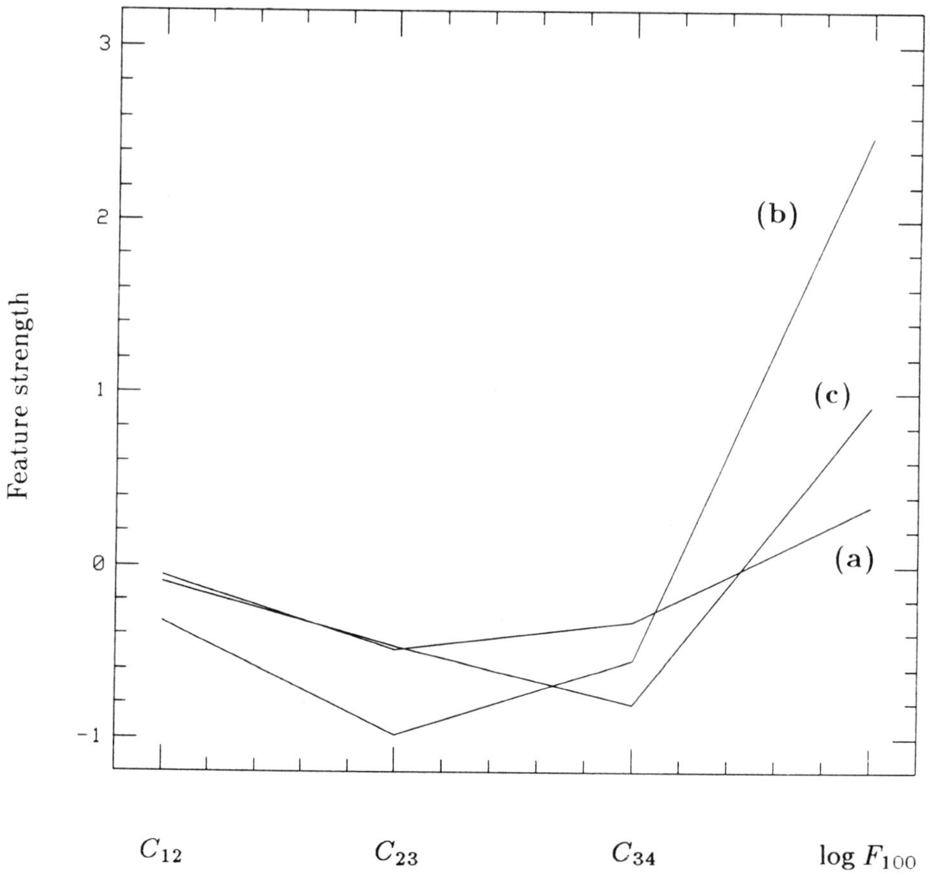

Fig. 4: The mean 'spectra' of the different source categories (a) *galaxies*, (b) *Thin Plane* and (c) *Cirrus*. representing the centres of the class-specific clusters in the parameter space.

STATISTICAL ANALYSIS OF C-RICH STARS IN THE LRS

R. Papoular[2], B. Pégourié[1,2], M. de Muizon[1]

1. Sterrewacht Leiden, Postbus 9513, 2300 RA Leiden (NL)
   Observatoire de Paris, Section de Meudon (France)
2. Service d'Astrophysique, CEN-Saclay (France)

ABSTRACT

Individual spectra of the LRS catalog rarely have a S/N high enough to allow a detailed and accurate analysis. We have circumvented this difficulty(1) by normalizing the spectra to 1 at 8 µm, then averaging the relative fluxes at each discrete wavelength over a few tens of spectra, grouped in sub-classes of the 10 main classes of the catalog. Each sub-class is defined by some secondary criterion: 8-µm brightness (indication of the distance), strength of the main feature, galactic location, etc...The size of the sub-classes is large enough to increase the S/N to acceptable values while retaining enough homogeneity so that the physical meaning of the average spectrum is not lost. This has made it possible for us to isolate three different shapes of the 11.3-µm (SiC) feature ; detect three features (about 8.6, 11.7 and 12.8 µm) previously not encountered in CS shells ; and measure galactic gradients of the strength of the 9.7-µm silicate feature.

The details of the methods and results for classes 1n, 2n, 4n, 6n are described elsewhere(1,2) ; here, we concentrate on C-rich stars.

1. PRINCIPLE OF THE METHOD

For each spectrum, $F(\lambda)$, we define the underlying continuum, $F_c(\lambda)$, by a power-law between two spectral points at λ_1, λ_2. Define the relative excess by $\mathcal{E}(\lambda) = F(\lambda)/F_c(\lambda) - 1$ and the shape of the spectral excess by $S(\lambda) = \mathcal{E}(\lambda)/\mathcal{E}(\lambda_o)$, with $\lambda_o = 11.3$ µm, the nominal location of the SiC feature. For each group of spectra we average $S(\lambda)$ over the group population to give $\overline{S(\lambda)}$. Both LRS spectral ranges (8-13 and 11-22 µm) are combined here in a single diagram ; this gives rise to a discontinuity at 13 µm. The second range is less reliable. It was found that the instrumental calibration left a common, weak, residual pattern superimposed upon even the smoothest spectra. To get rid of this, we smoothed each spectrum by means of a running average over three successive discrete wavelengths.

2. THE THREE SIGNATURES IN CLASS 4n (MOSTLY C-RICH STARS)

The spectral shapes in class 4n are found to fall in three sub-classes (fig. 1, 2, 3), all of which exhibit the SiC feature at ~ 11.3 μm but differ in details as follows:

- 4A: the largest sub-class ; the central (SiC) feature is narrow, extending from ~ 10 to 14 μm ; a narrower feature occurs on its red wing, peaking at ~ 11.7 μm (see also fig. 4) ; a third feature appears at ~ 8.6 μm ; the spectral excess is negative beyond ~ 15 μm. The objects in this group are selected by taking $\lambda_1 = 9.7$ μm, $\lambda_2 = 14$ μm and imposing $\mathcal{E}(8.6) > 0$.
- 4B: the central feature peaks at ~ 11.3 μm ; there is a low plateau on its blue wing and the spectral excess is positive beyond ~ 15 μm. These objects are selected by taking $\lambda_1 = 8.6$, $\lambda_2 = 15$ μm and imposing $\mathcal{E}(9.7) > 0$ and $\mathcal{E}(10)/\mathcal{E}(11.3) \leq 0.3$.
- 4C: the smallest sub-class ; the SiC feature is wide, extending from ~ 8.5 to ~ 15 μm, with a distinct shoulder at ~ 10 μm ; the peak is again at ~ 11.3 μm. These objects are selected by taking $\lambda_1 = 8.6$, $\lambda_2 = 15$ μm and imposing $\mathcal{E}(10)/\mathcal{E}(11.3) > 0.3$. There is a gap in the distribution of $\mathcal{E}(10)/\mathcal{E}(11.3)$ between about 0.2 and 0.4.

The average strength of the feature, $\overline{\mathcal{E}_0} \equiv \overline{\mathcal{E}(11.3)}$, as well as the range covered by $\mathcal{E}_0$ decrease as one goes from 4A to 4C through 4B. The corresponding std.dev. of galactic latitudes are ~ 9, 13.5 and 18° respectively.

3. Internal dispersion of shapes within sub-classes

Instrumental noise results in some spurious dispersion of shapes. However, sub-class A is large enough that one can still subdivide it in smaller groups whose average spectral shapes differ significantly (fig. 4). But the difference, now, is quantitative rather than qualitative. As $\mathcal{E}_0$ increases, the relative intensity of the narrow feature at 11.7 μm decreases continuously and the feature at 8.6 μm first decreases then increases.

The limiting shape for large $\mathcal{E}_0$'s (curve 6) is reproduced in fig. 5 (full line) together with the spectral extinction of (crystalline) α-SiC measured in the laboratory(3) (dashed line). The agreement is fairly satisfactory.

4. The 11.3 μm feature in class 1n

The "featureless" class 1n contains a number of noisy spectra with a weak feature at 11.3 μm, which escaped the test of the classification algorithm ; 120 such objects are found among the 445 brightest objects of 1n. Most of them (91) are of type 4C above (cf. figs. 6 and 3) ; the rest is of type 4A (cf. figs. 7 and 1). Note a new feature at ~ 12.8 μm in both figs. 6 and 7. A confirmation

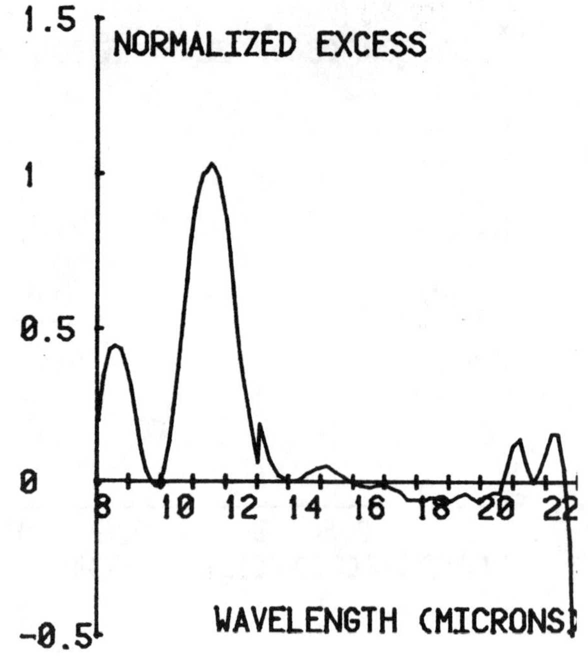

Fig. 1

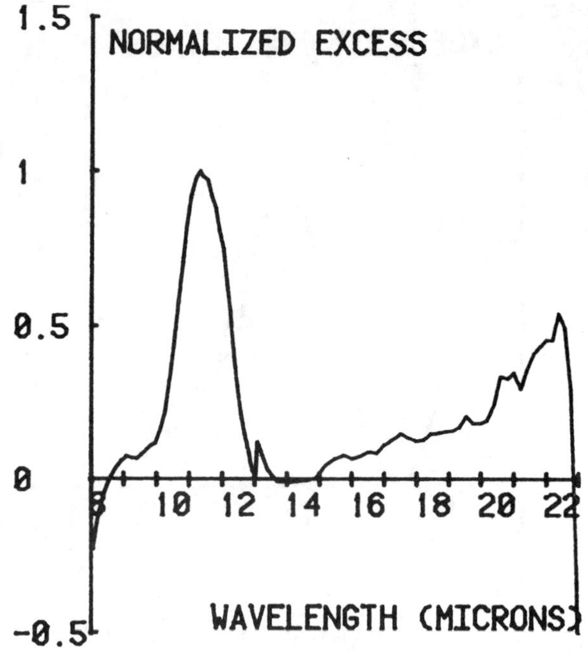

Fig. 2

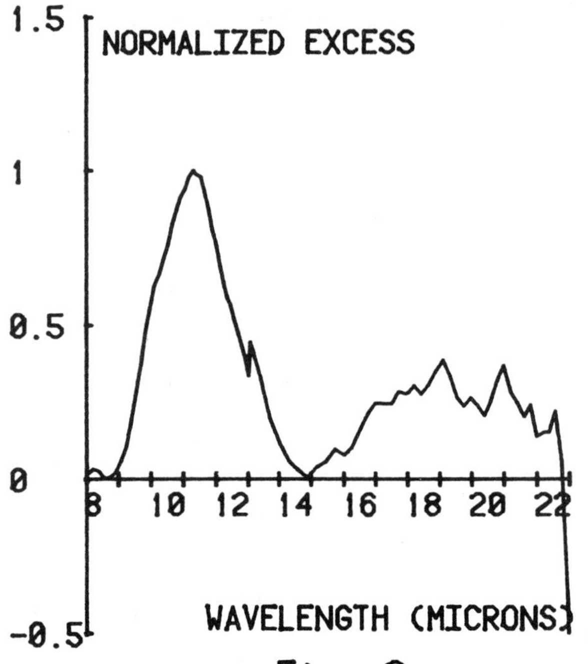

Fig. 3

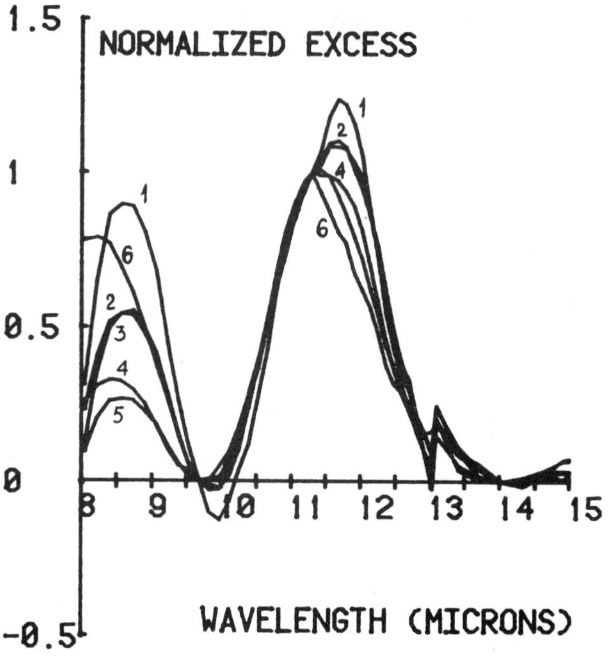

Fig. 4

AVERAGE SHAPES OF SPECTRAL EXCESS

1. Sub-class 4A: 305 objects ; note secondary features at 8.6 and 11.7 μm.
2. Sub-class 4B: 171 objects ; note plateau on blue wing.
3. Sub-class 4C: 46 objects ; note feature width.
4. Six groups in sub-class 4A with $\mathcal{E}(11.3)$ ranging between:
   1) 0.1-0.18 ; 2) 0.18-0.25 ; 3) 0.25-0.3 ; 4) 0.3-0.6 ; 5) 0.6-1 and
   6) 1-3. Number of objects in each group: 26, 57, 44, 127, 37 and 21.

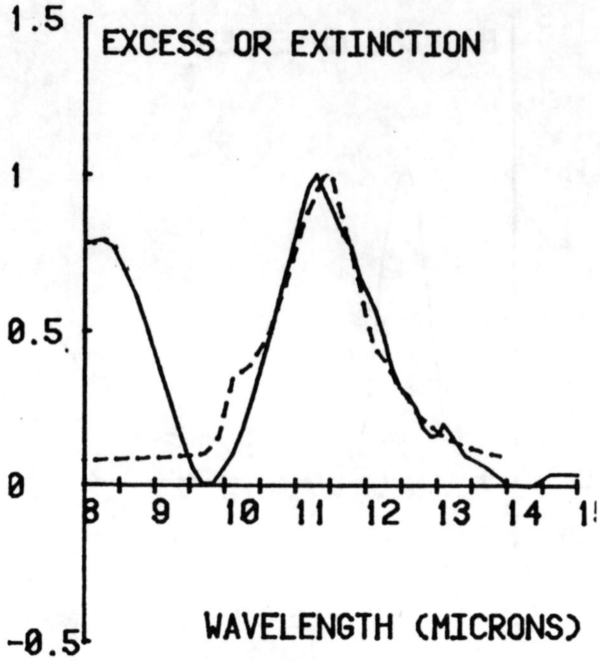

Fig. 5

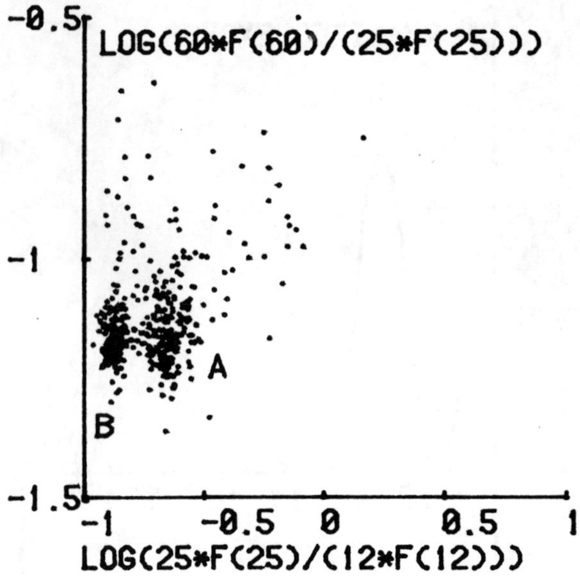

Fig. 8

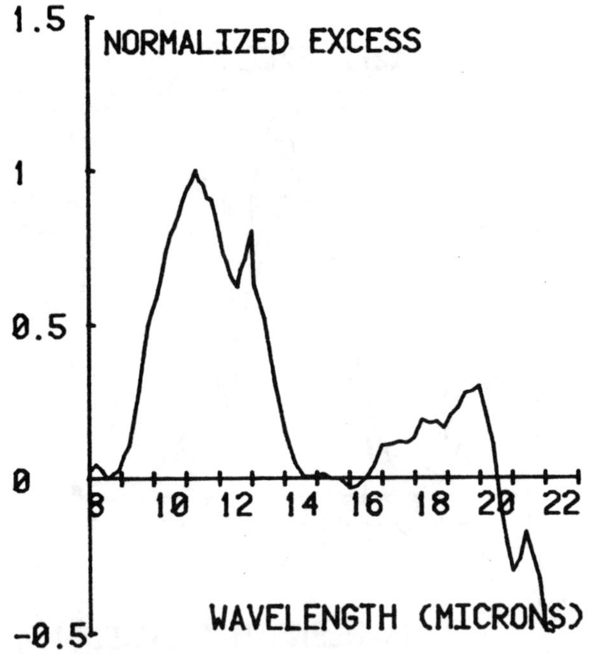

Fig. 6

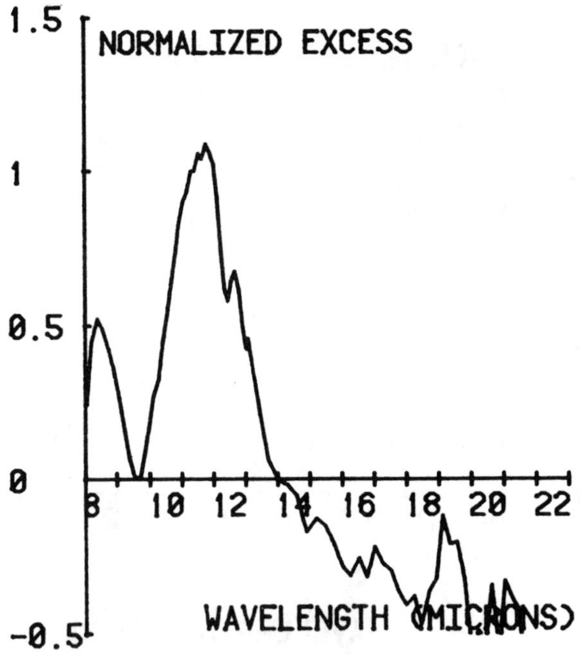

Fig. 7

5. Identification of the 11.3-μm feature. Dashed curve: laboratory cristalline SiC[(3)]. Full curve: "Asymptotic" feature of sub-class 4A (cf. fig. 4) ; here, the absence of shoulder at ~ 10 μm indicates an amorphous material.
6. The 11.3 μm feature in class 1n. Sub-class of 91 objects of type 4C (cf. fig. 3). Note feature at 12.8 μm.
7. The 11.3 μm feature in class 1n. Sub-class of 29 objects of type 4A (cf. fig. 1). Note feature at 12.8 μm.
8. IRAS-colours diagram for the brightest objects of class 1n. Objects of figs. 6 and 7 constitute the cloud of point A. Photospheres constitute cloud B.

of the circumstellar nature of these features is the location of these objects in the IRAS colour-colour diagram (fig. 8), away from the "cloud" of photospheres.

Although the discussion of the assignments and implications of the various features is outside the scope of this paper, the presence of amorphous hydrogenated carbon is strongly indicated by the features at 8.6 and 11.7 μm.

REFERENCES

(1) G al, O., de Muizon, M., Papoular, R., Pégourié, B., 1987, Astron. Astrophys., 183, 29.

(2) Baron, Y., de Muizon, M., Papoular, R., Pégourié, B., 1981, Astron. Astrophys. (in press).

(3) Borghesi, A., Bussoletti, E., Colangeli, L., de Blazi, C., 1983, Infrared Phys. 23, 321.

Statistical Analysis of Part of the IRAS Database: The Example of the Large Magellanic Cloud

J. Spicker [1] , U. Klein [2] , J.V. Feitzinger [3,1]

[1] Astronomisches Institut, Ruhr-Universität Bochum
Postfach 10 21 48, D-4630 Bochum, F.R.G.

[2] Radioastronomisches Institut der Universität Bonn,
Auf dem Hügel 71, D-5300 Bonn, F.R.G.

[3] Sternwarte der Stadt Bochum,
Castroper Str. 64, D-4630 Bochum, F.R.G.

1. Abstract

Some results of the first application of standard statistical methods to a variety of data on the Large Magellanic Cloud are presented. From an analysis of the two-dimensional autocorrelation and structure functions of the IR radiation at 60 μ and 100 μ, we derive the scale properties of dust in the LMC. The results are compared with a similar analysis of the HI, HII and radio continuum data.

2. Introduction

Throughout our analysis, we assume that the ISM in the LMC is in a state of turbulence, so that spatial fluctuations of velocity and matter occur and can be regarded as realizations of a stochastic process. The most commonly used statistics to investigate the behaviour of stochastic processes are the two-point correlation and structure tensors. In the present paper, some results of the first application of these statistics in its two-dimensional form to a complete stellar system are described, namely the analysis of the 60μ and 100μ IRAS data of the LMC. The same methods have already been successfully applied to the HI radial velocity field (Spicker and Feitzinger 1987b), the 13cm and 21cm cm radio continuum radiation and the optical polarization (Spicker and Feitzinger 1987c), and to the HI column density, 74cm radio continuum, colour excess of young stars, FUV and Hα radiation (to be published) of the LMC. The typical scaling properties of the ISM in the LMC could be derived from this analysis, and the state of turbulence of the matter could be proved for most of the components mentioned.

3. Statistical Methods

The full spectrum of the methods used in our analysis and the statistical vocabulary has been published in a seperate paper (Spicker and Feitzinger 1987a); we here only briefly summarize the basics.

The two-point covariance tensor defines the correlations of the fluctuations of an observable (radial velocity, column density, magnetic field pressure, ...) with another variable or itself at a certain displacement (lag) τ in form of deviations of mean values. It will reveal to what extent data-points separated by this lag are correlated. These statistics are only valid for stationary data, so that large scale trends or gradients eventually contained in the data have to be removed in order to extract the stationary components. An alternative to covariance tensors is given by the two-point structure tensor, which is particularly sensitive to small-scale fluctuations in suppressing large-scale trends, and which is also closely related to the Kolmogorov-Obukhov theory of isotropic incompressible subsonic turbulence. Astrophysical data normally consist observations of an observable N at discrete locations (x , y); only one finite realization of the ensemble of stochastic processes is therefore accessible (i.o.w., the ergodic hypothesis has to be fulfilled). With these assumptions, we are able to calculate the two-dimensional autocorrelation function (ACF) ρ of an observable N as a function of the discrete lag $(\Delta x, \Delta y)$

$$\rho(\Delta x, \Delta y) = \left\langle \left(N(x,y) - \langle N \rangle \right) \left(N(x+\Delta x, y+\Delta y) - \langle N \rangle \right) \right\rangle$$

where the brackets denote the spatial average, and the two-dimensional structure function

$$B(\Delta x, \Delta y) = \left\langle \left(N(x,y) - N(x+\Delta x, y+\Delta y) \right)^2 \right\rangle$$

Because of the limitations already mentioned (finite data extent, finite number of realizations of the stochastic process), only estimates of the true ACF and STR can be determined. The most commonly used estimates are the unbiased and the biased estimate (see Spicker and Feitzinger 1987a for a description of their properties), which were both computed.

The basic information contained in these functions are discussed by (Spicker and Feitzinger 1987a); we just mention, that the ACF of density fluctuations provide the typical sizes and separations of the clumps of radiating matter, whereas the STR is mainly used to calculate the Kolmogorov index of the observable investigated. Typical clump scales are derived from radial cuts through the two-dimensional ACF in different orientations and fitting parabolas to its central parts.

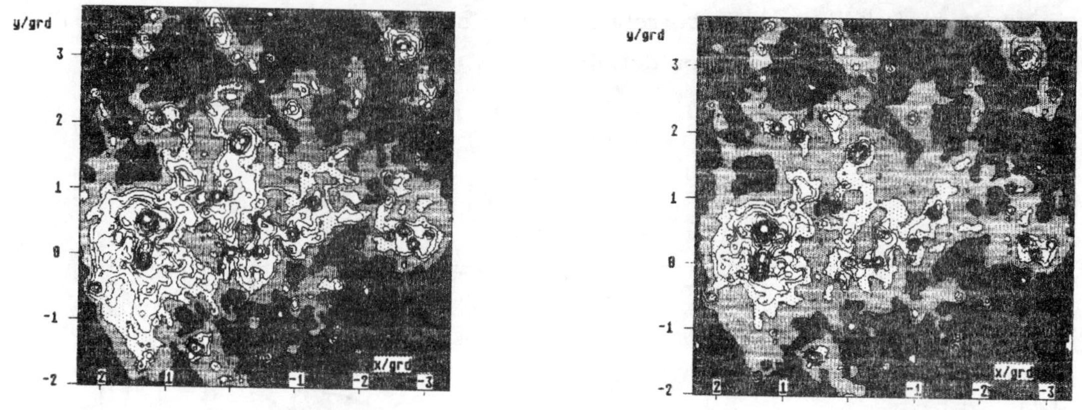

Fig. 1: Map of the 100μ (left) and 60μ (right) IR emission of the LMC as measured by the IRAS satellite in the standard tangential coordinate system. Contours are from 8 to 1400 in units of Jy/sr.

---

4. Data

The IRAS data used in the present paper are shown in Fig. 1. They were constructed from the all-sky survey-skyflux maps # 191 and 205 (see Israel and Schwering 1986 and references therein for a detailed description of the data) using BABSY, the Bonn standard image processing package and additional routines from the NOD-2 library. The 60μ and 100μ images were analyzed; the sky coverage of the maps corresponds to about 6^0 x 6^0 . The data have a resolution of 5' and a sampling of 2', so that features of about 80 pc size in the LMC are resolved. Already the plane appearance of the IR maps shows a pronounced similarity to the radio continuum maps (Haynes et al. 1986), a fact that is confirmed by the statistical analysis.

5. Results and Discussion

Fig. 2 shows the two-dimensional unbiased ACF of the 100μ IRAS data of the LMC. The map is given in the lag coordinate system ($\Delta y : \Delta y, \Delta x : \Delta x$) = (-5.°8 : 5.°8 , -5.°8 : 5.°8). Positive correlations are shown as full lines (0.8, 0.6, 0.4, 0.2, 0.1), the zero contour is enhanced, and negative correlations are drawn dashed (-0.1, -0.2). The main features visible are

1) There is a pattern of several recorrelations and anticorrelations at different lags and orientations. They are determined by the most intense IR sources in the LMC, namely 30 Doradus and its southern extension (the 'keyhole' structure), producing the two most prominent recorrelations. Some of the anticorrelations coincide with the positions of the HII supershells of the LMC, e.g. the 'hole' in the IR radiation at the constellation Shaple III at the northern edge. This hole has also been noted in the HI and radio continuum maps and is also strongly influencing the HI velocity field, as was proved by the ACF analysis (Spicker and Feitzinger 1987b);

Fig. 2: Two-dimensional unbiased autocorrelation function of the 100μ **IRAS** emission of the Large Magellanic Cloud. See text for details.

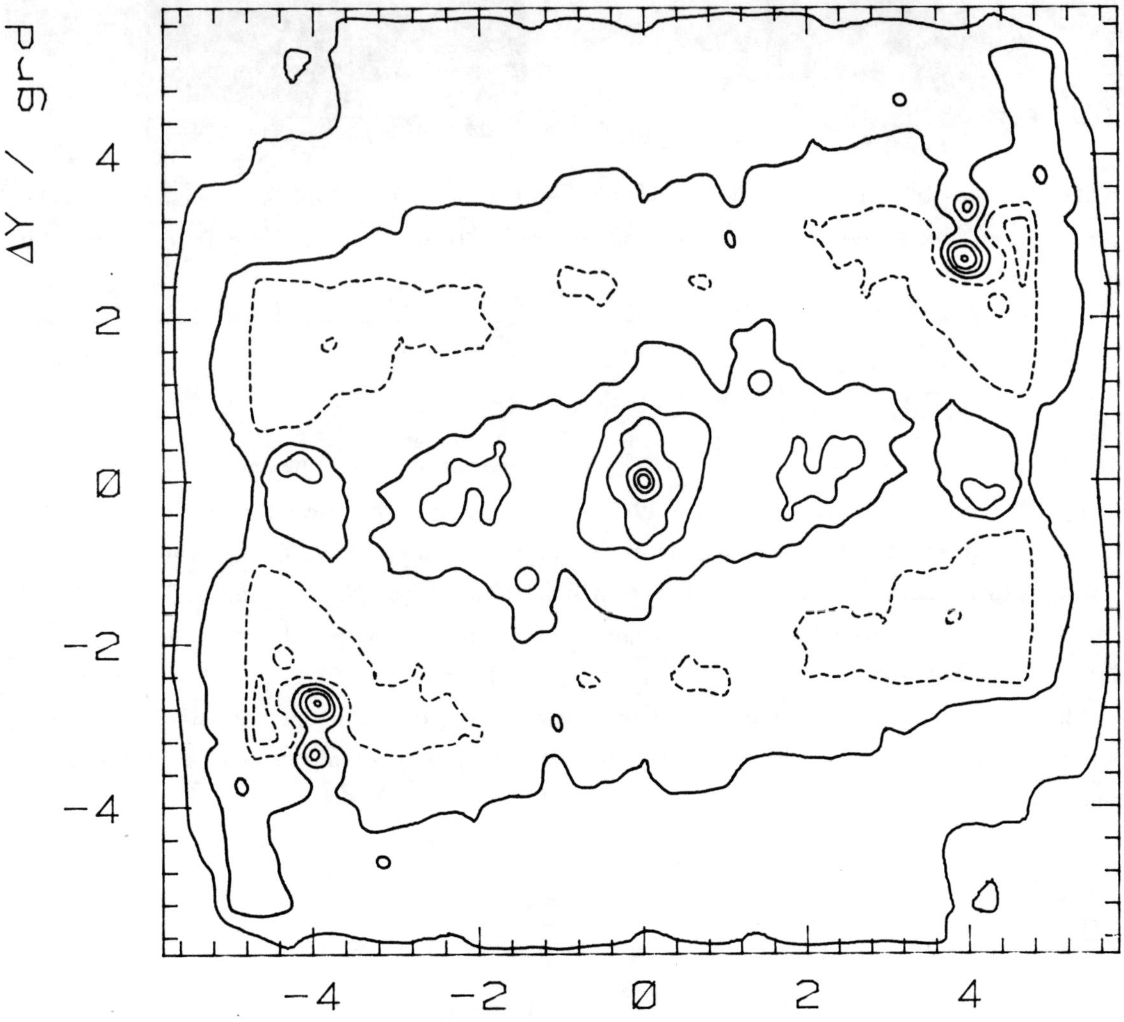

Fig. 3: Radial cuts through the autocorrelation function shown in Fig. 2 in the direction of the main recorrelation.

Top (3a): full cut

Bottom (3b): enlargement of the central part.

Note the double feature of parabolic shape near the origin, which can be associated with typical clump scales of the emission regions.

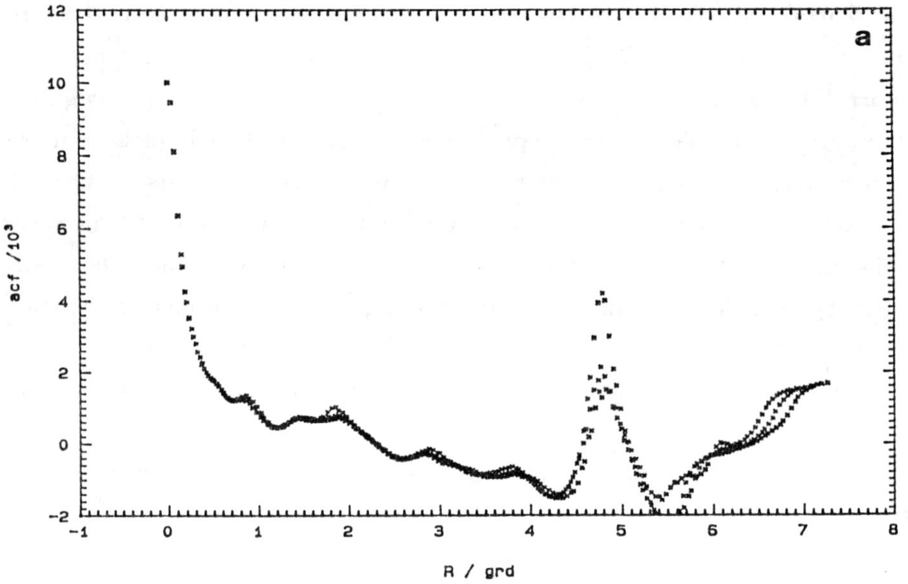

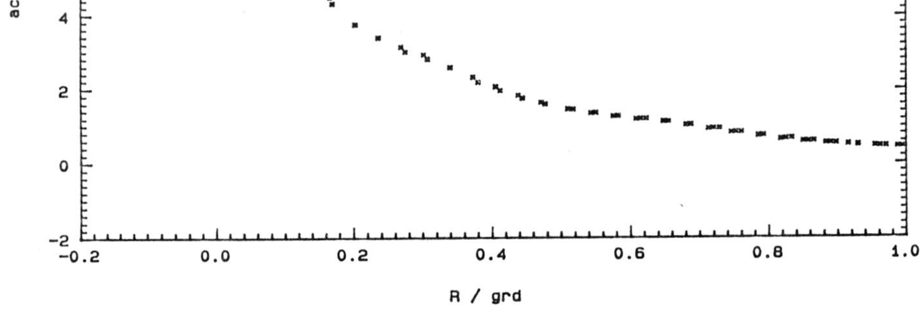

2) A central plateau of the ACF in about northern direction is visible, which is superposed on a weaker plateau in nearly perpendicular orientation. This pattern is obviously not determined by a few single sources, but more likely reflects the clumpy distribution of radiating matter, so that typical clumping scales can be derived.

Fig. 3 shows radial cuts through the unbiased ACF in the direction of the main recorrelation. Fig. 3a gives a complete cut, whereas Fig. 3b only extracts the central part. There is a double feature of parabolic shape near the origin, which is typical for emission coming from two distinct scale sizes of the emission regions. The pattern of several smaller recorrelations can be traced back to several larger emission regions located in the LMC bar and thus reflects the typical separations of the biggest emission regions. Other directions show even three distinct parabolic shaped forms of the ACF near the origin. The scale sizes derived from a variety of cuts through the ACF in several orientations give the three distinct scale sizes λ_1 = 200 .. 260 pc, λ_2 = 550 .. 600 pc, λ_3 = 900 .. 1000 pc. λ_1 is typical for the smaller HII regions, λ_2 corresponds to typical spheres of influence of the active star forming regions in the LMC, and λ_3 is the size of the LMC supershells. This is confirmed by the size of the anticorrelations found in their directions.

The ACF of the 60μ maps show essentially the same result, with scale λ_1 being even smaller. The analysis of the STR for both maps gives a similar distribution of scales. The characteristic Kolmogorov exponent is surprisingly small, s = 0.20 .. 0.29. Apart from the Kolmogorov exponent, the appearence of the IRAS ACF and STR maps is almost identical with the corresponding radio continuum maps (Spicker and Feitzinger 1987c). Because of the lower resolution (15') of the radio continuum maps, the scale λ_1 is not accessible, so that the smaller emission regions do not show up. The good agreement with the analysis of the HI radial velocity field (Spicker and Feitzinger 1987b) supports current theories of star formation processes in the LMC in revealing the scale sizes of cold, dense material not yet or no longer engaged in active star formation.

6. References

Haynes, R.F., Klein, U., Wielebinski, R., Murray, R.D., Astron. Astrophys. 159, 22 (1986)

Israel, F.P, Schwering, P.B., in: Light on Dark Matter, F.P. Israel (ed.), Dordrecht: Reidel 1986, p. 383

Spicker, J., Feitzinger, J.V., Astron. Astrophys. in press (1987a)

Spicker, J., Feitzinger, J.V., Astron. Astrophys. in press (1987b)

Spicker, J., Feitzinger, J.V., in: Interstellar Magnetic Fields, R. Beck and R. Gräve (eds.), Berlin: Springer 1987c, p. 82

The support of this research by the Deutsche Forschungsgemeinschaft (DFG) under grant Fe 196 / 3-2 is gratefully acknowledged.

STATISTICAL ANALYSIS OF STARCOUNTS IN REGIONS OF SOUTHERN DARK CLOUDS

J.A. Stüwe and J.V. Feitzinger

Astronomisches Institut
Ruhr Universität Bochum
Postfach 1Ø 21 48
D-463Ø Bochum 1
F.R.G.

Sternwarte Bochum
Castroper Str. 67
D-463Ø Bochum 1
F.R.G.

In the beginning of this century famous astronomers tried to solve the structure of our galaxy by inverting the basic formula of stellar statistics.

$$\mathcal{N}(m) = \omega \int_{-\infty}^{m} \int_{0}^{\infty} D(\vec{r}) \; \Psi(M;\vec{r}) \; r^2 \, dr \, dm'$$

Ψ : luminosity function
D : number density of stars
$\mathcal{N}$: cummulative number of stars brighter than m
ω : unit solid angle

Nowadays it is clear that this top-down commencement could not yield the desired result. Nevertheless an enormous amount of data has been accumulated in this phase of stellar statistics, which today is the basis of our knowledge about the structure of the Milky Way. A few years ago Bahcall and Soneira (198Ø) published a numerical model of our galaxy, which can be used to solve the basic equation of stellar statistics in a bottom-up manner. The interstellar extinction (A) is implicitely included in the equation through the conversion from absolute to apparent magnitude.

$$M = m + 5 - 5\lg(r/pc) - A(r)$$

That part of the extinction, that is due to the generaly distributed matter (not to dark clouds) has been taken into account when computing the cummulative starnumbers from the B.& S. model, so that, assuming this model to be correct, all observed deviations in the starcounts from that predicted by the model may be ascribed to density enhancements in the ISM, i.e. dark clouds.

This is, in short, the idea used in this paper to map the absorption of dark clouds, apparent on the SRC-J and ESO(R) photographic plates of the southern sky survey. In order to do so, one must be able to recognize all stars on the plates up to a given limit and register them with positions and magnitudes. Therefore microphotometric scans of the regions of interrest were done with a PDS 2000 GM, kindly made available by the "Astronomisches Institut der Westfälischen Willhelms Universität zu Münster". The digitalized data were processed with the "Image Handling And Processing" (IHAP) system (Middleburgh, 1983), installed at the "Astronomisches Institut der Ruhr Universität Bochum". Here the command SEARCH was used with option "Laplace" (written by P. Grosbol) to automaticaly find and register the stars. The parameters maximum density (D_{max}), integrated density (D_{int}) and surface (S) covered by the stellar image delivered by the program were used to construct one single photometric parameter.

$$p = \sqrt{S(1-\exp(-D_{max})) + D_{int}}$$

In the next step this parameter is converted to the apparent magnitude by comparing the cummulative star numbers as a function of this parameter with that predicted by the B.& S. model as a funtion of magnitude in an apparentely unobscured field. Calibration curves obtained in this way for the southern sky field No 369 are shown in figure 1.

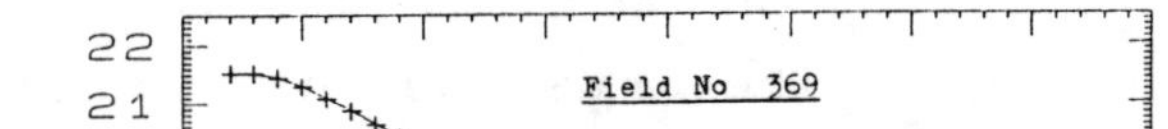

Fig. 1 : Calibration Curve

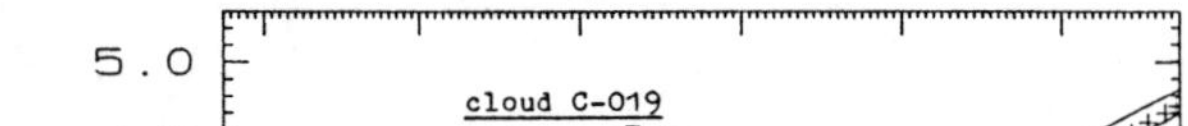

Fig. 2 : Wolf diagram

Figure 2 shows a "Wolf diagram", the construction of which is now straight foreward. The crosses show the observed data and the upper solid line is that predicted by the B.& S. model in the same area of the sky, but without extinction by a dark cloud (only the general obscuration is taken into account). The shift between the two datasets along the magnitude axis in the upper part of the diagram, where observed and model counts run parallel, is equal to the mean absorption in the analyzed area.

In a next step a dark cloud of this absorption and a user definable distance is switched on in the B.& S. computer model, and that distance which gives less deviation from the observed data (lower solid line in fig. 2) is assumed to be the distance of the dark cloud. The deviation betweeen model data and observed ones for several choosen distances is shown in figure 3.

Table 1 : Analyzed Dark Cloud Regions

Objekt	l	b	N^*_B	N^*_R	area cm^{-2}
C-Ø19	25Ø.96	-1.4Ø	758Ø	66Ø5	4
C-Ø15	249.Ø4	-3.23	7529	925Ø	12
C-Ø21	249.44	-2.22	2Ø466	24Ø36	15
G-Ø14/Ø1	251.87	-1.73	23927	22386	16
G-Ø14/Ø2	252.42	-2.Ø7	13327	14596	12
G-Ø14/Ø3	252.95	-2.24	12561	13634	12
G-Ø14/Ø4	253.36	-1.79	19224	21218	16
G-Ø1Ø	252.51	-3.52	41Ø1	785Ø	12
C-Ø24	253.22	-4.28	14986	1724Ø	15
C-286/Ø1	318.17	-3.52	4428	6549	2
G-192	32Ø.11	-2.11	17671	25459	1Ø
G-196	321.33	-1.96	31933	43ØØ7	12
C-289	318.99	-4.2Ø	25655	391Ø1	24
G-194	32Ø.62	-2.Ø6	3Ø182	415Ø1	16
G-195	32Ø.24	-3.46	52484	59644	16

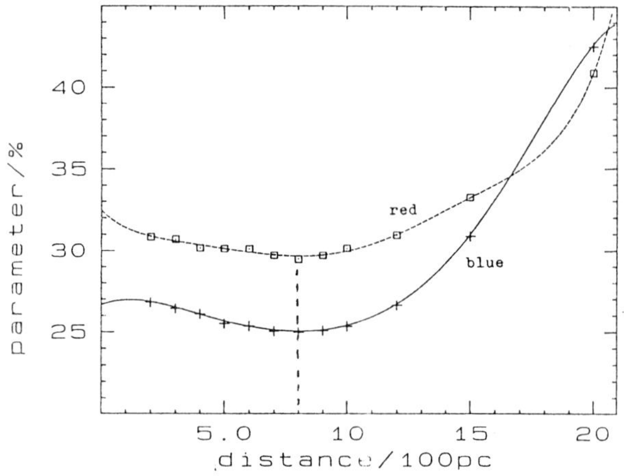

Figure 3 : Minimum Deviation Distance Fit

The minimum marks the dark cloud's distance. This procedure was done for 15 regions and the resulting distances are shown in table 1, together with area in the Schmidt-plates, the galactic coordinates of the regions, an error estimate from the distances derived seperately in two colours and the total number of registered stars.

Figure 4

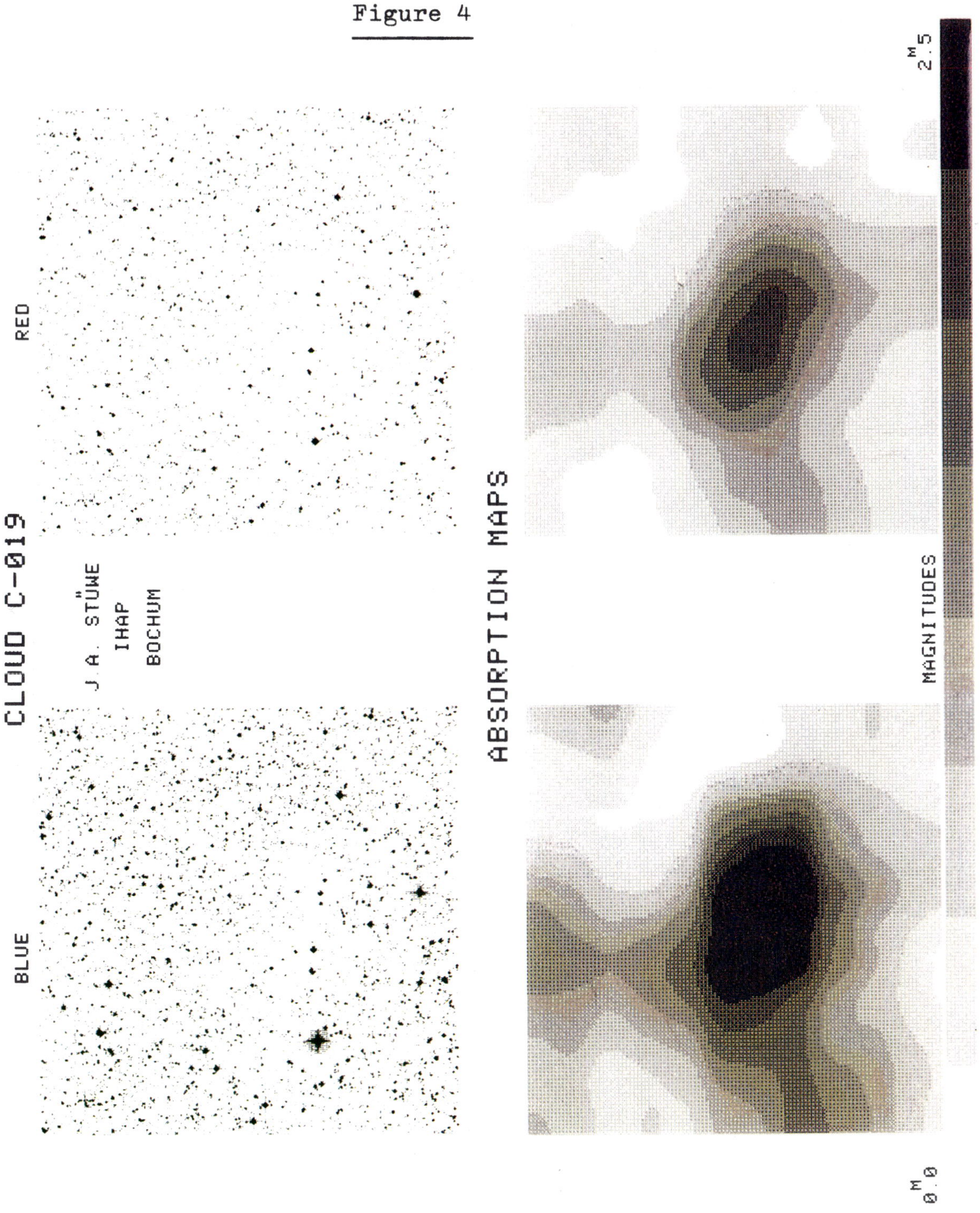
CLOUD C-019
J.A. STÜWE
IHAP
BOCHUM
BLUE
RED
ABSORPTION MAPS
MAGNITUDES
0.0 M
2.5 M

The sampled positions and magnitudes of the stars in the analyzed area can now be used to map the distribution of the absorption, in a way described by Dickman (1978), but now much more comfortable by a computer. Figure 4 shows the result for one example in a "bevore & after" manner. In the upper half one can see the star field of the dark cloud, as it is seen on the SRC-J and ESO(R) photographs. The lower half shows the resulting absorption maps in two colours (R,B). For all analyzed clouds the data are in agreement with the interstellar extinction law as given by Savage & Mathis (1979) within a +/- 0.2 mag limit, except for the most central regions of the densest clouds, where the observed absorption is already at the limit obtainable with the described method (Rossano, 1980). Together with the knowledge of the distance of the mapped regions the true cloud sizes and several physical paramters like mass and density (in the outer regions even density structure) may now be exploited.

Table 2 : Physical Dark Cloud Parameters

Nr	Object	R (pc)	σ_R (pc)	d (pc)	M (M)	M_J (M)	$\langle n_H \rangle$ (cm^{-2})
1	C-019	0.920	0.092	800	86	89	1065
2	C-015	0.872	0.175	500	82	85	1192
3	C-021 I	0.682	0.073	590	63	66	1910
4	C-021 II	0.793	0.081	590	67	77	1298
5	C-021 III	0.894	0.321	590	45	87	614
6	C-021 IV	0.709	0.073	590	20	69	529
7	G-014 I	0.695	0.056	450	75	68	2156
8	G-014 II	1.784	-----	450	200	174	340
9	G-014 III	-----	-----	520	---	---	----
10	G-014 IV	0.597	0.123	520	38	58	1712
11	G-014 V	0.612	0.120	520	47	60	1973
12	G-014 VI	0.622	0.120	520	42	60	1695
13	G-014 VII	0.658	0.078	620	26	64	868
14	G-014 VIII	0.734	0.114	620	19	71	464
15	G-014 IX	0.919	0.106	620	155	89	1926
16	G-014 X	0.633	0.100	620	21	62	785
17	G-014 XI	-----	-----	620	132	---	----
18	G-014 XII	1.063	0.163	590	167	103	1344
19	G-014 XIII	0.475	0.122	590	56	46	5005
20	G-014 XIV	0.602	0.084	590	25	59	1107
21	G-010 I	0.447	0.178	450	22	43	2402
22	G-010 II	0.583	0.113	450	40	57	1926
23	G-010 III	0.252	0.035	450	2	25	1208
24	G-010 IV	0.589	0.114	450	35	57	1636
25	C-024 I	-----	-----	540	80	---	----
26	C-024 II	-----	-----	540	---	---	----
27	C-024 III	-----	-----	540	220	---	----
28	C-024 IV	-----	-----	540	81	---	----
29	C-286	0.208	0.061	170	5	20	5155
30	G-192 I	1.485	0.192	675	169	144	499
31	G-192 II	0.881	0.111	675	44	86	626
32	G-196 I	1.230	0.147	880	226	97	1173
33	G-196 II	1.726	0.288	880	270	168	508
34	G-196 III	0.958	0.104	880	58	93	640
35	G-196 VI	0.429	0.045	880	3	42	404
36	C-289 I	1.263	-----	600	124	123	597
37	C-289 II	1.831	0.306	600	365	178	575
38	C-289 III	0.756	0.134	600	26	74	586
39	G-194 I	0.149	0.081	130	2	14	4674
40	G-194 II	0.138	0.074	130	1	13	5148
41	G-195 I	0.230	0.020	150	8	22	6354
42	G-195 II	0.168	0.022	150	2	16	3465
43	G-195 III	0.176	0.014	150	1	17	1950

The mass of a condensation in the maps is simply computed by summing up all the absorption belonging to the cloud and subtracting a background value derived from the surrounding area.

$$M = m_H \left(\widetilde{\Delta x} \cdot 67.2\,''/\mathrm{mm} \cdot d\right)^2 \left(1.9 \cdot 10^{21} \frac{\mathrm{cm}^{-2}}{\mathrm{mag}}\right) \left[\sum_{\mathrm{Surface}} A_V - A_{bg}\right]$$

$\widetilde{\Delta x}$: size of a pixel in the A_V-map

67.2 "/mm : scale of the ESO/SRC Schmidt plates

$1.9*10^{21}\ \frac{\mathrm{cm}^{-2}}{\mathrm{mag}}$: constant gas to dust ratio

d : distance of dark cloud

m_H : mass of a hydrogen atom

The obtained physical parameters of all mapped dark clouds are listed in table 2. Figure 5 shows a plot of the "Larson-Relation" (1981) between the size and the mean density of radioastronomicaly analyzed interstellar clouds (crosses). The boxes represent the data found in this work. The obvious agreement between the two completely differently obtained data sets shows that the dark cloud analyzing procedure developed by the authors is at least as accurate as the way of exploring the ISM with radio telescopes.

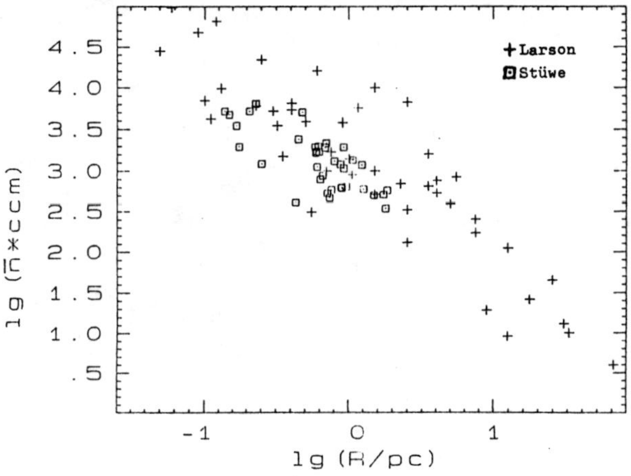

Figure 5 : "Larson" - Plot

References :

Bahcall J.N. & Soneira R.M. (198Ø) : Ap.J.S.S. 44, 73

Dickman R.L. (1978) : A.J. 83, 363

Larson R.B. (1981) : MNRAS 194, 8Ø9

Middleburgh F. (1983) : IHAP Manual, ESO, Garching

Rossano G.S. (1983) : A.J. 85, 1218

HIPPARCOS : SCIENTIFIC USES OF THE INCA DATA BASE

Catherine Turon, Ana Gómez, and Françoise Crifo
Observatoire de Paris, Section de Meudon
D.A.S.G.A.L.
F-92195 Meudon Principal Cedex (France)

ABSTRACT : The whole set of stars proposed for Hipparcos observation has been collected in a dedicated *INCA Data base.* This new data base includes all the data and information required for preparing the "Input Catalogue", final list of about 120 000 stars suited to Hipparcos observing possibilities and answering to a maximum of accepted observing proposals.

Some of the typical scientific uses of this data base are illustrated underneath :

- test of the global stellar content of all or part of the data base,
- test of the local stellar content, and detailed analysis,
- test of the specific stellar content for identified groups of stars.

1. INTRODUCTION.

The various aspects of the Hipparcos mission - the satellite, the preparation of the Input Catalogue, the data analysis - have been described in the literature on several occasions (see, for example, Perryman, 1987 ; Kovalevsky, 1984 ; Turon, 1987). The numerous activities involved in the construction of the precise Hipparcos observing programme, the Input Catalogue, have been extensively described in Turon and Perryman (1985) and more recently in Turon (1987), Gómez & Crifo (1987), Crifo et al (1987), and Réquième & Turon (1987).

The present paper will be concentrated on the illustration of some typical scientific uses of the INCA data base. The technical aspects are described in the paper by Arenou & Morin (this volume).

2. GLOBAL STELLAR CONTENT OF THE INCA DATA BASE AND OF THE SUCCESSIVE VERSIONS OF THE INPUT CATALOGUE.

In answer to the invitation for proposals released by the European Space Agency in 1981, more than 200 observing programmes have been presented by the world-wide astronomical community, including about 760 000 entries. These entries have been cleared out, i.e. the redundant appellations have been identified using successively : 1) the cross-identifications already known in SIMBAD*, 2) cross-identifications identified by a specific INCA search software, 3) "manual" identifications. The resulting list includes about 200 000 stars. A basic list of "bright" stars (called *Survey*) required for satellite operation and data reduction, has been selected in SIMBAD according to the following limits :

---

* Set of Identifications, Measurements and Bibliography for Astronomical Data, Observatoire de Strasbourg.

$(m_V)_{lim} = 7.9 + 1.1\ |\sin b|$ for spectral types earlier than or equal to G5,

$(m_V)_{lim} = 7.3 + 1.1\ |\sin b|$ for spectral types later than G5.

It includes about 57 000 stars. The *INCA Data Base* has been created to collect all the data required for preparing the Input Catalogue and to make it possible rapid and daily updatings, analyses, verifications, ... It now includes about 214 000 stars, the sky distribution of which is illustrated in Figure 1. *a)*.

The Input Catalogue is constructed from the INCA Data Base by successive iterations including numerical simulations of the mission and of the star observing strategy, analysis of proposal fulfilment, optimisation of the use of available observing time and detailed analysis of critical zones. The final Input Catalogue will be the result of sometimes conflicting requirements : 1) satisfy a maximum of high priority scientific programmes (which may already be conflicting), 2) respect the observational constraints inherent to the HIPPARCOS satellite operations. The magnitude distribution of the retained stars should be consistent with the available observing time. The magnitude distribution in two tentative versions is given in Table 1. *a)*. The sky distribution should be as regular as possible, with the exception that additional bright stars may be accepted in some zones (for example the galactic plane). This is illustrated in Fig. 1 *b)* and *c)*. As far as possible, Priority 1 and Survey stars should be retained. Preliminary results are given in Table 1. *b)*. Finally, a large percentage of the stars retained in the final Input Catalogue should be nearby stars. Estimated distance distribution of stars in *IC1* is given in Table 2.

Table 1

a) Magnitude distribution in IC1 and IC2 ; *b)* Preliminary results in IC1.

m_H	< 6	6-7	7-8	8-9	9-10	10-11	11-12	> 12	Total
IC1	4 400	8 300	22 500	42 600	26 300	7 200	2 100	950	114 500
IC2	4 400	8 700	23 000	43 100	27 600	8 600	2 700	600	118 700
% of successful priority 1 stars in IC1	98	97	97	95	91	80	80	57	93
% of successful Survey stars in IC1	95	95	91	91	87	-	-	-	9

Priority 1 stars : 72 000 in IC1, 72 600 in IC2
Survey stars : 53 000 in IC1, 54 000 in IC2

m_H is the magnitude in the Hipparcos band pass (wide band pass, intermediate between B and V, nearer to V).

Table 2

Estimated distance distribution of stars in IC1

d (pc)	≤ 100	100 - 500	500 - 1000	> 1000	Total
Nr of stars in IC1	29 000	65 700	9 800	3 500	108 000

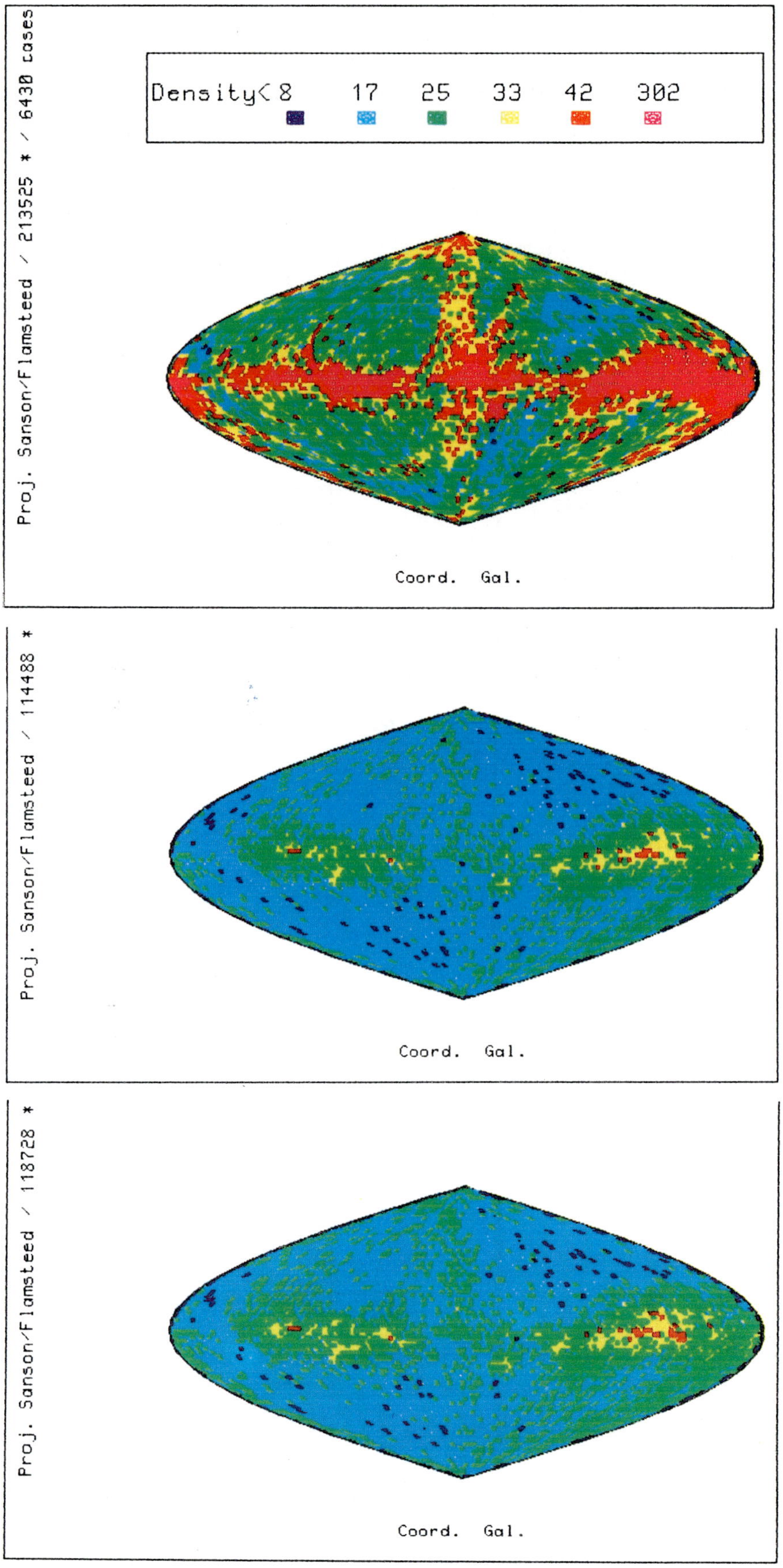

Figure 1

Sky distributions (in galactic coordinates) of :

a)
all proposed stars i.e. all stars in the **INCA** Data Base (~ 214 000 stars)

b)
all stars retained in the first version of the Input Catalogue, **IC1** (~ 114 000 stars)

c)
IC2
(~ 119 000 stars)
The inclusion of all stars brighter than V = 7 has been forced in this version.

3. TEST THE LOCAL STELLAR CONTENT.

In dense areas of the sky, the technical constraints due to the peculiarities of satellite operation (veiling glare effect due to bright stars, limited observing time due to the satellite scanning law, ...) turn out to be more and more imperative.

Figure 2

The area of the Pleiades' cluster. *a)* in SIMBAD. *b)* in INCA (all proposed stars). *c)* in IC1. *d)* in IC2.

Center : 3h 44m 00.0 +23° 54′ 00″

Dim +27° 00′ 00″ Scale : 1cm = +0° 08′ 00″

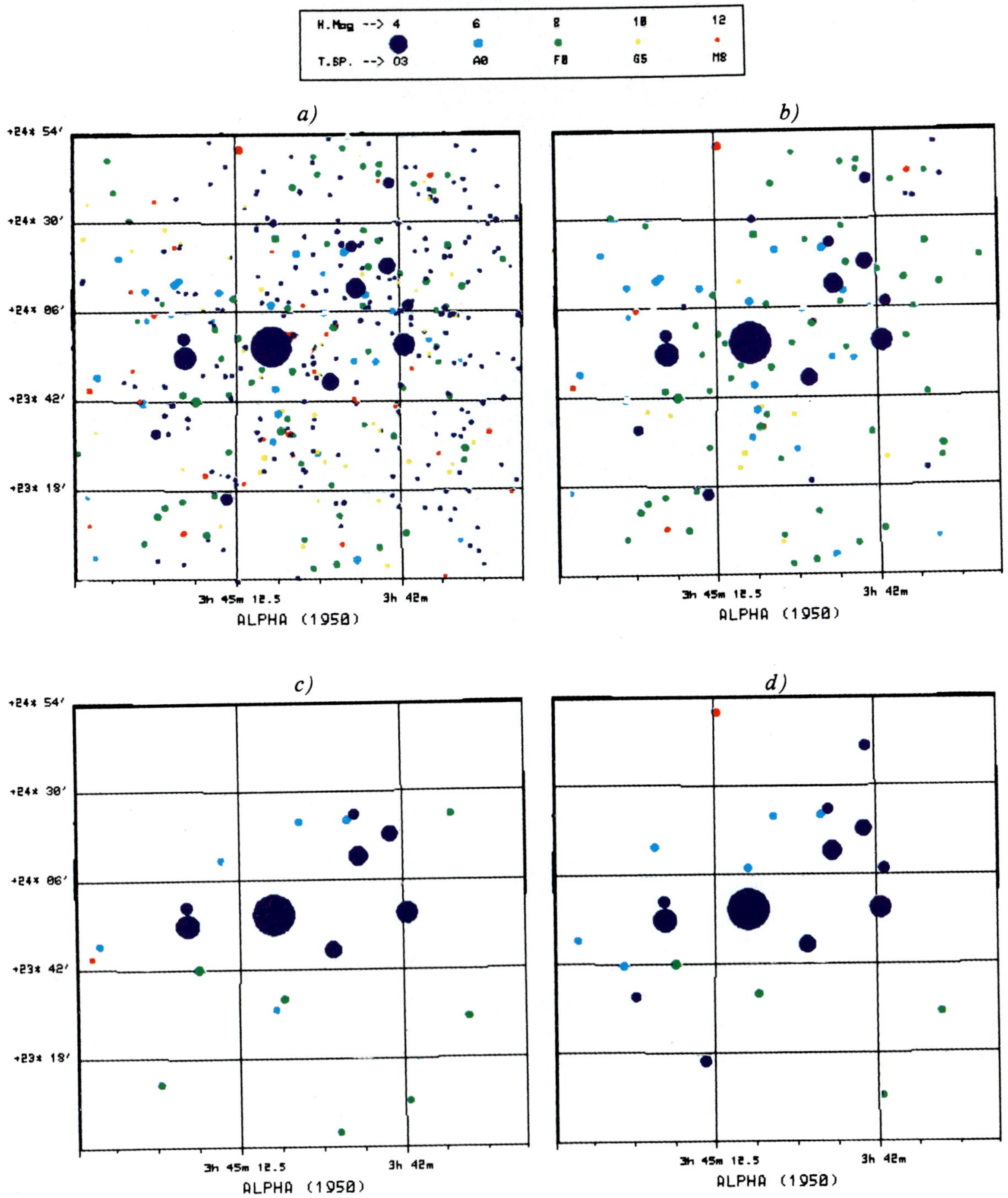

Two examples are illustrated here :

- **Figure 2** : the choice of stars in a galactic cluster, the Pleiades. Sampling by coordinates and by proposals (to discriminate between stars proposed as cluster members, and field stars also proposed) is required.
- **Figure 3** : a dramatic competition for observing time allocation between a nearby red dwarf star and a radiostar. The radiostar, basic object for linking the HIPPARCOS system to an extragalactic reference system (through the VLBI system), is retained.

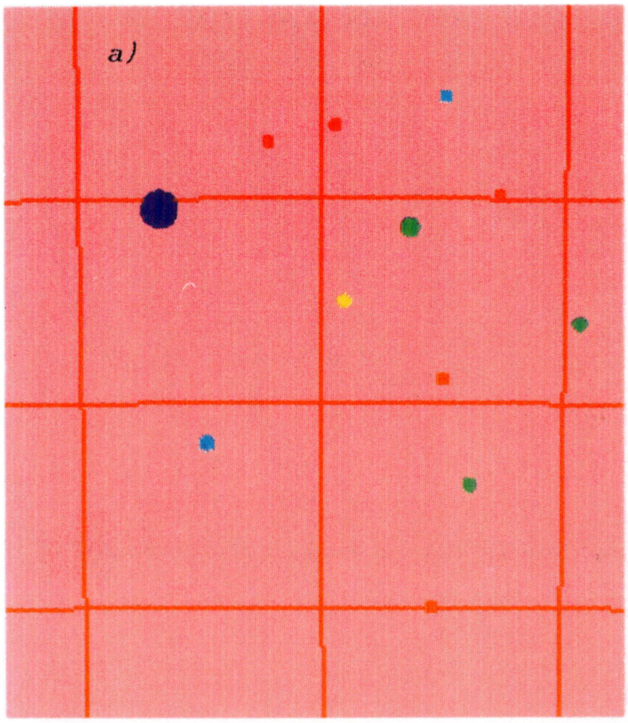

Figure 3

Illustration of a dramatic local conflict.
See the explanations in the text above.
a) all proposed stars (**INCA** Data Base)
b) **IC1** : 3 M dwarf stars, no radiostar.
c) **IC2** : the radiostar, 2 M dwarf stars.

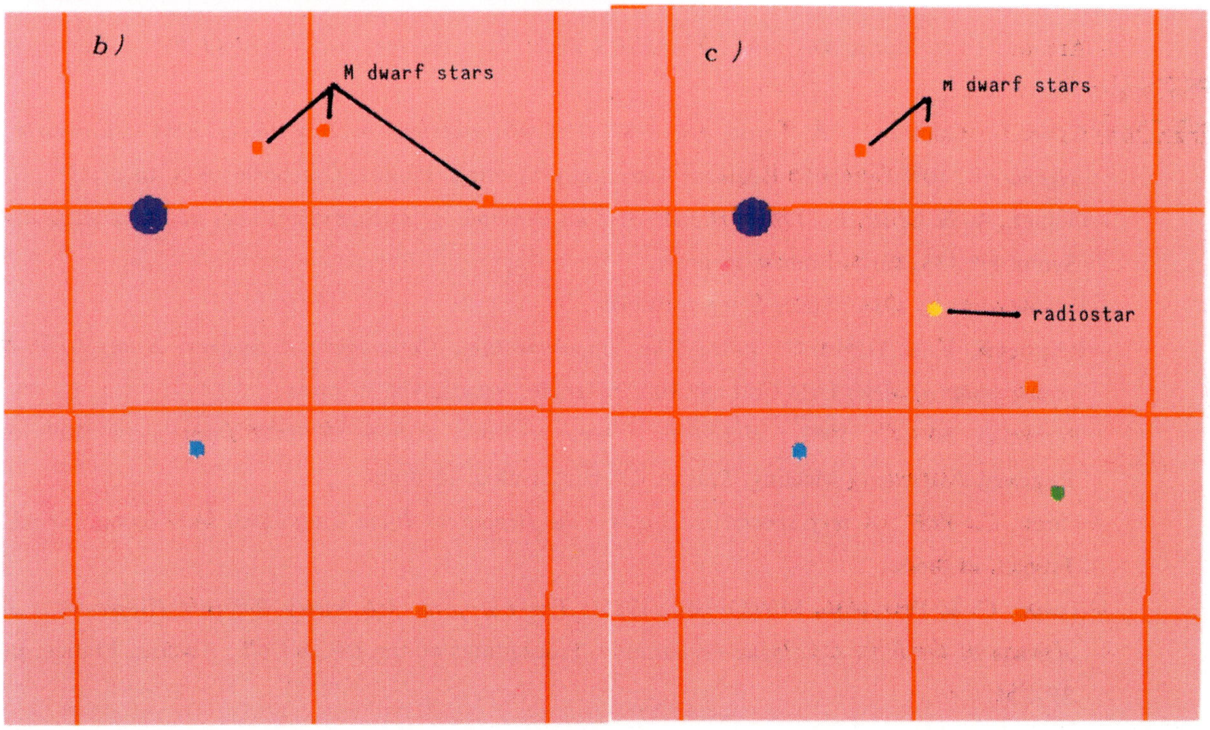

4. TEST THE SPECIFIC STELLAR CONTENT OF IDENTIFIED GROUPS OF STARS.

Sampling by spectral type, by proposal or set of proposals is currently used :

- to test the success of a given spectral type, of a proposal or a set of proposals ;
- to test the sky, magnitude, ..., distributions ;
- to compare the success of these groups in successive iterations ;
- to check the neighbourhood of high priority unsuccessful stars.

Some results of such samplings are illustrated in Table 3. The improvement of the content of the Input Catalogue with respect to these groups of especially interesting stars is clearly pointed out.

Table 3

Objects	Total Nr of proposed stars	Number of stars in IC0	Number of stars in IC1	Number of stars in IC2
Red dwarf stars	2330	1950	2130	2180
White dwarf stars	45	17	21	26
Cepheids	330	230	250	275
RR-Lyrae	390	190	210	215
Stars $\leq$ 25 pc	2000	1700	1800	1870
LHS ($\mu \geq 0''.5$/yr)	1140	970	1010	1045

5. CONCLUSION

As a conclusion, we would like to emphasize the constant kindness and helpfulness of the CDS staff who allows us to take benefit from their so efficient instrument : SIMBAD.

6. REFERENCES

- Arenou, F. & Morin, D., 1987 : *The INCA Data Base for the preparation of the Hipparcos mission"*, this volume.
- Crifo, F., Gómez, A. & Turon, C., 1987, *The astrophysical interest of the ESA astrometric mission Hipparcos*, 10th European Regional Astronomy Meeting, Prague, Czechoslovakia, in press.
- Gómez, A. & Crifo, F., 1987 : *Hipparcos : an example of astrophysical uses of astrometric data*, IAU Symp. N° 133, Paris, France, in press.
- Kovalevsky, J., 1984 : Space Science Review, *39*, 1.
- Réquième, Y. & Turon, C., 1987 : *The Hipparcos Input Catalogue : A progress report from an astrometric point of view*, IAU Coll. N° 100, Belgrade, Yugoslavia, in press.
- Perryman, M.A.C., 1987 : *Hipparcos : high precision astrometric data from space*, 10th European Regional Astronomy Meeting, Prague, Czechoslovakia, in press.
- Turon, C., 1987 : *A very special catalogue : the Hipparcos Input Catalogue*, IAU Symp. N° 133, Paris, France, in press.
- Turon, C. & Perryman, M.A.C. eds., 1985 : *The European Astrometry Satellite Hipparcos : Scientific Aspects of the Input Catalogue Preparation*, Proceedings of the INCA Coll., Aussois, France, pub. ESA SP-234.

Properties of AGN Host Galaxies - An example of statistics based on merged catalogues.

Stefan Wagner, Landessternwarte Heidelberg-Königstuhl, West-Germany

Abstract: A series of catalogues of extragalactic objects, available on magnetic tapes is presently assembled into a database of extragalactic objects. One of the tasks which are to be carried out is statistical analyses. An example described in this paper are the properties of AGN host galaxies. It is found that galaxies seen face-on or nearly so are overabundant among the near-by Seyfert galaxies. This indicates that the Seyfert nature of a galaxy is less likely to be discovered, if the host galaxy is seen edge-on. As a second result we found that Seyfert nuclei have a weak tendency to be overabundant in barred galaxies. Quantification of this trend is severly influenced by selection bias.

I. Introduction:

A merged compilation of galaxies is used to study properties of AGN host galaxies. An important problem of statistical analyses is biasing, which often misleads conclusions if the samples do not fulfill strict selection criteria. On the other hand, statistical studies may help to detect such selection bias as will be shown in this study.

The specific problem addressed here is to see whether the host galaxies of AGN have intrinsic differences compared to non-active objects. Such differences would indicate that environmental influences are important for formation and/or evolution of Seyfert nuclei.

II. The database:

A database of extragalactic objects is presently assembled at the Landessternwarte Heidelberg-Königstuhl, Germany. The purpose of the database is to help observers during preparations of observing programs and to enable statistical analyses. The database is compiled from essentially all catalogues available on magnetic tape at the Centre de Donnees Stellaires, Strassburg and several further catalogues of extragalactic objects which are available on magnetic tape. Addition of new catalogues to the database is easy.

The philosophy is to maintain the full information of all input catalogues. Each of those input catalogues thus forms a slice of a three-dimensional data cube with the three "dimensions" object - value of specific physical parameter - source catalogue. The database can be used in individual slices along any of the dimensions and in projections of the data cube on any two-dimensional plane. In the latter mode the arithmetical form of the projection can be specified (e.g. mean, sigma, maxvalue, medianvalue, means weighed by reliability of the input catalogue, etc.).

Here the distribution of the axis ratio andmorphological types for all Seyfert galaxies listed in the "Catalogue of Quasars and Active Nuclei (3rd edition)" by M.-P. Véron-Cetty and P. Véron are studied. The above mentioned parameters were taken from the Second Reference Catalogue of Bright Galaxies, G. de Vaucouleurs et al., 1976 (RC2), the Uppsala General Catalogue of Galaxies (UGC), P. Nilson, 1973 and The ESO Uppsala Survey of the ESO (B) Atlas (ESO), A. Lauberts, 1982.

III. Detection Bias

Biasing is the most severe problem of each statistical analysis in extragalactic research. This is particularly true if merged compilations are used which do not follow clear selection criteria. Even in such cases, however, it is possible to derive statistically meaningful answers to a number of questions. To some extend statistical analysis even enables the discovery of selection bias. The present study will give an example for both cases.

All galaxies listed by Véron-Cetty and Véron as Seyfert galaxies, and whose identification as AGN was found from optical observations were included in this study. The large majority of Seyfert galaxies are Spiral galaxies (see Sec. V). In a first part we excluded all those Seyfert galaxies whose morphology was classified as E, S0, Irr or Peculiar and examined the axis ratios of the Seyfert galaxies of types S or SB. The analysis was performed individually for all catalogues (RC2, UGC, ESO) and - for RC2 and UGC - axis ratios from red and blue plates were used. The aim of this study was to check the influence of host galaxy inclination on detection probability. The expected bias against objects seen edge-on was confirmed (see Sec. IV).

In a subsequent step the ratio of S:SB type host galaxies was examined. While there is no reason for the detection probability of an AGN to be dependent on the existence of bars one has to be sure that the classification of any galaxy as "barred" is not biased in any way correlated with the detection probability of an AGN. One also has to be sure that no Seyfert galaxies entered the catalogue which were found in studies biased to barred host galaxies.

IV. Axis Ratios

The different analyses from the individual catalogues and the two colours show no individual differences within the error bars. The following discussion is based on the sum over these individual studies.

If all S and SB systems were pure disks, the ratio of major axis to minor axis would allow a direct computation of the inclination. The existence of bulges however introduces a complication of this transformation. In order to account for the bulges the distribution of axis ratios in Seyfert galaxies is compared to the average axis ratio of all S or SB galaxies in the corresponding catalogue (Fig. 1). It is obvious from this diagram that Seyfert galaxies occur more frequently in galaxies with axis ratios approaching unity. This result is a confirmation of earlier results by Keel, which were based on a smaller sample. The easiest explanation for this result is that the Seyfert nature of active galaxies seen edge-on is more likely to be revealed by obscuration within the disk of the host galaxy.

It has to be pointed out, that a second-order effect might influence this study: Seyfert nuclei preferentially occur in massive spirals which themselves have larger bulge to disk ratios. Dominant bulges will increase the average axis ratio as compared to a randomly chosen comparison sample. It would therefore be appropriate to repeat the above study after subdividing the total group into the Hubble-Subtypes Sa-SBa, Sb-SBb, Sc-SBc. This would account for the bias mentioned above. Due to the small numbers of Seyferts, results based on this refined statistics are identical

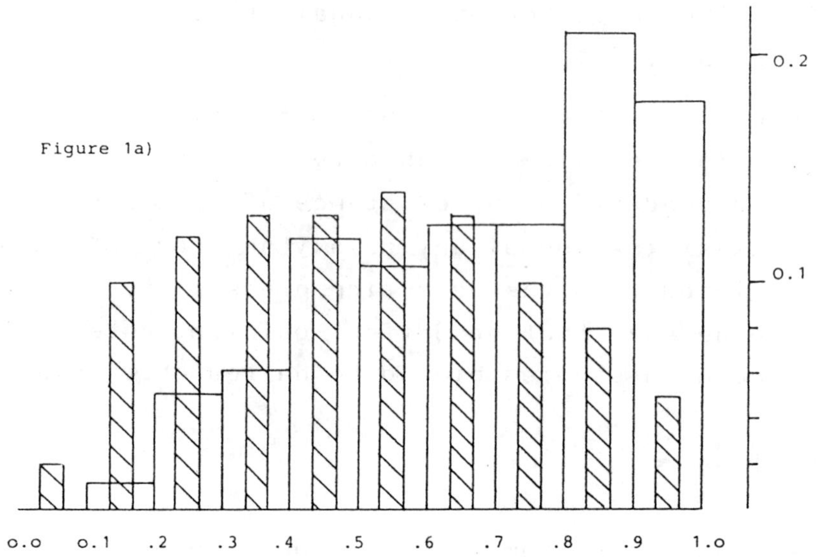

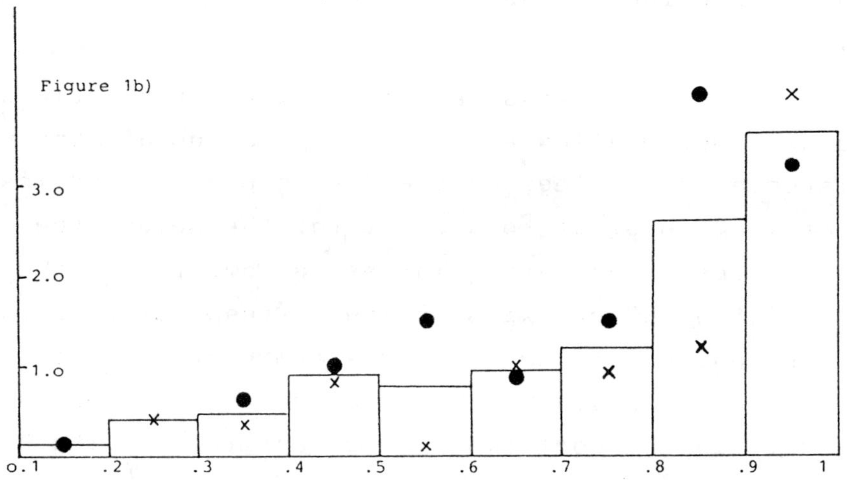

Fig.1a: Normalized distribution of the axis ratios of all spirals (hatched columns) and for host galaxies of Seyfert nuclei (blank columns).

Fig.1b: The ration of the normalized frequencies for active and normal spirals, derived in the different axis ratio bins. Seyfert 2 galaxies (x) are even more biased towards face-on galaxies than Seyfert 1 objects (•).

within the error bars.

V. The influence of perturbing bar potentials

Nuclear activity is often thought to be triggered or fed more easily in the presence of a perturbing non-axisymmetric potential. Many investigations claim an over-representation of barred galaxies among the hosts of Seyfert nuclei. To study this question quantitatively, we compare the distribution of Seyfert galaxies over the different morphological types with the corresponding distribution of all galaxies within the catalogues (Fig. 2). It is obvious that barred galaxies seem to be overabundant by a factor of 2 among the active nuclei. However, again the interpretation is not straightforward. As mentioned in Sec. III, four selection effects bias the control sample :

A) The frequency of spiral galaxies classified as barred decreases if the axis ratio differs from 1. Bars in edge-on galaxies are less likely to be identified. As was shown in Sec. IV, Seyfert galaxies are identified correctly in face-on galaxies which also have a higher chance to be classified as "barred".

B) The statistically higher absolute magnitude of active galaxies makes the volume covered in the magnitude limited catalogues RC2, UGC and ESO larger than for nonactive galaxies. Hence the apparent diameters are smaller in average. The ratio S:SB galaxies increases with decreasing apparent diameter since it is harder to detect a bar in a galaxy close to being angular resolved. Together with the on average smaller apparent diameter in Seyfert nuclei this weakens the conclusion found from Fig. 2.

C) As can be seen from Fig.2, the frequency of bars is much higher in early-type spirals. Comparisons with active galaxies give meaningful results only, if the sample differentiates between Hubble-subtypes.

D) There seems to be a statistically significant difference in bar-detection rates between the UGC and the ESO catalogues (Fig.3). This makes it necessary to subdivide the sample or to adjust the statisics with an appropriate normalization.

Fig.2: Ratio of barred/non-barred objects in each Hubble subtype. Hatched columns give the ratio for all spirals, the three blank columns in each subtype give the corresponding values for Seyfert 1, Seyfert 2 and all Seyfert galaxies (including LINERS and Seyfert candidates).

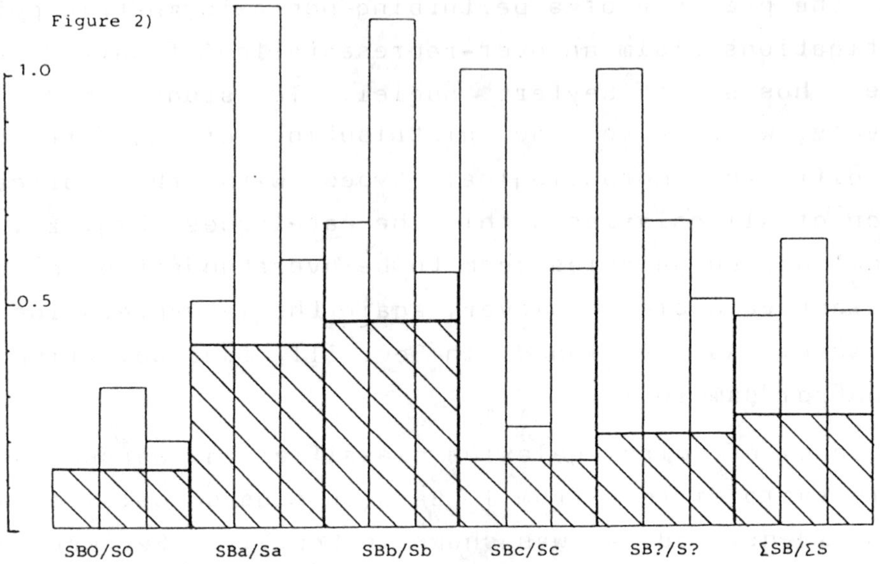

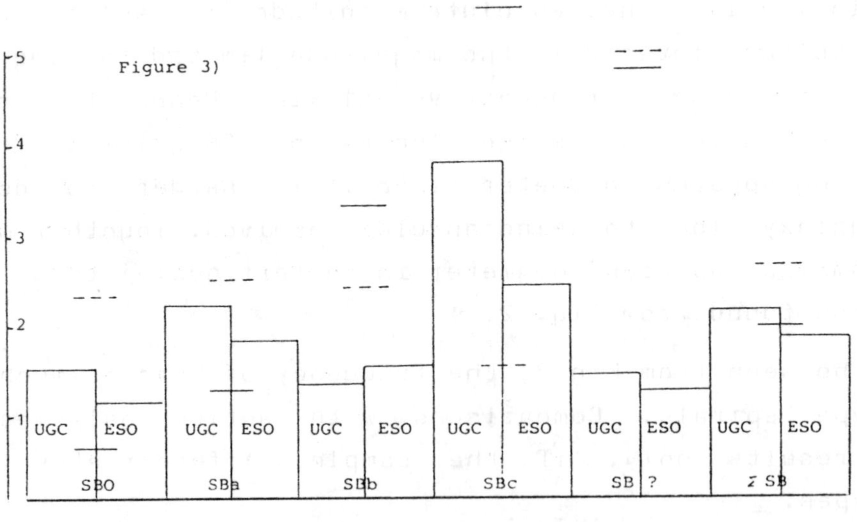

Fig.3: In each Hubble-subtype the excess of barred objects among the Seyfert galaxies is given. The two columns per bin give the values with the UGC and ESO catalogue being used as comparison sample of all spirals. (——) and (---) indicate the averaged values for Seyfert 1 and 2 galaxies respectively.
catalogues.

G E I S H A

P.R. Wesselius, Tj.R. Bontekoe, A.R.W. de Jonge, D.J.M. Kester

Space Research Department and Astronomical Laboratory "Kapteyn"
P.O. Box 800
9700 AV Groningen
The Netherlands

SUMMARY

The GEISHA (Groningen Exportable Infrared Survey High-resolution Analysis) system has as its two main goals to provide astronomers with direct access to the raw (uninterpreted) data and to develop high-spatial resolution algorithms. Exportability was formulated as a secondary design goal of the project.

The problem of calibration of detector read-outs in terms of surface brightness distribution over the sky involves a non-trivial, and possibly subjective, decomposition in detector physics and astrophysics. This led us to the decision to make un-interpreted detector readouts accessible, especially for attempting the highest possible resolution in imaging, but also for study of large-scale structures (tens of degrees).

GEISHA has two main output products for astronomers: sample files and sky maps. A sample file has all necessary information on detector positions and orientations, errors, and flux calibration. It can be used to make a sky map, or as input for the application of high-resolution techniques.

In order to be able to make maps, surface brightness and position calibrations are required. If (extra)galactic emission needs to be studied a zodiacal emission model (ZEM) is also needed. Such calibrations and a ZEM model are provided by GEISHA.

I IRAS DATA ACCESS

The basic material for the GEISHA project is the raw IRAS data, containing all telemetry data from the survey instrument detectors, and the pointing information. The data were originally taken in almost random scans over the sky, and presented in order of observation, distributed over 1000 tapes. To optimise the GEISHA system for detailed investigation of (not too large) sky regions both the detector data for the scans and all auxiliary data necessary for pointing and flux calibration were reorganised into a system of 363 plates. A plate is here defined as a coherent region of the sky, large enough to fill one magnetic tape in FITS format, which then forms a totally self-contained dataset.

For routine inspection of the data a set of programs is available, applying position and/or flux calibration to the raw detector outputs, with some variation in algorithm used. The end products of these programs are

- a file ("sample file") containing the sampled output of all IRAS detectors covering the region of interest during survey scans, each sample labeled with time, position and orientation of the detector, and errors in these quantities;
- an image of the region, formed by the surface brightness on an almost arbitrary pixel grid, as reconstructed from these samples.

All intermediate and end products of these programs are in FITS format to fulfill our export requirements. The subroutines used in these standard programs are available as a software library for special research projects.
Currently, we have the detector data distributed over the plate tapes, but not yet the auxiliary data. Auxiliary data and software are available for a flux calibration method based upon the zodiacal light, positional calibration according to IPAC methods, and imaging by co-adding.

II EXPORTABILITY

Exportability was formulated as a secondary design goal of the project, both for the data sets and for the access software. This would give the system better survival chances after the initial effort, enabling us to migrate to the best environment for funding/computing/access facilities.

The exportablity was achieved by:

- use of Fortran-77 as the commonest language in astronomical computing,
- isolation of computer-system dependencies into a small set of subroutines,
- use of magnetic tape in FITS-standard format for all data bases,
- use of two widely different computer systems in development, a DEC PDP-11/44 running UNIX V7 and a CDC Cyber 170 running NOS,
- a (present) transfer of the GEISHA user system to a VAX 8600 running VMS.

III CALIBRATION

III.1 Introduction

The GEISHA system will supply the astronomer with a choice of calibration routines and zodiacal emission models (ZEMs) to subtract if required.

All calibrations have a common structure: it is assumed that the raw detector data consist of a baseline (the detector bias) plus the 'true' infrared flux multiplied by a gain. Both the baseline and the gain might vary slowly due to photon and particle induced memory effects, instrumental degradation etc.

The main differences between IPAC (IRAS Processing and Analysis Centre, U.S.A.) and GEISHA are in philosophy. IPAC uses the instrumental response of the internal calibrator as their standard candle to determine the gain and the daily observation of a piece of sky near the ecliptic north pole (TFPR) as a monitor of the baseline. GEISHA has an external standard candle: the zodiacal emission. It tries to autocalibrate the IRAS database.

The first calibration schemes, both at IPAC and GEISHA, were quite simple: the changes in gain and/or baseline were slow and global. The later schemes involve some model on the behaviour of the detectors. Different instrumental models give rise to differences in calibration routines.

There exists a whole range of possible ZEMs, from a simple lower-envelope cubic spline fit to a full-fledged physical model with densities, emissivities, and temperatures as a function of position in the zodiacal dust cloud. Here again the choice is to the astronomer.

III.2 GEISHA calibration nr 1

The model for the first GEISHA calibration consists for each scan of a linearly changing baseline plus a constant gain. The zodiacal emission is the standard candle: data = baseline + gain * ZEM. A gain and a baseline is found for each detector in every scan. The ZEM itself is derived as the mean lower envelope to all scans.

As the intensity of the zodiacal emission changes also with the solar aspect angle, the ZEM can not simply be used as a standard candle. In this calibration procedure our assumption was that the ZEM could be separated into the product of two functionals, one in ecliptic latitude, ZE90(β), and one in solar aspect angle, f(θ). The latter is temporarily married to the gain. This product will be derived from a lower envelope fit of the scan to the ZE90. Afterward divorce takes place: the θ-dependent part moves into the ZEM = ZE90(β) * f(θ) and the remainder is the gain.

This calibration is operational since begin 1987.

III.3 IPAC calibration nr 1

In the first IPAC calibration every scan has a constant baseline. Monitoring the TFPR (total flux prime reference field) twice a day leads for every SOP (= Satellite Operating Plan = 1/2 day) to a new set of baselines, one for each detector. In flight it was discovered that the baselines at 60 and 100 μm were affected by the bias boost. A slow decay was included with timescales of 50 minutes typically. The gain is calculated from the response on the internal calibrator source at the begin of the scan and at the end. The gain was allowed to change linearly during the scan.

All presently available standard IPAC products are based on this calibration.

Note: this calibration will NOT be implemented in the GEISHA system.

III.4 GEISHA calibration nr 2

As a result of the first GEISHA calibration a ZEM was found that was a product of two functions, one in ecliptic latitude and one in solar aspect angle. The function in ecliptic latitude could be very precisely approximated by an ellipse. This led to the conjecture that the zodiacal emission as a whole could be approximated by the emission of a tilted oblate ellipsoid of uniform density and temperature. At the IRAS conference in London (1987) our geometric model was corroborated by a paper presented by W. Reach and C. Heiles who had made a physical ZEM consisting of ellipsoidal equidensities and had the model integrated up to a 'last equidensity contour'.

An ellipsoidal model can never fit the so-called sidebands of the zodiacal emission (Hauser et al., 1984). The sidebands have been identified with 3 asteroid families (Dermott et al., 1984). From considerations of celestial mechanics it seems reasonable to model them with a number of inclined tori fanned over all nodal angles. A side band model (SBM) incorporating these ideas is still under study by us.

Apart from the ZEM there is always a contribution of galactic emission. The galactic emission is so granulated that only a map of the whole sky can function as a model. A fairly coarse map is sufficient. We took an equal area projection with pixels of 1 by 1 degree. Each pixel is the mean of all overlapping detector values.

As a standard candle we take the sum: ZEM + SBM + GEM. We allow the baseline to change linearly over a scan. We assume that the slow exponential decay as found by IPAC will be sufficiently approximated by a straight line. The non-linear behaviour of the gain will be corrected according to a model developed in the report "IR Photon Detectors" May 7, 1986 by Do Kester.

The complete model, data = baseline + gain * (ZEM + SBM + GEM), contains as free parameters: 5 for ZEM, 6 for SBM, 40000 (pixels) for GEM, 4 for each scan (2 baseline and 2 starting conditions), and 4 for each detector (2 decaytimes and 2 increments). In total about 64000 parameters have to be estimated, still a tiny fraction of the complete database of 13 Gbyte.

This calibration and the pertaining ZEM will be available by middle 1988.

III.5 IPAC calibration nr 2

In IPAC's second model the baseline is a slowly varying function, now for the two shorter wavelength bands too. Its value at the time of the so-called bias boost decays on a typical timescale of 1000 seconds. For the two longer wavelengths (60 and 100 µm) a photon induced enhancement is incorporated in the gain. When the detector response exceeds a preset threshold the gain increases proportional to the detector response. That increase decays on timescales of 500 and 1500 seconds for 50 and 100 µm resp. The maximum correction is 10 to 20 %.

This calibration is under development and will also be made available to GEISHA users by middle 1988.

A physical ZEM is under development by J. Good.

III.6 References

Dermott, S.F., Nicholson, P.D., Burns, J.A., Houck, J.R. (1984), Nature, vol. 312, p. 505.
Hauser, M.G. et al. (1984), Astroph. J. Lett. vol. 278, p. L15.

IV SAMPLE FILES

In order to make the raw IRAS data in a computer-independent way available to the astronomical community a new product has been made: sample files. For a given region of the sky, a sample file contains all information present for one of the IRAS frequency bands. Each detector readout is present, leading to a long table of coordinates and brightnesses. This table has to be processed further to make maps of the region of interest. Presently, coadded maps can be made having a resolution equal to the sizes of the detectors: about 5 arcminute in the cross-scan direction, and 0.75 to 3 arcminute in the in-scan direction, depending on the frequency band. The overall accuracy of the coordinates is better than 3 arcseconds. The position of known point sources is reconstructed only to 3" in-scan and 20" cross-scan, due to the asymmetrical point spread function of the detectors.

The brightness can be calibrated against the zodiacal light emission (GEISHA calibration). Samplefiles can be used by an astronomer who wants to make his own (high-resolution) image of the sky.

V IMAGING

V.1 Linear Resolution Enhancement

In order to give an idea of the amount of data IRAS gave us: 1 square degree of the sky near the ecliptic equator has been sampled 40000 times in 12 and 25 μm (60 μm: 20000; 100 μm: 10000). Towards the ecliptic poles these numbers increase by an order of magnitude, due to the convergence of the scans.

A resampling of the 40000 detections evenly over the area, would give a linear sampling distance of 30 arcseconds. A linear least squares reconstruction technique has been devised which actually yields 30 arcsecond pixel images. However, the resolution cannot be better than the telescope quality permits, i.e. 25", 25", 60", 100" at 12, 25, 60, and 100 μm. The resolution will be worse because only some high spatial frequencies are present, but a resolution of 3 arcminutes should easily be attainable.

The method works as follows. If the area of interest is divided into small pixels, each detector readout covers a number of them, say 10. The measured brightness is then the sum of the unknown brightnesses of these 10 pixels convolved with the point spread function. This coverage can be described as one equation in many unknowns (= all pixels), for which the right-hand side is the one measured brightness. But the vast majority of these unknowns have a coefficient equal to zero, because most pixels are not seen by the detector under consideration. The next readout of the same detector now covers say 7 pixels of the previous 10, and 3 new ones as the satellite scans the sky. This new equation is partly coupled to the previous one through the 7 common unknown brightnesses (which are assumed to be constant).

The coverages of the area should interconnect all pixels, and the set of sparse equations thus formed can be solved for the unknown pixel values, in a least squares sense. A nessessary condition is that there are more equations (= readouts) than unknowns (= pixels). Therefore, in principle the number of pixels per unit area, and thus the resolution, will be higher near the ecliptic poles than at the equator.

V.2 Maximum Entropy

A further enhancement of the resolution is expected from Maximum Entropy Method (MEM) image restoration techniques. This method is still under study, but a further increase of the resolution to 30" seems possible (Skilling, private communication). The resolution, however, need not be constant over the map: areas with more coverages will be reconstructed with higher resolution than sparsely covered ones. A second map of the final resolution then has to be provided.

The advantage of MEM is that simultaneously a number of calibration problems can be solved for. Presently, when a strong source has been passed, the detector behaves as having got a new zero-point and a new sensitivity. This effect is not yet corrected for. Also slow drifts of zero-points and sensitivity, apart from the ones already accounted for in the calibration, can cause stripes over the map. The MEM permits parametrisation of these effects, and solves for the 'best' image and the 'best' determination of these parameters at the same time. The disadvantage, however, is the increase in computational cost by two orders of magnitude for the same area of sky.

VI CAPABILITIES OF GEISHA

The following tasks can at present be performed by the user/astronomer.

- Coordinate transformations between any of the following systems: Equatorial 1950, Equatorial 2000, Galactic, Ecliptic 1983.5, and Sunreferenced.
- Find the plate(s) where the data of a given region is stored. The output is one or more platenumber(s). Snip and/or glue a new plate with the desired region in its center.
- Calibrate the scans with GEISHA calibration nr 1.
- Subtract the zodiacal emission model.
- Plot a scan as a function of time or of ψ.
- Perform a one-dimensional coadd.
- Plot the run of all scans over a map.
- Coadd all scans with a circular beam into a map.
- Make a sample file, i.e. add position information.

- Coadd the sample file with detector sized/oriented beam into a map.
- Make a contour plot of a map.

Whenever a map or a projection is mentioned the astronomer can choose Stereographic, Gnomonic, Azimuthal equidistant, Azimuthal equal area, Orthographic, Cylindrical equal area, Mercator, Sinusiodal equal area, Hammer-Aitoff equal area or Cylindrical equidistant.

The GEISHA system can be used for a nominal fee. Call 31-50-634074 and ask for Wesselius, Kester or Arendz to arrange a visit.

Acknowledgements:

We thank P.C. Arendz, P.D. Duyst, W. v. Oosterom, and J.E. v. Weerden for their continuous efforts in creating software and running it.

THE EXOSAT RESULTS DATABASE

N.E. White
EXOSAT Observatory/ESTEC
2200 AG Noordwijk
The Netherlands

The EXOSAT database will contain the principal results obtained from a standard analysis performed on every observation. The objective is to provide a computer accessible overview of each EXOSAT observation. This will include a summary of the results from each instrument, as well as the option of obtaining the data in a reduced form (e.g. compressed images, background subtracted spectra, lightcurves etc). A data base management system will allow manipulation of the results summary files (e.g. cross- correlating parameters). The database will in many cases circumvent the need for astronomers to analyse the final observation tapes (FOTs). This system will be accessible by remote users via the SPAN network and will include the possibility of file transfer, both of data and programs. A preliminary version of the results summary data base has been generated using the output of the first version of the EXOSAT automatic analysis of the data from the low energy telescope. The reprocessing of all EXOSAT data using second generation software is now well underway and the full database system is expected to be available by the end of 1988.

I. Introduction

Over the past twenty years X-ray astronomy has evolved from a field where a few specialists were studying a hundred poorly understood X-ray sources, to the rich field it is today where virtually every known astrophysical object is at some level an X-ray source. This transformation has resulted in non-specialist astronomers using the more recent X-ray astronomy observatories such as *Einstein* and EXOSAT. One major problem for a newcomer to data obtained from an X-ray observatory like EXOSAT is the complexity of the data acquisition modes used and the rather specialized knowledge required to perform relatively basic tasks such as background subtraction. To undertake the analysis of this data requires a considerable investment in time to write the appropriate software and to gain experience in the nuances of a particular instrument. Such an effort is only worthwhile at institutes expecting to analyse a fairly large fraction of EXOSAT observations. Up to now the *non-specialist* has been dependent on the EXOSAT interactive analysis system provided by ESA, which has meant travelling to ESTEC/ESOC to access the data. To improve this situation an EXOSAT results database is being setup that will contain the following:

(i) the basic results from each instrument for every observation, so providing an immediate overview of the results from each EXOSAT observation.

(ii) a database management system for cross-correlating those results against each other as well as against other astronomical catalogues and databases.

(iii) the data in a reduced form e.g. background subtracted spectra and images.

(iv) A suite of analysis programs to operate on the reduced data.

(v) remote access, via SPAN, with the possibility to extract the reduced data and analysis software.

The output from the LE automatic analysis has been used to create a preliminary version of (i) and (ii). This has proved extremely successful and the full database is now being implemented. This has involved an updating and rewriting of the automatic analysis programs to take into account the latest calibrations and analysis techniques. The first reprocessing of the ME data is now over 90% complete (as of October 87). The GSPC reprocessing started in June 87 and the LE in September 87. The database will be ready for public consumption by the end of 1988, with some limited access possible prior to that date.

II. The Data Base Management System

There are four *instrument summary files* which contain the principal results from each instrument (L1, L2, ME and GS) for the entire mission. A fifth *slew summary file* contains a record of all X-ray sources detected by the ME experiment while the spacecraft was slewing from one target to the next. In addition background subtracted grating spectra will also be available for the approximately 20 observations where it was used. For the instrument summary files there will be one entry per source, with up to 220 parameters stored for every entry. The summary files are accessed by a database management system (DMS) developed at the EXOSAT Observatory. This software package has been specifically designed to manipulate large amounts of astronomical data. The system, besides providing all the basic functions supported by the majority of commercially available packages, also allows a number of statistical tests frequently used in astronomy. The main facilities are:

1. The possibility to display to any terminal the basic results information for every entry.
2. Retrieve information on sources located in a specified area of the sky.
3. Select subsets using any relation of the Boolean algebra.
4. Cross-correlation of the four summary records, or of any subset with catalogues of cosmic sources.
5. Definition of new parameters. These can be virtually any function of the primary parameters i.e. parameters directly stored in the database. Parameters defined in this way are called derived parameters.
6. Possibility to sort the database (or any subset of it) by any primary or derived parameter.
7. Production of ASCII tables of primary or derived parameters.
8. Generation of histograms of any primary or derived parameter.
9. Plotting of parameters against each other.
10. Possibility to fit analytical functions to the data.
11. Statistical analysis e.g. Calculation of mean value, variance, regression analysis, χ^2 etc.

Each of the LE records is cross-correlated with well known catalogues of astronomical objects and any coincidences are also written into the record. These catalogues include the Bright Star catalogue, the IRAS catalogue, the revised catalogue of Quasi Stellar Objects, plus several other listings of extragalactic objects. It is also planned to include the catalogues available at the Centre de Donnees Stellaire at the Observatoire de Strasbourg.

III. The Reduced Data Files

The *reduced data files* created by the automatic analysis are used to produce the instrument summary files by operating on them with a standard set of analysis routines e.g. source detection and spectral fitting programs. These files are saved on an off-line medium (currently magnetic tapes, but eventually optical disks) and include such things as background subtracted spectra, images, lightcurves etc. These files will be given in a format that will be compatable with a suite of interactive programs to provide further analysis of these data. These programs will include:

1. An image processing package to detect, or set upper limits to sources in the field of view.
2. A spectral fitting program.
3. Fast fourier transform, folding and general plotting facility (the latter may be system dependent).
4. A program to convert the files to FITs format.

These programs and the data will be made available to be run on Vax systems.

IV. The Automatic Analysis

The revised LE automatic analysis software performs a detailed standard processing of LE data, taking into account the final calibration of the telescopes and detectors. For every observation a *compressed* image file is created for each filter used. A compressed image is a file containing all the information necessary to create a standard EXOSAT X-ray image and, because each image is a sparse array, is written in way that enables a factor of 10 saving in disk space with respect to more common methods of storing images. The images are searched for sources and for every one detected its position, count rate, significance of the detection and several other parameters of scientific interest are calculated and written to the database. In addition for each source a file containing all the relevant timing information is created and analysed by timing analysis software (FFTs etc). The compressed images (for each filter used) and the time series files for all detected sources plus the associated background are saved.

The ME automatic analysis system has been designed to produce, in an automatic fashion, as many high quality spectra and lightcurves as possible. In many cases the quality is such that users need not access the original FOT. These files are operated on by a standard set of analysis programs and the results written to the summary file.

Great care has been taken to design a system that can cope with the myriad of OBC modes, observing configurations and types of source observed by the ME. One of the most difficult areas of ME analysis is the choice of background subtraction technique which can be dependent on how each observation was performed, solar activity and the performance of individual detectors. It is envisaged that a second pass through some 20-30% of the observations will be required to 'fix' problems due to poor background subtraction. For many observations the automatic analysis gives acceptable spectra for sources as faint as 1 UFU. With similar constraints, the limiting sensitivity for detection is $\sim$ 0.25 UFU. Any slew data associated with the observation is processed to produce rates files for the aligned and offset data as well as a file containing pointing positions against time during each slew. If data is available with detector or quadrant ID then rates files are produced separately for each offset quadrant since they point in different directions.

The following reduced data files will be archived:

1. Time series files with the highest PHA time resolution available at energies of 1-3, 3-6 and 1-6 keV.
2. Time series files for 1-6 keV with a time resolution of 30 sec for both source and background.
3. Image files from the FFT analysis.
4. Slew time series files
5. The history of the spacecraft pointing during slews.
6. Spectra from the offset quadrants if there is evidence for background contamination.
7. Spectra for each observation plus a summary file with results from the automatic spectral analysis

Only 2, 4 and 7 will be kept permanently available. The others are either too large or not of sufficient interest to keep on disk, but they can be be restored on request. The GSPC analysis will primarily consist of a background subtracted spectrum for each observation where the ME count rate was greater than 5 ct/s/half. For weaker sources the signal to noise is not sufficiently high to justify analysis. For both the ME and GSPC the results of an automatic spectral analysis will be written to the instrument summary files. In the case of the ME a timing analysis is also made and the results stored in the instrument summary files.

V. Remote Access

The DMS will reside on both the observatory micro-vax (known as Vivean) and the HP1000 computers. These will be available for use by the community at ESTEC. The five summary files and the DMS will be kept on line. Because of disk space limitations it is not currently possible to keep all the reduced data files permanently available on disk for instant access. With the advent of optical disks this may become possible in future.

The micro-vax can be accessed via SPAN and this will form the main route for the community to interact with the DMS. Because of the differences in network protocols not all members of the community have the same capabilities, e.g. IBM machines will only be able to transfer files back and forth to Vivean (mail access), while VAXes and most other small systems may have the capability for remote interactive access, depending on their network links. In the oncoming years it is expected that the richness of these links will increase greatly so that almost anyone on an academic VAX in Europe, North America or elsewhere should have interactive access. Those users could, in principle, be allowed access equivalent to members of the Observatory on site, although graphics will be slow (image displays in particular would take a *long* time) unless a high speed link is available.

Access to Vivean will occur on four levels as follows:

(i) Mail access - The request account.

A special account (EXOSAT::REQUEST) has been setup that acts as a general mailbox. This account is checked daily to process incoming mail and would be similar to the Observatory BITNET account on Profs (MAILEX@HNOESA10). The following items would be handled by this account:

1. Requests for information or archive requests.
2. Requests for restoration of reduced data files.

3. Batch processing requests. This would allow mail-only users to submit jobs to access the DMS. The output would then be returned by computer mail (or normal mail).

(ii) Direct Access

Any user on another SPAN node may access the DMS and any activity allowed to DECNET can be performed. The prime purpose of this is to allow free use of the DMS to interrogate the summary records. In this way astronomers will be able to obtain a concise summary of each observation as well as being able to manipulate the various parameters. It is also anticipated that a complete bibliography of EXOSAT publications will be made available as part of the database.

(iii) Visitor accounts

In those cases where a more detailed analysis is required it will be possible to grant individual user IDs to observers, as occurs when observers physically visit the observatory. Then the full interactive facility available on Vivean could be used to further analyse the reduced data files. This will include a more detailed spectral analysis and specialized timing and plotting programs. Because some of the reduced data files will not be permanently mounted this will require at least one days notice for the relevant files to be restored. Eventually it is envisioned that the most frequently used reduced data files will be written to permanently mounted optical disks.

(iv) File Transfer

Users with SPAN access will also be allowed to transfer to their local computer some of the information in the database. This will include images, lightcurves, spectral buffers (and the associated detector response matrix) plus other assorted useful items. In this way it will be possible to fit more complex models to the data that may not be available on Vivean, or directly compare the EXOSAT data with those from other observatories. Also the executables of the interactive software required to operate on these files will also be made available for use on other Vaxes. The transfer out of data for extended analysis will be encouraged since the capacity of the micro-vax is limited.

VI. Present Status

The database is currently only available on the HP1000, but is being ported to the observatory microvax. This task should be complete in early 1988. The automatic processing is now well underway and will be complete for all instruments by mid-1988. Then the emphasis will concentrate on quality control. For the LE this will involve carefully checking for spurious source detections, while for the ME it will be concerned with manually correcting observations where the background subtraction was not optimum. Quality control will take a subatantial effort and will probably not be complete until well into 1989. However, it is anticipated that the database will be made available to the community by the end of 1988, with the proviso that some of the entries may be spurious. In this way feedback from the use by outside users will provide another mechanism for quality control.

Acknowledgement

The establishment of the database is a major effort involving all members of the EXOSAT observatory team. Key personnel in this effort have to date been P. Giommi, L. Chiappetti, A. Parmar, M. Gottwald, L. Stella and J. Osborne.

Object Classification Problems

RECENT RESULTS IN CLUSTERING VERY LARGE DATA SETS

M. Bruynooghe

Laboratoire d'Astronomie Spatiale
Traverse du Siphon , Les trois Lucs
13012 Marseille , France

Abstract .– Astronomical image analysis may require the clustering of very large data sets (VLDS) , composed of more than 10000 objects . Up to now none clustering method was able to deal with such a VLDS , with a reasonable computing time , especially in the case of a high dimensional representation space . In this paper a clustering strategy for VLDS is presented . It overcomes the limits of the classical strategy that is based on the clustering of the so called "strong patterns" . This new clustering strategy is based on the complementary utilization of an accelerated partitioning algorithm generating a large number of small clusters and then of an accelerated hierarchical clustering algorithm . It is about 25 times faster than the classical strategy and leads to results of higher statitical quality . Thus , it is now possible to classify large data sets as those encountered in astronomical image analysis , for object classification in bidimensional photometry , for example , or for analysis of the large amount of data to be send down by the Space Telescope ...

Keywords : Hierarchical Clustering – Partitioning Algorithms – Minimum Variance Criterion – VLDS Clustering – Astronomical Image Analysis.

I. Introduction.

Clustering algorithms are generally used to provide a reduction of multidimensional data . In this paper , we will be solely concerned with unsupervised clustering , which does not require a learning set to classify the data set .

As in the case of factorial analysis , the objects or individuals to be classified have numerical values defined on a set of variables or attributes . These objets may then be represented as points in a multidimensional space . The main interest of a preliminary factorial analysis is to reduce the dimensionality of the representation space , the ten to twenty first principal factorial coordinates beeing generally sufficient to reconstitute most of the non random part of data .

This paper deals with clustering algorithms which are based on the minimum variance criterion . The preference for this clustering criterion is motivated by the consistency it offers with factorial analysis , which also involve a decomposition of variance .

Clustering methods , hence may be used in addition to factorial analysis : principal components analysis or correspondence analysis .

Condiderable literature exists on clustering methods . A distinction should be made between hierarchical algorithms and partitioning algorithms . About 75 per cent of all published work on clustering has employed hierarchical algorithms , according to Blashfield and Aldenderfer (1978) . Hierarchical methods generate results of greatest interest for interpretative purposes . Neththerheless , for the clustering of a large data set , partitioning algorithms have been prefered up to now , because of their better computational performances .

However , the experimental results presented in this paper show that this last approach applied to image analysis is very time consuming : the clustering of 10011 pixels represented by the first 37 factors of a correspondence analysis requires about six hours of computing time on a VAX 11/780 !

Hence , to deal with very large data sets as those encountered in astronomical image analysis , it is necessary to elaborate a more efficient clustering strategy . This is the aim of this paper , in which accelerated clustering algorithms are presented .

The results we have obtained with these accelerated algorithms show that it is now possible to classify very large data sets with a relatively small amount of computing time and a high statistical quality .

II. Introduction to a Classical Methodology for Clustering Large Data Sets .

As pointed out by Murtagh and Heck (1986) , a large number of partitioning algorithms have been proposed : "k-means algorithm" (Hall and Ball 1965) , "variable centers algorithm" (Forgy 1965 , Mac Queen 1967) , "dynamic cluster algorithm" (Diday 1971) ...

Many of these algorithms are iterative techniques for improving a crudely derived initial partition . They are generaly characterized by the random choice of a **small** number of cluster seeds , fixed in advance , from which the centers of the classes of minimum internal variance are sought by means of an iterative procedure : these partitioning algorithms proceed by alternately generating clusters around cluster seeds and then selecting cluster centers as new cluster seeds , until a stable partition is obtained .

As a non hierarchical strategy , the variance criterion makes it possible to evaluate the quality of the partition obtained after the random selection of a **small** number of cluster seeds .

The partition so obtained is generally no more than a locally optimal solution to the clustering problem and depends on the initial choice of cluster seeds . Consequently the operation has to be repeated several times choosing cluster seeds at random , producing hence a number of locally optimal partitions equal to the number of initialisations of the clustering procedure .

A practical remedy for this situation , proposed by Diday (1972) , is to seek for the intersection of the different locally optimal partitions . The resulting elementary clusters are the so called "strong patterns" , which are stable groups of individuals that are always assigned to the same cluster in each of the partitions .

The hierarchical clustering of a reduced number of "strong patterns" may then be used in the context of clustering large data sets .

For example , Molliere (1985) has proposed to suppress in the data set atypical individuals which can be isolated in very small "strong patterns" , especially in those reduced to one individual . The application of this clustering methodology to a data set consisting of one thousand individuals , represented by the first five factors of a preliminary correspondence analysis , led to 467 "strong patterns" of one individual among the whole set of 571 "strong patterns" resulting of 20 different partitions into 8 clusters .

III. Critical Evaluation of the Classical Strategy for Clustering Large Data Sets .

The locally optimal partitions into a **small** number of clusters may be caracterised by a very low statistical quality , as illustrated by the following results related to the clustering of a set of 10011 pixels represented by the 37 first factors of a correspondence analysis .

This set of pixels was partitioned into 37 clusters corresponding to the background of a sensitometry image and to 36 areas of approximately constant optical density . The table I presents the results related to 20 different partitions into 37 clusters .

Although 37 clusters were initially required , a smaller number of classes has been obtained at each iteration , varying from 28 to 36 clusters , making it more difficult to compare the statistical quality of the various partitions .

Our results show that this classical strategy for clustering large data sets results in :

1 - an extreme variability of the partition quality , as measured by the percentage of variance which is explained by each partition .

2 - a total computing time of about 6 hours on a VAX 11/780 , which is the result of the great number of iterations required to determine the "strong patterns" with a minimum of statistical pertinence .

IV. An Efficient Strategy for Clustering Very Large Data Sets.

In order to overcome the limits of the classical strategy for clustering large data sets , we propose :

1 - to generate one , and only one , locally optimal partition in a **large** number of small clusters , using an accelerated partitioning algorithm , described in detail in Bruynooghe (1988) .

2 - to build then a binary tree on the set of elementary clusters , previously obtained , using one of our accelerated hierarchical clustering algorithms , Bruynooghe (1977 , 1978 , 1980 , 1988) .

Then , by cutting the binary tree at different stratification levels , several partitions of high statistical quality may be obtained .

The accelerated partitioning algorithm we have implemented incorporates a fast nearest neighbour algorithm (Bruynooghe 1977) , which is a very efficient generalisation of other nearest neighbour algorithms (Fukunaga and Narendra 1975 , Friedman , Bentley and Finkel 1977) .

The first fast hierarchical clustering algorithms , i . e . the "reciprocal nearest neighbour clustering algorithm" and the "reducible neighbourhoods clustering algorithm" , were proposed by Bruynooghe (1977 , 1978 , 1980) . These algorithms have overtaken the limits of the traditional hierarchical clustering algorithm by generating an exact hierarchy on a large data set . The theoretical justification of these algorithms is the so called "reducibility property" or "neighbourhood contracting property" discovered by Bruynooghe (1977 , 1978) .

A variant of the "reciprocal nearest neighbour algorithm" has been dubbed "bachelor algorithm" by De Rahm

(1980) and Juan (1982), due to the fact that each agglomeration leaves certain points unmatched . Furthermore , a theoretical improvement has been obtained by Benzecri (1982) with the "nearest neighbour chain algorithm" whose time complexity increases at most as the square of the number of objects to be classified . Other time complexity results and an excellent survey of these last clustering algorithms are to be found in Murtagh (1983 , 1986) .

These fast hierarchical clustering algorithms of the early 1980 ' s have been accelerated by incorporating :

- a fast nearest neighbour algorithm in the case of the "reciprocal nearest neighbour clustering algorithm" .
- and an efficient branch and bound algorithm for the generation of a set of neighbourhoods in the case of the "reducible neighborhoods clustering algorithm" .

The utilization of such efficient searching algorithms in high dimensional spaces , which is described in detail in Bruynooghe (1988) , is able to greatly reduce the computing time required for clustering , by eliminating the necessity of calculating many pairwise distances .

The table II shows the results we have obtained , with our VLDS strategy , using 1024 initial cluster seeds , randomly choosen among the 10011 pixels to be classified . After convergence of the partitioning algorithm , the number of clusters obtained has been reduced from 1024 to 711 , due to the elimination of cluster seeds around which no cluster has been constructed .

It is to be pointed out that the percentage of variance explained by the locally optimal partition in 711 clusters is rather important , greater than 99 per cent . This result is an experimental demonstration of the high quality when based on a large number of cluster seeds . The computing time of this first step is equal to 618 seconds .

At the next step , the computing time of the hierarchical clustering of the 711 elementary clusters , previously obtained , is equal to 231 seconds . Thus the total computing time for the clustering of the 10011 pixels amounts to 850 seconds . That is to say 25 times less than for the classical strategy , which is based on the clustering of the "strong patterns" . Furthermore , the statistical quality of the results have been significantly improved .

V. Conclusion.

The complementary utilization of an accelerated partitioning algorithm generating a large number of small clusters and then of an accelerated hierarchical clustering algorithm has given very satisfactory results . This new strategy is about 25 times faster than the classical approach and leads to results of higher statistical quality . Our approach may hence be considered a very efficient one for clustering very large data sets , as those encountered in astronomical image analysis . Thus , the computational performance of our clustering strategy is such that it is applicable to a wide class of astronomical problems : for image segmentation , object classification in bidimensional photometry with star-galaxy separation , prediction of spectral type from photometry , taxonomy construction for stars , galaxy clustering , gamma and X-ray astronomy , or for analysis of the large amount of data to be provided by the Space Telescope ...

Acknowledgements.

Thanks are due to our friends and colleagues who encouraged this work and suggested examples for us to consider ; amoung them are A.Llebaria , A.Vuillemin from Spatial Astronomy Laboratory and Pr. J.P.Benzecri from Paris VI University .

References.

Ball, G.H. Hall, D.J. : 1965, "*Isodata , a Novel Method of Data Analysis and Pattern Classification*", Techn. report 5 RI. Project 5533 , Stanford Research Institute , Menlo Park , California USA .

Ball, G.H. Hall, D.J. : 1967, "*A Clustering Technique for Summarizing Multivariate Data*", Behavioral Sciences no 12 , pp. 153-155 .

Benzecri, J.P. : 1973, "*L'analyse des données*", Tome 1 ,"*La taxinomie numérique*"; Tome 2 ,"*L'analyse des correspondances*", Paris, Dunod

Benzecri, J.P. : 1982, "*Construction d'une Classification Ascendante Hiérarchique par la Recherche en Chaine des Voisins Reciproques*", Les Cahiers de l'Analyse des Donnees , Vol .VII , no 2 , pp . 209-218 .

Blashfield, R.K. Aldenderfer, M.S. : 1978, "*The literature on cluster analysis*", Multivariate Behavioral Research , 13 , 271-295 .

Bruynooghe, M. : 1977, "*Methodes nouvelles en classification automatique de donnees taxinomiques nombreuses*", Colloque International de Taxonomie Numerique , Orsay . Statistique et Analyse des Donnees ,no 3 / 1978 .

Bruynooghe, M. : 1978, "*Hierarchical Clustering of large data sets and geographical pattern recognition*", The Fourth International Conference on Pattern Recognition, Kyoto, Japan

Bruynooghe, M. : 1978, "*Classification Ascendante Hiérarchique des Grands Ensembles de Données*" . *Un algorithme rapide fondé sur la construction des voisinages reductibles* . Les Cahiers de l'Analyse des Donnees , Vol .III , no 1 , pp . 7-33 .

Bruynooghe, M. : 1980, "*Large Data Set Clustering Methods using the Concept of Space Contraction*" . *COMPSTAT 1980 , 3rd Symposium held in Leiden , The Netherlands .*

Bruynooghe, M. : 1988, These de Doctorat d'Etat es Sciences Mathematiques , Paris VI University , (in preparation) .

De Rahm, C. : 1980, "La Classification Hiérarchique Ascendante selon la Methode des Voisins Réciproques", Les Cahiers de l'Analyse des Donnees , Vol . V , no 2 , pp . 135-144 .

Diday, E. : 1971, "Une nouvelle methode en classification automatique et reconnaissance des Formes", Revue de Statistique Appliquee , Vol . XIX , no 2 . pp.61-95 .

Diday, E. : 1972, "Optimisation en Classification Automatique et Reconnaissance des Formes", Revue Francaise de Recherche Operationnelle , 6th year , pp.61-95 .

Forgy, E.W. : 1965, "Cluster Analysis of Multivariate Data : Efficiency versus Interpretability of Classifications", Biometrics , 21 , pp. 768-769 .

Friedman, J.H. Bentley, J.L. and Finkel, R.A. : 1977, "An algorithm for finding best matches in logarithmic expected time", ACM Transactions on Mathematical Software 3 , 209-226 .

Fukunaga, K. and Narendra, A. : 1975, "A branch and bound algorithm for computing k-nearest neighbours", IEEE Transactions on Computers C-24 , 750-753 .

Juan, J. : 1982, "Le Programme Hivor de Classification Hiérarchique Ascendante selon les Voisins Réciproques et le Critere de la Variance", Les Cahiers de l'Analyse des Donnees , Vol .VII ,no 2 , pp . 173-184 .

MacQueen, J. : 1967, "Some Methods for Classification and Analysis of Multivariate observations", Proc. 5th Berkeley Symp. 1965 , pp. 281-297 .

Molliere, J.L. : 1985, "What is the real numeber of clusters", Proceedings of the 9th Annual Meeting of the Classification Society , Karlsruhe 26-28 June .

Murtagh, F. : 1983a, "A survey of recent advances in hierarchical clustering algorithms", The Computer Journal 26 , 354-359 .

Murtagh, F. : 1983b, "Expected time complexity results for hierarchic clustering algorithms which use cluster centers", Information Processing Letters , 16 , 237-241 .

Murtagh, F. : 1985, "A cluster-based strategy for determining nearest neighbours", Proceedings of the 9th Annual Meeting of the Classification Society , Karlsruhe 26-28 June .

Murtagh, F. : 1986, "Clustering Techniques and their Applications", International Workshop on Data Analysis and Astronomy , Erice , Sicily , 20-30 April 1986 . To appear in Data Analysis and Astronomy , Plenum Press .

Murtagh, F. and Heck, A. : 1986, "Multivariate Data Analysis" D.Reidel Publishing Company, Dordrecht , Holland .

Ward, J.H. : 1963, "Hierarchical grouping to optimise an objective function", J. Am Statist. Ass. , 59 , 236-244 .

Table I. — *Classical Strategy Results*

Iteration of Number	Number of clusters after convergence	CPU in Seconds on a VAX 11/780	Mean Number of distances computed per individual	Explained Percentage of variance
1	28	1077	9.83	41.65
2	29	977	8.33	52.17
3	31	1058	9.64	48.61
4	31	1251	11.65	39.26
5	32	1046	9.33	47.49
6	32	1127	10.12	44.52
7	32	1064	8.97	48.70
8	32	1022	8.59	55.28
9	33	975	8.04	59.57
10	33	1096	9.76	45.89
11	33	1065	9.44	51.85
12	34	1016	8.56	54.55
13	34	1064	9.32	50.71
14	34	1152	10.54	44.89
15	35	981	8.58	60.67
16	35	933	7.79	61.29
17	36	968	8.17	62.18
18	36	1247	11.31	46.47
19	36	1189	10.64	50.14
20	36	911	7.63	62.85

Table II. — *VLDS New Clustering Strategy Results*

Iteration of Number	Number of clusters after convergence	CPU in Seconds on a VAX 11/780	Mean Number of distances computed per individual	Explained Percentage of variance
1	711	618	61.43	99.40

Extraction of morphological parameters from images of galaxies

Preben Grosbøl
European Southern Observatory
Karl-Schwarzschild Straße 2, D-8046, Garching, FRG.

Summary

A number of methods and algorithms for extraction of morphological parameters from surface photometry maps of galaxies are discussed. Automatic methods to determine position and inclination angles for galaxies using Fouries techniques are given with a discussion of the main sources of errors. These techniques can be used for the morphological classification of elliptical galaxies. For spiral galaxies, extraction of additional parameters for their spiral structure is discussed.

1 Introduction

The spiral density wave theory (Lin and Shu, 1964) can now explain the dynamics of grand design spiral structures in galaxies. Recent developments (Haass *et al.*, 1982; Bertin *et al.*, 1984) make it possible to predict the main spiral modes in a disk galaxy from its distribution functions of mass, angular momentum and velocity dispersion. This opens the possibility to deduce general properties of these functions from the distribution of parameters of spiral structure in galaxies. In order to perform such investigations, a large set of quantitative parameters for spiral structures in disk galaxies is needed.

The present paper discusses methods to estimate such parameters in an automated way. The determination of orientation and axial ratio for galaxies is discussed in section 2 while a procedure for extraction of spiral parameters is given in section 3.

2 Projection parameters

The orientation of the galaxies with respect to the line of sight must be determined before parameters for their internal structure can be extracted. To obtain position and inclination angles, it is assumed that the outer parts of spiral galaxies can be approximated with an axisymmetric thin disk. The methods can also be applied to early type galaxies; although the interpretation of the derived parameters is different.

2.1 Position angle

To estimate the position angle *p.a.* of a galaxy, azimuthal intensity profiles (AIP) are extracted along circles centered on the galaxy and with radii corresponding to its outermost isophotes. Foreground stars and other sharp features, considered as gross errors, are removed from the profiles by non-linear filters (*e.g.* a median filter (Pratt,1978) or a recursive filter (Grosbøl,1980)). Finally, the profiles are Fast Fourier Transformed (FFT) to obtain the amplitudes A_m and phases θ_m of the mth harmonics. For the determination of position angle only the harmonics $m = 0, 1, 2$ and 4 are used. The position angle of the galaxy, relative to a reference line, is given by $p.a. = \theta_2/2$.

Several checks can be performed to ensure that the estimated value of $p.a.$ is unaffected by gross errors:

- One of the main problems in estimating the $p.a.$ is spiral structures and warps in the outer parts of the galaxies. In order to determine if such features have effected the estimate, the $p.a.$ should be computed for at least three radii in the outer region of the galaxies. A correlation between the radius r and θ_2 indicate that such systematic errors have effected the result.

- Normal galaxies should have $\theta_4/4 \approx \theta_2/2 \pmod{\pi/2}$ If they differ significantly problems associated with spiral structure or warps can be expected.

- For undisturbed galaxies, it is expected that $A_1 < A_2$. However, if $A_1 > A_2$ and $\theta_2/2 \approx \theta_1 \pmod{\pi}$, it is likely that an object close to the galaxy affect the measurements significantly.

- The expression $A_0 + A_2 + A_4$ gives a lower limit of the isophotal level as which the $p.a.$ is determined.

A main advantage of this procedure is that it is less sensitive to asymmetric isophotes and companions than other methods since only the even harmonics are used.

2.2 Inclination angle

With known position angle, the inclination angle i can be determined in a similar manner using FFT. AIP's are extracted along ellipses with different inclination angles ϕ_k and major axes defined by the previously determined $p.a.$ Each of these profiles are Fourier transformed giving the corresponding amplitudes $A_{2,k}$ and phases $\theta_{2,k}$. Changing ϕ_k from 0 to $\pi/2$, the phase $\theta_{2,k}$ will shift from 0 to π at $\phi_k \approx i$. For a galaxy with a thin exponential disk the inclination angle i is determined (see, Grosbøl, 1985) as the root of the regression line through the points

$$(x, y)_k = (\cos(\phi_k), A_{2,k}\cos(\theta_{2,k})). \tag{1}$$

The scatter around the regression line gives the accuracy of the determination. A large uncertainty indicates that spiral features or warps effects the estimate. Also here the method is little sensitive to asymmetries in the isophotes since they mostly affect the odd harmonics which are not used.

Projection angles for open spirals are in general difficult to determine due to the pertubation of their disks. If the spiral pattern covers an azimuthal angle of more than π, the smoothness of the deprojected arms can be use to find the best values of $p.a.$ and i (Danver, 1942). This may not be possible for very open patterns. In such cases, an estimate of the projection angles can be obtained by minimizing $|\theta_2 - \theta_4/2| \pmod{\pi}$ along the deprojected spiral. Although such methods can be used they are very difficult to make automatic.

3 Spiral parameters

The extraction of parameters for the spiral structure in galaxies is often done by visually outlining the pattern and measuring it (Danver, 1942). In order to compare the results with the density wave theory it is however necessary to get objective measures for all spiral modes in the galaxy. This has been done by applying two-dimensional Fourier transforms to deprojected images of the galaxy in polar coordinates (*e.g.* Kalnajs (1975) and Iye *et al.* (1982)). A $(\log(r), \vartheta)$ coordinate system is normally used because most spiral arms can be approximated with logarithmic spirals. Two main problems are associated with the use of two-dimensional FFT:

- the axisymmetric intensity of the galaxy must be removed to avoid to get spurious frequencies.
- the radial region, within which the FFT should be performed, must be specified. From a theoretical point of view, it would be reasonable to do the analysis between the main resonances in the galaxy. However, they are normally not known since the angular speed of the pattern is unknown. Furthermore, there may exist several modes with different pattern speed which makes it impossible to define a common region to analyze.

These problem can to some degree be avoided by performing a one-dimensional FFT of each radius and analyzing the radial variation of the phases. This is also much faster in computer time than using a two-dimensional transform. The main difficulty with the one-dimensional approach is that it only determines the dominant wave number k in a region and not the full spectrum.

3.1 Pitch angle

AIP's are extracted along ellipsis corresponding to circles in the plan of the galaxy and a one-dimensional FFT is performed. This yields amplitudes $A_m(r_j)$ and phases $\theta_m(r_j)$ for the mth harmonics at the radius r_j. The radial increment should be of the order of the resolution element times $\sec(i)$ to avoid correlation between the individual AIP's along the minor axis. Spiral patterns with mode m in the galaxy can then be found by analyzing the histogram of

$$\psi_{n,j} = \theta_n(r_j) - \int_{r_0}^{r_j} k(r)dr \pmod{2\pi} \qquad (2)$$

for all radii r_j. The bin size of the histogram depends on the number of radii used and the desired resolution in the wave number k. For a logarithmic spiral with mode m and pitch angle ι, the wave number is $k = m/r\tan(\iota)$.

The value of k or ι is varied over the whole definition range and a χ^2-test is performed to determine if the corresponding histogram can be drawn from a uniformed distribution. A spiral may be present for values of k for which the χ^2-test shows the histogram to be inconsistent with a uniformed distribution. k gives the wave number of the possible detected spiral pattern. The radii r_j for which $\psi_{n,j}$ are in the mode bin of the histogram then define the radial region where a spiral is present. To verify that these radii constitute a region, a run-test can be perform. If this test shows that the radii corresponding to the mode cannot be taken form a random distribution the existence of a pattern is confirmed. When radii in the same region pass both tests for several k, the value of k which gives the highest mode is chosen.

4 Conclusions

The described methods for extraction of morphological parameters for galaxies provide an easy and robust way of obtaining not only position and inclination angles for galaxies but also structural parameters for spiral patterns. They were applied to a sample of 605 spiral galaxies studied by Grosbøl (1985) for which also amplitudes and pitch angles for the spiral modes $m = 1, 2, 3$ and 4 were derived.

By visual inspecting, it was found that less than 15% of the estimated projection angles were in error. The main reasons for such errors were spiral structures and warps in the outer regions and companions. Both of these problems are indicated by the test mentioned above. Thus, the described procedure can be used for automatic extraction of projection and morphological parameters for large samples of galaxies. Although the methods have been derived for spiral galaxies, they can also be used to determine the position angle and ellipticity of early type galaxies.

References

Bertin,G., Lin,C.C. and Lowe,S.A. : 1984, *Proc. of Course and Workshop on Plasma Astrophysics* **ESA SP-207**, 115.

Danver,C.-G. : 1942, *Lund Ann. Obs.* No. **10**.

Grosbøl,P. : 1980, *SPIE* **264**, 118.

Grosbøl,P. : 1985, *Astron. Astrophys. Suppl. Ser.* **60**, 261.

Haass,J., Bertin,G. and Lin,C.C. : 1982, *Proc. Natl. Acad. Sci. USA* **79**, 3908.

Iye,M., Okamura,S., Hamabe,M. and Watanbe,M. : 1982, *Astrophys. J.* **256**, 103.

Kalnajs,A.J. : 1975, in *La Dynamique des Galaxies Spirales, Colloque Intern CNRS*, No. **241**, ed. L. Weliachew, 103.

Lin,C.C. and Shu,F.H. : 1964, *Astrophys. J.* **140**, 646.

Pratt,W.K. : 1978, *Digital Image Processing*, Wiley-Interscience, New York.

MASS ANALYSIS OF TYCHO PHOTOMETRIC DATA FOR VARIABLE STARS

A. Heck, G. Jasniewicz, C. Jaschek

C.D.S., Observatoire Astronomique, Strasbourg, France

A methodological strategy is prepared to analyze the photometric data of the Tycho space experiment with the following objectives:
- detection of variable stars,
- classification of lightcurves,
- prediction of variable types.

The classification process is intented from scratch (i.e. without an a priori knowledge of the type of variability). Simulations will be carried out on real lightcurves of known types of variability to test the algorithms.

THE SEARCH FOR STRUCTURE: OBJECT CLASSIFICATION IN LARGE DATA SETS

Michael J. Kurtz

Harvard-Smithsonian Center for Astrophysics
60 Garden Street
Cambridge, MA 02138 USA

Those lovely patterns in the heavens are decorations in the visible world, so they may be thought the most beautiful and perfect of visible things; but they fall far short of those which are true...

Plato, The Republic

I. Introduction

A. Classification in General

Classification is the ordered idealization of observational experience; to be successful a classification scheme must also be an ordered idealization of physical reality. Object classification schemes provide the nouns for the formal languages required to describe phenomena.

Classification is not measurement, rather it is an organizational synthesis of many measurements. These multitudes of measured values fit into patterns, which represent the classes; these patterns then fall into patterns of patterns, which represent the classification schemes.

The usefulness of a classification is at first determined by the ease with which the multivariate observations can be put into the classes. The long term usefulness is determined however by whether the classification scheme can serve as an organizing principle to allow an understanding of the processes which gave rise to the classes.

Classifications are functions of the multivariate observations from which they were defined. They are not properties of the objects which gave rise to the observations; thus one may speak of a G2V spectrum, but not a G2V star. When one attempts to determine the class of an object by using parameters other than those which define the classification one is not actually classifying, rather one is predicting the classification of the object. In most of what follows the distinction between prediction and classification will be ignored.

B. Astronomical Classification

Astronomers have been using classification as a tool to organize and understand their data since they started having data complicated enough to warrant it. A. Secci 1867 developed the first classification scheme in astronomy, and published the first catalog of classifications (for stellar spectra).

Since then classification schemes have been developed for almost every type of astronomical data and/or object. There are schemes to classify the spectra of normal stars, peculiar stars (of several different types), of galaxies, of QSOs. There are schemes to classify the light curves of variable stars, of binary stars, of novae, and of supernovae. One can classify point distributions as different types of star or galaxy clusters. One can classify optical pictures of galaxies, x-ray pictures of clusters of galaxies, and radio pictures of galactic jets.

Crude classifications are often used to classify objects into the two categories "interesting" and "uninteresting". Examples might be classifying stars and galaxies on direct plates, and QSO spectra and non-QSO spectra on objective prism plates. The reader is left to decide which are the "interesting" objects in these cases.

II. Classification Techniques

A. Syntactic

Since the pioneering work of Chomsky in essentially founding the field of analytical linguistics the burgeoning field of computer science has fully embraced the syntactic/linguistic approach. It is, for example, quite probable that no beginning student in computer science today does not know of the existence and use of parsers.

Within the last two decades a great body of work has laid the basis for the engineering discipline now known as syntactic pattern recognition. Led, until his recent untimely death, by K.-S. Fu, this development has proved enormously fruitful for a large number of applications.

Basically the method begins by segmenting an image into pattern primitives, which I'll call morphemes. These morphemes are organized into groups by means of a pattern grammar; I'll call these words. The words are then organized into sentences by the pattern grammar, or the pattern description language. The sentences then represent the objects to be classified.

The sentences are analyzed by parsing them into words or phrases according to the pattern grammar, and looking up the words and phrases in a dictionary. The technology used in analyzing the sentences is quite similar to that used in writing compilers, save that it is usually permitted for a compiler to fail when a grammar or a spelling error is encountered.

B. Decision Theoretic

Decision theoretic methods of classification are by far the most common currently. They are the only mathematical classification techniques to be widely used in astronomy. Included are such methods as Principal Components Analysis, Cluster Analysis, and Multiple Discriminant Analysis. These topics have been reviewed by Bijaoui 1979, Kurtz 1983, and especially Murtagh and Heck 1987.

The great power of these relatively simple methods continues to show much promise for astronomical classification, especially when dealing with new types of data, where established classification schemes do not exist. The recent work of Heck, *et al.* 1986 demonstrates with great vigor the power of this approach.

In many datasets in astronomy the underlying distributions of the multivariate observations are unknown. Also our data often have "outliers", objects which are not really members of the group of objects we wish to study. The analysis of this kind of data often requires the robustness of non-parametric, distribution-free statistical methods. Some non-parametric techniques which could be of use in astronomical classification problems include non-metric multidimensional scaling (see Kurtz 1983 for review), Spearman's coefficient of rank correlation, and Kendal's coefficient of concordance (*e.g.* Gibbons 1985).

C. Fuzzy Sets

Fuzzy techniques are of considerable interest because they seem to have very much in common with the human thought process (Zadeh 1984). As Di Gesu is reviewing fuzzy methodologies at this meeting I will confine myself to a few general remarks.

Whereas decision theoretic techniques reflect the probabilistic nature of the data, fuzzy techniques reflect the fuzzy nature of the definitions (*e.g.* the set of all red stars is clearly a fuzzy set); this leads to substantially different methodologies. Fuzzy thinking is a requirement for some artificial intelligence applications, the development of natural language parsers might be one example. Because of the very rapid development of A.I. technologies, one may expect fuzzy methods to be even more widely used in the near future.

D. Stochastic/Fuzzy Grammars

The syntactic methods provide excellent means of describing complex data structures. Modifications on the basic method are however required when one is dealing with data of less than infinite signal to noise, or with morphemes which are not always perfectly formed. Stochastic grammars (Fu 1982) and fuzzy grammars (Kandel 1982) are designed to address this issue.

Essentially these techniques are algebras which permit the combination of uncertain morphemes into uncertain words into uncertain sentences, which can then be compared with patterns which may or may not be very well defined. Stochastic and fuzzy languages, especially in association with error correcting parsers (Fu 1982), can be extremely powerful tools for the analysis of complicated multi-element datasets. They are very regularly used now, as a brief perusal of the cybernetics or pattern recognition literature will attest.

III. Current Problems in Astronomical Classification

A. Large Data Sets

The theme of this conference points toward the overwhelming problem being faced by some astronomers, the enormous explosion in our ability to collect data in digital form. The sheer mass of data is causing insurmountable problems for some fields already. To take the most extreme example, the VLA will soon be capable of producing images which are 8192 x 8192 in the spatial dimension, 512 in the spectral dimension, and 4 in the polarization dimension, with dynamic ranges high enough to require 32 bit real numbers.

As I am not a radio astronomer I can perhaps be forgiven for not attempting to solve the VLA data problem here. Whole sky digital optical surveys will only contain an equal amount of data to a single VLA image, should the VLA be able to be used to its capacity, but the optical surveys will still contain huge amounts of data.

It is clear that whenever one is dealing with large numbers of objects some form of classification is essential. The type and quality of the science to be derived from huge datasets with large numbers of objects is directly linked to the type and quality of the classifications, or data characterizations performed. For example if one chooses to reduce an objective prism survey into a two class classification, possible QSO spectrum and non-QSO spectrum, one is clearly limiting the type of science to be gleaned from the data. The quality of any work based on these data will clearly be dependent on the quality of the classifications.

B. Spectra

i. General

Spectra have provided the key to astrophysics ever since the work of Kirchoff and Bunsen. With typically 100 to 1000 independent data elements spectra require a rich assortment of languages for their description. Beginning with Secci the languages and dialects of spectral classification have grown greatly. Nearly every subfield of observational astrophysics where spectra are a major data source has created at least one classification scheme for its spectra.

Three examples will serve to demonstrate the breadth of current research involving spectral classification. Lawrence 1987 has reviewed the situation with regard to the classification of spectra from active galactic nuclei, and has suggested a number of improved methods. Burg 1987 has investigated the properties of a complete magnitude limited sample of over 2000 galaxy spectra using statistical classification techniques. McMahan 1986 has classified white dwarf spectra by comparing them with grids of spectra from his theoretical model atmosphere code.

ii. Normal Stars

Certainly with the term spectral classification one normally conjures up the idea of spectral classification of stars. This line of research has played and continues to play a central role in what is perhaps the greatest achievement of modern astronomy: understanding the physics of stars. Spectral classification continues to play an important role in understanding the chemical and dynamical evolution of our galaxy.

Several active programs are underway where low dispersion spectra are classified as a step in investigations of galaxies or QSOs. As several papers detailing the exact state of this work are being presented at this conference, I shall leave a discussion of this interesting work to them.

Classification at MK dispersion has until now been almost entirely by classical visual means. A full discussion of the classical spectral classification methodology has recently been given by Jaschek and Jaschek 1987. The limits of classical classification may be seen by noting that N. Houk will have spent most of her life in the HD re-classification when she finishes, and will "only" have classified a quarter of a million spectra. It is not unusual for a single objective prism plate to have over ten thousand classifiable spectra on it; several such plates may be observed in a single night.

iii. The Automation of MK Classification

There has been as yet no automated MK classification; not a single spectrum. To the best of my knowledge only LaSala 1987 is working on this problem. Very little work is being done at MK dispersion, or near it. The only recent project of any size has been Einasto, *et al.* 1985 who have obtained automated classifications for 3000 Abastamauni Schmidt F star spectra.

Because of the great importance of spectra at this dispersion, and of the MK System and its revisions (Johnson and Morgan 1953, Morgan and Keenan 1973, Keenan 1987) to stellar astronomy, I should like to discuss some of the requirements to automate it, despite the apparent current dearth of interest.

I have already (Kurtz 1985) suggested an algorithm which I believe can automate the system, essentially an iterative application of multiple linear discriminant functions. One feature of this algorithm, the need for a large number of standard spectra in each classification box has proven controversial. However the requirement for many standards is not just caused by the requirement of this particuliar algorithm to know the variance (cosmic scatter) of every spectral feature within a classification box. The requirement is caused by the nature of the homomorphism which is the classification process.

No classification scheme which attempts to classify (map) continuous functions to points in a two-dimensional continuous space can possibly succeed. The probability of one of these continuous functions (spectra) falling exactly on one of the points (standard star spectra) in the 2-D space (classification grid) is exactly zero. In practice the visual classifier, as a result of experience, establishes a complex set of fuzzy rules for every spectral feature, at every point in the classification grid.

The need for many standards reflects the need to quantify this experience for the machine, and is independent of the algorithm employed. Any computer program which assigns spectral types to spectra must have rules to define "nearness" to a spectral type; the only question is whether these rules will be *ad hoc* numerical ones, such as equal weighting of fluxes, or will be based on standard stars, embodying the collected experience of generations of visual classifiers.

Knowing the variance of each spectral feature at each spectral type would provide important information for the continued improvement of the definition of the standard star based classification system; for an exact determination of the resolution of the system, both in terms of the standards, and in terms of the signal-to-noise in the unknown; for the establishment of local additional dimensions in the classification space; and for the direct determination of spectral peculiarities.

The standard star system could be improved by rejecting as standards any spectra which have spectral features significantly more different than the mean spectrum at its type, when compared to the normal variance at that type of the other standards. The standard grid could also be improved by requiring self-consistency, and that all standards be classified by the program as the types they represent.

The resolution of a classification system could easily be determined by noting at each type the difference between the mean values of the spectral features at that type and adjacent types, and comparing that with the variance of these features at the type (the cosmic scatter). One could at once see the degree of discrimination between spectral types due to each spectral feature, and from this one could easily calculate the signal to noise necessary to make the discrimination.

At any spectral type the existence of a large variance in a particular spectral feature implies directly that the classification scheme cannot explain that feature. There are two possible reasons: either the grid is not fine enough, or a "third" dimension is required to explain the data. In these cases Principal Components Analysis, or similar statistical classification techniques, would be quite successful in developing attached local "third" dimensions to the classification space. One can certainly expect such schemes to be necessary to describe abundance differences, just as they have been for visual classifiers (for a review see Keenan 1987).

The principal advantage of spectral classification over photometry is that one can discover the unexpected. Thus it is incumbent on any classification scheme to include the ability to discover "peculiar" spectra. In order to determine what is "peculiar" however, one must know what is "normal". The variance information for each spectral feature at a spectral type provides this information. Once one has a measure of "peculiarity" one can obtain a measure of degree of "peculiarity" for each spectral feature in an unknown spectrum, and can develop secondary classification schemes to analyze the spectrum further.

C. Galaxies

i. Systematic Errors

One of the most worked on of current classification programs is to classify all objects on a plate (or in a CCD frame) which are brighter than a certain magnitude limit and which are galaxies as galaxies. This simple sounding data can be very useful in studies of clusters of galaxies, and of the large scale structure of the universe. Unfortunately as the size of the survey grows, and as the uses of the data become more subtle, the possibility of some systematic error contaminating the results becomes more and more real.

The statistical moments of the spatial distribution of galaxies are of substantial theoretical interest (Peebles 1980). Geller, de Lapparent, and Kurtz 1984 have derived the stringent limits on the ability of photometric surveys to measure the second moment, or angular two point correlation function, as a function of the systematic error at the corresponding physical scale on the plate: the square root of the amplitude of the correlation function should be much greater than the systematic error (in magnitudes) at the spatial scale under consideration.

Because the amplitude of the correlation function decreases as a power law with physical scale, and the statistical noise in its determination is a function of the number of galaxies within the scale radius, it has either been necessary to measure galaxies over a large angular scale, or to make very deep measurements. Both strategies invoke the specter of substantial systematic error.

Another measurement of considerable interest to cosmology is the relation between galaxy number density and apparent magnitude. Of particular interest to studies of galactic evolution is the difference between the measurement at faint magnitudes and the $N \sim 10^{0.6m}$ relation expected for a universe of standard candles in a euclidean space. Differences in the behavior of this function as a function of galaxy color are also quite important.

This measurement, as well as studies of individual clusters of galaxies, can be adversely affected by any systematic tendency in the magnitudes as a function of local galaxy density. This could be an especially pernicious systematic error because of the known properties of the galaxy distribution. Namely that galaxies cluster, and that this tendency is a strong function of galaxy color (e.g. Postman and Geller 1985).

ii. Blended Objects

Much work has been put into the problem of classifying and measuring objects in crowded fields. For the case of crowded star fields the problem has essentially been solved by the use of point spread fitting techniques, as in DAOPHOT (Stetson 1987). In the case of crowded galaxy fields the problem is much more difficult, in that there is no *a priory* knowledge of the exact shape of the objects.

TABLE 1

Some workers and their methods for dealing with blended images

Cawson	Pixel Migration	Cawson, *et al.* 1986
Chan	Eye	Chan, *et al.* 1986
Dickey	Eye	Dickey, *et al.* 1987
Irwin	Iterative Object Subtraction	Irwin 1985
Kron	Eye	Koo, *et al.* 1986
Kruszewski	Octant Search plus Iterative Object Subtraction	Kruszewski 1987
Kurtz	Eye; Non-Linear Image Restoration plus Model	Kurtz, *et al.* 1985; Falco and Kurtz 1988
Lasker	Correlation with Image Library	Lasker 1987
Malagini	Ascending Isophote Along Projection plus Eye	Malagini, *et al.* 1985, Malagini 1987
Peterson	Descending Isophote plus Model	Peterson 1987
Sebok	Eye	Sebok 1986
Stobie	Unpublished Algorithm by Steven Beard	Stobie 1986, 1987
Valdes	Ascending Isophote plus Assisted Eye	Valdes 1982, 1987

The importance of solving this problem is indicated above. Currently it is difficult to classify faint stars and galaxies correctly in crowded fields; getting the photometry correct is very difficult. That this is so is recognized by all workers in the field, the authors of the best current algorithms describe them as "crude". In Table 1 I have listed a number of workers, and the methods they are using to deal with the problem of blended objects. I would especially like to thank those who shared with me the details of their methods in advance of publication for the purpose of making this table.

iii. Galaxy Types

Another outstanding current problem is the automated classification of single galaxies. Some work is being done now, *e.g.* Thonnat *et al.* 1985, Mussio 1985, Pasian *et al.* 1986, Postman 1987, Grosbol 1987, but work in this area is just beginning. The existence of large samples of typed galaxies would be of substantial importance to all branches of extragalactic astronomy; many current questions could be addressed with these data, but I prefer to think of those problems which will only be able to be asked as a result of having it.

The classification scheme possible for galaxies is directly related to the number of resolution elements available. The classical visual classification schemes (Sandage 1975) are based on very high resolution photographs; surely one cannot classify a galaxy SBb(r) without several thousand resolution elements. For this sort of data a syntactic approach, using a stochastic or fuzzy grammar, would seem to be the best technique. While conceptually straight forward, the practical implementation of this scheme can be expected to be quite arduous.

Were it only to be used with ground based data, where the effects of seeing severely limit the number of well resolved galaxies, it would perhaps not be worth the effort to create such a program. The launching of the Hubble Space Telescope, with the Wide Field/Planetary Camera, will create an immediate requirement for the classification of large numbers of faint, well resolved galaxies. How these objects will be classified has yet to be determined.

For less well-resolved data, wherein the overwhelming majority of galaxies in a digital sky survey would be found, there is no well defined classification scheme at present. Certainly fitting functions, such as an exponential or power law, to the luminosity profile of the object would seem to be a good way of obtaining parameters which could form the basis for the classification. What exactly those parameters would be, and how well the derived classifications could be mapped to the physical parameters known from the high resolution data remain as outstanding problems.

D. Clusters of Galaxies - Large Scale Structure

Clusters of galaxies represent an area where classification schemes have yet to lead to substantial new insights into the underlying physical situation. This is not surprising due to the complicated nature of the objects, and the paucity of data. The only type of data for which thousands of observations are present are optical Schmidt plates. Struble and Rood 1987 have performed the most extensive work with these data; they have visually classified all the clusters in the Abell 1958 catalog into the Rood-Sastry 1971 system.

Galaxy clusters are also luminous in x-rays. It may well be argued that the x-ray maps better represent the physical state of the cluster, as the x-rays trace the intra-cluster gas. At present only a few dozen good x-ray maps exist; many of these have been classified by Jones and Forman 1984 into their system.

The distribution of redshifts of the galaxies in the cluster also provides important information on the physical state of the cluster. While no classification on this basis has yet been proposed studies of individual clusters incorporating redshifts, magnitude limited photometry, and x-ray maps have begun to appear with some regularity (*e.g.* Beers *et al.* 1982, Bothun *et al.* 1983, Kurtz, *et al.* 1985, Fabricant *et al.* 1986, Postman *et al.* 1986, Chapman *et al.* 1987). The danger of classifying clusters on the basis of the spatial distribution of x-rays and galaxies alone can be seen by comparing the Coma cluster (Kent and Gunn 1982) with the very similar appearing Abell 2256 (Fabricant, *et al.* 1987).

In the future multiple object spectrographs and x-ray detectors capable of measuring the spatial variation of the temperature of the x-ray emitting gas will improve the situation greatly. One might dare to hope for a cluster classification scheme which could explain the x-ray luminosity--velocity dispersion diagram (Quintana and Melnick 1982, Zabludoff 1987), or some other cluster equivalent to the HR diagram.

Just which point distributions are classified as clusters is a question of substantial interest, both for the study of the clusters, and for studies at larger scales. Abell 1958, Zwicky *et al.* 1962 and Shectman 1985 all have different ways of doing this. Obtaining complete, well defined catalogs of clusters, which correspond well to the actual three dimensional galaxy distribution, is one of the most important tasks of a digital sky survey.

Certainly our knowledge of structures at larger scales does not permit a detailed characterization of them at this time. The recent discovery by de Lapparent *et al.* 1986 of galaxy-filled surfaces surrounding large empty regions similar to the void in Bootes (Kirshner *et al.* 1987) can give rise however to some wild speculation. Perhaps sometime in the next century astronomers will have distinguished several different kinds of surfaces and voids, and will be able to attempt a comparison of the distribution of the various types with galaxy clusters, with each other, and with the distribution of fluctuations (by then discovered?) in the cosmic microwave background. From this, knowledge about the VERY large scale structure of the universe may result.

IV. Conclusions

A. The Challenge - New Data

We are already being swamped with data from the measuring engines and radio interferometers. We are beginning to be swamped with data from CCD surveys. We will soon be swamped with data from the next generation of space missions, HST, ROSAT, AXAF, LAMAR, SIRTF, *etc*.. The clear challenge which these data present is to develop feasible techniques to reduce the huge data streams into sets of descriptors which are compact enough to be manageable, yet rich enough to contain most of the available information. This is the greatest challenge and the greatest opportunity for astronomical object classification since the invention of the photographic plate.

One way to avoid the problem of too much data is not to take most of it; to take only the minimum data necessary to answer specific questions. Some examples of surveys of this sort are the Arizona CCD Transit Instrument Survey of McGraw *et al.* 1987, the Palomar Deep Cluster Survey of Gunn, *et al.* (Postman 1987), and the CfA-Dartmouth "Century Survey", a combined photographic, CCD, and redshift survey, of Geller, *et al.* 1988.

B. The Methods - New Algorithms

The recent work of the measuring engine groups will attest to the fact that these new huge datasets require new, clever algorithms to reduce them. This is nothing new; all new types of data require new methods for their proper reduction.

Some examples of areas requiring algorithmic development are: "three"-dimensional classification of stellar spectra, classification of peculiar spectra, classification of well resolved-galaxies, classification and photometry of blended galaxies, and classification of point patterns as clusters of galaxies. Many more examples can be found in the proceedings of this conference.

C. The Tools - New Machines

When one pauses to think of the several orders of magnitude increase in the amount of data to be reduced, and the increasingly complex algorithms which our ever more demanding applications require, one might get but a wee depressed. One, however, need not worry; computers are coming to the rescue!

The e-folding time for increases in available computer capacity is about three years, and shows no sign of slowing down. Indeed current trends in computer science, toward multiple parallel processors, seem ideally suited to the upcoming multi-dimensional data reduction needs of astronomy. These developments may solve the VLA data reduction problem; whether they can solve the VLBA data reduction problem is not yet clear.

D. The Results - New Knowledge

We are at the beginning of the information age. For generations our data has lay dormant in plate collections, occasionally giving up a bit or two to a determined investigator. Now our archives have become alive, our new instruments produce mountains of data that are immediately useful. In many cases the problem is not how to get the data, but what to do with it.

Within our lifetimes we will have spectra for tens of millions of objects, positions, colors, and magnitudes as functions of time for hundreds of millions. We will have huge hordes of gamma ray, x-ray, ultra-violet, optical, infra-red, and radio data. The science to be gleaned can only be understood via the dramas and discoveries of the next half century of astronomical research. I can only give this advise as to what to do: *a chacun son gout!*

Acknowledgements

I should like to thank the many who discussed their work with me in preparation for this talk. For partial financial support I thank NAGW-201 and the Smithsonian Scholarly Studies Program. I dedicate this paper to the memory of King-Sun Fu, who blazed the trail which all of us will eventually follow.

References

Abell, G. O. 1958, *Astrophys. J. Suppl.,* **3**, 211

Beers, T.C., Geller, M.J., and Huchra, J.P. 1982, *Astrophys. J.,* **257**, 23

Bijaoui, A, 1979, in *Classification Stellaire,* Compte Rendu de l'ecole de Goutelas, ed. D. Ballereau, Meudon: Observatoire de Meudon, p. 1

Bothun, G.D., Geller, M.J., Beers, T., and Huchra, J. 1983, *Astrophys. J.,* **268**, 47

Burg, R. 1987, *Thesis*, Massaschusetts Institute of Technology

Cawson, M.G., McGraw, J.T., and Keane, M.J. 1986, *S.P.I.E. Proceedings,* **627**, 79

Chan, R., Pellegrini, P.S., and da Costa, L.N. 1986, *Rev. Mex. Astron. Astrophys.,* **12**, 393

Chapman, G.N.F., Geller, M.J., and Huchra, J.P. 1987, *Astron. J.,* **94**, 571

Dickey, J.M., Keller, D., Pennington, R., and Salpeter, E. 1987, *Astron. J.,* **93**, 788

Einasto, Ya., Malyuto, V.D., and Kharchenko, N.B. 1985, *Astron. Tsirk.,* **1394**, 1

Fabricant, D., Beers, T.C., Geller, M.J., Gorenstein, P., Huchra, J.P., and Kurtz, M.J. 1986, *Astrophys. J.,* **308**, 530

Fabricant, D., Kent, S.M., and Kurtz, M.J. 1987, *Astrophys. J.,* to appear

Falco, E.E. and Kurtz, M.J. 1988, *in preparation*

Fu, K.-S. 1982, *Syntactic Pattern Recognition and Applications*, Englewood Cliffs: Prentice Hall

Geller, M.J., de Lapparent, V., and Kurtz, M.J.1984 *Astrophys. J. Lett.,* **287**, L55

Geller, M.J., Kurtz, M.J., Huchra, J.P., McMahan, R., Schild, R.E., Hamwey, R., Thorstensen, J.R., and Wegner, G. 1988 *in progress*

Gibbons, J.D. 1985, *Nonparametric Statistical Inference, 2nd ed.,* New York: Dekker

Grosbol, P.J. 1987, *these Proceedings*

Heck, A., Egret, D., Nobelis, p., and Turlot, J.C. 1986, *Astrophys .and Space Sci.,* ***120**, 223*

Irwin, M.J. 1985, *Mon. Not. Roy. Astron. Soc.,* 214**, 575**

Jaschek, C. and Jaschek, M. 1987, *The Classification of Stars,* Cambridge: Cambridge University Press

Johnson, H.L. and Morgan, W.W. 1953, *Apstrophys J.,* **117**, 313

Jones, C. and Forman, W. 1984, *Astrophys. J.,* **276**, 38

Kandel, A. 1982, *Fuzzy Techniques in Pattern Recognition*, New York: Wiley

Keenan, P.C. 1987, *Pub. Astron. Soc. Pac.,* **99**, 713

Kent, S.M. and Gunn, J.E. 1982, *Astron. J.,* **87**, 945

Kirshner, R.P., Oemler, A. Jr., Schechter, P.L., and Schechtman, S.A. 1987, *Astrophys. J.,* **314**, 493

Koo, D., Kron, R., Nanni, D., Trevese, D., and Vignato, A.1986, *Astron. J.,* **91**, 478

Kruszewski, A. 1987, *personal communication,* to appear in *Acta Astronomica*

Kurtz, M.J. 1983, in *Statistical Methods in Astronomy*, ed. E. Rolfe, European Space Angency Special Publication **SP-201**, p47

Kurtz, M.J. 1985, in *The MK Process and Stellar Classification,* ed. R.F. Garrison, Toronto: David Dunlap Observatory

Kurtz, M.J., Huchra, J.P., Beers, T.C., Geller, M.J., Gioia, I.M., Maccacaro, T., Schild, R.E., and Stauffer, J.R. 1985, *Astron. J.,* **90**, 1665

de Lapparent, V., Geller, M.J., and Huchra, J.P. 1986, *Astrophys. J. Lett.,* **302**,1

Lasker, B.M. 1987, *personal communication*

LaSala, J. 1987, *these Proceedings*

Lawrence, A. 1987, *Pub. Astron. Soc. Pac.*, **99**, 309

Malagini, M.L. 1987, *personal communication*

Malagini, M.L., Pasian, F., Pucillo, M., and Santin, P. 1985, in *Data Analysis in Astronomy,* eds. V. Di Gesu, *et al.,* London: Plenum, p. 57

McGraw, J.T., Cawson, M.G.M., and Keane, M.J. 1987, *these Proceedings*

McMahan, R. 1986, *Thesis*, Dartmouth College

Morgan, W.W. and Keenan, P.C. 1973, in *Ann. Rev. Astron. Astrophys.,* **11**, 29

Murtagh, F. and Heck, A. 1987, *Multivariate Data Analysis,* Dordrecht: Reidel

Mussio, P. 1985, in *Data Analysis in Astronomy,* eds. V. Di Gesu, *et al.,* London: Plenum

Pasian, F., Pucillo, M., and Santin, P. 1986, *Mem. Soc. Astron. Italia,* **57**, 251

Peebles, P.J.E. 1980, *The Large-Scale Structure of the Universe,* Princeton: Princeton University Press

Peterson, B.A. 1987, *personal communication*

Postman, M. 1987, *personal communication*

Postman, M. and Geller, M.J. 1985, *Astrophys. J.,* **291**, 85

Postman, M., Geller, M.J., and Huchra, J.P. 1986, *Astron. J.,* **92**, 1238

Quintana, H. and Melnick, J. 1982, *Astron. J.,* **87**, 972

Rood, H.J. and Sastry, G. 1971, *Pub. Astron. Soc. Pac.,* **83**, 313

Sandage, A. 1975, in *Stars and Stellar Systems* ***IX****, Galaxies and the Universe,* eds. A. Sandage, M. Sandage, and J. Kristian, Chicago: University of Chicago Press, p. 1

Sebok, W.L., 1986, *Astrophys. J. Suppl.,* **62**, 301

Secci, A. 1867, *Memorie della Soc. Italiana delle Scienze,* **1**, 67

Shectman, S.A. 1985, *Astrophys. J. Suppl.,* **57**, 77

Stetson, P.B. 1987, *Pub. Astron. Soc. Pac.,* **99**, 191

Stobie, R.S. 1986, *Pattern Recognition Lett.,* **4**, 317

Stobie, R.S., 1987, *personal communication*

Struble, M.F. and Rood, H.J. 1987, *Astrophys. J. Suppl.,* **63**, 543

Thonnat, M., Granger, C, and Berthold, M. 1985, in *Proceedings I.E.E.E. Computer Soc. Conf. on Computer Vision and Pattern Recognition,* Silver Spring: I.E.E.E. Computer Soc. Press

Valdes, F. 1982, *S.P.I.E. Proceedings* **331**, 435

Valdes, F. 1987, *personal communication*

Zabludoff, A. 1987, *Senior Thesis,* Massaschusetts Institute of Technology

Zadeh, L.A. 1984, *I.E.E.E. Spectrum*, August 1984, p. 26

Zwicky, F., Herzog, E., Wild, P., Karpowicz, M., and Kowal, C. 1961-1968, *Catalog of Galaxies and of Clusters of Galaxies,* Vol **1-6**, Pasadena: California Institute of Technology

A PROGRAM FOR AUTOMATIC TWO-DIMENSIONAL SPECTRAL CLASSIFICATION OF OBJECTIVE-PRISM SPECTRA

Jerry LaSala
University of Southern Maine
Portland, Maine, USA

Abstract: The classification of objective-prism spectra is an area of astronomy that would benefit greatly from the successful application of automated techniques. I report on the current status of a project to develop an automated classification system which has been underway since 1983. At this point, a series of programs is being refined and tested using some 3200 spectra digitized from the original objective-prism plates obtained as part of the Michigan Spectral Catalog survey. The programs use cross-correlation with standard spectra to obtain initial classifications and wavelength calibration, then determine a final two-dimensional classification by a combination of pattern-recognition and criterion-evaluation techniques. I describe here the data, data reduction, the classification programs, and plans for further work.

I. INTRODUCTION

Many authors (e.g. West 1972, 1976; Houk 1976, 1984; Keenan 1987) have pointed out the desirability of a fully-automated system of two-dimensional spectral classification, preferably one based on the MK system used for visual classification. The principal advantages of such a system would include increased speed and durability over human classifiers, as well as greatly improved homogeneity: an automatic classifier would be able to classify large numbers of spectra on a self-consistent system. Large, homogeneously classified samples are especially important for statistical studies of stellar distributions. Such studies represent a valuable potential use of the enormous existing data base of objective-prism spectra. The largest homogeneously-classified body of spectral data is the Michigan Spectral Catalog (Houk and Cowley 1975; Houk 1978, 1982); this work-in-progress currently offers a data base of about 100 000 classified spectra. When completed it will contain some 225 000 classifications and will have taken 40 years to finish. Clearly larger homogeneous samples, for example complete samples to some faint magnitude limit, will be impossible to obtain without automatic techniques.

It seems clear that automated techniques offer the greatest possible advantage when applied to objective-prism spectra. The enormous amount of data obtained on even a single exposure results in a very unfavorable ratio of reduction (classification) time to observing time, equivalent in some cases to ten years of classifier time for one night's observations. The potential speed of automated methods would improve this situation greatly.

Several attempts at developing more or less fully automated classification systems have been undertaken in the past two decades. The most recent review of previous work is that by Kurtz (1984). It is only in the last few years, however, that the availability of high-speed, low-noise microdensitometers (*e.g.*, APM,

COSMOS) along with improvements in computing and new developments in the fields of image processing and pattern recognition have made a large-scale automatic classification system feasible.

Techniques which have been applied to automatic spectral classification programs fall generally into two classes which may be called criterion evaluation and pattern recognition respectively. Criterion evaluation methods involve the measurement of certain prespecified quantities (such as equivalent widths of certain lines, ratios of certain line strengths, synthesized color indices, etc.) and calibration of these measurements in terms of desired parameters such as spectral type and luminosity class. Pattern recognition techniques, which include cross-correlation and various multivariate statistical methods, effect classification by determining a suitable defined similarity measure (such as correlation coefficient or distance in the space defined by the standard spectra considered as vectors) and assigning the spectrum to the class defined by the standard it most closely resembles.

Most previous work in automatic spectral classification has made use of criterion evaluation methods. Most notable among these are the work of Schmidt-Kaler et al. (1976), of Malyuto (1984), and of Zekl (1984). Advantages of this approach include relatively rapid computing time, easy recalibration, and the fact that it is not necessary to have large numbers of standard spectra stored on-line for comparison. Disadvantages include the ignoring of large parts of the spectra (i.e., everything but the specified criteria), the necessity for precise wavelength and intensity calibration, and the fact that the selection of criteria is often based on physical assumptions about the processes going on in the star, thus making the classification theory-dependent.

The most important work in the pattern-recognition approach has been done by Kurtz (1982, 1984). He applied a variety of multivariate statisitcal methods to low-dispersion digital spectra and was able to achieve classifications with mean errors slightly smaller than those typically achieved by visual classifiers working at similar resolution. Advantages of pattern recognition methods include the use of the information contained in the entire spectrum, elimination of the need for precise wavelength and intensity calibration, avoidance of physical assumptions in defining the classification scheme, and presumably closer correspondence with the process of the visual classifier. Disadvantages include relatively large computing time requirements and the need for large numbers of standards available for comparison.

Since 1983, I have directed a project extending the work and ideas of Kurtz (1984) to higher-dispersion objective-prism spectra, with the goal of achieving a fully-automated two-dimensional classification system with accuracy comparable to that of visual classifiers working with similar data. This work was begun at the

College of Wooster, Ohio, USA, and is now being continued at University of Southern Maine. I report here the current status of this work.

II. CLASSIFICATION PROGRAMS

Kurtz (1982) found that of the various pattern-recognition methods he tested, the best results were achieved with a weighted metric distance technique. In this technique the digitized spectra are considered as vectors, each wavelength resolution element representing one dimension of the vector. The standard-star spectra then define a classification space. The distance of each spectrum to be classified from each standard spectrum is computed according to a specified metric, and the unknown is assigned to the class of the standard from which its distance is a minimum. The magnitude of the distance from the standard gives an estimate of the reliability of the classification. If a spectrum does not fall within some specified threshhold distance of any standard, it can be flagged as peculiar. In addition, individual features may be flagged as peculiar if they differ from the corresponding features of the standard by several standard deviations. Thus a classification program based on this technique will not only classify normal stars, but will provide an efficient method for detecting peculiar spectra as well.

The metric used in the current classification programs is as follows. The general metric distance between two vectors $\vec{X}$ and $\vec{S}$ may be defined as:

$$d^2_{XS} = \frac{1}{m}\sum_{i=1}^{m} \alpha^2(i)[X_i - S_i]^2,$$

where $\alpha(i)$ is a weighting factor defined for each component of the vector. If $\alpha(i) = 1$ for all i, the result is the ordinary Euclidean metric. Kurtz (1982) experimented with a weighted metric in which $\alpha^2(i) = \sigma^2(i)$, the variance of feature (i) within a particular error box, thus weighting most heavily the features with the largest variance, which presumably carry the most information. This method gave Kurtz the best results of all methods tested.

In 1984, Kurtz proposed an improved weighting factor,

$$\alpha^2(i) = \frac{\Sigma^2(i) - \sigma^2(i)}{\sigma^2(i)},$$

where $\Sigma^2(i)$ is the variance within the error box associated with some initial approximation and $\sigma^2(i)$ is the variance within the final classification box associated with each standard spectrum. Thus this metric weights most heavily those features which have little variance within a given final classification box, but large variance within the initial approximation box, that is those features which are most sensitive to final classification. This is the metric adopted in this work.

The actual classification programs are designed to work as follows. One standard spectrum from each of the principal spectral classes B-K is chosen as the basis of an initial rough classification. Each unknown spectrum is cross-correlated with each of these standards, and is assigned as an initial approximation the spectral type of the standard with which it has the highest correlation coefficient. A rough wavelength calibration is then performed by shifting each unknown spectrum by the amount required to produce the maximum correlation coefficient. Thus within each approximation group, all spectra will be shifted so that their wavelength scales correspond to that of the standard for the group. If necessary, a more precise wavelength calibration may be performed at this point by locating prominent features in the spectrum on the basis of the initial approximate classification and fitting a polynomial to the pixel-wavelength relation. However, I do not believe at this time that calibration to this precision will be necessary, at least for homogeneous data.

Final classification is then accomplished by the weighted metric distance method described above. The weighted distance of each unknown spectrum from each standard spectrum (one for each spectral type and luminosity class) in the neighborhood of the approximation box is calculated, and the unknown is assigned to the class defined by the nearest standard. Peculiar spectra and peculiar features are flagged as described above.

Selection of standard stars and calculation of weighting factors $\alpha(i)$ are discussed in Section IV. below.

III. TEST DATA AND DATA REDUCTION

To provide standard stars, a basis for determining the variances $\Sigma^2(i)$ and $\sigma^2(i)$, and sample data for testing the programs, I obtained eight Curtis-Schmidt objective-prism plates from N. Houk. These plates were originally obtained as part of the Michigan Spectral Catalog Survey. They were chosen to provide a set of data with homogeneous accepted classifications, namely Houk's classifications in the Michigan Catalog itself. The plates are all 20-minute exposures on IIaO emulsion, obtained with the 10° prism on the Curtis Schmidt and provide a resolution of 2A over the entire MK classification region of the spectrum. The plates used were selected from the region of the galactic plane to maximize both total number of stars and representation of various spectral classes.

The spectra on the plates were digitized using the APM flying-spot laser scanner (Kibblewhite *et al.* 1984) at the Institute of Astronomy, Cambridge. The spectra were located automatically on the plate using their SAO Catalog coordinates. A 128 x 1024 pixel two-dimensional raster scan was made of each spectrum was made, with a pixel size of 15.2 microns. Scanning an entire plate required about 6 hours of APM time.

The two dimensional scans were reduced to one-dimensional, 1024-element spectra by applying a median filter across a "slice" through the actual spectrum on the scan (about 60 pixels wide). Several possible filters were tested for this step, and the median was found to be the most successful at eliminating plate flaws and scratches while preserving spectral features. The center of the filter "slice" was shifted as the filter progressed along the length of the spectrum to correct for the skewness of the spectra on the plate. When two or more spectra overlapped (which occurred for over 50% of the spectra in these crowded fields), the center of the "slice" was moved to the center of the non-overlapping region and the width of the "slice" was adjusted accordingly. Overlap was determined automatically from SAO coordinates and known spectral dispersion and widening.

The resulting digitized spectra contain 1024 resolution elements corresponding to the 15.2-micron pixel size, providing a spectral resolution of 1.6A. Altogether, 3200 digital spectra were thus obtained.

The final data reduction step before applying the classification programs is to rectify the spectra to eliminate continuum and instrumental response effects. Rectification is accomplished using the Fourier division technique described by LaSala and Kurtz (1985) and Malyuto and Pelt (1981).

IV. STANDARD STARS, WEIGHTING FACTORS, AND TESTING

For this testing and refining project, two types of standard spectra will be tried. First and simplest, a relatively bright star without known spectral peculiarities is chosen from each spectral type and luminosity class to serve as a standard. Second, a synthesized spectrum, the mean of all stars in the sample with the same Houk classification, can be used as the standard for that class. For future work, it would be desirable to obtain rectified spectra of the actual MK standards for each class, although the synthesized mean spectra may prove more reliable even than these for machine classification.

After the initial cross-correlation, wavelength calibration step described above, the variances Σ^2 in the initial approximation group and σ^2 among all stars with the same Houk classification can be calculated and used to determine the weighting factors $\alpha^2(i)$ defined above.

The final classification program can then be run and results tested for both internal consistency and agreement with the accepted Houk classifications.

When the programs are performing satisfactorily, they will be made available on request. All programs are written in VMS FORTRAN and will run on a VAX 11/780 or a MicroVAX.

V. FUTURE WORK

Several future projects for improving and applying the classification programs are planned. These include expansion of the classification programs to take as data matrices of the ratios of all spectral elements to each other rather than simply vectors of the elements. This will require much more computing time, but should produce classification which are reliable over a much wider range of exposure conditions. Other plans include exploring the possibility of incorporating Zekl's criterion evaluation methods as a supplementary or alternative final classification method, and obtaining new objective-prism data optimized to the characteristics of the APM, so that scanning and reduction time may be further reduced. I also hope to apply these methods to spectra taken with other telescopes, and to begin a survey of galactic clusters, determining distances by main-sequence fitting with the automatic classifications provided by these programs.

I wish to thank N. Houk for providing plate material; the Astronomical Data Center at the NASA Goddard Space Flight Center for providing the Michigan Catalog in machine-readable form; the staff of the APM for valuable advice and assistance; M. J. Kurtz and H. Zekl for discussions and encouragement; the Institute of Astronomy for research facilities and assistance while I was a Senior Visiting Fellow there; the College of Wooster and the University of Southern Maine for travel support.

REFERENCES

Houk, N., and Cowley, A. P. 1975, University of Michigan Catalog of Two-Dimensional Spectral Types for HD Stars, Vol. 1 (Ann Arbor: University of Michigan).

Houk, N. 1976, remarks following West (1976).

Houk, N. 1978, 1982, University of Michigan Catalog of Two-Dimensional Spectral Types for HD Stars, Vols 2, 3 (Ann Arbor: University of Michigan).

Houk, N. 1984, remarks following Kurtz (1984).

Keenan, P. C. 1987, Publ. Astr. Soc. Pacific, Vol. 99, p. 713.

Kibblewhite, E. J., Bridgeland, M. T., Bunclark, P., and Irwin, M. J. 1984, in Proceedings of the Astronomical Microdensitometry Conference, 1985, ed. D. A. Klinglesmith (NASA Conf. Publ. 2317).

Kurtz, M. J. 1982, Thesis, Dartmouth College.

Kurtz, M. J. 1984, in The MK Process and Stellar Classification, ed. R. F. Garrison (Toronto: David Dunlap OBservatory) p. 131.

LaSala, J., and Kurtz, M. J. 1985, Publ. Astr. Soc. Pacific, Vol 97, p. 605.

Malyuto, V. 1984, in Astronomy with Schmidt-Type Telescopes, ed. M. Capaccoli (Dordrecht: Reidel) p. 287.

Malyuto, V., and Pelt, J. 1981, Acad. Sci. Eston. SSR Preprint A-1 (1981).

Schmidt-Kaler, Th., Diaz-Santillana, G., Rudolpf, R., and Unger, H. 1976, Forschungsbericht des Landes Nordhein-Westfallen, Vol. 2595 (Opladen: Westdeutscher Verlag).

West, R. M. 1972, in I.A.U. Symposium 50: Spectral Classification and Multicolor Photometry, eds. Ch. Ferenbach and B. E. Westerlund (Dordrecht: Reidel) p. 109.

West, R. M. 1976, in Proceedings Third European Astron. Meeting, ed. E. K. Karadzi (Tblissi: Abastumani Astrophysical Observatory) p. 23.

Zekl, H. 1984, in The MK Process and Stellar Classification, ed. R. F. Garrison (Toronto: David Dunlap Observatory) p. 175.

Statistics, Pattern Recognition and Expert Systems

Towards Heterogeneous Distributed Very Large Databases

H.-M. Adorf[1], *R. Albrecht*[1,2], *M.D. Johnston*[1,3], *R. Rampazzo*[1]

[1]Space Telescope - European Coordinating Facility
European Southern Observatory
Karl-Schwarzschild-Str. 2
D-8046 Garching b. München, F.R. Germany

[2]Affiliated to the Astrophysics Division
Space Science Department
European Space Agency

[3]Space Telescope Science Institute
Homewood Campus
Baltimore, MD 21218, U.S.A.

Abstract: With the availability of advanced storage facilities very large data archives will come into existence which will contain astronomical data collected by space missions and by ground-based observatories. Maintenance considerations will render the centralization and standardization of these archives difficult, with the result that remote query and retrieval processes across such distributed archives will become prohibitively cumbersome. Recent developments in the area of artificial intelligence seem to make it attractive to support sophisticated archive operations through expert systems. This paper highlights some of the techniques and implemented systems found in the literature, and describes some of the commercially available software tools.

Introduction

Database and expert system technology are both areas in which such rapid progress is being made that it is difficult to make firm predictions about future developments. It is important, however, to survey the possibilities and trends from the perspective of an astronomer who will need access to one or more of the growing number of remotely accessible astronomical databases. While it is not possible here to provide a comprehensive survey of the topic, we will highlight recent developments which — according to our present view — are likely to influence the development of next generation systems (cf. R. Albrecht et al. 1986). We also discuss our best estimates of the most promising developments with application to astronomical databases.

Our considerations should be viewed in the context of the proposed "European Space Information System" (ESIS) project at ESRIN Frascati/Italy (M. Albrecht et al. 1987) and the "Distributed Access View Integrated Database" (DAVID) project at NASA's Goddard Space Flight Center (Jacobs 1985). ESIS, for example, will in its first phase connect via a wide-area network the IUE archive (Villafranca, Spain), the EXOSAT archive (Noordwijk, The Netherlands), the Hubble Space Telescope archive (Garching, F.R. Germany), Centre de Donnees astronomiques (Strasbourg, France), Rutherford Appleton Laboratory (Chilton, Didcot, United Kingdom) and ESRIN (Frascati, Italy). The databases at these geographically distributed sites are independently created and administered and they differ in various aspects. The user will have access not only to raw and calibrated scientific data, but also to data analysis results, engineering data, and bibliographic references.

Conceptually the user should be able to perceive a distributed database system like ESIS as if it were a single integrated classical database, supplemented by tools allowing queries in astronomical rather than database terms, easy communication between users, file transfer, etc. Establishing such a perception for the user of a network of heterogeneous vary large databases will be a very difficult task (see Chu 1987). It is only through advanced software techniques, particularly those arising out of several fields of "artificial intelligence" (AI) research, that a solution of this problem appears possible. In the following we will first give a broad overview of the benefits one might hope to gain from the application of AI techniques in general, and then take a closer look at the use of an expert system as a 'front end' to one or more databases. We will also review the options for advanced user interfaces, in particular the possible use of natural language for this purpose.

Artificial intelligence programming techniques

Astronomical software development can profit from work done in AI research in four ways: Firstly, AI gives us *powerful languages* like LISP, which makes it easy to implement even more abstract and thus more powerful languages. Secondly, AI workers have developed a variety of *data structuring techniques* useful for knowledge representation (see e.g. Fikes & Kehler 1985). Thirdly, AI research has contributed to the development of *methods* and *concepts*, e.g. the logic programming method, or the concept of cognitive acts (Wahlster & Kobsa 1986), which is important for the construction of man-machine interfaces. And finally, AI research has brought us powerful integrated *programming environments* on workstations, which have been shown to considerably speed up the software development cycle. See the article by Bobrow & Stefik (1986) or the three volumes by Barr & Feigenbaum (1981) for general perspectives on AI programming.

Benefit from applying AI techniques will be most obvious in the first place not so much to the astronomer as to the software developer. Dedicated AI workstations based on the LISP language have successfully integrated the programming language, operating system, editor, and file system into a powerful development environment. The emergence of Common LISP as a standard (Steele 1984) has encouraged the development of portable applications. Modern compilers have nearly eliminated the execution speed difference between LISP and conventional languages such as FORTRAN and C. The LISP language is rich, extensible, and offers the advantages of both interpreted and compiled modes. Very readable introductions to the Common LISP language can be found in the books by Winston & Horn (1984) and by Tatar (1987).

As extensions to the LISP language most development systems provide *object-oriented programming* capabilities. This allows the user great economy and flexibility of problem representation since objects can be arranged hierarchically in order to inherit properties from their ancestors; objects have 'slots' to hold attributes, and can also have attached 'methods' which are procedures activated by sending a 'message' to an object.

Expert systems form one of major subfields of AI. They can be broadly defined as systems which incorporate, in some limited domain, the knowledge and reasoning ability of a human expert. Expert systems have been developed for numerous domains, and several are in routine commercial use. See Hayes-Roth et al. (1983) for a review of the development and operation of several representative expert systems. Most commonly the ability to reason with 'rules', chained in either a forwards or backwards mode, is associated with expert system programming. *Expert system shells* have been developed which provide integrated toolkits for building expert systems. These shells incorporate rule processing modules, but they have productively included object-oriented programming techniques as well, along with various user interface tools. Commercially available shells of this type include ART (Automated Reasoning Tool from Inference Inc.) and KEE (Knowledge Engineering Environment from IntelliCorp Inc.). See the articles by Cross (1986), Gevarter (1987), Gilmore et al. (1986) and Richer (1986) for overviews on and comparisons of commercial expert system shells.

At present, development using these shells can only be carried out on fairly high-powered workstations. Delivery systems are gradually appearing that will allow expert systems developed on LISP machines to run on high-end Personal Computers, such as the Macintosh II and IBM PC/AT.

Accessing astronomical databases — the goal

One of the early considerations within the "Artificial Intelligence Pilot Project at the ST-ECF" (Adorf & Johnston 1987) was the support of sophisticated archive operations through expert systems. A "Network and Database Assistant" was conceived as follows (Adorf et al. 1986):

> The database system currently being installed for the joint Space Telescope and ESO data archive puts the astronomical community into a novel situation. Various aspects present a challenge to current database technology: (1) The data will be *inhomogeneous*, so the database management system may have to be enhanced by some 'fuzzy' query capability to allow the retrieval of data which come 'close' to those actually requested. (2) The data volume will become *very large*; within 15 years — the expected lifetime of the Space Telescope — it will receive some Terabytes of images, spectra, calibration data etc. (3) The database must be *remotely accessible* by astronomers, largely unfamiliar with database systems, via a computer network from all over Europe. Also, there are plans to integrate the HST/ESO database into an international network of remotely accessible distributed databases.
>
> The purpose of the "Network and Database Assistant" would be to navigate a remote user through the unfamiliar and changing issues of networking, operating systems, and relational database query languages. Ideally the assistant would be equipped with a versatile user interface possessing natural language capability to accept (restricted) English queries. On request it should transparently access the network, query the database and, depending on the quantity, either transmit the data retrieved and display them formatted on the user's screen or submit a retrieval request.

A useful model for the Network and Database Assistant would therefore contain two major components: (1) an expert system with knowledge of the available databases and their contents, and (2) a user interface through which the astronomer formulates queries and views the results.

The expert system component of a basic system may consist of a data dictionary which would provide meta-information on all available databases, including information on network access paths and retrieval costs. Such a system would aid the astronomer in performing ad-hoc feasibility checks to keep him/her from (possibly inadvertent) unreasonable operations. A more advanced system would allow the scientist to formulate a request in an abstract, astronomy-oriented manner to the expert system, residing locally, which would in turn remotely consult the different archives, taking into account the idiosyncrasies of different query languages, of different operational procedures and of computer network connections.

The user interface could consist of a natural language query system, a graphical object-oriented display with a pointing device for user input, or, most likely, a combination of the two (see e.g. Hanne et al. 1986; Hayes 1986).

Expert systems and databases

The problem of database access by users who need information but do not necessarily care how it is stored is, of course, not peculiar to astronomy. This problem is widespread in business applications, and consequently possible couplings between expert systems and database management systems have received a good deal of attention (see Al-Zobaidie & Grimson 1987; Kerschberg 1987; Rosenthal 1987 and references therein). A number of custom-built expert systems have been coupled to (single) external databases, e.g. the Hubble Space Telescope proposal transformation expert system (Rosenthal et al. 1987) but the only products which are near to general availability are KEEconnection and IntelliScope (Kehler 1987) from IntelliCorp Inc.

IntelliScope can be considered as a knowledge base browser to be used within KEE. Its features include scatter plots, histograms and table displays. The main purpose of IntelliScope is to be used in conjunction with KEEconnection, which is used to interface KEE with any

SQL-driven relational database. In conjunction with KEEconnection, IntelliScope can be used for query refinement and subquery formulation. KEEconnection consists of three pieces: (1) a graphical editor for setting up the translation of database relations into a KEE's frame structure, (2) a query translator for translating queries into SQL, and (3) networking software for the connection of the AI machine to the database management system (DBMS). KEEconnection can handle several relations which may reside in different databases — even on different machines — including joins between the relations. KEEconnection can be interfaced to Oracle, Ingres and Britton-Lee DBMSs, using the DECnet or TCP/IP protocol for communication.

KEEconnection is used in the following way: The database administrator sets up the translations from DB relations to KEE frames. Thus a data dictionary containing meta-knowledge about the DB is constructed. This need only be done once. The knowledge engineer can construct knowledge bases as usual using any required facts, regardless of whether they are stored in the knowledge base or in the database. If needed, parts of the database will be transferred into the expert system. The user may then interact with the DB through IntelliScope or any other KEE application.

At present, KEEconnection only works on one database at a time. This may not be a severe limitation for the browser, but it is a certainly a problem for a more general interface to a heterogeneous network of databases. The general problem of access to multiple databases with different query languages is not likely to receive much attention in the commercial world. Most current work is targeted for databases which support SQL. This is a strong argument for migrating existing custom-built astronomical databases to the SQL standard whenever this is feasible.

Advanced user interfaces to database systems

It will become increasingly important for astronomical database queries to be made in terms familiar to the astronomer, which are not necessarily those used by the database designer. This is especially true for networks of heterogeneous databases since there are no general standards for nomenclature, units, subject classifications, etc. The possible use of Natural Language (NL) interfaces is therefore well worth investigating. (See Rich (1984) for a recent review on NL interfaces.) NL front ends to database systems are likely be most useful to casual users; the experienced user may prefer the more concise expressiveness of a formal language. Jarke & Vassiliou (1985) systematically address the question of which query language to choose for which user.

Constrained NL interfaces have successfully been built and used as front-ends to database systems because of their restricted domain of discourse. A natural language user interface called LIFER to the distributed database system LADDER was operational as early as 1978 (Hendrix et al.). At the same time an English question answering system for a large relational database has been constructed by Waltz (1978). CO-OP, an experimental portable NL database query system has been developed by Kaplan (1984). The HAM-ANS system (Wahlster 1985) especially concentrated on the cooperativeness aspect. Lehmann et al. (1985) describe a portable and even *multilingual* interface to a relational database system. TACOS, a constrained but extensible natural language system, has been interfaced to the Hubble Space Telescope time allocation committee database (Hornick et al. 1987). Also, an expert system may be used to assist the user in the choice of the right vocabulary for consulting a database as described by Shoval (1985).

Today, various database systems with built-in restricted NL interfaces are commercially available, e.g. dBase III, Q&A, TEAM, and INTELLECT, which was launched in 1980 and is probably the most prominent. There also exist toolkits for constructing NL interfaces, the best known of which probably is Language Craft from Carnegie Group Inc. For comprehensive information on NL systems se the survey by Guilfoyle (1987).

Impressive as the advancements in the area of natural language front ends to — even *distributed* — database systems may be, we feel that natural language will catch on as a natural means of man-machine interaction only after speech recognition and voice synthesis have matured and been integrated with natural language understanding systems. After all, natural language is for speaking and listening, not for typing.

Talking to databases is not out of question. A working voice recognition system has been demonstrated by Hewlett Packard at the International Joint Conference on Artificial Intelligence at Milano this year. Voice recognition systems until recently have had a vocabulary of a few hundred words, but a system like the Kurzweil Voice System 3000 can now recognise 1000 words or phrases with 97 % accuracy (Bramley 1987).

Until some of these developments reach the marketplace it is likely that NL interfaces will play a relatively minor rôle in database access systems. The most promising alternatives center on the further development of graphical interfaces where the user navigates by 'pointing and clicking' (with a mouse or other pointing device) on pop-up or pull-down menus and other 'active regions' on a multi-window bitmapped screen. These techniques can be used to minimize the need for typing commands and queries, and can provide intuitively clear signals to the user of what functions are available and how to access them. Graphical interfaces of this type were originally developed on and for AI workstations; they are now widely available on general purpose workstations and personal computers. A particularly interesting example of this is the recently released HyperCard system from Apple Computer Inc., which provides a simple way for Macintosh PC users to browse through large quantities of related information. Embedded in HyperCard is an object-oriented programming language which allows the creation of very flexible applications.

Summary

Artificial intelligence techniques will come to play an increasingly important rôle in astronomical software systems, although the initial advantages will be realized primarily by system developers. Workstations with high-resolution displays, window systems etc. will play an important rôle in prototyping and developing complex software systems. Large expert system shells are providing a rich environment for prototyping and testing new ideas: the facilities to connect these and other shells to databases will improve dramatically as spinoffs from commercial applications. Delivery of expert systems on inexpensive personal computers will greatly lower the cost of widely disseminating expert knowledge. Restricted natural language interfaces will increasingly attain importance, in particular for casual users, although progress will not be too rapid. Eventually we will see speech recognition and voice output added to our systems and use it as normal means for man-machine communication like we use terminals and keyboards today.

References

Adorf, H.-M., Johnston, M.D.: 1987, "The Artificial Intelligence Pilot Project of the ST-ECF", *ST-ECF Newsl.* **8**

Adorf, H.-M., Albrecht, R., Fosbury, R.A.E.: 1986, "Artificial Intelligence Applications within the Space Telescope – European Coordinating Facility", *ST-ECF internal document*, Nov. 1986

Albrecht, R., Adorf, H.-M., Richmond, A.: 1986, "Next Generation Software Techniques", in: *"Instrumentation in Astronomy IV", D.L. Crawford (ed.), SPIE Proc.* **627**, 225–230

Albrecht, M., Russo, G., Richmond, A., Hapgood, M.: 1987b, "European Space Information System Pilot Project, User Requirements Document, Working Paper, 2nd Draft – October 1987", 1–40

Al-Zobaidie, A., Grimson, J.B.: 1987, "Expert systems and database systems: how can they serve each other?", *Expert Systems* **4**, 30–37

Barr, A., Feigenbaum, E.A.: 1981, *The Handbook of Artificial Intelligence*, Pitman Books Ltd., London

Bernold, T., Albers, G.: 1985, *Artificial Intelligence: Towards Practical Applications*, Elsevier Science Publishers B.V., North Holland

Bobrow, D.G., Stefik, M.J.: 1986, "Perspectives on Artificial Intelligence Programming", *Science* **231**, 951–957

Bramley, I.: 1987, "Talking to Databases", *Systems Intern.* **Jan**, 47–48

Chu, W.W. (ed.): 1987, "Distributed Database Systems, Special Issue", *Proc. IEEE* **May**

Cross, G.R.: 1986, "Tools for constructing knowledge-based systems", *Opt. Engineering* **25**, 436–444

Fikes, R., Kehler, T.: 1985, "The Role of Frame-based Representation in Reasoning" *Commun. ACM* **28**, 904–920

Gevarter, W.B.: 1987, "The Nature and Evaluation of Commercial Expert System Building Tools", *Computer* **May**, 24–41

Gilmore, J.F., Howard, Ch., Pulaski, K.: 1986, "A Comprehensive Evaluation of Expert System Tools", *SPIE Proc. "Applications of Artificial Intelligence IV"* **657**, 194–208

Guilfoyle, Ch.: 1987, "Natural Language for Smarter Systems", *Expert Systems User* **2**, 14–15

Hanne, K.H., Hoepelman, J.Ph., Fähnrich, K.P.: 1986, "Combined Graphics / Natural Language Interfaces to Knowledge Based Systems", in: *Proc. "Artificial Intelligence and Advanced Computer Technology Conference, Wiesbaden*, Sept. 1986

Hayes, Ph.J.: 1986, "Providing Expert Systems with Integrated Natural Language and Graphical Interfaces", in: *Proc. 1986 Fall Joint Computer Conference, Dallas*

Hayes-Roth, F., Waterman, D.A., Lenat, D.B. (eds.): 1983, *Building Expert Systems*, Addison-Wesley Publishing Company, Reading, Massachusetts

Hendrix, G.G., Sacerdoti, E.D., Sagalowicz, D., Slocum, J.: 1978, "Developing a Natural Language Interface to Complex Data", *ACM Trans. Database Syst.* **3**, 105–147

Hornick, Th., Cohen, W., Miller, G.: 1987, "A Natural Language Query System for Hubble Space Telescope Proposal Selection", in: *Proc. "Goddard Conference on Space Applications of Artificial Intelligence and Robotics"*

Jacobs, B.E.: 1985, "Heterogeneous Distributed Database Management — The DAVID System", *NASA/Goddard Space Flight Center*

Jarke, M., Vassiliou, Y.: 1985, "A Framework for Choosing a Database Query Language", *ACM Computing Surveys* **17**, 313–340

Kaplan, J.S.: 1984, "Designing a Portable Natural Language Database Query System", *ACM Trans. Database Syst.* **9**, 1–19

Kehler, Th.P.: 1987, "KEEconnection and IntelliScope: A Strategic Direction for IntelliCorp", Intellinews **3**, No. 2

Kerschberg, L.: 1987, "Expert database systems", *Computer Bull.* **June**, 7–16

Lehmann, H., Ott, N., Zoeppritz, M.: 1985, "A Multilingual Interface to Databases", *IEEE Database Engin. Bull.* **8** No. 3

Richer, M.H.: 1986, "An evaluation of expert system development tools", *Expert Systems* **3**, No. 3, 166–182

Rosenthal, D.A.: 1987, "Applying Artificial Intelligence to Astronomical Databases — A Survey of Applicable Technology", *this volume*

Rosenthal, D., Monger, P., Miller, G., Johnston, M.: 1987, "An Expert System for Ground Support of the Hubble Space Telescope", in: *Proc. "Goddard Conference on Space Applications of Artificial Intelligence and Robotics"*

Shoval, P.: 1985, "Principle, Procedures and Rules in an Expert System for Information Retrieval", *Inform. Proc. & Management* **21**, 475–487

Steele, G.L.: 1984, *Common LISP*, Digital Press, Bedford

Tatar, D.G.: 1987, *A Programmer's Guide to Common LISP*, Digital Equipment Corporation, Maynard

Wahlster, W.: 1985, "Cooperative Access Systems", in: *Artificial Intelligence Toward Practical Applications, T. Bernold & G. Albers (eds.)*, pp. 33–45

Wahlster, W., Kobsa, A.: 1986, "Dialog-Based User Models", *Berichte des KI-Labors am Lehrstuhl für Informatik IV, Saarbrücken* **3**, 1–34

Waltz, D.L.: 1978, "An English Language Question Answering System for a Large Relational Database", *Commun. ACM* **21**, 526–539

Winston, P.H., Horn, K.P.: 1984, *LISP*, Addison-Wesly Publishing Company, Reading, MA

Intelligent Access to a Bibliographical Full Text Data Base

H.-M. Adorf[1], *E. K. Busch*[2]

[1]Space Telescope – European Coordinating Facility
European Southern Observatory
Karl-Schwarzschild-Str. 2
D-8046 Garching b. München, F.R. Germany
[2]Information Access Systems, Inc.,
Suite 202, 3340 Mitchell Lane
Boulder, Colorado 80301

Abstract: Searching for and evaluating relevant literature is an integral part of scientific activity. The availability of large database facilities created e.g. within the Hubble Space Telescope project allows us to consider their use for storage and retrieval of astronomical bibliographical information. We analyze the storage and processing requirements for a bibliographical full text base and show that user requirements for information retrieval can largely be met by today's commercial 'intelligent' text retrieval systems.

Introduction

Almost every astronomer will have experienced the increasing difficulties of keeping up with the amount of published literature in his field of interest. This difficulty is enhanced when research topics require information from outside astronomy, e.g. from computer science, applied mathematics, pattern recognition etc. Given the ubiquous increase in computer storage and processing power it seems natural to consider using computerized systems for storage retrieval of bibliographical information.

Progress towards computer-readable bibliographical databases has already been made on small scale:

- Personal bibliographical text bases have been used at ESO since 1985. Key features include: Simple ASCII format, so easy transfer of database records to other machines/operating systems. Database records provide fields for author(s), title, journal, keywords and brief personal comments. Retrieval via free text pattern matching either on specific fields or on whole records. Multiple text strings can be conjuncted (*and*ed) for the search. Retrieved references can be cast into various formats (e.g. *Astronomy & Astrophysics, Monthly Notices* or *P.A.S.P.*). Output can also be obtained in a form suitable for further processing with the TEX (Knuth 1984) and LATEX (Lamport 1986) text processing software, which has increasingly been used by scientists and engineers for preparing their publications (Adorf 1986).
- ESO is currently implementing a machine-readable preprint service on its Britton Lee IDM/500 database machine (Ochsenbein & Sachtschal 1987), which will be be remotely accessible over the computer network similar to the other scientific databases stored in the database system.

However, the most important source of contemporary astronomical bibliographical information is the well known *Astronomy and Astrophysics Abstracts* series, (Böhme et al. 1987), which quite comprehensively documents and organizes the literature in all fields of astronomy. *Astronomy and Astrophysics Abstracts (AAA)* has been published semi-anually since 1969 and allows to access full text abstracts – which form the bulk of *AAA* – by means of three indices: one for authors, one for subjects and one for celestial objects. Starting with Vol. **33** of 1983 computers have been used for the preparation of these abstracts and beginning with Vol. **38** of 1984 the complete volumes have been recorded on machine-readable media (Schmadel 1987).

On-line access to Astronomy and Astrophysics Abstracts?

So far, *AAA* is available to the astronomical community in printed form. The advantages of being able to access this important source of bibliographical information also via computer are obvious and storage of *AAA* within database systems is currently being considered by the Astronomisches Recheninstitut, Heidelberg (Schmadel 1987; Rey-Watson 1987). At the time where large astronomical databases are already in use and very large databases are foreseen e.g. within the Hubble Space Telescope project (Russo et al. 1986a, 1986b), it is due to take the position of a potential user and to analyze the requirements which would make a computerized bibliographical information system beneficial to the astronomical community.

The requirements for storing a single complete volume of *AAA* are modest: For instance, volume **42** of *AAA* contains references to more than 10,000 papers. Being based on the original author's abstracts, the size of the text entries varies, of course, but generally stays between 500 and 1500 characters. So the pure text of one volume of *AAA* would need roughly 10 Mbyte of storage space — essentially the same space as occupied by a single large (2000×2000) CCD-frame. All published volumes of *AAA* would certainly fit on a single optical disk and thus could be kept on-line.

The problem with full-text bases is not so much storage but retrieval. Astronomical image databases can be described as *very large but rather simple* structured. After all the relational data model is believed to be sufficient for data representation. Bibliographical information in text bases, on the other hand, is characterizable as *modest in size but complex*. So the challenging problem is 'understanding a request and retrieving relevant information', and a computer solution presumably requires both, processing power and some sort of built-in intelligence.

Getting this intelligence into place is a non-trivial undertaking and can be reasonably considered only for a valuable text base, in other words for *Astronomy & Astrophysics Abstracts* itself.

User requirements for text retrieval

Let us now consider the expectations which a potential user might have towards a bibliographical information system. We assume that the basic unit for retrieval would be a document containing the full text of an astronomical abstract and a number of formatted fields, which would allow a user to specify author(s), date, journal, existing classification number etc. Documents would then be accessible on the basis of formatted fields only or in conjunction with other retrieval strategies to be detailed below.

Query processing can be considered on a number of levels:

- A *simple query* would consist in a request for 'completion': some information on the values of formatted fields is entered and the system retrieves all bibliographical entries which are extensions of the incomplete specification. Such a service would be valuable for the tedious task of constructing reference lists.
- The next level of queries is one using *classificatory keywords*. As a subject classification is provided within *AAA*, such a query capability would present no additional technical problem to the text management system. However, the current classification system may be too coarse with the effect that a query based on a few classificatory keywords only may retrieve too many references.
- A high level query would use *free text search* within the full text of the abstracts. One could conceive a "query-by-topic" or a "query-by-example" where the user would submit a *natural language request* or a *sample abstract* to the system which would retrieve all similar entries, rank-ordered for relevance. A system being capable to perform such a service would have to 'understand' the request to some extend and use an internal representation for testing similarity between request and stored bibliographical entry. It

Quick response

The ITMS retrieval response time is a direct linear function of the data base size. Response time can be tuned to meet specific requirements by adding to the number of concurrent parallel processors. This modular growth approach provides maximum flexibility in controlling hardware costs while achieving user satisfaction.

ITMS is providing several levels of query processing: (a) *Boolean.* Version 4.0 of ITMS will provide the basic boolean capabilities of 'and' and 'or'. (b) *Query by topic.* The ITMS allows a user to enter a request in conversational language (or languages) by specifying particulars about a given subject matter or topic. There as no system limitations to the length of the request. (c) *Query by example.* The user may submit an existing document or sub-document to the systems. If the document already has been indexed into the system, the user identifies the document as a request. If not, the user must enter the document by the keyboard or by a page scanner.

Implementation

Prior to installation of a expert system like ITMS, an astronomy knowledge base would have to be developed by knowledge engineers working with selected experts of the astronomical community. The steps to be taken are the following:

(1) Analysis of a representative set of astronomical articles by the knowledge engineers.
(2) Preliminary selection of system vocabulary terms and sub-topic areas.
(3) Finalization of system vocabulary terms and sub-topic areas by experts from the astronomical community.
(4) Collection of subject matter judgement data from experts within the astronomical community. (From previous experience it can be estimated that this phase would take approximately 320 hours of experts' judgement time.)
(5) Analysis of the judgement data by knowledge engineers.
(6) Formulation of an astronomy indexing structure.
(7) Integration of astronomy indexing structure with the text retrieval system.

The implementation of a computerized indexing system is facilitated by the existence of an established astronomical classification system already in use for *AAA*.

Problems and issues

There are a number of open issues which would have to be addressed if a system like ITMS is to be be used to access a computerized *AAA* full text base:

- *What form should the output take?* One would probably want to get the bibliographical information retrieved into an ASCII file so that for instance a full screen editor may used for reading and searching.
- *How should one deal with special (e.g. greek or mathematical) characters? AAA* is currently using a special coding scheme based on three-letter sequences enclosed in escape characters. One might consider the possibility to use a preprocessor to convert the special character coding to the de facto TeX-standard.
- *How will a user enter his natural language (NL) query?* Typing is conceivable for short NL queries, but ideally one would like to talk to the system. Research in coupling speech recognition systems to databases is underway at various places over the world (see e.g. the October 1987 issue of *Systems Intern.*).
- *How will a user enter a sample document?* Very often a suitable document will already exist in the computer system, either as a text file in the user's personal workspace or in the text base and could be presented to the system as the sample to be used. Otherwise the document would have to be entered via keyboard or a page scanner.

is questionable whether such advanced retrieval requests could be processed for a large text base in real-time.

A very important requirement is of course that the service is *cheap* such that all institutes linked to the astronomical computer networks (STARLINK, ASTRONET, SPAN, ASTRODOMAIN, etc.) could participate.

In the following we will illustrate our considerations with a specific database system which can be considered to represent the state of the art in the field of automated text retrieval. We will show that the requirements outlined above can be fulfilled by today's commercial information systems.

Intelligent text retrieval

The Intelligent Text Management System (ITMS)* provides expert-system based document classification, cross-referencing and retrieval for very large text bases. It is designed to provide conversational language access to very large document collections.

The general retrieval procedure operates with formatted fields, plain text (narrative fields), or both. Formatted fields retrieval selects documents on the basis of specific terms (e.g. author's name) or a range of values (e.g., between January 1, and March 30, 1987).

For plain text (narrative fields), key words, string search and "intelligent retrieval" are available. Intelligent retrieval is accomplished either by specifying the subject matter of the documents to be retrieved, or by providing a sample of text and requesting documents similar to the example in regard to subject matter.

Results of key word, string search and intelligent retrieval are combined and the documents are ordered in terms of their subject matter relevance to the request.

Salient functional features of the ITMS include: (a) user definable entry and retrieval screens, (b) conversational language query capability, (c) support for submitting full or partial documents as a request, (d) sub-document indexing and retrieval capability, (e) rank-ordering of retrieved documents with respect to relevance to the query and (f) retrieval of documents on the basis of subject matter content.

ITMS is based on ideas from descriptive psychology (Jeffrey 1981). It consists of the *Intelligent Text Management Engine* and an *Expert Judgement Indexing Module* covering the subject matter the system is to manage, i.e. in our case, astronomy.

The Expert Judgement Module is a functional and modular element which captures the judgemental perspective of a human expert. In application, it provides the human expert's judgemental perspective to the research process. This is like having that expert available to the researcher as a research assistant. The researcher gives the 'captured expert' his query, for instance: "Give me recent articles on supernovae. I am particular interested in spectroscopic observations in the ultraviolet." The 'captured expert' then goes through the available document base(s) and returns documents ranked in order of relevance to the request.

Distributed access

The ITMS provides a unique parallel distributed architecture which addresses the performance issues of very large document collections while allowing access to document bases located on different processors or in different geographical areas. This architecture consists of several modules which communicate utilizing the DECnet protocol. The DECnet protocol allows several of the modules to exist in the same processors or on distributed processors. The capacity to support mutiple concurrent processing modules allows parallel processing on data line queries to occur, providing significant performance improvements over commonly used data base technologies. In addition, the processing modules can support networked or true distributed processing.

---

* marketed by Information Access Systems of Boulder, Colorado, USA

- *How can the back volumes* **1** *to* **37** *of* AAA *be transformed into machine-readable form?* This is a laborious task and would certainly require using page scanners.
- *Where should the information retrieval system be located?*
- *How can the installation and maintenance of such a bibliographical service be funded?*

Conclusion

It has been shown that an intelligent full text base system containing astronomical bibliographical information is technically feasible. The obvious target for such a system is the *Astronomy and Astrophysics Abstracts* series. Having access to the *Astronomy and Astrophysics Abstracts* textbase in machine-readable form might already be useful, but a larger benefit can be expected from an integration of the textbase into an intelligent text retrieval system. The implementation of such a system would require a major investment in terms of expert judgement time and would also need major computer processing capacity so that it would be feasible only at a major data center. The information service would then have to be distributed using the established wide-area computer networks.

References

Adorf, H.-M.: 1986, "TEX— Erfahrungen nach einem Jahr", *Proc. Digital Equipment Users Soc. Symp., Stuttgart 1986* **9**, 423–426

Böhme, S., Esser, U., Hefele, H., Heinrich, I., Hofmann, W., Krahn, D., Matas, V.R., Schmadel, L.D., Zech, G. (eds): 1987, *Astronomy and Astrophysics Abstracts* **42**, 1–14

Intelligent Text Management System, ITMS, Version 3.1, Software Product Description, Information Access Systems, Inc., 3340 Mitchell Lane, Boulder, CO 80301, USA

Jeffrey, H.J.: 1981, "A New Paradigm for Artificial Intelligence", *Adv. Descript. Psychology* **1**, 177–194

Knuth, D.E.: 1984, *The TEXbook*, Addison-Wesley Publishing Company, Reading, Massachusetts

Lamport, L.: 1986, *LATEX, A Document Preparation System, User's Guide & Reference Manual*, Addison-Wesley Publishing Company, Reading, Massachusetts

Ochsenbein, F., Sachtschal, E.: 1987, European Southern Observatory, *pers. comm.*

Rey-Watson, J.M.: 1987, "Access to Astronomical Literature Through Commercial Databases", *this volume*

Russo, G., Albrecht, R., Ochsenbein, F.: 1986a, "Optical mass storage for a 10 Terabyte archive: system integration problems and user interface", *Proc. SPIE 30th Ann. Intern. Techn. Symp. "Optical and Optoelectronic Applied Sciences and Engineering"*, San Diego, California, 17–22 Aug. 1986

Russo, G., Richmond, A., Albrecht, R.: 1986b, "The European Scientific Data Archive for the Hubble Space Telescope", in: *Data Analysis in Astronomy II, V. Di Gesù et al. (eds.)*, 193–200

Schmadel, L.D.: 1987, Astronomisches Recheninstitut, Heidelberg, *pers. comm.*

A TEST FOR SPATIAL CLUSTERING IN THE PRESENCE OF AUXILIARY INFORMATION

Jeff Banfield

Department of Statistics, University of Washington

ABSTRACT

A technique is proposed to test for the presence of clusters when both spatial and auxiliary information is available. The test is based on randomizing the auxiliary information over the spatial locations. It requires no model assumptions (such as a Poisson process) for the spatial distribution. The procedure may be used to screen data for possible clustering.

1. Introduction

In the fields of research where cluster analysis is used, it is often the clustering mechanism, not the clusters themselves, that is of interest. For example, in acid rainfall studies the questions of interest involve the physical mechanisms that are causing the clustering. Similarly, in quasar clustering the current work deals with the relation between clustering and the evolution of structure (Shaver 1987).

One type of problem encountered in spatial cluster analysis is that of testing for the significance of a cluster after the cluster has been found. This is an area of cluster analysis that Dubes and Jain (1979) term "cluster validity". This paper will look at this particular problem in cluster validation with an eye toward addressing the deeper question, the existence of a mechanism that is causing the clustering.

In this paper we will consider data composed of a spatial component and auxiliary information which is measured at each of the spatial locations. For example, in acid rainfall studies the latitude and longitude of the collection stations would define the spatial locations and the chemical composition of the rainfall at each station would be the auxiliary measurements associated with that station. More formally, let us define the observations, $\mathbf{x}_1, \ldots, \mathbf{x}_n$, by $\mathbf{x}_i = (\mathbf{s}_i, \mathbf{a}_i)$ where $\mathbf{s}_i$ is the vector of spatial coordinates and $\mathbf{a}_i$ is the vector of auxiliary measurements associated with the location given by $\mathbf{s}_i$.

A common technique for measuring the validity of a spatial cluster, is to generate a large number of random data sets from a hypothesised homogeneous process and then calculate the extent of clustering found in these sets. If the amount of clustering is small, then we would conclude that there is evidence that the original cluster did not occur by chance. Although cluster validation does not directly confirm the existence of a clustering mechanism (it can only measure the probability of the observed cluster having occurred under certain assumptions) the procedure proposed here allows the researcher to limit the alternatives.

2. The Procedure

If the original cluster was composed of k data points then it is compared to the "best" cluster of size k in each of the random data sets. The usual statistic for this type of test is the percent of random data sets that had a "better" cluster of size k. We will denote this percentage, for a cluster of size k, as p_k. Under the null hypothesis, that the original data set was generated by the same random process that was used to generate the random data sets, the distribution of p_k is uniform between 0 and 1. Thus p_k is the significance level for testing the null hypothesis. Small values of p_k suggest that the null hypothesis should be rejected.

The sequence of p_k values, $\{p_k\}$ $k = 2, \ldots, n$, contain information about the clustering structure in the data. We can use this information as an exploratory tool. Define a function over k by $P(k) = 1 - p_k$. A graph of $P(k)$ can be used to examine clustering structure and as an indicator of the presence of a cluster. If there is a cluster of size k present, then the graph of $P(k)$ should have a high sustained peak around the value k. If there is no cluster present, then there should be no pattern in the graph of $P(k)$. Figure 1a shows a typical plot of $P(k)$ for a Poisson data set without a cluster. Figure 1b shows a graph of the $P(k)$ function for a Poisson data set that also has a cluster of size six.

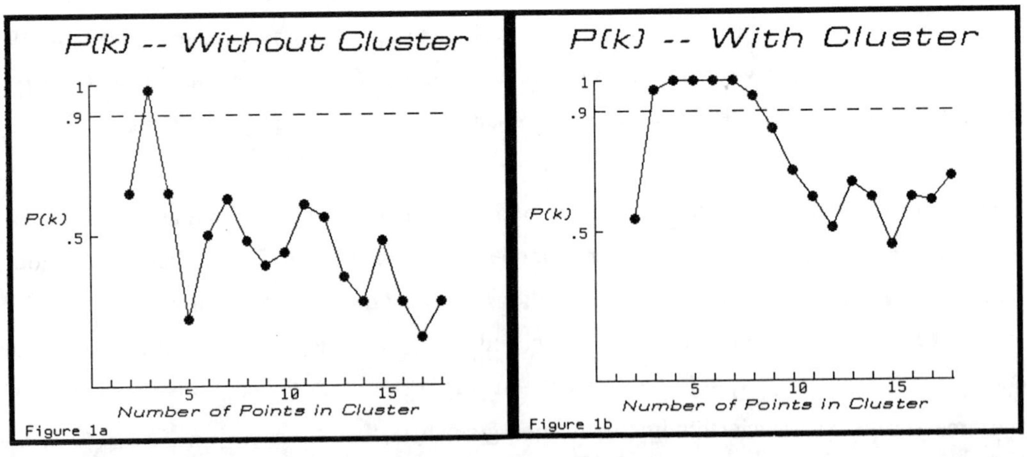

It should be noted that p_k, for a single value of k, may be used as a statistical hypothesis test. The value of p_k being the probability of a type I error. However, the graph of $P(k)$ may not be used as a hypothesis test. There are problems with multiple comparisons and dependence between the observations ($P(k)$ is not independent of $P(k+1)$). The $P(k)$ function is only proposed as an exploratory device.

Before calculating p_k two key points must be addressed. First, how do we generate the random data sets? What null distribution should we use? The second point is how do we determine when one cluster is "better" than another? What measure of clustering should be used? The following two sections look at these questions.

3. Generating Random Data

We need to generate random samples in which there is no clustering. One way to do this is to generate spatial locations from a random homogeneous distribution, such as the Poisson distribution, then randomly assign the auxiliary measurements to those locations. Several problems can arise when using a hypothesised homogeneous distribution to generate random spatial locations There may be sampling biases that would not be accounted for. Vignetting in photographic plates for example. There may be physical constraints on the spatial locations that would be difficult to model. For example, in tracking whale migration routes, coastlines, ice floes and ocean floor geography must be accounted for in any set of random spatial locations. Dubes and Zeng (1987) point out that problems with edge effects, because of the finite sampling area, could produce erroneous results. Finally, if we reject the null hypothesis, the only conclusion that we may draw is that the cluster we observed in the original data is unlikely to have occurred if the original data came from the hypothesised distribution. This doesn't address the question of a clustering mechanism.

For the type of data we are considering, spatial with auxiliary information, there is a way to generate random data sets without resorting to a hypothesised homogeneous distribution. Permute the auxiliary measurements over the spatial locations. This produces random data sets with the same spatial distribution as the original and the same marginal distribution for the auxiliary measurements. Ko, et al. (1986) successfully used this spatial permutation technique to test the significance of observed whale migration patterns.

This permutation test corresponds to a null hypothesis of independence between the auxiliary measurements and the spatial locations. An advantage in using it is that by rejecting the null hypothesis we may conclude that the type of clustering we observed is unlikely if the auxiliary information is independent of the spatial locations. This comes closer to addressing the question of the existence of a clustering mechanism than simply concluding that, for instance, the data does not appear to be Poisson. It still does not answer the question directly. As Diggle (1979) points out, without additional information, especially in observational studies, definitive conclusions cannot be expected.

One problem that can occur with this method is loss of sensitivity when the spatial area of a cluster is large relative to the area being studied. This is because most of the clustering is in the auxiliary measurements. This may be remedied by generating additional auxiliary measurements. These measurements may come from either a hypothesised distribution or from additional auxiliary measurements made at other spatial locations.

4. Clustering Criteria

Dubes and Jain (1979) present an excellent survey of validation techniques for cluster analysis. Ripley (1977, 1979), and Diggle (1979, 1983) discuss testing randomness in spatial point patterns. None of these techniques allow the researcher to specify the type of cluster to be validated. An advantage of the p_k method is that it allows the researcher to bring his insight and special knowledge into the testing procedure. He does this by designing the clustering criterion. Because of this, specific types of clustering may be tested.

A clustering criterion is a function that operates on subsets of the data. It provides a measure of how "good" a cluster each subset would be. Gordon (1981) discusses many of the clustering criteria that have been developed. Many of these criteria are based on the L_2 norm, sums of squared distances. Friedman and Rubin (1967) discusses invariant L_2 criteria. Scott and Symons (1971) derive the conditions under which several L_2 criteria produce maximum likelihood estimates of clusters. Murtagh and Raftery (1984) propose decomposing the sum of squared distances and downweighting various parts of the decomposition. This method provides flexibility in assigning relative influence to various features in the data. The Murtagh and Raftery downweighting scheme resulted in elongated elliptical clusters that they used in problems involving character recognition. Other decompositions have been used in the location of ice floes in LANDSAT images (Banfield 1987).

5. An Example

Crampton, Cowley and Hartwick (1986) (hereafter CCH) speculated on the presence of a cluster of 6 quasars in the 1338+27 field of their CHFT survey. They ran a Monte Carlo test using a Poisson distribution to simulate the spatial locations. Redshifts were randomly drawn from all the observed redshifts in the survey and assigned to the simulated spatial locations. In only .1% of the random data sets did they find a cluster of 6 objects, located within a .5 deg^2 field, which had a redshift spread less than that observed in the original cluster.

Figures 2 and 3 show $P(k)$ functions based on the data in the CHFT survey. The results support the findings in CCH. Redshifts were drawn from all the observed redshifts in the survey and randomly assigned to the observed spatial locations in the 1338+27 field. Figure 2 shows the $P(k)$ function for a single-link clustering criterion that was designed to emphasize the redshift differences. Figure 3 shows the $P(k)$ function based on co-moving separations. The co-moving separations were calculated according to the formulas in Osmer (1981). The clustering criterion was the maximum separation between the center of the cluster and each of the quasars in the cluster.

As the graphs in figures 2 and 3 show, there is strong evidence that the redshifts are not independent of the locations. One interpretation of this is that there is a cluster present. The exact size of the cluster is not clear from the $P(k)$ functions, but the shapes indicate that there is a cluster. The results are independent of whether the redshift is considered a measure of distance, using a cosmological model, or simply an auxiliary measurement. Technically, we may not consider the value for p_6 (the cluster found in CCH was of size 6) as a significance level for a statistical test of independence between the redshifts and the locations. This is due to the fact that the cluster found in CCH was found by inspection of all the fields in the study. The simulations used in this paper considered only the field where the cluster had been found. In order to duplicate, exactly, the procedure used in CCH to locate this cluster, the simulations should have considered the "best" cluster from all the fields. However, it is felt that this would not have made a significance difference in the results.

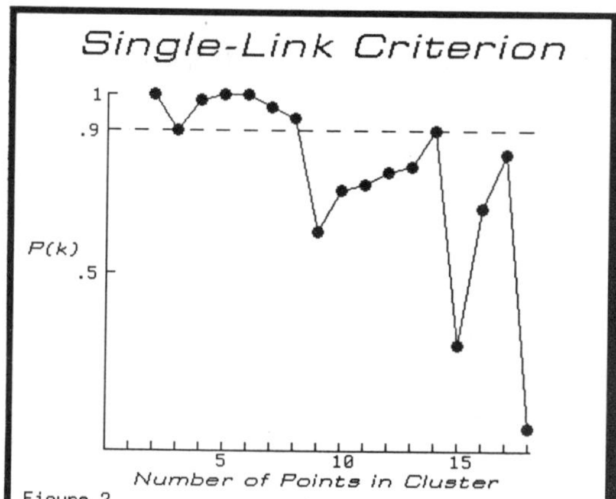

Figure 2

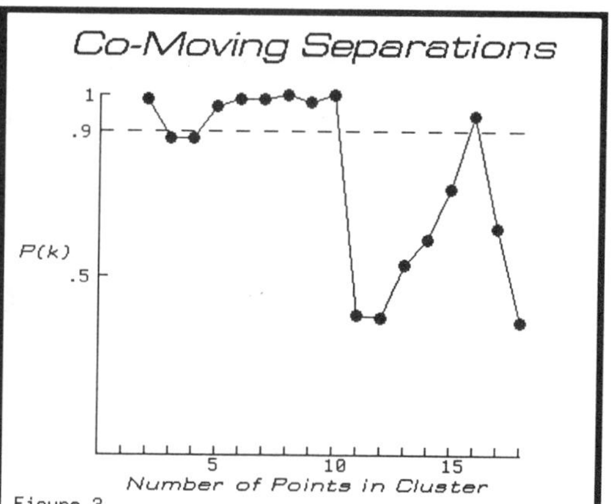

Figure 3

Acknowledgments

It is a pleasure to thank David Crampton, Dominion Astrophysical Observatory, for providing me with the data, and Adrian Raftery, University of Washington, for his advice and comments. Research for this work was supported in part by DOE grant DE-FG06-85ER25006 and NASA grant NAGW-412.

References

Banfield, J.D. (1987) Unpublished dissertation. Department of Statistics, University of Washington, Seattle Washington.

Crampton, D., Cowley, A.P. and Hartwick, F.D.A. (1987) "The Space Distribution of Faint Quasars From the CFHT Survey" *Ap. J.* **314,** 129-144.

Diggle, P. J. (1979) "On Parameter Estimation and Goodness-of-fit Testing for Spatial Point Patterns" *Biometrics* **35,** 87-101

Diggle, P. J. (1983) *Statistical Analysis of Spatial Point Patterns,* New York: Academic Press.

Dubes,R. and Zeng,G. (1979) "Validity Studies in Clustering Methodologies" *Pattern Recognition* **11,** 235-254

Dubes,R. and Zeng,G. (1987) "A Test for Spatial Homogeneity in Cluster Analysis" *Journal of Classification* **4,** 33-56

Friedman,H.P. and Rubin,J. (1967) "On Some Invariant Criteria for Grouping Data" *J. Amer. Statist. Ass.* **62,** 1159-78

Gordon, A.D. (1981) *Classification, methods for the exploratory analysis of multivariate data.* New York: Chapman and Hall

Ko,D., Zeh,J., Clark,C., Ellison,W., Krogman,B. and Sonntag,R. (1986) "Utilization of Acoustic Location Data in Determining a Minimum Number of Spring-Migrating Bowhead Whales Unaccounted for by the Ice-Based Visual Census" *Rep. Int. Whal. Comm.* **36,** 325-38

Murtagh,F. and Raftery,A.E. (1984) "Fitting Straight Lines to Point Patterns" *Pattern Recognition* **17,** 479-483

Osmer, P.S. (1981) "The Three-Dimensional Distribution of Quasars in the CTIO Surveys" *Ap. J.* **247,** 762-773

Ripley, B.D. (1977) "Modelling Spatial Patterns" *J. R. Statist. Soc. B* **39,** 172-192

Ripley, B.D. (1979) "Tests of 'Randomness' for Spatial Point Patterns" *J. R. Statist. Soc. B* **41,** 386-374

Scott,A.J. and Symons,M.J. (1971) "Clustering Methods Based on Likelihood Ratio Criteria" *Biometrics* **27,** 387-97

Shaver,P.A. (1987) Preprint, "Quasar Clustering and the Evolution of Structure"

ANALYSIS OF RADIAL PROFILES

A.Bijaoui Nice Observatory B.P.139 F-06003 Nice Cedex
E.Slezak E.S.O. La Silla 213832 La Serena (Chile)
G.Wlérick Paris Observatory F-92195 Meudon PP Cedex

I-Interest of the radial profile.

Many strategies have been applied for faint object photometry: integration of fixed size spots (Kurtz et al. 1985), isophotal magnitude, well adapted to real time detection (Mac Gillivray and Stobie 1984), 2D profile fitting used for stars, or derived from radial profiles (Le Fèvre et al. 1986).

The use of the radial profiles is based on the idea that faint astronomical sources are mainly described as peaks. Almost all the information is contained in i/the position of the maximum intensity, ii/the ellipticity of the isophots in the region around the maximum, iii/ the radial intensity profile.

There are some advantages in using the radial profile: easy computations, accurate determination of the faintest part by summation of an increasing number of pixel intensities, accurate determination of the background even in a crowded field, rejection of defects or neighbouring objects with a simple test. All astronomical objects can be measured, especially stars and faint galaxies.

II-Profile estimation.

The determination of the profile is made iteratively. At each step, the program computes the mean profile and the corresponding dispersion, thus allowing us to reject pixels with an intensity too far different from the mean at such a distance from the center. Only two iterations are needed to reject major defects or objects which are too close.

As the number of pixels used is proportional to the radius, we can compute the mean profile up to a fainter level than the one determined for isophotal photometry.

In the algorithm, we compute a distance to the object center which takes into account the ellipticity and the orientation. That allows us to improve the pixel rejection.

III-Parameter extraction.

Two main parameters must be extract at first: the background level and the object maximum extension.

A complete background map is always computed and subtracted in the first steps of the image processing. Generally that is not sufficient, a small correction is necessary to obtain accurate magnitudes. We have developed algorithms to compute this parameter and the best one results from the estimation of the mode of the local intensity histogram. That histogram is easily computed with the profile. That algorithm works with Schmidt images, but does not work with deep fields, like the ones obtained with the large field electronic camera at the Cassegrain focus of the CFHT (Wlérick et al. 1984). In that case the density of objects is so high that the algorithm often gives too great a value, leading to an integrated magnitude which is too small.

As the total number of objects is small (about 600), we have interactively measured the background. This measure is performed graphically on the integrated flux, instead of the profile. That allows us to obtain a better determination of the effect of the neighbouring objects.

The determination of the object extension is harder to obtain. That is more sensitive to the noise and to the neighbouring objects.So we developed an algorithm which works well on Schmidt plate objects, but fails for deep fields. For scuch images, interactive determination on the integrated flux allows us to obtain a good value. That determination requires a careful study of the field.

After the determination of these two essential parameters, the background and the extension, we extract a set of parameters which are independant of the model: integrated flux, equivalent area, flux for given magnitudde threshold, equivalent radius, Le Fèvre's Q parameter, invariant moments, etc.. Using these secondary parameters we have developed a method to differentiate between stars and galaxies, based on Bayesian classifiers (Slezak et al. 1987). For Schmidt plates the number of misclassified objects depends on the magnitude, and it is small for $m < 18$. One of the major problem is the separation between double stars and galaxies and no extracted parameter from the radial profile works: we solve that point by using the angular profile.

For the deep fields, the number of resolved pixels is small but we use the equivalent radius and only a separation with the colours really works for $m > 21$ which corresponds to the most interesting objects.

IV-Model fitting and fuzzy classifier.

For well defined galaxies, the data permits us to fit models. On a field of 1°5 × 1°5 measured in the Coma Supercluster we have detected 1048 objects classified as galaxies. We fitted the radial profiles with 4 laws:

1/de Vaucouleurs's law for elliptical galaxies:

$$p(r) = A\exp(-b(r/r_e)^{1/4})$$

2/Exponential disk model:

$$p(r) = A\exp(-b(r/r_e))$$

3/King's law:

$$p(r) = A(1/(1 + (r/a)^2)^2$$

4/Gaussian law:

$$p(r) = A\exp(-0.5(r/\sigma)^2)$$

For each model we obtain a length scale and a deviation σ_i $(i = 1, 4)$. From the set of values σ_i we derive an object classification.

First we have considered the ratio σ_i/σ_j. The corresponding histograms show generally 3 regions where: i/the value is significantly greater than 1, ii/the value is significantly little than 1, iii/the ratio is around the value 1.

As we cannot have a clear classification for a set of galaxies, we introduce a fuzzy classifier (Kauffman 1975). For each model i we define a weight as:

$$w_i = \sigma_i^{-2}$$

If s if the sum of the weights for a given object, the fuzzy parameter is given by:

$$p_i = w_i / s$$

p_i can be understood as a probability belonging to a given class.

If the measurement dispersion is small, a weight w_i is greater than the other and no doubt is possible, but if the studied object is faint, with few resolved pixels, all the weights are almost the same, and it is impossible to distinguish between the profiles.

We applied that approach to a set of 1048 detected galaxies. We plot the diagram in which:

$$x = w_2 + w_3 \qquad \text{and} \qquad y = w_3 + w_4$$

The origin corresponds to a pure de Vaucouleurs' profile, the point $(1, 0)$ the exponential one, $(1, 1)$ King's one, and $(0, 1)$ the gaussian one.

The diagram shows that it is very difficult to classify an object. A large void exists. It is due to the fact that if the weight is the highest for the first fit, the other ones are in decreasing order. That means that there is a continuity between the model profiles.

For faint objects, almost all the information is the radial profile and the parameters of the elliptical contour fitting. For the other ones, an anisotropy can exist (spiral or irregular galaxy). A criterion is derived if we compare the original galaxy image (after taking into account the background and the extension) and a synthetic image obtained with the radial profile and the elliptical parameters. The correlation coefficient is a good parameter for that comparizon. We have only test that criterion on a few objects (computing time, size). Coefficients of correlation around 0.9 are generally obtained.

V-Use of the crosscorrelation.

A classical idea in profile comparison is to use the crosscorrelation between the profile (Karhunen-Loève expansion for cross correlation). Difficulties in that case are:

1/All the profiles are not defined with the same number of pixels;

2/Two astrophysical similar objects may belong to the same class, but the distance are not the same, so the profiles are different;

3/The data dispersion can be different from one profile to another.

We take into account these features in the following procedure:

1/We compute an equivalent radius r_e with the profile;

2/We resample the magnitude profile, taking into account this radius;

3/We give to each profile a weight w as:

$$w = r_e / \sigma^2$$

where σ is the minimum dispersion value, previously computed.

4/We computed the crosscorrelation matrix.

5/We derived the corresponding eigenvalues λ_k and eigenvectors V_k.

If we use all the profiles, we do not obtain significant results. Too many profiles corresponds to too few pixels and the diagram (V_2, V_3) shows strips, which disappear if we plot only the large objects (more than 10 pixels in radial profile).

VI-Conclusion.

The data processing applied on the radial profile shows that it is a good tool for the analysis of detected objects in astronomical images; it can can be used for:

1/Determination of an accurate magnitude, taking into account the near objects;

2/Computation of some object characteristics without modeling, which allows us, for example, to separate the stars and the non- stellar objects;

3/Model fitting which leads to a fuzzy classification, simple to implement;

4/A 2D Correlation coefficient between observed image and restored image, leading to detection of anisotropic objects;

5/Crosscorrelation which could indicate new classes of profile.

References:

Kauffman,A. (1975) Introduction à la théorie des sous-ensembles flous (tome 1) Masson Paris.

Kurtz,M.J., Huchra,J.P., Beers,T.C., Geller,M.J., Gioia,I.M., Maccaro,T., Schild,R.E., Stauffer,J.R. (1985) Astron.J. **90** 697.

Le Fèvre,O., Bijaoui,A., Mathez,G., Picat,J.P., Lelièvre,G. (1986) Astron. Astrophys. **154** 92.

MacGillivray,H.T., Stobie,R.S. (1984) Vistas in Astronomy **27** 433.

Slezak,E., Bijaoui,A., Mars,G. (1987) Astron. Astrophys. (submit.).

Wlérick,G., Renard,L., Horville D., Gex,F., Munier,J.M., Lelièvre, G., Llebaria, A., Bijaoui, A., Dubout,R., Petit,M. (1984) in "Very Large Telescopes, their instrumentation and Programs" p.639 ed.E.S.O. Garching.

STATISTICAL AND NUMERICAL PROBLEMS IN POOLING ASTROMETRICAL DATA. APPLICATION TO THE PARIS ASTROLABE DATA.

M. L. Bougeard
Université Paris X, I.U.T. "Techniques aérospatiales"
1 Chemin Desvalliéres, F-92410 Ville d'Avray, France.

Abstract: *Up to now, Astrolabe observations have been regularly used mostly for latitude and Time work. Yet, for the purpose of future revision of these long astrometrical series for proper motion determinations or catalog work, it may prove useful to analyze carefully the model of reduction. In this paper, through multivariate statistical methods, we try to identify nonhomogeneous data, then to detect and separate some outstanding errors in the usual day-reductions as well as in the global reductions. The methods are demonstrated in their application to the visual Paris-Astrolabe data, and may be extended to many other astrometrical data.*

1. *INTRODUCTION*

Since 1956, visual Astrolabe observations of stars have been regularly used for latitude (ϕ) and Time determinations. In the case of the Usual Astrolabe (UA) model, the equation of condition of an observed star i can be written, in matrix algebra notation and usual notations, as:

$$y_i = (1\ ,\ \cos Z_i\ ,\ \sin Z_i\ )\ \beta + f_i + \varepsilon_i$$

$$f_i = -\sin Z_i\ (\ \cos\phi .\ \Delta\alpha_i) + \cos S_i(\Delta\delta_i) \qquad \text{(UA)}$$

where Z_i is the azimuth of the observed star, β is a (3,1) vector of differential corrections in parameters (instrumental zenith distance, latitude and Time), ε_i denotes data noise and is the error term. As for $\Delta\alpha_i$ and $\Delta\delta_i$, they stand for the errors on the adopted coordinates of the star at the time of the observation. The term y_i is a difference between the observed value and a calculated value of the star's time of passage; it includes standard corrections. Usually, the vector β is assumed constant over an interval of about 90 minutes ; the equations obtained for all n observed stars of a given group g (g=1,..12 and $n \leq 30$ for the Paris-astrolabe) are solved by the method of Ordinary Least Squares (OLS). In such a so-called day-group-reduction, the term f_i is omitted because it is non-determinable and usually estimated by an a-priori fixed correction- the so-called Internal Correction (CI) of the star in the given group g- so that the error term in (UA) is assumed to be a term of pure error.

The Paris-Astrolabe data considered here were obtained between 1956 and 1982; since 1983, the instrument has been being modified and adapted to the solar observation (see:Mangombi, 1986). This period 1956-1982 concerns 7352 group (day) reductions and about 180 000 STAR-OBSERVATIONS, which represents a large astrometrical database. These data are homogeneous insofar as they are all computed in the I.A.U. system of constants 1964-84 and with a fixed system of internal corrections computed in 1964 from an iterative weighted mean of the star OLS-residuals.

Nevertheless, these data may be non-homogeneous since:

(a) There are two instrumental periods :
- -1956-1970 (4382 group-reductions) which corresponds to the Danjon's prism astrolabe (APD) with an optical prism working in transmission,
- -1971-1982 (2990 group-reductions), characterized by the full-pupil astrolabe (APP) with an optical system, a priori more stable, in reflection.

(b) The observer team has varied. We recall that in the case of visual astrolabes, a star observation consists in noticing (by eye) the coincidence of two images of this star.

(c)The estimations in Time and latitude obtained by group-reductions are related to reference frames which depend on the observed stars of the group. Thus, they are directly affected by particular errors on coordinates of some stars in a given group. This had led to the use of group-corrections obtained by the so-called " chain-method" (Débarbat,Guinot:1970) in several catalog works.

Nowadays, modern stable numerical techniques are available for solving large linear systems,even in the case of sparse overdeterminated systems of linear equations (see: Golub,Meurant,1983 and their references). Therefore, proper motion determinations and catalog works are being performed by pooling the equations of condition over large periods of time. In this paper, we shall deal with the preliminary step which consists in the detection -through multivariate statistical methods- of sources of significant non-homogeneity and of perturbations (bias) in the numerical estimations obtained by solving the system of equations of condition by the OLS fit as usually performed.

2. OBSERVER AND GROUP EFFECTS ON THE ESTIMATED VARIANCES

Let s^2 be the OLS estimated error-variance in a group-reduction (linear regression) and $P:=0.1/s^2$ the so-called weight of the reduction. It may be seen in Paper I (Bougeard 1987a;Tab.2, Fig.1) the variations of these variances per year and per observer: a non-constant effect of the observers can be noticed. Moreover, an "apprenticeship curve" has been detected (observers Nr.47,49); it is characterized by a significant decrease in the mean variances at its begin-

ning. Actually, since these variances (or weights) have to be statistically compared through intervals of confidence, we have chosen two test-periods, distant of 12 years such as to avoid some periodic terms of the polar motion in the comparisons: - 1977-79 (APP/ 9 observers/17258 star-observations) which is a period situated in a stable part of the above curves,
- 1965-67(APD/14 observers/ 21465 star-observations).

Then, in a first step, the observer effect on the reduction-weights (more precisely on 5 classes of weights: P1=(P<1.2) to P5=(P ≧3.6)) has been analysed through a factor analysis of correspondences which has shown a first component clearly correlated with the scale of weights with a small loading on most observers for both test-periods with the exceptions of :

- period 1977-79: a contrast between the observer Nr49 (near the smallest weights P1) and observer Nr 19 (near the highest weights),
- period 1965-67: 2 observers (NR 4, 24) near weights (P1-P2).

In a second step, we have performed a factor analysis of multiple correspondences on the 3 characters (observer codes/ weights P1-P5/ group codes g=1,..12) which has proved that- except for the observers already mentioned- the chosen test-periods were homogeneous enough in terms of observer and group effects on the estimated variances (small correlation with the first axis "scale of weights") with, nevertheless, a tendancy of the groups (NR 11,12,1) observed in summer to have higher error variances.

3. *OUTLIERS AND INFLUENTIAL OBSERVATIONS*

The detection of outliers and suspicious observations in a linear regression is an old problem in astronomy which leads to the question: has such an observation to be or not to be deleted from the reduction? The current Paris-Astrolabe method consists in deleting an equation as soon as the related OLS residual exceeds (in absolute value) 2.8 times the estimated OLS standard error of the reduction. As proved in Paper I and Paper II (Bougeard 1987b), this method seems to be not severe enough since it may remain in the reductions some outliers which destroy the normality of the residual distribution and consequently affect directly usual statistical tests. Furthermore, we proved in Paper I (Sect. 3.5) that the presence of some remaining outliers can :

- introduce a perverse time-effect ("rank error") in a reduction,
- prevent the detection (through stepwise regressions) of some real effects in magnitude and colour, by destroying the gaussian distribution of the errors in a given linear regression.

Nowadays, the notion of outlier can be advantageously replaced by the notion of influential observation through available "influence diagnostic statistics" such as the Belsley-Kuh-Welch's statistics or the Cook's D influence statistic which were introduced in years 1977-80. The aim of these statistics is to quantify the extent to which an observation or a set of observations has affected the OLS estimates and standard errors. Some of these statistics were introduced in Paper I and applied to the reduction of selected observations. They proved to be a very efficient tool to bring into evidence undetected errors still contained in the observational data studied.

In Paper III (Bougeard 1987c) -see also (Bougeard 1987e)- the systematic use of these influence diagnostic statistics led to the detection of 9 "atypical stars" in the Paris-Astrolabe programme observed during both test-periods. These stars are "atypical" for the UA model insofar as they appear as:

* very influential -high value of the Cook's D statistic- on the OLS estimate of the vector β,
* with ill-behaved residuals (significant nonzero mean of these residuals for the test-periods, especially the period 1977-79) in the fixed group where these stars are observed,

which has probably introduced bias and additional variations in the Paris-Astrolabe determinations in Time and latitude. The list of these stars is given in the following Table: it may be seen that they are mainly stars of the FK4 supp. catalogue.

TABLE 1: Atypical stars of the Paris-Astrolabe Programme

n° FK4, FK4 supp.	738	2319	2889	3185	3320	3594	3656	3722	3733
group	12	2	7	9	11	10	1	1	10
rank in the group	27	3	26	24	11	13	12	18	28
visual magnitude	4.6	4.9	5.7	4.9	5.3	5.5	5.6	4.9	4.7
double star	yes	.	yes	.	.	.	.	.	.

No atypical effect has been detected in groups Nr 3,4,5,6,8.

4. CORRELATED ERRORS AND HETEROSCEDASTICITY

The next problem we come to is the implicit assumption that in the OLS regression, the variance-covariance matrix of the errors is a <u>scalar matrix</u>; this means that, in particular, the errors ε_i i=1,..n are

(a) mutually uncorrelated,

(b) with common variance σ^2.

Hypothesis (b) is known as homoscedasticity and its violation as <u>heteroscedasticity</u>; (b) is also said, in the case of gaussian errors, as spherical disturbances. When assumptions (a)(b) are violated, the use of conventional OLS formulae to measure the <u>sampling variance of the OLS estimators</u> is potentially <u>misleading</u>; other methods than OLS have to be used to solve the system of the equations of condition.

To test that autocorrelation -first order serial correlation- is zero in a linear regression, the common test is the Durbin-Watson's d statistic. We have shown in Paper II (Sect.3) that zero-autocorrelation can generally be accepted, with although some exceptions, in the case of the usual Paris-Astrolabe group-reductions. Since autocorrelation of the errors indicates some systematic lack of fit of a linear model, the analysis of the data with non-zero error autocorrelation has led us to the detection of possibly omitted variables in the current model. In a first step, the search for significant variables possibly omitted in the linear model -such as Sidereal Time (TS) of passage of the star, visual magnitude (mv) or B-V colour index (BV) of the star- has been performed in Paper I (Sect.3.2) (see also:Bougeard,1987e,h) through factor analysis in principal components. Such an analysis can bring into evidence factors of importance among regressors of a given linear model.

In a second step, the use of stepwise regressions , at a 0.10 F-level for entry and maintenance of a variable in the model, has proved (Papers I,II) that the variables TS,mv,BV could be highly significant in addition to the usual regressors (sin Z, cos Z); neglecting them introduces a source of correlation of the errors. Furthermore , we have proved (Paper II, Table 2,reduction 6607) that the internal corrections, a priori added , could introduce a B-V colour effect, not present in some data of the test-period 1977-79 ; indeed these internal corrections -evaluated in 1964 from observations made with the Paris prism astrolabe (APD)- used for the period 1977-79, are highly correlated, in some groups, with the B-V colour index of the stars (see:Bougeard ,1987 d , h). This surprising result can be interpreted in terms of difference in the instrumental transmission of the APP and the APD.

The last point we come to is the assumption of homoscedasticity. We have proved through classical tests (Paper II, Sect.4) that, in the case of Paris Astrolabe data, heteroscedasticity holds especially in pooling the equations of condition : - over several days for a given group,
- over several different groups,

even if it is the same observer. In such situations, methods other than OLS should be used (at least, iterative Weighted Least Squares with a priori estimates of the variance-covariance of the errors); such a thing seems to have been ignored in all the global astrolabe reductions we known at this time, even when they have been performed through stable numerical methods using the so-called Q-R algorithms for solving the large sparse (global) linear systems.

5. *CONCLUSION*

In this paper, we have been concerned with possible sources of nonhomogeneity in astrometrical data and we have proved that some problems still occur in global or group reductions of astrolabe data so far performed. After having considered in Sect.2 possible significant non-constant effects of observers and groups on the error variances in the reductions of two test-periods (1965-67, Danjon prism astrolabe / 1977-79, full pupil astrolabe), we have been interested in Sect.3 in the detection of potential outliers. Through influence diagnostic statistics, we have seen that some observational errors and"atypical" stars with ill-behaved residuals still remain in the usual OLS group reductions. In Sect.4, the detection of a non-scalar variance-covariance matrix of the errors in the reductions of some data sets has led us to the search of variables missing in the common model, such as sidereal time of passage,visual magnitude and B-V colour index of the star; this point has been performed through factor analysis in principal components and stepwise regressions. Finally, heteroscedasticity (non common variance of the errors) has been proved to significantly occur in pooling astrolabe equations of condition over several days and we suggest to take, at least, into account the differences in the variances in global reductions, in addition (of course) to stable numerical algorithms of resolution such as the so-called Q-R methods.

NOTE: This work was supported in part by the Paris Observatory, CNRS(UA 1125) and the Center C.I.R.C.E (Orsay- France).

6. REFERENCES

Bougeard M.L. 1987a: Astron. Astroph., 173, 1916203 (jan.)

b: Astron. Astroph., 183, 156-166 (sept.)

c:"Nonnormality of the errors and suspicious observations" Astron. Astroph. Supp. (in press)

d:"Paris full pupil astrolabe and Danjon's prism astrolabe:a statistical comparison in terms of time-magnitude-colour equation", Colloquium "Mapping the sky",Paris,(in press)

e:"Are double star observations the most suspicious ones in the Paris-Astrolabe data?",Coll. IAU 97,Bruxelles (press)

f:"Statistical detection and separation of disturbing effects in astrometrical observations",Coll. ICIAM,Paris

g:"Statistical problems with the OLS use in astrometry. I: general problems, II:detection of catalog errors", Coll. IAU 100, Belgrade (sept.)

h:"Time-magnitude-colour equation in visual astrometrical data" (in preparation

Débarbat S. Guinot B. 1970;"La methode des hauteurs égales",Gordon and Breach

Golub G.,Meurant G. 1983:"Résolution numérique des grands systémes linéaires",Eyrolles

Mangombi D.I.J. 1986: thèse de docteur ingénieur, Observatoire de Paris.

AN AUTOMATIC METHOD FOR THE REDUCTION OF LONG-SLIT SPECTRA

M. Bruynooghe , D. Burgarella , A. Llebaria

Laboratoire d'Astronomie Spatiale
Traverse du Siphon, Les Trois Lucs
13012 Marseille , France

Abstract

We present an automatic reduction software of stellar and nebular spectra, which automatically corrects the geometrical and optical distortions of long slit spectra and more specifically of bidimensional photographic image-tube spectra .

Any photographic plate bordered with two comparison spectra can be processed. Our spectral reduction method is fully independent of the spectrograph and of the tube-image characteristics : no a priori knowledge of the different distortions beeing required .

The so called "SPECTRA" software we developped on VAX 11/780 has been included into the ESO's image processing system (MIDAS) . This automatic reduction software can be used in a very simple way. It was successfully applied to plates originating from the Haute Provence Observatory (OHP), the ESO Observatory at La Silla and the Canada-France-Hawaii telescope (CFHT).

I. Introduction.

This work does not deal with the echelle spectrograph reduction that provides high dispersion spectra, since several works have correctly resolved this problem (Moreno and al 1985, Gehren and Ponz 1986, Mc Keith and al 1987).

The goal of the present paper is the reduction of long slit spectra of extended objects and the software package we have developped is optimized for photographic image-tube spectra .

In this case , it is well known that two kinds of distortions characterize spectral images :

- The optical distortions due to the spectrograph : field curvature and non linearity of the dispersion function .
- The so called "S" geometrical distortion , due to the inhomogeneities of the electronical optics (Bijaoui 1984) , in the case of an image-tube or of an electronographic camera . This geometrical distortion limits the useful spectral range . This latter geometrical distortion may be ten times more important than the

optical distortions .

Thus , with respect to the nature and to the extent of the different distortions which are present in spectral images , our automatic reduction method of photographic image-tube spectra is composed of two independent processes :

- A first geometrical distortion correction process .
- And a second optical distortion correction process .

This reduction method of long slit spectra needs not a priori informations about the different distortions . It therefore appears as a tool of general use regarding its independence versus the spectral data acquisition techniques .

II. The Geometrical Distortion Correction Process .

II.1 *The geometry of recorded spectra* .

On each spectral image the spectrum of an astronomical object is surrounded by two comparison spectra both coming from the same calibration lamp , which will be both used to determine the spectrograph dispersion function .

These spectra represent three horizontal stripes above a plate background , each stripe beeing composed of contiguous rows .

Furthermore, each stripe is vertically structured , the spectral lines having a vertical orientation , i .e . perpendicular to the direction of dispersion .

These informations , related to the structure of the spectral image, will be used by the geometrical distortion correction process to automatically identify the spectral lines by a pattern recognition approach (Bruynooghe 1988) .

II.2 *The "S" geometrical distortion correction process* .

The geometrical distortion correction is based on a statistical approach which enables the processing of noisy images. It is composed of three steps, fully automatized.

In the first step, the astronomical spectrum and the two calibration spectra are automatically recognized , using jointly correspondence analysis (Benzecri 1973) and automatic classification of the rows profiles of the spectral image . The results of this classification are interpreted using simple rules for an automatic pattern recognition purpose .

The second step is related to the evaluation of the spectral line positions in the two comparison spectra , using a new statistical edge-points detection algorithm , described in detail in Bruynooghe (1988) . This algorithm is based on the simultaneous utilization of classical edge detection operators (Sobel and local contrast) and of standard statistical tests (Suk and Hong 1984) . The results of these different tests are submitted to a preliminary multiple correspondence analysis . A partitioning algorithm is then applied in order to bipartition the set of pixels in the factorial space , previously generated , and to therefore automatically identify the searched edge points . Then , the different spectral lines can be automatically recognized , by sorting the different segments where edge points are accumulating , after an horizontal projection . With this statistical method of spectral line detection , images with a non-uniform background and a low signal/noise (S/N) ratio may be efficiently processed .

Finally , the third step modelizes the geometrical distortion of the two calibration spectra and thus determines the vertical correction to be applied at each pixel . Bidimensional spectrographic plates can then be automatically corrected of the "S" geometrical distortion .

In this last step , vertical offsets of the different spectral lines have been independently estimated using a standard correlation technique , the "S" geometrical distortion beeing then modelized by an orthogonal polynomial of order three or five .

The geometrical correction process , described in this paragraph , presents an obvious interest, when spectral photographic images are acquired from a spectrograph equipped with an image tube or an electronographic camera . In the case of spectral images collected with a spectrograph equipped with a CCD camera , like in the CFHT plates , no correction of the "S" geometrical distortion is needed . Nethertheless , the automatic detection of spectral lines remains of interest and may be used for the next step of field curvature correction and of wavelength conversion .

III. The Optical Distortion Correction Process .

Corrections of the field curvature and of the non linearity of the dispersion function are now to be considered .

Three steps are to be used by the row by row approach suggested by Pelat (1986) and implemented by Burgarella (1987) .

In the first step , spectral lines of the calibration spectrum are automatically identified using the algorithm of Pelat (1986) , the different rows of the spectral image beeing independently processed . In the case of calibration spectra containing a great number of rows , averaging groups of 5 to 7 contiguous rows for example , may increase the S/N ratio before the identification of spectral lines . After this preliminary row by row processing , positions of the identified lines are extended to the whole bidimensional spectral image by fitting orthogonal polynomials (Peterson 1979) modelizing thus the field curvature of the spectral lines .

The second step estimates the parameters of the non linear dispersion function , row by row , using once more the Peterson's method . Thus , the dispersion function of each spectral image row is modelized by an orthogonal polynomial , whose order is automatically determined using a standard F-test .

The third step is a wavelength conversion . A cubic interpolation function (Keys 1981) is used for the wavelength re-sampling .

In this row by row approach the different monodimensional fittings are independently performed , spectral line by spectral line . The accuracy of such an approach may be somewhat limited when the number of pixels related to each spectral line is too small .

Thus , a bidimensional approach , based on the simultaneous modelization of the field curvature and of the dispersion function , would provide more statistically significant results , beeing based on the analysis of a larger number of data . Nethertheless , at the 0.2 pixel level of precision which may be obtained , the interest of such a bidimensional approach remains to be proved , due to the relative importance of possible local defaults .

IV. Experimental Results.

In order to test the efficiency of the previously described method we applied it to spectra from three observatories :

- The D spectrograph of the 1m93 OHP telescope .
- The Boller and Chivens spectrograph of the ESO (Chile) where several subdwarf spectra were obtained .
- Observations of two interactive galaxies at the spectrograph Casshawec of the CFHT were recorded by a CCD camera.

The annexed table summarizes the characteristics of the three spectra used to test the reduction method presented in this paper .

This reduction method is versatile , considering the wide variety of spectra reduced , giving high quality results in the case of the ESO and CFHT samples .

V. Conclusion

Our software package for automatic reduction of long slit spectra was optimized for photographic image-tube spectra. Satisfactory experimental results has been obtained with this first version .

The adaptation of this first software to CCD spectra from BC , CASPEC and ESOFOSC will be the subject of future developments . The related package will be implemented into MIDAS as a general spectral reduction facility .

Acknowledgements.

Thanks are due to our friends and colleagues who encouraged this work and suggested examples for us to consider ; amoung them are M.Viton , T.Hua , A.Vuillemin from Spatial Astronomy Laboratory and Pr. J.P.Benzecri from Paris VI University .

References

Benzecri J.P. : 1973, *"L'analyse des données"*, Tome 1, *"La taxinomie numérique"*; Tome 2, *"L'analyse des correspondances"*, Paris, Dunod

Bevington, P.,R. : 1969, *"Data Reduction and Errors Analysis for the Physical Sciences"* Mc-Graw-Hill Book Company, New York

Bijaoui, A., Marshall, J., Ounas, C. : 1978, *Astronomy and Astrophysics*, **65**, 71

Bruynooghe, M. : 1978, *"Hierarchical Clustering of large data sets and geographical pattern recognition"*, The Fourth International Conference on Pattern Recognition, Kyoto, Japan

Bruynooghe, M. : 1988, Thése de Doctorat d'Etat es sciences mathématiques, Paris VI University (in preparation) .

Burgarella, D. : 1987, *"Automatic Corrections of Bidimensional Spectra Distorsions"* Laboratoire d'Astronomie Spatiale, Marseille (Thèse de Doctorat).

Gehren, T., Ponz, D. : 1986, *Astronomy and Astrophysics*, **168**, 386

Keys, R. G., : 1981, *IEEE trans. on Acous., Speech, and Signal Proc.*, vol. ASSP-29, **6**, 1153

Moreno, V., Llorentes de Andres, F., Jimenez, J. : 1982, *Astronomy and Astrophysics*, **111**, 260

Mc Keith, C. D., Bates, B.,Catney,E., Jorden, P. ,R., van Breda, I., G. : 1987, *Astronomy and Astrophysics*, **173**, 204

Pelat D. : 1986, Private Communication.
Peterson, D. N., : 1979, *Publications. of the Astron. Soc. of the Pacific*, **91**, 546
Peterson, R. C., Title, A. M., : 1975, *Applied Optics*, **14**, 2527
Pratt, W.,K. : 1978, "*Digital Image Processing*" John Wiley and sons, New York
Rossi, C., Lombardi, R., Gaudenzi, S., De Biase, G.A. : 1985, *Astronomy and Astrophysics*, **143**, 14
Spite, M. : 1980, *Astronomy and Astrophysics*, **81**, 365
Suk, M., Hong, S. : 1984, *Comp. Vision., Graph., and Image Proc.*, **25**,24

table 1 : Characteristics of the spectra used to test the SPECTRA chain.

instruments	1m93 OHP D spectrograph	1m52 ESO Boller Chivens spectrograph	3m60 CFHT spectrograph Casshawec
observers	R. Augarde	M. Viton	C. T. Hua & B. Lelievre
receptor	IIaO plate	IIIaJ plate	CCD RCA
intensificator	tube 33063 BP2	tube EMI	none
studied object	IC 4593	HD 39194	Markarian 131
object type	planetary nebulae	subdwarf	interacting galaxies
spectral range	3600 - 5300 A	3500 - 4700 A	6300 - 6900 A
dispersion	92.6 A/mm	39.3 A/mm	30 A/mm
comparison lamp	Fer - Neon	Helium - Argon	Fer - Neon
slit of the microdensit.	20*50 microns	16*40 microns	none
Size of the image (pixel)	90 * 1000	93 * 2048	320 * 512
number of comparison rows	20	20	300
distortion correction	7 min	15 min	0
conversion	6 min	15 min	1 h 55 min
Accuracy of the vertical correction	0.3 pixel	0.3 pixel	
Mean precision of the conversion on one line	0.72 angtroem	0.06 angtroem	0.09 ansgtroem

figure 1: Bidimensional spectra showing "S" distorsion before correction

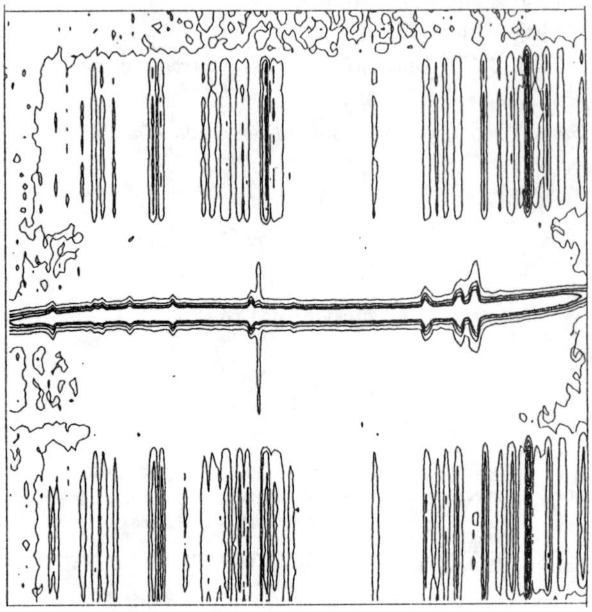

figure 2: Bidimensional spectra after correction

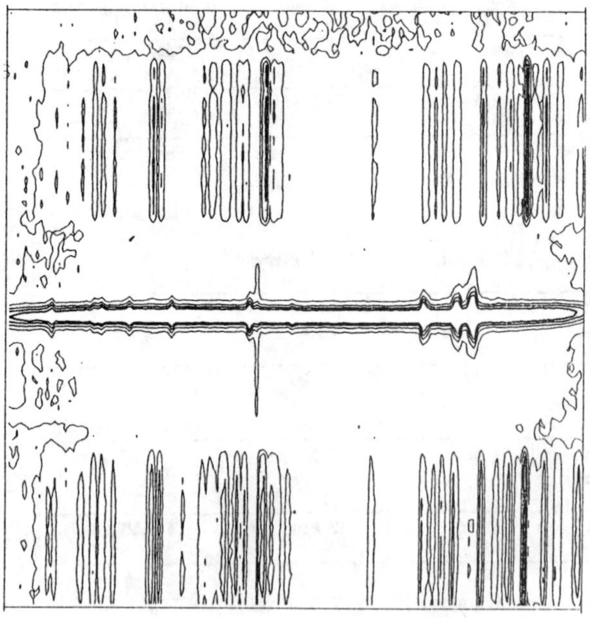

AUTOMATED QUASAR DETECTION (AQD)

Roger G Clowes
Royal Observatory, Blackford Hill, Edinburgh EH9 3HJ, Scotland

Abstract

AQD, for Automated Quasar Detection, is the generic name for the system of software and procedures at ROE that allows quasar candidates to be discovered from measurements of objective-prism plates. It was developed because of the great advantages of replacing the original visual searches by an automated procedure. The visual searches had many defects, including unknown and varying selection criteria, wastage of information, and tedium, all of which may be overcome by automation.

The databases that AQD searches for quasars are among the largest used in astronomy, the largest so far for AQD being ~ 312 Mbytes for ~ 150,000 objects from a single Schmidt plate. Each record of such a database corresponds to one object, and contains a set of identifiers and an objective-prism spectrum.

This paper describes the present version of AQD and the plans for its further development. The plans have two major components, which are intended to lead to purer samples of quasars and to generalisation to other classes of objects. The first component is the interpretation of selected objects by an expert system, and the second is a restructuring of the AQD processes as a set of specialised and routine database operations.

1. Introduction

AQD, for Automated Quasar Detection, is the generic name for the system of software and procedures at ROE (Royal Observatory, Edinburgh) that allows quasar candidates to be discovered from measurements of objective-prism (or similar) plates. It is based on the COSMOS fast plate-measuring machine (MacGillivray & Stobie 1984) at ROE, and typically uses plates from the UK Schmidt Telescope (UKST).

AQD was developed because of the great advantages for statistical work on quasars of replacing the original visual searches by an automated procedure. The visual searches had many defects, including unknown and time-varying selection criteria, wastage of information, and tedium, all of which could be overcome by automation. For detailed descriptions of the initial and present versions see Clowes, Cooke & Beard (1984) and Clowes (1986) respectively.

Typically the quasars are selected by one or both of emission lines and ultraviolet excess, and other options are also available for special cases. Any of these options can, of course, be used to select classes of objects other than quasars.

A few simple statistics readily illustrate that AQD is one of the most efficient ways of finding quasars in large numbers over large areas of sky. The Hewitt & Burbidge (1987) compilation lists 3681 quasars that have been published in a 23 year period since the first quasars were discovered. In a period of just eight months the ROE/ESO large-scale AQD survey (Clowes, Iovino & Shaver 1987) yielded ~ 1500 high-grade candidates from the five best plates (~140 deg^2); these are candidates for which the success rate is ~ 80%. Including other surveyed fields the total number of high-grade candidates is ~ 3000. Additionally ~ 3000 quasars are expected to be present in the lower grades, for which confirming observations will be required.

Quasars selected by AQD are being used in a wide range of applications. The following list gives some examples, the items in square brackets being the features of AQD which are of particular importance: (i) clustering of quasars with redshifts in the range 1.8–2.4 (Clowes, Iovino & Shaver 1987) [large samples, large-area coverage, objective-prism redshifts]; (ii) isotropy of the universe [large samples, large-area coverage]; (iii) common and associated absorption in close pairs on the sky [large samples].

2. AQD – the present version

This section gives a brief description of the present version of AQD. Many complexities have been omitted, but full details can be found in Clowes (1986).

2.1 The databases

The emphasis here is on the databases for AQD. For input, there are two, formed from measurements of direct and objective-prism plate-pairs. The direct-plate database, which may be regarded as the master, contains a set of parameters (eg x, y, RA, Dec, magnitude, ellipticity etc.) for each recognised image. The prism-plate database, which may be regarded as dependent, contains the corresponding spectra. This latter database also repeats a set of identifiers (eg object number, RA, Dec) from the direct-plate database so that a link between the two can always be established.

The prism-plate databases can be very large, the largest so far being ~ 312 Mbytes for ~ 150,000 objects from a single prism plate. Each record corresponds to one object, and most of the bytes are for the spectrum. The spectrum is intrinsically two-dimensional but is unravelled into a one-dimensional array for inclusion in the database.

AQD can test the spectra for one or more of the following features:
(i) emission lines, (ii) absorption lines, (iii) continuum discontinuities, (iv) ultraviolet excess, and (v) red excess. Emission lines and ultraviolet excess are the most productive for quasars, and the other options are intended for special cases – BAL and high-redshift quasars in particular. Any of these options can, of course, be used to select classes of objects other than quasars.

The output from AQD for the selected objects comprises a new master database and two new dependent databases. The master contains parameters from the direct plate and new parameters calculated from the spectra (wavelengths of emission lines, equivalent widths, colours etc). The dependent databases contain the two-dimensional spectra and a set of one-dimensional-spectra (intensity-converted, continuum, and noise). As before, all databases can be linked by identifiers.

2.2 Interpretation of the output database

At this stage objects have been selected very simply by features that exceed specified limits. The selection is therefore objective, but needs further interpretation for several reasons. The goal of this interpretation is to assign a number to each selected object which is ideally the probability that the object is a quasar, but, if that is not possible, is a probability index which, it is hoped, increases as the probability.

A variety of objects may be present in the output database. Some are obvious quasars, for example, with very strong emission lines, which generally need no further confirmation. Some rather less obvious quasars, faint with ultraviolet excess, for example, may appear indistinguishable from the white dwarfs which are also present. There are non-quasars that genuinely satisfy the selection criteria – intergalactic HII regions and white dwarfs. There are non-quasars that satisfy the selection criteria because of limitations of the data – the continuum between absorption troughs in F stars can mimic emission lines. And of course there are non-quasars that are entirely spurious detections caused by specks of dirt, emulsion holes etc. The need for further interpretation should now be clear.

Simple interpretation of the parameters in the output database is remarkably successful, and, as will be described in the following section, a more sophisticated interpretation should be still more successful.

In the present version of AQD the quasar candidates are interpreted or graded using only conventional database operations. Probabilities are unknown, and so a probability index is produced by a scheme for assigning points to a candidate – positive for quasar-like properties and negative for non-quasar-like properties; the index is the sum of the points. Thus, emission and ultraviolet excess contribute postive points, and the contribution for emission is further increased according to the strength of the strongest line. Classification as a galaxy or the characteristics of F stars result in negative contributions.

High-grade quasar candidates are defined as those for which the probability index exceeds the value, H, that gives a success rate of $\sim$ 80%. Note that the surface densities from just these high-grade subsets typically exceed those from the visual searches. The low-grade candidates are collectively those for which the probability index is less than H: confirming observations are generally required for these. Some examples of high-grade candidates are shown in Figure 1. Each plot is of relative intensity against pixel number, and also shows the fitted continuum superimposed. Increasing pixel number corresponds to decreasing wavelength, but the correspondence is not linear.

Figure 1. Examples of high-grade quasar candidates.

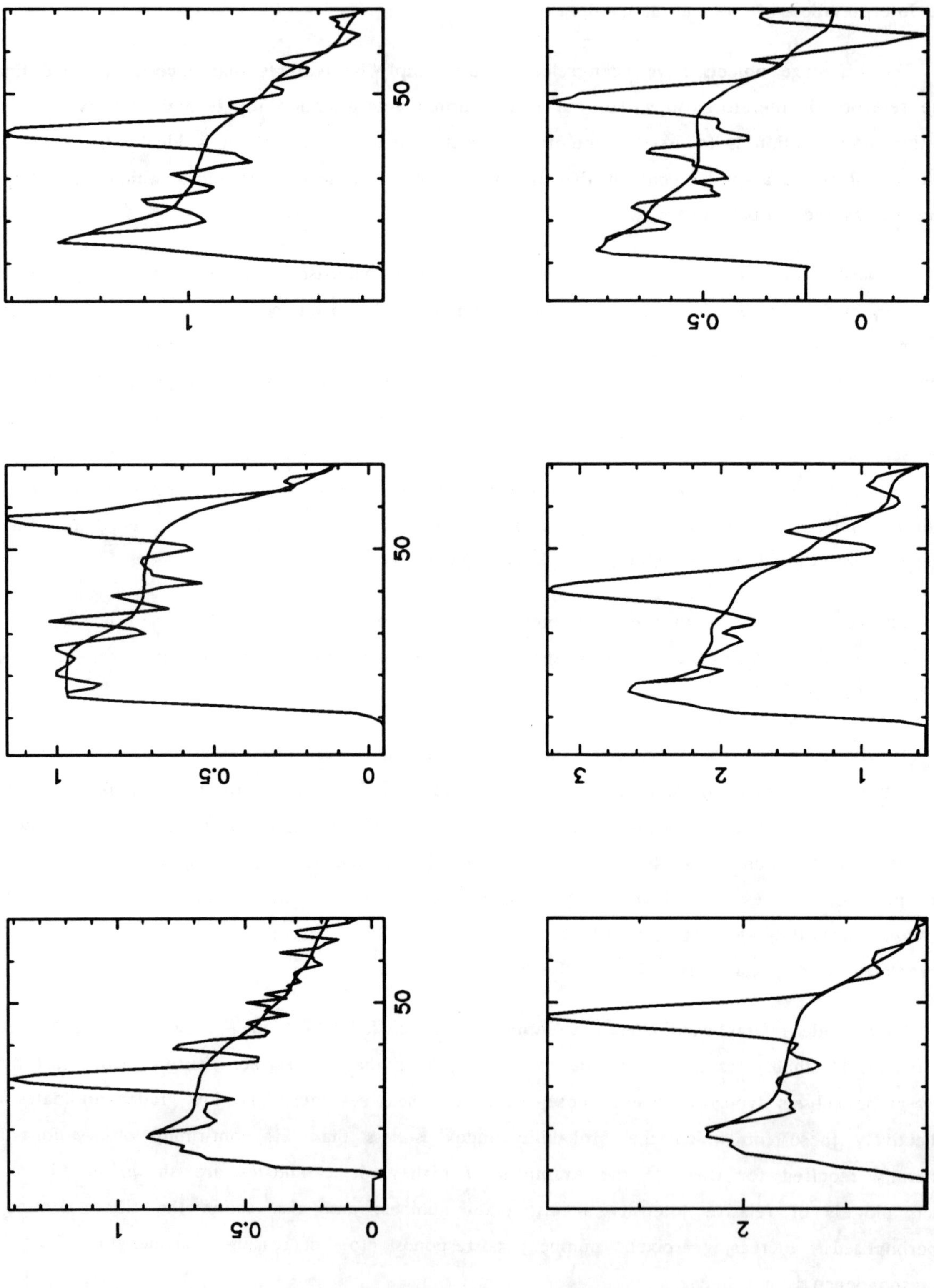

3. AQD - plans for further development

This section describes the plans for the further development of AQD. The plans have two major components, which are intended to lead to purer samples of quasars and to generalisation to other classes of objects. The first component is the interpretation of selected objects by an expert system, and the second is a restructuring of the AQD processes as a set of specialised and routine database operations.

Both components have now become feasible because of external factors - (i) inexpensive AI software and support from the Artificial Intelligence Applications Institute of the University of Edinburgh, and (ii) the availability of the powerful STARLINK package, SCAR, for processing astronomical databases.

3.1 Interpretation using an expert system

The grading scheme described in section 2.2 was inspired by the expert system approach used in artificial intelligence (AI): each point that is assigned depends on the application of rules. The limitation of this scheme is that the rules must be simple or else the database operations become too complicated and unwieldy.

Work is now beginning on the replacement of the grading scheme by a proper AI expert system, "Blackboard Shell", which is written in Edinburgh Prolog, and supplied by the Artificial Intelligence Applications Institute of the University of Edinburgh. One particularly impressive feature of Blackboard Shell is its great flexibility, which is due to the coexistence of the shell and Prolog. This means, for example, that the method for propagating certainty factors can be specified by the builder of the knowledge base, rather than being imposed by the shell.

The use of an expert system is, of course, experimental. The minimum requirement is that the success of the grading scheme is continued. The anticipated enhancement is an expansion of the knowledge base, made possible by the ease of introducing both more and more complicated rules, such that the probability index more accurately reflects the probability of a candidate being a quasar. This enhancement is particular important for the low-grade candidates, where quasars are presently being lost unnecessarily.

The expert system shell is not committed to quasars only. A different knowledge base, together with the proposed developments that are described in section 3.2, could be used for other classes of objects.

3.2 Restructuring AQD for database operations

The present version of AQD is a mixture of conventional database operations and special-purpose programs, the latter treating the databases as ordinary files. In principle all of the AQD processes could be restructured as a set of specialised and routine database operations. Most advantageous would be the restructuring of the processes that process the spectra in the prism-plate database: this would lead to the flexibility to easily construct whatever database operations might be required, and the generality for applications other than quasars.

Until recently, with the distribution of the STARLINK package, SCAR, there was no possibility of processing spectra with a database package. In fact, it is not yet possible with SCAR, but investigations by Dr B.D.Kelly at ROE have shown that only minor modifications are needed to allow applications for spectra. Spectrum operations should therefore be introduced into SCAR soon.

Examples of the flexibility and generality include: (i) non-standard selection operations - eg construction of a new parameter and selection on it, (ii) the use of applications for non-quasar projects - eg galaxy redshifts (Cooke et al. 1986), and IRAS spectra (iii) ease of connection to existing software for multivariate data analysis (eg Murtagh & Heck 1987), and (iv) ease of connection to an expert system for interpretation of the data.

REFERENCES

Clowes,R.G., Cooke,J.A. & Beard,S.M., 1984. Mon. Not. R. astr. Soc., 207, 99.

Clowes,R.G., 1986. Mitteilungen der Astron. Ges., 67, 174.

Clowes,R.G., Iovino,A. & Shaver,P.A., 1987. Mon. Not. R. astr. Soc., 227, 921.

Cooke,J.A., Beard,S.M., Emerson,D., Kelly,B.D. & MacGillivray,H.T., 1986. Mon. Not. R. astr. Soc., 219, 241.

Hewitt,A. & Burbidge,G., 1987. Astrophys. J. Suppl., 63, 1.

MacGillivray,H.T. & Stobie,R.S., 1984. Vistas Astr., 27, 433.

Murtagh,F. & Heck,A., 1987. "Multivariate Data Analysis", Reidel.

THE APPLICATION OF A MAXIMUM LIKELIHOOD ANALYSIS TO DETECTION OF SOURCES IN THE ROSAT DATA BASE.

R.G.Cruddace(NRL),G.R.Hasinger(MPE) and J.H.Schmitt(MPE)

1. Introduction

During the ROSAT mission a large X-ray telescope (0.1-2 kev) will perform a six-month long all-sky survey, followed by at least twelve months of pointed observations. We estimate that the number of sources detectable by ROSAT is of order 100,000. A maximum likelihood (ML) algorithm has been developed for detecting sources in the data, in the presence of variable background produced by particles, cosmic X-rays and scattered solar X-rays. The ROSAT scientific data are basically a string of x-ray events characterised by the photon energy, arrival time, and position. The starting point for the algorithm is a provisional list of sources, identified by search procedures which look for regions of high signal-to-noise ratio. The algorithm examines each candidate and determines its position, its intensity expressed in counts per second, and if it is an extended source its angular diameter. Statistical criteria are used to reject unlikely candidates, and to reject unsatisfactory measurements of angular extent.

In the survey mode the telescope field sweeps continuously across the sky, making the task of the algorithm more difficult in two respects. First, the telescope collection efficiency and point-spread function with respect to this source are changing continuously. Second, the background will vary significantly, not just because the sources of the background are variable but also because the background count-rate is not distributed uniformly over the detector field. The algorithm takes these factors into account.

2. Construction of the Likelihood Function.

We consider a field of N "sky" pixels at the positions x_i, y_i (i=1,2,.....,N). A pixel records n_i events, which may be a mixture of background and X-ray source events. We assume that the field contains one source only, but the results are applicable to fields containing numerous sources, because the preliminary search procedures break the field down into small regions, in each of which a candidate has been isolated. Regions containing several confused sources require special techniques. The detector field is divided into J concentric annular rings. Photons recorded in such a ring correspond to a particular off-axis angle of the source. During the observation the source spends a fraction η_j of the time in ring j(=1,2,....,J), so that,

$$\sum_{j=1}^{J} \eta_j = 1 \tag{1}$$

We define ϵ_j as the ratio of the telescope efficiency at the off-axis angle of the ring j to the on-axis efficiency. Thus the expectation value for the source count λ_s is,

$$\lambda_s = \lambda_{so} \sum_{j=1}^{J} \eta_j \epsilon_j, \tag{2}$$

where λ_{so} is the count expected if the source remained on-axis.

The probability p_{ik} that the kth event is recorded in pixel i is,

$$p_{ik} = \frac{\lambda_{Bj} P_{Bik} + \lambda_{so} \eta_j \epsilon_j P_{Sik}}{\lambda_{Bj} + \lambda_{so} \eta_j \epsilon_j}, \tag{3}$$

where j is the detector ring associated with event k (=1,2,..,M), λ_{Bj} is the expectation value for the background count in ring j, P_{Bik} is the probability that event k is a background event in pixel i, and P_{Sik} is the probability that event k comes from a source located at position x_o, y_o.P$_{Sik}$ is derived using the instrument point-spread function. For simplicity we identify from now on the probabilities P_B and P_S with the index k only, assuming that information about the event location in detector and celestial co-ordinates is stored for each k. The maximum likelihood technique can incorporate any point-spread function for the telescope and detector. We assume here that this function is a two-dimensional Gaussian, so that,

$$P_{Sik} = \frac{1}{2\pi\sigma^2(R_k, E_k)} exp(-\frac{r_{ko}^2}{2\sigma^2(R_k, E_k)}), \tag{4}$$

where $r_{ko}^2 = (x_k - x_o)^2 + (y_k - y_o)^2$ and x_o,y_o are the source co-ordinates. The variance σ^2 depends both on the distance R_k of the event from the optical axis, and the photon energy E_k. In addition, when the source is extended a model of the surface brightness distribution must be convolved with the point-spread function, in order to estimate the probability P_{Sk}. We examine a simple case, where the brightness has circular symmetry and the radial distribution is Gaussian, with a variance ${\sigma_e}^2$. We make the substitution, $\nu = 2\sigma^2$. The overall distribution is therefore Gaussian and $\nu_k^* = \nu(R_k, E_k) + \nu_e$.

The background terms λ_{Bj} and P_{Bk} are obtained from an independent analysis and modelling of the detector background, which establish the count-rate per detector pixel as a function of time and position in the detector field. The probability, P_{Bk}, may be expressed approximately in the form,

$$P_{Bk} = \frac{1}{A_j}\frac{\delta t_i}{\tau} f(t) \tag{5}$$

where A_j is the area of annulus j measured in sky pixels, τ is the observation time, δt_i is the time pixel i spends in annulus j, and $f(t)$ is the ratio of the instantaneous total detector count-rate to the average count-rate. This is an approximation, because the sum of the three background components is not separable into a product of a time-dependent and a spatially-dependent function.

Consider a subset, n_j, of the M events, namely those recorded in detector ring j. They comprise an array of n_{ij} events(i=1,2,.....N) , where $n_j = \sum_{i=1}^{N} n_{ij}$. The probability of this configuration, P_j, is given by,

$$P_j = p_j \frac{1}{\prod_{i=1}^{N} n_{ij}!} \frac{\prod_{k=1}^{n_j}(\lambda_{Bj} P_{Bk} + \lambda_{so}\eta_j\epsilon_j P_{Sk})}{(\lambda_{Bj} + \lambda_{so}\eta_j\epsilon_j)^{n_j}}, \tag{6}$$

where p_j is the probability of recording n_j events in ring j, when the expectation value is $\lambda_{Bj} + \lambda_{so}\eta_j\epsilon_j$. Taking the product over all rings we obtain the likelihood, $L = -2log_eP$, in the form,

$$L = 2\sum_{j=1}^{J}(\lambda_{Bj} + \lambda_{so}\eta_j\epsilon_j) - 2\sum_{k=1}^{M} log_e(\eta_j \frac{\lambda_{Bj}}{A_j} f(t) + \lambda_{so}\eta_j\epsilon_j P_{Sk}) + constant \tag{7}$$

3. Determination of the Source Characteristics.

Differentiating equation (7) with respect to the x-coordinate of the source position, x_0, and using the point-spread function in equation (4), we obtain upon setting the derivative to zero,

$$x_o = \frac{\sum_{k=1}^{M} \alpha_k x_k}{\sum_{k=1}^{M} \alpha_k}, \tag{8}$$

where,

$$\alpha_k = \frac{\epsilon_j P_{Sk}}{\nu_k^*(\frac{\lambda_{Bj}}{A_j} f(t) + \lambda_{so}\epsilon_j P_{Sk})} \tag{9}$$

An iterative calculation is performed to solve equation (9) for x_0, and a similar calculation yields the other co-ordinate, y_0.

In order to obtain λ_{so}, the source count-rate on axis, and the source extent, measured by ν_e, we obtain the derivatives of L in equation (7) with respect to these parameters, and then perform an iterative calculation to find a minimum of L.

In order to obtain estimates of the probable errors, we wish to establish confidence intervals for the parameters x_o, y_o, λ_{so} and ν_e, such that,

$$L(x_o, y_o, \lambda_{so}, \nu_e) - L_{min} = A \tag{10}$$

where A is the desired confidence level and L_{min} is the likelihood of the best-fitting solution. An error estimation, in which the parameters (r_l) were altered systematically in order to deflect L from L_{min}, would be prohibitively expensive in computer time. Instead we use the technique proposed by Strong (1985), in which L is expanded in a Taylor series about L_{min}, and the errors may be obtained from the eigenvalues, λ_m, and the matrix of eigenvectors, Q_{lm}, of the "information matrix", H_{lm}, where,

$$H_{lm} = \frac{\partial^2 L}{\partial r_l \partial r_m} \tag{11}$$

The dependence of an error on all the parameters may be taken into account, and in the case where it is calculated allowing all other parameters to vary freely the errors δ_l in r_l may be expressed as,

$$(\delta_l)^2 = \delta L \sum_m \frac{Q_{lm}^2}{\lambda_m} \tag{12}$$

where δL is the specified confidence interval (for a Gaussian distribution a 1σ error corresponds to $\delta L = 1$). In our algorithm the errors in x_o, y_o and λ_{so} are estimated in this manner. As only a small fraction of the ROSAT sources will have detectable angular extent, in order to economise on computing time the error in the extent is estimated assuming no dependence on the other parameters.

The calculation of the source characteristics is performed in stages, each terminated when the fractional change in likelihood falls below a set threshold, usually set at $\sim 10^{-5}$. The first obtains the position x_o, y_o under the assumption that λ_{so} and ν_e are zero. The second maintains the condition $\nu_e = 0$, and using the calculated position as a starting point we repeat the position and flux calculations alternately, until again the fractional change in likelihood falls below the threshold. At this point the derivative of L with respect to ν_e is calculated, and if it is positive then the source is judged to be a point source and the results are stored. If the derivative is negative then a third stage starts, in which position, flux and extent are estimated in sequence, until for the last time the fractional change in likelihood falls below the threshold.

Two likelihoods are calculated. The first is an overall likelihood, which must lie above a threshold if the solution found by the algorithm is to be accepted as a real source. The second likelihood is used to judge whether the result for an extended source is a significant improvement in comparison with a point source. The likelihood threshold is established during detailed simulations of telescope observations and has a value of ~ 10.

4. Testing the Algorithm.

The algorithm has been tested in complete simulations of ROSAT observations. A catalog of sources is created using known luminosity functions, and then the resulting "sky" is observed with the telescope in either the survey or the pointed observation mode. Background is introduced into the observations, taking into account orbital variations of the background. The resulting data, stored in the ROSAT telemetry format, is then treated just like flight data, and is analysed using the search procedures mentioned earlier. The last step uses the maximum-likelihood algorithm. Such simulations have been performed successfully, using catalogs of stars, active galactic nuclei and clusters of galaxies. Due to space limitations we defer discussion of the results to a paper being prepared for publication. Instead we present a few results of simpler and faster tests of the algorithm, in which the field contains only one source. Simulation of the source, background and telescope characteristics is complete, but the preliminary search procedures are not invoked and the data are not put in telemetry format.

In Figures 1-4 we show results obtained using a source with a power-law spectrum ($\alpha = 1.7, N_H = 3.10^{20} cm^{-2}$), observed in the survey mode. The survey field has a radius of 45 arc min and the total exposure time is 675 sec. The particle and cosmic xray background components are 3 and 9 counts/sec respectively. All results are the averages of 100 search attempts. The ability of the algorithm to correct for the variations in sensitivity as the source moves across the field of the xray optics is demonstrated in Figure 1. At low fluxes the detection rate falls, because the higher fluctuations in the count dominate in the detected sample, and so the flux is overestimated. In this case the detection efficiency reaches 50 percent when the expected count is 11 and 99 percent at a count of 32. Figure 2 shows the dependence of the position error on the source strength, and Figures 3 and

4 compare the average measured deviations from the expected position and flux with the estimated errors. At fluxes sufficient to yield a significant detection rate, the estimated errors are reliable.

Reference

Strong, A. W. 1985, Astronomy and Astrophysics, vol. 150, p. 273.

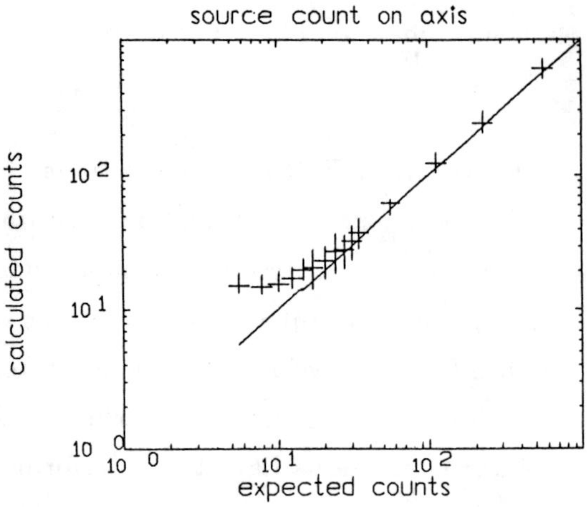

Figure 1

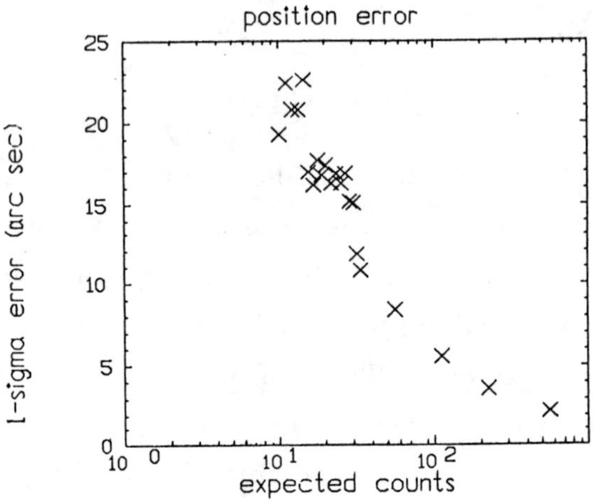

Figure 2

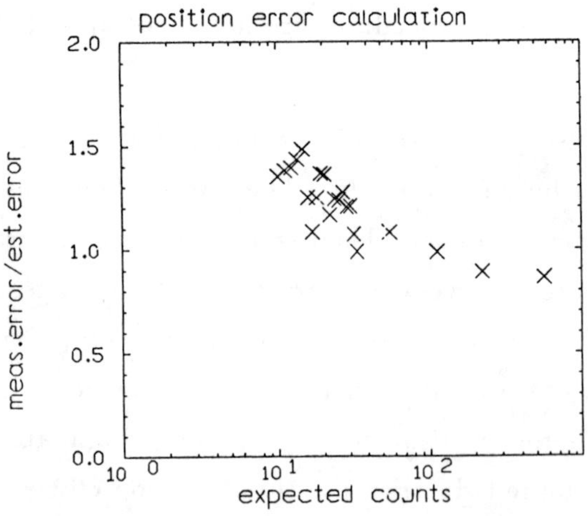

Figure 3

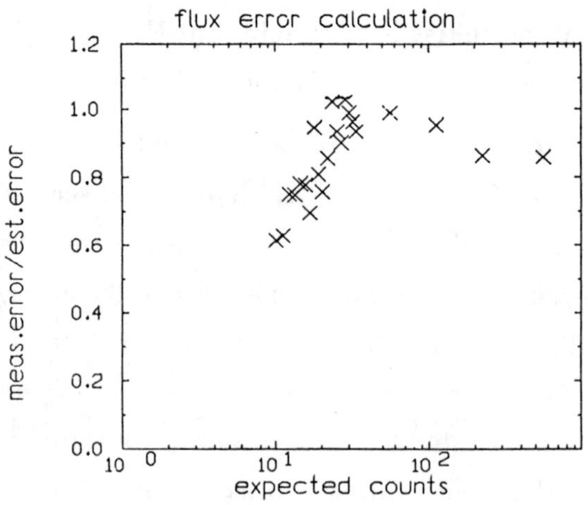

Figure 4

FUZZY SETS AND DATA ANALYSIS

Vito di Gesu'

Dip. di Matematica ed Applicazioni, Univ.di Palermo, Palermo, ITALY
and
I.F.C.A.I./C.N.R., Palermo, ITALY

ABSTRACT

Aim of the paper is to give an overview of the theory of *fuzzy-set*. Methods and their applications to data analysis problems are presented. *Fuzzy-sets* theory seems to be a powerful tool to model uncertainty and vagueness present in the data and to represent the human thinking in a more natural way.

Key words: *fuzzy-sets, data analysis, image processing, clustering.*

1. INTRODUCTION

Sciences as physics, engineering, chemistry built exact models of phenomena, by starting from experimental data, and use these models to make predictions. A very crucial point of this process is, of course, the correct analysis and evaluation of the experimental data. The models are mainly based on the algebra and the probability theory.

Data analysts are also using mathematics and probability. However the *human experience* may plays a foundamental role whenever inexactness and vagueness are inside the data. Moreover *cloudy* quantities exist which are not describable in terms of probability distributions, as the *beauty of some thing* or the *tallness of a man.* In all these cases, education, fashon and global knowledge may play a crucial role in making decisions.

In 1962 L.Zadeh (Zadeh 1962) introduced the concept of *fuzzy-set,* as an extention of the classic ones. In the theory of *fuzzy-set* the characteristic function, χ_A, of a subset **A** of a universe **X** range in the interval [0,1] instead in the set {0,1}.

Given an element $x \varepsilon \mathbf{X}$, $\chi_A(x)$ is also said the belonging degree of x to the *fuzzy-set* **A**. The universe **X** could represent the measured data and χ_A could be an evaluation of them respect to the hypothesis **A**.

Since *fuzzy-sets* have been introduced, their mathematical foundation has been widely developed. Measure theory has been applied in order to extend on them topological spaces (Katsaras 1981). An extension of the information theory has been studied (De Luca and Termini 1972, Dimitrov 1983). *Possibility* theory has been developed in connection with the probability theory (Yager 1980).

Fruitful applications have been made in the fields of the artificial intelligence (Negoita 1985), of the image analisys and processing (Dutta Majumder 1980), of the linguistic analysis (Yager 1980) and multivalent logic (Kandel 1976, Zadeh 1985).

In particular their application to image-analysis seems to be promising, if the data are *sparse* and/or the probability model is not available or difficult to issue (Di Gesu' 1986).

For example the hinner analysis (structure, morphology) of faint and extended celestial objects could make profict from a *fuzzy set* approach whenever is hard to answer questions like:

*which is the probability that the shape of the object **X** is a spiral with two arms and a kernel of size d?*

or

which are the most relevant parameters in order to represent possible clusters of radiopulsar? (the number of parameters ~10, number of samples 336).

Fuzzy-sets could be a good basis in order to handle such kind of problems and the combination of probability and possibility theory should be used in order to analyse data, as a matter of fact that complementary information come out from both sides.

Section 2 introduces some basic definitions. Section 3 gives an overview of the main fields, where *fuzzy-sets* have been succesfully applied. Some remarks on the use of *fuzzy-sets* in the analysis of astronomical data are made in Section 4. Concluding remarks are made in Section 5.

2. BASIC DEFINITIONS

Fuzzy-sets may be formally defined in the framework of the *measure theory* .

DEF 1. Given a subset **A** of a universe **X** and a function χ:**X**->[0,1] then the set of ordered pair (x, $\chi_A(x)$) is called a *fuzzy-set* of **X**, in symbols:

$$\mathbf{A} = \{x \mid (x, \chi_A(x)),\ x \in \mathbf{X}\}$$

$\chi_A(x)$ is termed the grade of membership of x to **A**. The elements with grade 1 are called the *ideal members* of **A**, those with grade 0 are the *not-members* of **A**.

The interpretation and the values of χ depends on the a priori knowledge of the system under study (physical model, subjective thinking) and on the goals of the analysis (data transformations, classification, presentation). Figure 1 shows two *fuzzy-sets* defined on the real numbers.

Set operations may be, also, defined on the fuzzy sets. Let **A** and **B** be two fuzzy sets, with membership functions $\chi_{\mathbf{A}}$ and $\chi_{\mathbf{B}}$ respectively, then:

DEF 2. Union of **A** and **B**:

$$\chi_{\mathbf{A} \cup \mathbf{B}}(x) = \max\{\chi_{\mathbf{A}}(x), \chi_{\mathbf{B}}(x)\}$$

DEF 3. Intersection of **A** and **B**:

$$\chi_{\mathbf{A} \cap \mathbf{B}}(x) = \min\{\chi_{\mathbf{A}}(x), \chi_{\mathbf{B}}(x)\}$$

DEF 4. Complement of **A**:

$$\chi_{\bar{\mathbf{A}}}(x) = 1 - \chi_{\mathbf{A}}(x)$$

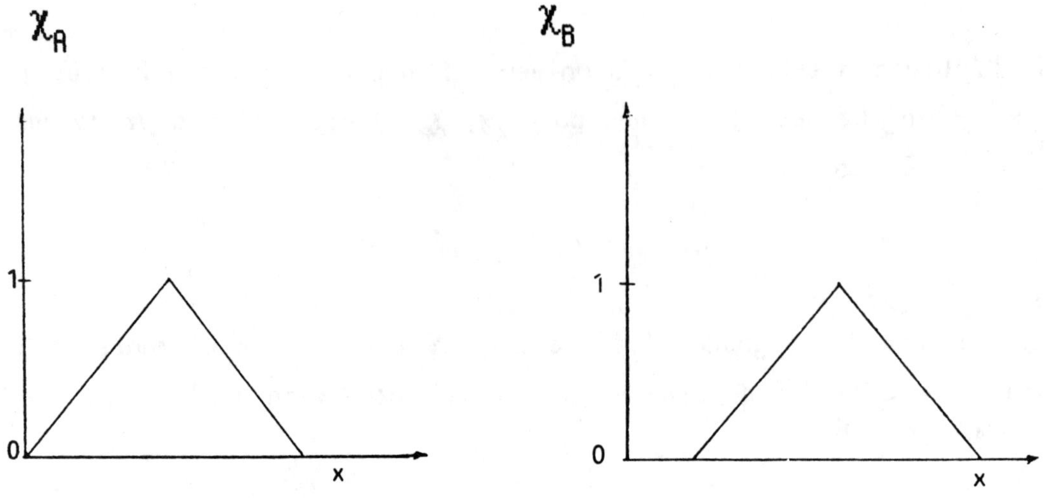

Fig.1

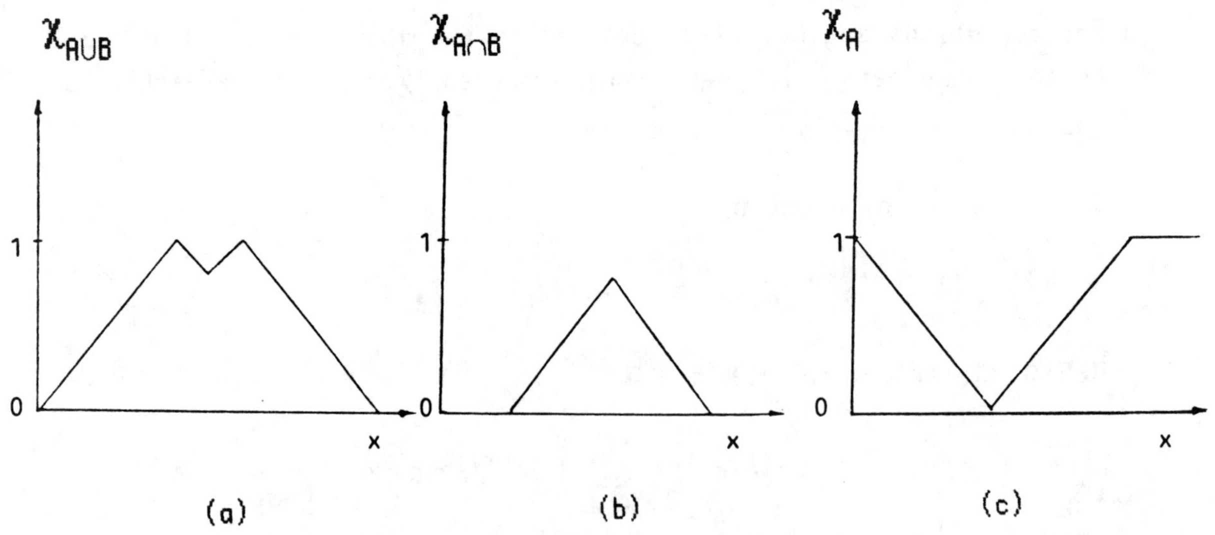

Fig.2

DEF 5. Difference of **A** and **B**:

$$\chi_{A-B}(x) = \min\{\chi_A(x), 1-\chi_B(x)\}$$

It is easy to see that the union and the intersection are associative, commutative and mutually distributive. The complement is involutive ($\bar{\bar{A}} = A$).

However, on the basis of its definition, must be noted that:

$$A \cap \bar{A} \neq \varphi \quad \text{(empty set)}$$
$$A \cup \bar{A} \neq X \quad \text{(universe)}$$

These operations are equivalent to the classic ones if applied to a characteristic function. Figures 2a, 2b and 2c give the union, intersection and complementation of the *fuzzy-set* in Figure 1.

Authors have introduced new set of operations, usefull in the design of decision systems, such as the product and the exponentiation (Kaufmann 1980).

The reader may find a complete exposition of the algebra of the *fuzzy set* in A.Kandel's book (Kandel 1986).

Multigray levels digital images may be seen as *fuzzy sets*, after opportune normalization of the pixels values. From this point of view, all the operations previously introduced, could be interpreted as transformations on images.

For example all the operation of the mathematical morphology (Serra 1982), well defined on binary images, can be extended to the case of multigray levels images by using *fuzzy-sets* representation of the image.

The definition of *fuzzy graph* has been also introduced by Rosenfeld and Zadeh:

DEF 6. Given a graph **G**=<**V**,**E**>, where **V** is the set of nodes and **E** is the set of arcs, a *fuzzy-graph*, $\mathbb{G}$, is a pair <$\mathbb{V}$, $\mathbb{E}$>, where $\mathbb{V}$ is a fuzzy set on **V and** $\mathbb{E}$ is a fuzzy relation on **V**x**V** such that:

$$\chi_{\mathbb{E}}(x,y) \leq \min\{\chi_{\mathbb{V}}(x), \chi_{\mathbb{V}}(y)\}$$

This definition is usefull for the solution of *fuzzy-clustering* problems, in which the degree of connectivity between two nodes is determined by computing the length of the shortest path between them.

DEF 7. A fuzzy-matrix $\mathbb{M}$ of dimension nxn is a set:

$$0 \leq m_{ij} \leq 1 \qquad i,j=1,2,\ldots,n$$

Given two fuzzy-matrix $\mathbb{A}$ and $\mathbb{B}$, the following algebra may be defined:

(Summ) $\mathbb{A} + \mathbb{B} = ||\max(a_{ij}, b_{ij})||$

(Product) $\mathbb{A} \times \mathbb{B} = ||\max_k(\min(a_{ik}, b_{kj}))|| \quad k=1,2,\ldots,n$

The definition of fuzzy-matrices and the relative algebra is widely applied to fuzzy-clustering problems and to implement queries in relational data base.

Theorem. Given a fuzzy matrix $\mathbb{M}$, an integer p exists such that:

$$\mathbb{M}^p = \mathbb{M}^{p+1}$$

The last Theorem allows to extabilish a convergence criterium in fuzzy clustering, whenever $\mathbb{M}$ is interpreted as the adiacency matrix of a *fuzzy graph*. In fact each entry of $\mathbb{M}^p$ is the degree of connection of each couple of nodes througth a path of length p.

3. APPLICATION FIELDS

The theory of *fuzzy-sets* has been applied to a wide class of problems such as: the fault tollerance analysis in digital systems, the optimization in decision making (Negoita 1981), the pattern clustering (Ruspini 1977), the search trees (Flowers and Kandel 1985), the relational data base (Zemankova-Leech and Kandel 1984), the image analysis (Dutta Majunder 1986).

In the following are scketched three applications strictly related to data analysis problems.

3.1 Expert Systems

In the design of the rule-based expert systems the concept of decision-table (DT) has been introduced in order to achieve the independence from the applications.

DTs are made up of four major parts:

- the condition **stub** and **entry**;
- the action **stub** and **entry**.

The condition stub contains a row for each condition to be evaluated, the action stub contains a row for each action.

The condition and the action entry sections are divided into colums called rules. Each column specifies values for conditions and resulting action to be taken, whenever the conditions meet the specified values.

In classic knowledge based systems the condition entries have truth values in the domain {T,F}.

Fuzzy-logic gives a mathematical basis to manage uncertainty in knowledge based systems. In this case the values of the conditions entry are non-deterministic, and may contains linguistic variables (low, small, high, large, very large,...) and/or modal operators (possible, necessary).

Whithin this formalism it is easier and more natural to represent the user's subjective evaluation of the conditions.

Knowledge based systems, that include *fuzzy logic* and its application to the *inference engine*, have been designed and realized.

The system CLASSIC (Thonnat 1985) has been realized for the automatic classification of celestial objects, by using also the fuzzy evaluation of the rule-set. The system RIMA (Di Gesu' and Maccarone 1987), designed for the analysis of *sparse images*, will apply the resolution-principle in the context of the *fuzzy-logic*.

3.2 Pattern Clustering

Given a data set **X**, the clustering problem may be simply described as one of finding a natural grouping or partition of the elements of **X**.

The space **X** is usually a vector space of dimension k, on which a norm or a similarity function, δ, is defined.

The clustering problem is often vague, in spite of the semplicity of its definition. A good knowledge of the probability distribution of the element in **X** is required, and gaussian assumption, often considered as valid, is wrong. Linear separation between classes is usually a rough approximation. The features or parameters representing the elements of **X**, may have non-numerical values so that their interpretation become subjective and does not follow any statistical model.

Graph-theoretical approach has been applied to group elements of **X**. Single-linkage or multiple-linkage algorithms select clusters of elements on the basis of a choosen threshold $\delta_o>0$.

For example if single-linkage algorithm is used then two samples α and β are said similar if a chain $\alpha=\alpha_1,\alpha_2,\ldots,\alpha_n=\beta$ exists, such that $\delta(\alpha_i,\alpha_{i-1})>\delta_o$ for i=1,2,...n-1. If a complete-linkage algorithm is used all samples in a given cluster must be similar to one other.

Similarity matrices have been introduced to perform cluster analysis, they may be defined as follow:

$$S_{ij}=\begin{cases}1 & \text{if } \delta(i,j)>\delta_o\\ 0 & \text{otherwise}\end{cases}$$

Fuzzy similarity matrices, may still be used to represent *fuzzy-graphs*, whenever the clustering problem is applied to data with inexact knowledge on the model or it is usefull represent non-numerical features (linguistic variables). In this case the columns and the rows are the nodes (samples) and each entry is the function $\mathbb{E}$).

As an example let be $\mathbf{X}=\{x_1,x_2,x_3,x_4\}$ and $\mathbb{M}$:

	x_1	x_2	x_3	x_4
x_1	1	0.	0.5	0.3
x_2	0.	1	0.9	0.6
x_3	0.5	0.9	1	0.3
x_4	0.3	0.6	0.3	1

The minimized form of $\mathbb{M}$ is obtained by using the theorem mentioned in the previous section.

Then:

	x_1	x_2	x_3	x_4
x_1	1	0.5	0.5	0.5
x_2	0.5	1	0.9	0.7
x_3	0.5	0.9	1	0.7
x_4	0.5	0.7	0.7	1

and we have the partitions:

$$
\begin{aligned}
C(\delta_o=1) &= \{[x_1],[x_2],[x_3],[x_4]\} \\
C(0.7<\delta_o\leq 0.9) &= \{[x_1],[x_2,x_3],[x_4]\} \\
C(0.5<\delta_o\leq 0.7) &= \{[x_1],[x_2,x_3,x_4]\} \\
C(0<\delta_o\leq 0.5) &= \{[x_1,x_2,x_3,x_4]\}
\end{aligned}
$$

Two main problems must be solved in order to complete the analysis: the choise of the best value δ_o and the validation of the clusters obtained.

3.3 Image Analysis

Fuzzy sets provide also methods for the analysis and processing of images. Their application cover most of the analysis, from low to high level vision.

Image analysis needs for non-standard techniques whenever the distribution of the intensities and/or the instrument answer are unknown, the ratio signal/noise or the statistics (number of counts/pixel) are very low. Images of such type often may be treated as a random graph and are termed *sparse images* (Di Gesu' 1986, Di Gesu' and Maccarone 1986).

The transformation of the image, $\mathbf{I}$, to be analysed in a fuzzy image, $\mathbb{I}$, is preliminary to the whole analysis. The values of $\mathbb{I}$ may be interpreted as the belonging degree of its pixels. The operators used to perform this operation are called *fuzzifiers*. The choise of the appropiate *fuzzifier* is related to the specific problem.

Low Level Vision.

Image enhancement algorithms, based on *fuzzy-sets* operations, have been designed (Pal and King 1980).

Edge detection may be expressed in terms of local operators as:

$$\square_{rk} = \max_{Q}\{ \square_{ij}\} + \min_{Q}\{ \square_{ij}\} , \quad i,j \varepsilon Q$$

Where Q is a spherical neighbour of the pixel (r,k).

Fuzzy-convolvers have been designed in image analysis (Dutta Majumder 1980,), so that their kernels reduce the edge and the salt pepper effects often introduced by the classical ones. In the definition of the kernel values are considered both the answer of the detector and the local pixel-relations.

Thresholding algorithms have been developed by several authors (Frei 1977, Deravi and Pal 1983).

Region labeling and segmentation methods have been settled by using relaxation operation (Rosenfeld, Hummel and Zucker 1982).

Classification.

Classification of shapes falls in the general domain of the decision problem (Sebestyen 1962, Duda and Hart 1973). Fuzzy approach seems to be promising whenever probability model is not available and the classes are not well separated (Bezdek 1977, Bezdek and Dunn 1975, Ronbens 1978).

Concept of fuzzy-decision tree is usefull (Chang and Pavlidis 1977) in pattern classification problems. It is defined in such a way that each non-leaf node, x, has a k-tuple decision function:

$$\mathbf{d}_x : \mathbf{X} \rightarrow [0,1]^k$$

The decision algorithms, designed to evaluate the function $\mathbf{d}_x$, define a membership function, $\chi_j(x)$, associated with each pattern $x \varepsilon \mathbf{X}$ for the Jth class. χ_j is often expressed in terms of a similarity or distance function (Ruspini 1977, Dunn 1974, Bezdek 1977).

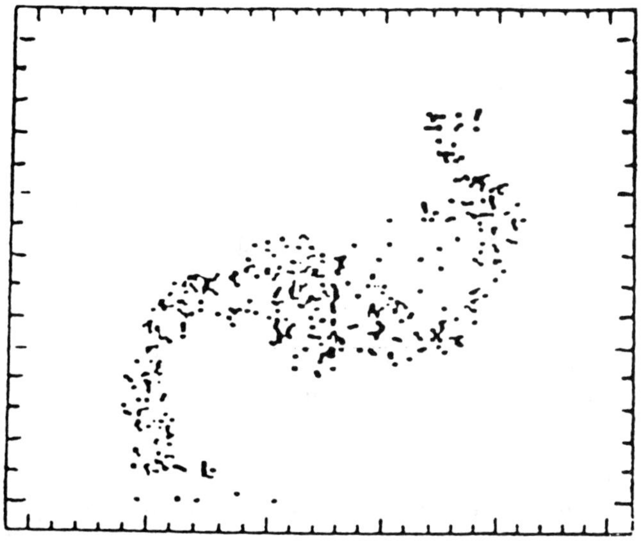

Fig.3

Shape analysis.

Shape analysis consists in both the structural description of the objects detected (number and position of the harms and concavities, orientation,...) and the evaluation of numerical shape parameters (area, perimeter, elongation).

Classical methods are not, usually, applicable to sparse images. Contour analysis may be, for example, performed by using the concept of convex-hull (Touissant 1982), with its geometrical properties, instead of the classical contour analysis algorithms. The skeleton of an image may be retrieved by applying graph-algorithms (Di Gesu' and Maccarone 1984).

4. FUZZY SETS AND ASTRONOMY

Astronomy and astrophysics as well as the high energy physics are spring of data (images, time series, spectra), analysis of which is very influenced by the a priori knowledge of the researcher (physical models, previous results, intuition). Not all these aspects of the analysis can be described in terms of deterministic or probability model.

Some time the information contained in the data is so poor that it could be impossible to apply probability theory to retrieve it.

Gamma and X ray astronomy images of the sky often are at the limit of the visibility and the signal is embedded in a structured noise background.

Figure 3 shows an example of sky image, it is a degraded version of the galaxy NGC5247 (contained in the MIDAS test file).

The preprocessing phase, carried out by using single link clustering algorithm, detects the relevant clusters (De Biase, Di Gesu' and Sacco 1986).

The selected clusters are then classified in shape with respect to a predefined set of prototypes by using a possibility function which take in account the objects simmetry. The Normalized Fuzzy Entropy (De Luca and Termini 1972) has been used as fuzzy discriminant function.

SHAPE CLASSIFICATION

Prototypes	Membership degree
CIRCLE	0.63
ELLIPS. 1.5	0.68
ELLIPS. 2	0.69
ELLIPS. BAR	0.60
SPIRAL-2-KERNEL	0.76
SPIRAL-2-BAR	0.63

Object classified as a spiral with kernel and two arms.
Normalized Fuzzy Entropy used as possibility function.

(a)

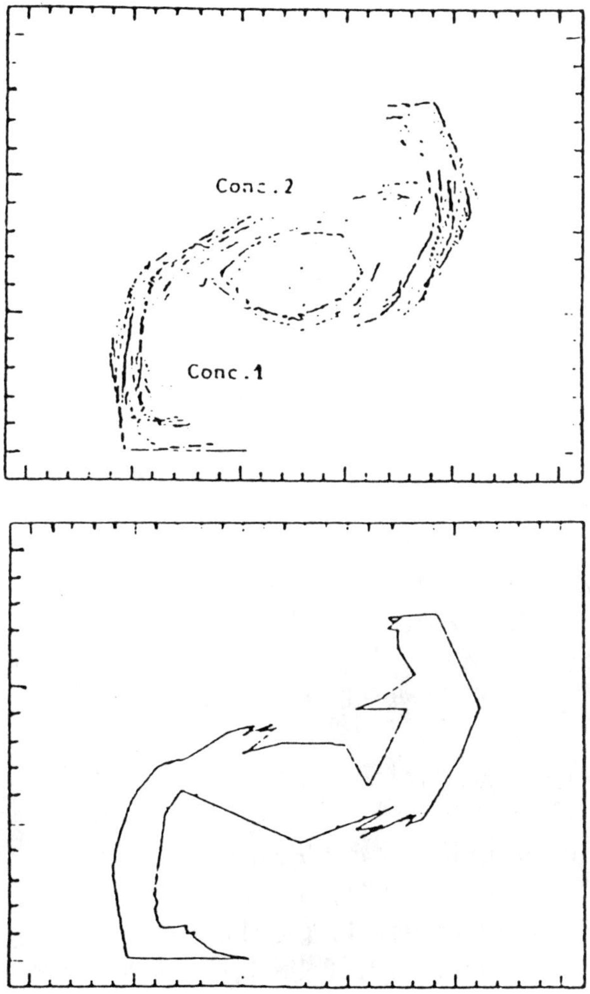

MORPHOLOGICAL STRUCTURAL ANALYSIS

First External Convex Hull	
Perimeter	670.
Surface	25579.
Free mean path	42.
Cut threshold	84.
Num. concavities	2

Concavities

	Conc.1	Conc.2
Perimeter	552.	607.
Surface	8087.	7090.
Size	31 %	27 %
Direction	322°	136°

Final Contour

Num. points	83
Perimeter	1257.
Surface	10743.
Size	42 %

(b)

Fig.4

Goals of the morphological analysis are to determined both the internal structure and the global shape parameters (concavities, orientation, area, perimeter,...).

The internal structure is evaluated by defining the two *fuzzy-sets* named **full** and **empty** (Di Gesu' and Maccarone 1986). The global shape features are computed by using the properties of convex hull.

Figures 4a and 4b show the results of the analysis for the object in Figure 3.

This short example must be considered, only partially as a paradigm for the analysis of fuzzy-images. The semantic interpretation of the results (consistency with physical model) should be also included.

5. CONCLUDING REMARKS

This paper may be considered a short introduction to *fuzzy-set* theory and its applications. *Fuzzy-set* theory is young and not completely well understood. It has been introduced to overcomes some problems present in the theory of probability and to quantify some axpects of the human reasoning. It must be experimented to discover its limit and bugs and to develop new ideas.

REFERENCES

Bezdek J.C. (1977), *IEEE Trans.Syst.Man.Cyber.,* Vol.SMC-7, pp.87-92.

Bezdek J.C. and Dunn J.C. (1975), *IEEE Trans.Comput.,* Vol.24, pp.835-838.

Chang R.L. and Pavlidis T. (1977), *IEEE Trans.Syst.Man.Cyber.,* Vol.SMC-7, pp.28-35.

De Luca A. and Termini S., (1972), *Information and Control,* Vol.20, pp.301-312.

De Biase G., Di Gesu' V. and B.Sacco (1986), *Pattern.Rec.Letters,* Vol.4, pp.39-44.

Deravi F. Pal S.K. (1983), *Patt.Rec.Letter,* Vol.1, pp.417-422.

Di Gesu' V. (1986), *NATO ASI series,* Vol.F 30, P.Devijver & Kittler (eds.), Springer Verlag.

Di Gesu' V. and Maccarone M.C. (1986a), *Proc.Int.Conf.* I.A.P.R., Parigi.

Di Gesu' V. and Maccarone M.C. (1984), *Proc.Int.Conf.* I.A.P.R., Montreals.

Di Gesu' V. and Maccarone M.C. (1987), *Proc.Int.Conf.* I.A.P.R. (Italian Chapter),Cefalu'.

Dimitrov V. (1983), *Fuzzy sets and Systems,* Vol.9, No.1, pp.25-40.

Duda R. and Hart P. (1973), *Pattern Classification and Scene Analysis,* Wiley, NY.

Dunn J.C. (1974), J.Cybernetics, Vol.4, pp.1-15.

Dutta Majumder D. (1980), *Int.J.Systems Sci.,* Vol.11, No.12, pp.1435-1445.

Dutta Majumder D. (1986), *Halsted Press,*NY.

Flowers P. and Kandel A. (1985), *Fuzzy sets and Systems,* Vol.16, No.1,pp.1-24.

Frei W. (1977), *Comput.Graph. and Image Proc.*, Vol.6, pp.286-294.

Kandel P.F.A. (1976), *IEEE Trans.Syst.Man.Cyber.,* Vol.SMC-6, pp.215-219.

Kandel P.F.A. (1986), Fuzzy mathematical techniques with applications, *Addison Wesley.*

Katsaras A.K. (1981), *Fuzzy sets and Systems,* Vol.6, No.1, pp.100-111.

Kaufmann A. (1980), *Academic Press,* NY.

Negoita C.V. (1981), *Fuzzy sets and Systems,* Vol.6, No.3, pp.261-270.

Negoita C.V. (1985), *Expert Systems and Fuzzy Systems,* Benijamin/Cummings, Menlo Park.

Pal K.S. and King R.A. (1980), *Electron.Letter,* Vol.16, pp.376-378, 1980.

Rosenfeld A., Hummel R.A. and Zucker S.W. (1976), *IEEE Trans.Syst.Man.Cyber.,* Vol.SMC-6, pp.420-423.

Ronbens M. (1978), *Fuzzy sets and Systems,* Vol.1, pp. 239-253.

Ruspini E.H. (1972), *IEEE Trans.Syst.Man and Cyberns.* Vol.SMC-2, p.541.

Sebestyen G.S. (1962), in *Pattern Recognition,* The Macmillan Co.,NY.

Serra J. (1982), *Academic Press.*

Thonnat M. (1985), *Int.Report,* INRIA, No.387, March.

Touissant G.T. (1982), in *Pattern Recognition Theory and Applications,* J.Kittler, K.S.Fu, L.F.Pau (eds.), pp.73-91.

Yager R.R. (1980a), *J.of Cybernetics,* Vol.10.

Yager R.R. (1980b), *J.of Cybernetics,* Vol.10.

Zadeh L. (1962), *Proc.Institute of Radio Engineers,* Vol.50, pp.856-865.

Zadeh L. (1985), *IEEE Trans.Syst.Man.Cyber.,* Vol.SMC-15, No.6, pp.754-763.

Zemankova-Leech M. and Kandel A. (1984), *Fuzzy Relational Data Base - A Key to Expert Systems,* Verlag.

NEGLECTED PARAMETRICAL EXTRAGALACTIC CLUSTERING

A. Heck
C. D. S., Observatoire Astronomique, Strasbourg, France

Abstract

A survey of the literature shows that the clustering methodology is applied to extragalactic data generally to study only the spatial distribution of the objects (spatial clustering). In most cases, the potentialities of the methodology for studying the parametrical similarities (parametrical clustering) are neglected.

Statistical clustering

Maybe somewhat paradoxically, astronomers have not been among the quickiest to realize the potentialities of the multivariate statistical methodology. It is obvious however that its importance is continuously increasing in our science (see e. g. Heck et al., 1985), be it only because ever larger amounts of data are being collected. And the rate will continue to accelerate. It will be therefore necessary to work on larger samples if full advantage is to be taken of the information contained in the data accumulated. And one way to work effectively on large samples is to apply, and if necessary to develop, an adequate statistical methodology.

Among the various multivariate statistical methodologies, the cluster analysis (or clustering) is quite a popular one. It is used to determine the inherent or natural groupings in the data, or to provide a convenient summarization of the data into groups. Clustering methods can also be used to pick out anomalous or peculiar objects. A description of various algorithms can be found in Chap. 3 of Murtagh & Heck (1987a), together with a list of astronomical examples and annotated bibliographical references.

In other words, cluster analyses may be used:

- for determining groups of objects, using some criterion, or,
- without prior knowledge of what groups are present, for obtaining a range of groupings and for using this to assess the true group structure.

Statistical clustering in astronomy

Cluster analyses, as other multivariate statistical methods, are particularly well appropriate to the variety and heterogeneity of astronomical data. A compilation of multivariate statistical methods in astronomy has been carried out (Murtagh & Heck, 1987b) and is continously updated.

The range of astronomical clustering applications covers a wide variety of problems such as star-galaxy separation; classification of spectra and images; taxonomy of asteroids, stars, cosmic sources, lightcurves, etc.; study of planetary surfaces, lunar rocks, rare earths in stellar spectra, groups of galaxies, gamma and X-ray data, photometric colour-colour diagrammes, etc. Among the few applications to extragalactic data, it is striking that practically none goes beyond a mere study of the spatial clustering of the objects (groups or clusters of galaxies, superclusters, etc.).

Clusters and clusters

The application of the clustering methodology should however not be restricted only to that aspect. The other potentialities, calling for other reference frames than the geometrical one, should not be neglected. Thus the clusterings in multiparametrical spaces, where the axes can be directly observed quantities or already reduced variables (by a preliminary principal-component analysis, for instance), should be investigated.

This is specially true for fields as extragalactic astronomy where the physics is not yet as advanced as in stellar astronomy. Physical groups could be pointed out on the basis of physical parametrical similarities (in the statistical sense), as it has been the case in some instances for stellar data.

Statistical methods are not intended to replace physical analysis. These should be seen as complementary. Statistical methods can be used to run a rough preliminary investigation, to sort out ideas, to put a new ("objective" or "independent") light on a problem, or to point out aspects which would not come out in a classical approach. Physical analysis is necessary to subsequently refine and interpret the results, and to take care of the details.

References

Heck, A., Murtagh, F., Ponz, D.J. 1985, The increasing importance of the statistical methods in astronomy, Messenger 41, 22-25

Murtagh, F., Heck, A. 1987a, Multivariate data analysis, Astrophys. Sp. Sc. Library 131, D. Reidel Publ. Co., Dordrecht, xvi+210p.

Murtagh, F., Heck, A. 1987b, An annotated bibliographical catalogue of multivariate statistical methods and of their astronomical applications (magnetic tape), Astron. Astrophys. Suppl. 68, 113-115

MRSP - AUTOMATED DATA ACQUISITION AND THE USE OF DATA BASES IN OBSERVATIONAL COSMOLOGY

H. Horstmann, P. Schuecker, V. Gericke
Astronomical Institute of Muenster University, FRG

1. Introduction

The Muenster Redshift Project (MRSP) is a study of large-scale distribution of luminous matter in the Universe and astrophysical/cosmological implications. The data obtained from direct and objective prism Schmidt-telescope plates are of photometric accuracy $m_J \lesssim 0.1$ and redshift accuracy $dz \lesssim 0.01$. They lead to a homogeneous data base and a "fair" sample of galaxies and quasars. The distances reached with galaxy redshifts range from $z = 0$ to $z = 0.3$ (with a programme extension to possibly $z = 1.0$) and with quasar redshifts from $z = 0.2$ to $z \lesssim 4$. The sky coverage (at present restricted to the southern hemisphere) is projected to grow plate by plate to a total of 12 000 square degrees.

A single Schmidt plate increases the number of known galaxy redshifts by 20 percent (observed), the number of known quasars by 10 percent (extrapolated), and the observed volume occupied by normal field and cluster galaxies with $m_J \lesssim 20.5$ by a factor 100.

The sequence of processes from fully automated data reduction to object interpretation requires the compilation, control and use of large internal and external data bases. The organization of data analysis follows the scheme presented by Teuber (1987).

2. Data bases and organization of data analysis

Table 1 presents permanent and non-permanent data bases used in the MRSP for the reduction and evaluation of photographic data.

3. Examples

a) Low, medium and high level vision for individual objects on direct Schmidt plate (Horstmann, 1987)

Object segmentation and intensity calibration are examples of **low level vision** of the direct plate. They are used to reduce redundant image information (sky background) and to enhance the object quality.

For segmentation objects are defined as connected pixel areas on the digitized photographic plate showing significant increase of density over the background. The fact that all relevant objects (stars and galaxies with $15.0 \leq m_J \leq 21.5$) have sizes $\leq$ 20"x20"

Table 1: Data bases and organization of data analysis

DATA BASE	PROCESS	DESCRIPTION	DATA FOR A SINGLE PLATE
primary data base		direct and objective prism Schmidt plates	350 mm x 350 mm photographic plate
		film copies permanent storage	5×10^9 pixels of non-uniform size (grains)
	DIGITIZATION	direct Schmidt plate with the Muenster PDS 2020GM microdensitometer	step size aperture area covered 300 mm x 300 mm
1st temporary data base		used for on-line reductions	0.8×10^9 pixels of uniform size
		no permanent storage	
	LOW LEVEL VISION	on-line image segmentation	1.5×10^5 objects with $m_J \leq 21.5$ in high galactic latitude fields
	MEDIUM LEVEL VISION	calculation of object data: positional parameters, number of pixels, sky background, noise	
1st MRSP data base		images and image information extracted from the 1st temporary data base permanently stored	
	MEDIUM LEVEL VISION	calculation of object features: moments, effectice object radius, object central density	
	HIGH LEVEL VISION	object classification: star/galaxy separation	1×10^5 starlike objects 5×10^4 galaxies
2nd temporary data base		spectra from digitized objective prism Schmidt plate	dispersion 246nm/mm at Hγ
		selected parameters from the 1st MRSP base relevant for processing objective prism data	
		used for on-line reductions no permanent storage	
	LOW LEVEL VISION	on-line segmentation	4×10^4 spectra of starlike objects 1×10^4 galaxy spectra
2nd MRSP data base		images and image data extracted from the 2nd temporary data base permanently stored	

(cont.)

Table 1 (continued)

	LOW LEVEL VISION	object enhancement: wavelength calibration, intensity calibration, rectification	
	MEDIUM LEVEL VISION	feature evaluation: correlations, least-squares fits, moments, line strengths	
	HIGH LEVEL VISION	line identification spectral classification quasar separation redshift determination	
3rd MRSP data base		permanent storage	7×10^7 galaxy redshifts 5×10^3 quasar redshifts
	MEDIUM LEVEL VISION	data evaluation and compilation from the 1st and 3rd MRSP data bases for astrophysical/cosmological application colours, spectral types, absolute magnitudes, 2D and 3D densities	
4th MRSP data base		2D and 3D galaxy and quasar distributions	
	LOW LEVEL VISION	image segmentation: clusters of galaxies	
	MEDIUM LEVEL VISION	calculation of cluster properties: density, mass, luminosity function, cluster profile (Schuecker et al. 1987)	
	HIGH LEVEL VISION	evaluation of object relations: subclustering, superclustering, voids (Schuecker et al. 1987)	
5th MRSP data base		data and results from the 4th MRSP data base supplemented by data from selected external data bases, organized for common access	
	LOW LEVEL VISION	target selection for HST, VLT, new generation telescopes source verification and classification from existing data bases source identification for EXOSAT, ROSAT, IRAS, other surveys	
	MEDIUM LEVEL VISION	derivation of general object properties	
	HIGH LEVEL VISION	astrophysical/cosmological interpretation and comparision with models of space geometry/matter distribution obtained from evolutionary models	

leads to top down segmentation. Objects are stored as quadratic segments of the scan data ("picture frames").

External intensity calibration is performed with artificial intensity standards, e.g. the calibration spots on the UKST plates. Internal calibration is based on the method of Hoemberg (1987).

In **medium level vision** object parameters are computed: object position, object size, maximum and central intensity, magnitude, moments and related features (effective radius, total area).

In **high level vision** the object features are used for star/galaxy/artefact classification. For bright objects the total area and the effective radius, for faint objects the central intensity and the effective radius are used for star/galaxy separation. Artefacts are isolated using the maximum intensity, central intensity and position of the center of the image at various intensity levels.

Fig.1 shows a 3D-presentation of the two dimensional galaxy distribution versus number density in the SRC field No 411. The total number of galaxies found is approximately 42 000. All isopleths with densities lower than twice the background density are set to a constant value.

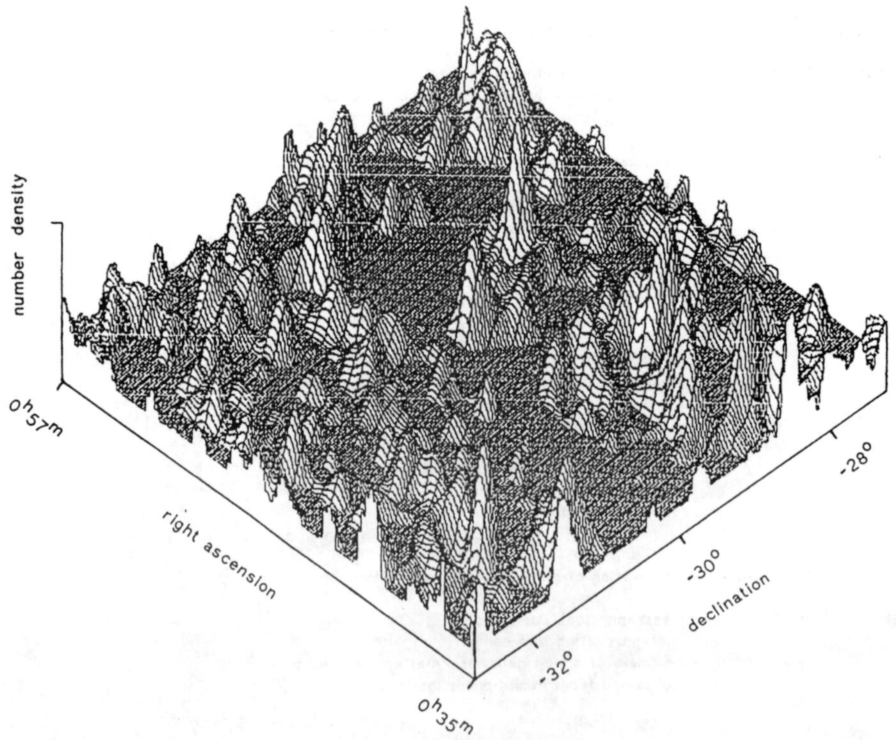

Fig. 1: 3D-presentation of the two dimensional galaxy distribution versus number density in the SRC field No 411

b) Low, medium and high level vision for individual objective prism spectra (Schuecker, 1987)

Wavelength calibration and rectification are typical examples of **low level vision**. They are used to enhance the object quality, i.e. to show off spectral features relevant for subsequent recognition tasks.

The wavelength calibration uses the transformation of object coordinates from the direct to the objective prism plate (Tucholke et al., 1987). The calibration objects are automatically selected G-type stars, where the position of the CaII (H,K) blend serves as the reference point on the objective prism plate.

For rectification an adaptive fuzzy set filter was developed. It adjusts filter parameters with respect to different criteria, relevant for rectification (e.g. short continuum, clear protrusion of spectral features, low RMS deviations between spectrum profile and continuum). Each criterion is represented by a fuzzy set. The filter parameter which optimizes the preset criteria best is determined by fuzzy set algebra.

Evaluation of features relevant for subsequent spectrum classification and redshift measurement is the major task of **medium level vision**. Several features are computed. For galaxy redshift determination correlation functions and least squares fits between the programme spectra and the standard spectra are used. The position of the correlation maximum and the position of the least squares miminum determine the galaxy redshift.

Other procedures detect spectral features directly. Positions and strengths of the individual features are obtained from 1st and 2nd order derivatives of the spectrum profile. For galaxy redshift determination the position of the CaII blend and the G-band are measured.

For the identification of blue galaxies and quasars the center of gravity of the spectrum profile is used. Another identification criterion is the presence of emission lines.These are also used for redshift determinations.

Classification of objective prism spectra (necessary for wavelength zero point determination in redshift measurements) and identification of individual spectral features are **high level vision**. Here fuzzy set classifiers, Bayes classifiers (classification of M-type stars, Schuecker et al., 1986) and rule-based decision algorithms are applied. The trainability of the classification procedures ensures quick adjustment to further classification problems.

Fig. 2a shows a wedge diagram with 6 312 galaxy redshifts obtained fully automatically for the SRC field No 411. The redshifts are calculated with the three procedures of galaxy redshift measurement mentioned above. Galaxies with low absolute magnitudes ($-14 > M > -18$) are found at redshifts $z < 0.1$, whereas absolutely bright galaxies ($M < -18$) are detected at redshifts $0.1 < z < 0.3$ (Malmquist bias).

At z = 0.1 a prominent supercluster is outlined by 550 galaxy redshifts (Fig 2b). Additional redshifts at z = 0.16 identify the supercluster as member of a super-supercluster. More details are given in Schuecker et al. (1987).

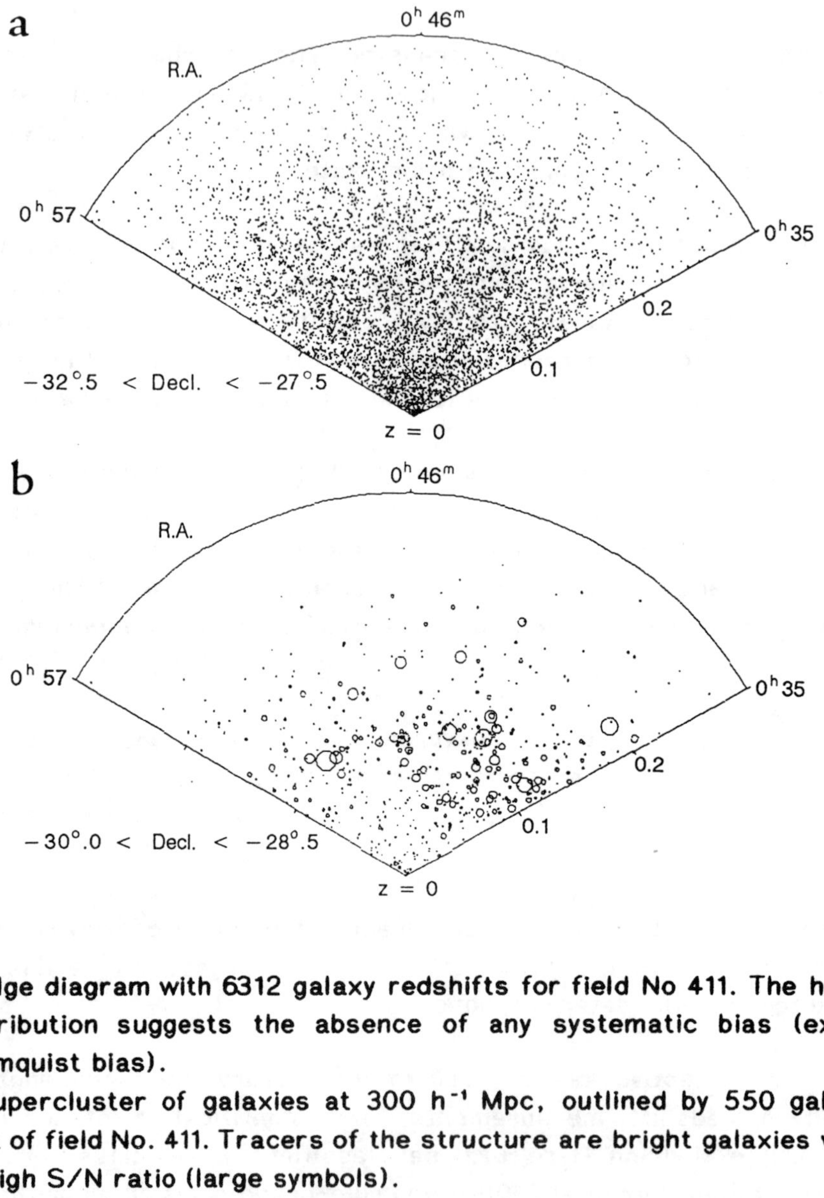

Fig. 2: a) Wedge diagram with 6312 galaxy redshifts for field No 411. The homogeneous distribution suggests the absence of any systematic bias (excluding the Malmquist bias).

b) A supercluster of galaxies at 300 h^{-1} Mpc, outlined by 550 galaxies, for a part of field No. 411. Tracers of the structure are bright galaxies with spectra of high S/N ratio (large symbols).

For the same SRC field 188 quasar candidates (75 with high probability) are found using the blue excess criterion (Gericke et al., 1987). It is expected that the introduction of the emission line criterion and the subsequent automated line identifications and redshift measurements will noticeably increase the number of high probability quasars.

It seems possible that no follow-up observations are needed if the quasar data are used for statistical purposes. This is supported by the good agreement between the radial velocities obtained from the MRSP and corresponding catalogue data.

4. Present status and outlook

While 3 direct plates and 1 objective prism plate have been fully analyzed during the course of programme development, the status in October 1987 is 9 direct plates - 210 square degrees and 3 objective prism plates - 70 square degrees, including incomplete reductions.

The future perspective is presented in the following.
All plate fields ($b \leq -30$):

		2×10^{12}	non-uniform pixels
		3×10^{11}	uniform pixels
estimated	›	6×10^{7}	objects
		2×10^{7}	spectra
	›	4×10^{7}	starlike objects (images)
		2×10^{7}	galaxies (images)
		2×10^{7}	stellar and quasistellar spectra
		4×10^{6}	galaxy spectra
		3×10^{6}	galaxy redshifts
		2×10^{5}	quasar redshifts

Space extension by a factor of 40 000 for normal galaxies (not restricted to clusters) of $m_J \leq 20.5$

10^4 - 10^5 clusters of galaxies expected

5 h^{-1} Gpc limiting size for super-structures of normal galaxies

References

Gericke, V., Schuecker, P., Seitter, W. 1987, in *Workshop on Astrophotography*, ed. Marx, S., Lecture Notes in Physics, Springer, Heidelberg (in press).

Hoemberg, D. 1987, in *Workshop on Astrophotography*, ed. Marx, S., Lecture Notes in Physics, Springer, Heidelberg (in press).

Horstmann, H. 1987, in *Workshop on Astrophotography*, ed. Marx, S., Lecture Notes in Physics, Springer, Heidelberg (in press).

Schuecker, P., Horstmann, H., Volkmer, C.C. 1986, in *Second Workshop on Data Analysis in Astronomy*, eds. Di Gesu, V., Crane, P., Friedman, H.J. and Levialdi, S., Plenum Press, New York and London, p. 109.

Schuecker, P. 1987, in *Workshop on Astrophotography*, ed. Marx, S., Lecture Notes in Physics, Springer, Heidelberg (in press).

Schuecker, P., Horstmann, H., Seitter, W. 1987, in IAU Symp. No. 130, *The Structure of the Universe*, eds. Audouze, J. and Szalay, A., Reidel, Dordrecht (in press).

Teuber, D. 1987, these proceedings.

Tucholke, H.-J., Schuecker, P., Horstmann, H., Seitter, W. 1987, in IAU Colloquium No. 100, *Fundamentals of Astrometry*, eds. Eichhorn, H.K., Pakvor, I. (in press).

STATISTICAL ANALYSIS OF SYNTHETIC STELLAR PHOTOMETRY WITH HST FILTERS

M.L. Malagnini
Dipartimento di Astronomia, Universita' degli Studi di Trieste, via G.B. Tiepolo 11, I-34131 Trieste, Italy
C. Morossi
Osservatorio Astronomico, via G.B. Tiepolo 11, I-34131 Trieste, Italy
L.Rossi
Istituto di Astrofisica Spaziale del C.N.R., PO Box 67, I-00044 Frascati, Roma, Italy.

ABSTRACT

The Wide Field Camera of the Hubble Space Telescope is provided with an ample collection of photometric filters carefully devised for many different astrophysical purposes. It is obvious that, due to the limitation in the ST observing time allottable to any project, the maximum of information must be obtained from a minimum number of preselected filters, optimized for the specific research program. Therefore, with the aim of deriving the fundamental atmospheric stellar parameters effective temperature, gravity, and metallicity from HST spectrophotometric scans, we investigated the possibility of selecting the minimum number of observations required. For this analysis, we adopted Kurucz' models (Kurucz, 1979; Model atmospheres for G,F,A,B, and O stars: "New convective models" from the magnetic tape version with unpublished models) for simulating the energy distributions emitted by stars, and synthetic photometry has been derived by convolving the simulated fluxes with WFC filter transmissions and overall efficiency. The 42 WFC filters describe almost completely the stellar spectral energy distribution in the observable HST wavelength range. By applying principal component analysis to the total set of normalized scans, it results that a minimum of 3 components must be used in order to obtain an adequate

representation of the overall bulk of data.
The analysis of the correlation matrix of the 41 colors obtained through normalization to F569W (HST V magnitude) and the physical parameters effective temperature, gravity, and metallicity permits us to select a set of colors strongly dependent on these parameters, while almost uncorrelated one among each other.
Next step will be the study of different combinations of the selected colors through multiple regression techniques in order to obtain linear relationships relating them to the physical parameters effective temperature, gravity, and metallicity.

GRID OF THEORETICAL FLUX DISTRIBUTIONS FOR MODEL ATMOSPHERES
(from Kurucz, 1979; Model atmospheres for G,F,A,B, and O stars: "New convective models" from the magnetic tape version with unpublished models)

No. of models	294
No. of wavelength points	342

Parameter	Range	Step
Effective temperature	5500 - 8500	500 (K)
Surface gravity	1.5 - 4.5	0.5
Metallicity	-3.0 - 0.0	0.5

SELECTED COLORS (normalized to F569W) :

F284W F336W F368M F375N F850LP

DATA ANALYSIS, DATA BASES AND EXPERT SYSTEMS : The COMMON INTERFACE.

H. RALAMBONDRAINY*, E. DEMONCHAUX**, G. JOMIER***

* INRIA
Domaine de Voluceau BP 105
78153 LE CHESNAY Cedex.
France.

** SEMA-METRA.
16-18 Rue BARBES
92126 MONTROUGE Cedex.
France.

*** Université PARIS SUD.
ISEM
91405 ORSAY.
France.

Abstract

An analysis information system needs to be powerful, user-friendly and intelligent. In this paper, we show how a system which combines a data base management system and a statistical system is efficient in data management, and data analysis. We use Artificial Intelligence tools to make complicated statistical systems more accessible for novice users and we propose a learning method in order to generate rules for expert systems.

1. INTRODUCTION

In most scientific areas (medicine, biology, astronomy, ...) users collect vast sets of objects and face the problem of data management and data analysis (D.A.). Different kinds of solutions have been proposed. The existing systems which provide statistical analysis and data management facilities may be classified into three categories [Kobayashi and Ikeda, 1981]:

1) Statistical packages whose data management functions are strengthened by incorporating data base management techniques;

2) Data Base Management Systems (D.B.M.S.) whose functionalities are extended by statistical analysis;

3) Systems which integrate an existing data base management system and a statistical package.

The first kind of system can be efficient in statistical analysis but has not the data management power of a D.B.M.S. On the other hand a D.B.M.S. of the second category can supply only a small number of statistical programs. The main reason for these limitations is that a statistical system was not initially conceived to be a D.B.M.S. and vice versa. For instance an efficient organization file for a statistical method will not be suitable for a management procedure.

Most of the D.B.M.S. provide interface with others systems which moves data from the base to A.S.C.I.I. files but in this case, the exchanges between the two systems are made by the user. We present and discuss in section 2, a system which combines a relational D.B.M.S. PEPIN [Bouchet and al., 1981] and a statistical system SICLA [Ralambondrainy, 1983] .The integrated system is seen by the user as a unique system via a menu and the user has the ability to move from one system to another without having to program anything.

For a long time, researchers have been emphasizing efficiency of software and looking for systems like described above i.e. *powerful* in data management and in data analysis. Nevertheless such systems are oriented towards professional statisticians. Recently researchers have been interested in making them easier to use to naïve statistical users. Statisticians frequently notice that current statistical packages are open to misuse. Novices have difficulty in determining the suitable method to analyze their data and to interpret the results. Recently a growing interest in Artificial Intelligence (A.I.) technology has appeared among statistical researchers [Gale and al., 1986] . They have discovered that A.I. tools, such as Expert Systems (E.S.), help users to select suitable methods and to prevent misuse by checking automatically the validity of statistical methods assumptions.

Several systems, like REX [Precibon and Gale, 1984], have demonstrated the utility of A.I. techniques in statistics. We present in section 3, an E.S. called CLAVECIN [Demonchaux and al., 1985] [Ho Tu Bao, 1987], which is an intelligent assistance for users to analyze data using the Data Analysis system SICLA .

Systems need to be powerful, user-friendly but also to be *intelligent*. We mean by intelligent system, a system able to learn and improve by observing and analyzing the problems that it must solve. D.A. methods, such as discrimination, regression analysis, A.I. tools such as "Learning techniques" [Michalski and al., 1983] are techniques which generate decision rules from data. They will be main components for such intelligent systems called *Learning systems*. For this problem, the statistical approach differs from the A.I. approach. D.A. techniques focus on efficient methods applied to data that have numerical attributes. But D.A. results are not easy to understand. A.I. techniques deal with symbolic representations of knowledge (qualitative data) and try to find clear descriptions of concepts. We propose in section 4 a learning algorithm which combines the two points of view.

2. DATA BASE MANAGEMENT AND DATA ANALYSIS

2.1 Introduction

Two converging trends are appearing in data bases and in data analysis. As data base management systems are used to store, manipulate and query large data bases, it is common to have access to some elementary statistical functions (minimum,maximum,mean,...) called "*aggregates*". But the designers of those systems are discovering more and more the help that statistical tools [Lefons and al., 1983] and especially *multivariate* data analysis (factorial analysis, clustering methods, ...) can provide for data base users : they allow a deeper and fuller data knowledge, the possibility of taking into account several parameters simultaneously, and offer a more synthetical view of data. Moreover the statistical systems designers are more and more interested in using D.B.M.S. rather than traditional file systems to manage the data which are to be analyzed. This is explained by D.B.M.S. power : it makes the update and consultation of data easier, allows a more powerful control of their coherence and assures their consistency maintenance in a multi-user context, even when crashes occur. A lot of work has been done on statistical data bases and the paper of Ozsoyoglu and al. (1985) is particularly interesting as a survey on existing and proposed statistical database management systems.

In this section, first we point out the interest of relational DBMS for statistical analysis and we present a system which integrates the interactive data analysis system SICLA and the relational D.B.M.S. PEPIN.

2.2 Relational model and data analysis

The relational systems are the D.B.M.S. second generation, the zero generation designating the file management systems and the first generation, the hierarchical and network D.B.M.S. The relational model appears particularly well fitted to D.A.: the rectangular arrays (objects-variables) may be seen as relations (tuples-attributes) and the relational operations(selection,join,...) correspond to the data management needed by the statisticians. On the other hand from a data base user point of view, data analysis techniques present the following advantages:

1) They offer a synthetic way to represent multidimensional data ;

2) They bring out underlaying variables which may be integrated as new attributes in the database and used ;

3) Statistical and relational queries are complementary and it is interesting to be able to link them together.

2.3 PEPIN and SICLA

The relational D.B.M.S. we use PEPIN, has some very useful qualities in a data analysis context: automatic data coherence controls, existence of atomic transactions, possibilities of multi-user, data consistency maintenance even when crashes occur, dynamically evolving schema... PEPIN, written in PASCAL, was developed in the ISEM laboratory (Paris-Sud University). It is operational and available on different computers.

SICLA is an interactive and general purpose package for data analysis. It offers the speed, reliability, flexibility, and ease of use needed by the user to perform multivariate analysis. The package contains statistical procedures, varying from elementary statistics (means, standard deviations, chi-square,...) to multivariate analysis (factorial, clustering, hierarchical and discriminate analysis methods,...).

Special efforts are devoted to prevent misuse of methods. Systematically a statistical procedure, before starting to run, verifies the data type (data array, dissimilarity matrix,...) and the variables type (quantitative,binary,qualitative,...). For instance a clustering method for quantitative data will not run on data which are made up of quantitative and qualitative variables. Data structures, as partitions, hierarchies, factorial coordinates,... generated by D.A. procedures are stored and can be reused for deeper analyses.

The SICLA system has been developed in the "classification and pattern recognition" project at INRIA. It is written in FORTRAN 77 and operational on different computers, from IBM PC to VAX.

2.4 The PEPIN-SICLA integrated system

The system features

The PEPIN-SICLA system [Jomier and al., 1986] in its operational version offers the data management functions of the relational D.B.M.S. PEPIN and the data analysis methods of SICLA.

Before their analysis the data, stored in relations, must be prepared, "*statistically normalized*", for multivariate analysis i.e. the data are numerically coded and statistically described. The data coding is an important and complex step. It depends on the data type and will affect analysis results. For instance, if we have to code variables dealing with the absence or presence of one feature, the feature presence can be represented by 1 and the absence by 0. This choice gives a larger weight to presence while the coding (1,0) and (0,1) balances both possibilities. Multivariate analyses need statistical variables descriptions : the type (binary, qualitative, quantitative,...), text associated,... All these operations are processed by PEPIN using relational operations and dialogue with the user (figure 1). Statistically normalized relations are built and sent to SICLA to be analyzed.

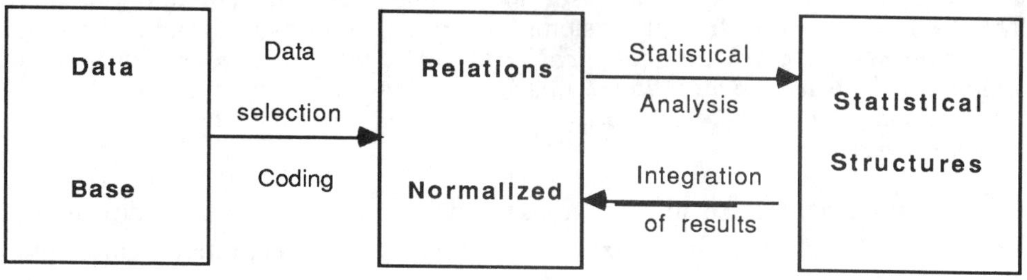

Figure 1.

On the other hand, it is possible to integrate D.A. results in the D.B.M.S. PEPIN. Any D.A. results which can be represented by matrices, may be seen as relations. For instance a partition is a relation having two attributes : the objects identification and the class number. This possibility of results integration in the data base allows the D.A. system SICLA to extend a PEPIN data base.

The system design.

The integrated system is seen by the user as a unique system via a menu. It is interactive and self-documented. Figure 2 shows the general architecture of the system. The general menu is subdivided into three groups. Each one corresponds to a hierarchical sub-menu. The first group is composed of the D.B.M.S. commands. We essentially find relational algebra operations. The second group deals with SICLA statistical commands and the third group concerns interface specific commands. These last commands execute and control the data exchange between PEPIN and SICLA : extraction and normalization of data from the relational base and integration of statistical structures.

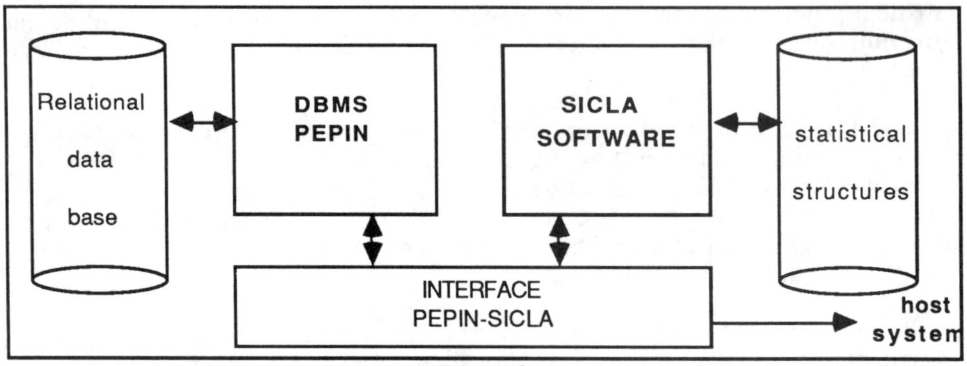

Figure 2.

2.5 Discussion

Different solutions have been proposed to interface a D.B.M.S. and a D.A. system. The first one is to allow the D.B.M.S. to generate A.S.C.I.I files to export relations. This solution presents the disadvantage that the user must make all the preparations of the data for the D.A. system. Another solution consists in a complete integration of both, D.A. system and D.B.M.S.. If the system resulting of such integration is supposed to be as powerful as a "complete" D.B.M.S. and to provide the users with all the data analysis facilities, it would be enormous, with a lot of difficulties to build it, to debug, to maintain and to make it evolve.

3. EXPERT SYSTEMS AND DATA ANALYSIS

3.1 Introduction

The statistical systems are not only required to be powerful in data management or in data analysis but also to be user friendly. Generally, the user doesn't know the most suitable method to solve his problem. The need for more user friendly systems is generally admitted, as pointed out by Chambers (1981): "*Here is the moral imperative: Statistical system in its present form, made widely available by cheap computing, will participate much uninformed, unguiding and simply incorrect data analysis. We are obliged to do something to help ....I have argued here that expert system for data analysis would respond to an important need.*"

Several expert systems, related to standard statistical system, have been developed.The REX system [Precibon and Gale, 1984] is well-known. It provides guidance, interpretation and instruction in regression analysis with the system S. The EXPRESS system [Carlsen and Heuch,1986] gives help in processing the BMDP system. An example of an expert system shell for SAS System (executing sequences of statistical analysis) has been described by Darius (1986).

The next section describes the basic design features of the CLAVECIN expert system related to the SICLA system.

3.2 CLAVECIN overview.

An expert system is usually made up of two parts:

- the inference engine.
- the knowledge base.

In this case, CLAVECIN uses the inference engine KX with a knowledge base to help the user to choose and to call different procedures of the D.A. interactive SICLA system.

The Inference Engine: Knowledge eXpert.

KX is a tool kit for knowledge engineers and A.I. system builders developed by SEMA-METRA.

The knowledge base is a set of rules, which represents part of an expert knowledge in a given domain, and a set of facts, which describes the particular situation to be analyzed.

The inference engine is domain-independent, an interactive rule interpreter using combined forward and backward reasoning and a special strategy able to ask the question which informs the most, by computing a criterion similar to the Shannon entropy. Two of the activation rule mechanisms are the "modus ponens" and "modus tollens".

Propositional logic is used in KX and three kinds of rules are available:

- Production rules *(if ...then...).*
-.Summarizing rules*(if ...write..).*
- Exclusion rules *(Exclusion : subject: p1,p2,.., pn).*

It is possible to associate to each proposition used in the rules special types: "to be considered first in the evaluation process", "askable","visible if the evaluation is true", etc. These types are taken into account in every possible strategy. Experts can modify the defective strategy by using other options. The name of every interactive command can be introduced in the rules.

The rules integrate calculus and comparison modes and can process any external program.

The system asks the user such questions as:

- Do you know....?
- Do you think .... is true?
- Choose between.. ?

In this case the possible answers are "yes", "no", "a choice number" or "!" (I don't know).

Example:

Question to ask: **Kx Consultant**

What kind of help do you need ?
Choose one :
1: What you have to know before starting.
2: Data management guide.
3: Data premilinary description guide.
4: Data Analysis guide.
5: Data editing guide.
6: What you have to do if problems occur in a SICLA session.
7: Data exchange with SPAD system (Interface).
8: Perform a data preliminary analysis (Evaluation).
9: Strategy guide for a data array study.
10:Run procedures.

Figure 3.

The system design

CLAVECIN provides a convenient help to SICLA users in data analysis, data management, and processing errors. The system, (figure 4), can be seen as a piloting of SICLA system by the KX system using the knowledge base.

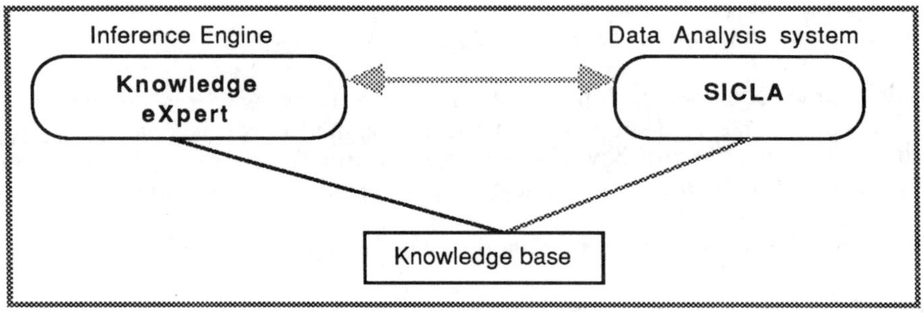

Figure 4.

A CLAVECIN session consists of the following steps :

1. The initial generation of the fact base of the SICLA's environment by transmitting the information of the structured data from SICLA to CLAVECIN

2. The determination of the user's problem and needs by dialogue. The user interacts with KX by a dialogue for the acquisition of facts.

3. After having chosen one or more SICLA's procedures, the user can execute them without leaving the expert system (figure 5).

Knowledge base :

The base deals with the knowledge as follows:

- Knowledge on the preliminary steps in data analysis: description, edition, management and evaluation of data, coding, ...

- Knowledge on SICLA methods: The methods are grouped in themes (clustering, discrimination, factorial analysis,...) according to their purposes. The conditional part of each rule expresses the running context of methods (kind of data needed and purpose).

An example of a rule :

IF	type of statistical problem is clustering
AND	type of clustering wanted is partition
AND	all variables are quantitative and homogeneous
AND	clustering variables is wanted
THEN	
	USE **TRANSP** to transpose data
AND	USE **MNDQAN** to classify.

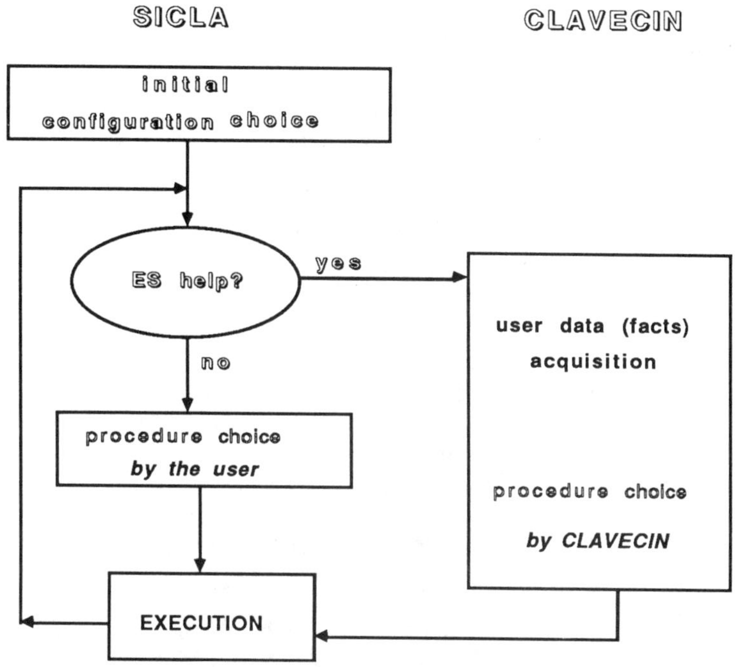

Figure 5.

- Knowledge on data analysis strategies: for each kind of data : quantitative, qualitative, heterogeneous data, matrix of similarity, ... a selection of suitable procedures is proposed. It covers the different steps of a data analysis session.

- Knowledge on processing errors: actions are proposed to correct errors which occur during a SICLA session.

The knowledge is represented in two forms :

1) A file of the production rules, the main source of expertise.

2) A file of commentaries where purposes and parameters of methods are detailed. This solution allows the insertion of texts or graphics to explain the conclusions. It is easy to modify these files.

Discussion

In most statistical expert systems, the inference engine has been specially developed for the application. In this case the inference engine is necessarily limited and domain dependent. Our solution takes advantage of the features of the inference engine KX : user friendly, powerful, and domain independent. Different knowledge bases and not only statistical knowledge bases can be processed. The problem of diffusion and maintainability is simplified.

We have chosen this tool for many reasons :

1: the KX formalism provides facilities to write a knowledge base.

2: the system is interfaced with the external environment and is able to run load modules which produce new facts that KX will take into account in the future expertise.

3: it is a powerful system which can explain the logical reason of a question but can give more information (precisions) from an expert point of view.

4: it is easy to translate the expertise into other foreign languages : expert explanations, texts related to propositions, messages are in files which are different to the knowledge base file and accessible to the users.

5: it also runs on micro computers such as IBM-PC compatible.

4. LEARNING SYSTEMS AND DATA ANALYSIS

4.1. Introduction

A great problem in designing E.S. is knowledge acquisition. Most of the time, it is difficult to get knowledge from domain expert (doctor,chemist,...) due to the communication problem (we don't speak the same language, it is not easy to express knowledge) . For that reason, A.I. techniques, what we call *learning systems* [Michalski and al., 1983] is nowadays developing for automatic knowledge acquisition. Learning methods may be viewed as forms of exploratory data. The purpose of both learning and data analysis techniques is to detect structures within data. Nevertheless, they greatly differ on the following points :

1) *Data representation and mathematics tools* : D.A. techniques deal with numerical representations of data. The mathematics tool is linear algebra. Metric spaces play a great part in D.A. techniques. On the other hand A.I. methods deal with symbolic data as assertions, rules, qualitative data. It is possible to deal with logical or hierarchical relations between variables. The mathematics tool is the logical theory.

2)*Efficiency or understandability :* D.A. methods are efficient and powerful to process large arrays. Resulted structures (clusters,factorial axis,...) are represented by numerical summaries. A usual complaint among data analysts is that D.A. results are difficult to interpret. For its part, A.I. techniques emphasize the ability to understand the results. These techniques are devoted to finding concepts and symbolic descriptions of structures which are easy to understand.

For instance, the methods of numerical taxonomy (clusters, hierarchical analysis) give an extensional description of clusters : the objects which belong to each cluster of the partition are listed. The clusters are characterized by elementary statistics . With this information, the data analyst has to find out the underlying conceptual structure. The conceptual clustering methods [Michalski and Stepp, 1983] enlarge the numerical taxonomy methods. In conceptual clustering, a cluster is not represented as a simple enumeration of objects but by rules which define membership. A concept will be defined as a collection of rules. For instance the concept of "high consumer" is characterized by the following rules:

"IF (cigarettes smoked > 2 packets) AND (beer drunk > 2 liters) OR
IF (sex=woman) AND (sleeping pill=yes) THEN "high consumer"

In both numerical and conceptual clustering, we distinguish an aggregation step (how to find a good partition) and a characterization step : how to describe the clusters. We are interested in the second point, namely to find intentional rules to characterize clusters or structures resulting from data analysis methods.

4.2 An inductive learning algorithm

The techniques which look for recognition rules for clusters are called in A.I. learning techniques from examples. More precisely, the objects of one cluster are called examples and the other objects counter-examples. These methods induce recognition functions which describe every example and none of the counter-examples. In D.A. the decision tree approach applied to qualitative data gives recognition rules.

The proposed algorithm combines Data Analysis and Artificial Intelligence points of view. It looks for easily understandable descriptions of clusters and uses hierarchical techniques to reduce the order of algorithm complexity.

Data representation

We note Ω a set of n objects and $X_1,\dots, X_p$ a set of variables : functions defined into Ω, $X_j : \Omega \rightarrow D_j$, for $j=1,\dots,p$ where D_j is a set of finite elements. An object of Ω has p components : $\omega=(d_1,\dots d_p)$ and $X_j(\omega)=d_j$ with $d_j \in D_j$. A couple (X_j,d_j), defines an elementary assertion $A = (X_j=d_j)$ which is a boolean function defined into Ω:

$$A = (X_j=d_j) : \Omega \rightarrow \{false, true\}$$
$$\omega=(d_1,\dots,d_p) \rightarrow A(\omega)=true \text{ if } X_j(\omega)=d_j \text{ and false if not}$$

We shall consider assertions as conjunctions of elementary assertions. For instance :

$A=\wedge\{ (X_j=d_j \mid j\in J'\}$ where $J \supset J'$

We note ω_A the set of objects recognized by the assertion A. The length of A is the number of elementary assertions which compose A: card(J')=length(A).

If $A=\wedge\{ (X_k=d_k \mid k\in J\}$ and $B=\wedge\{ (X_l=d_l \mid l\in J\}$ are two assertions, if $\omega_A \supset \omega_B$, we have: $B \Rightarrow A$ and we say that A is more general than B.

Let M be the indexes set related to the common terms of A and B: $M=\{m \mid d_k=d_l, k\in J$ and $l\in J\}$. If $M\neq\emptyset$ then the *generalization* of A and B is the longest assertion more general than A and B which is defined as following :

$Gen(A,B)=\wedge\{ (X_m=d_m \mid m\in M\}$

Example :

X_1: nationality — D_1: { French, Belgian, European, American, African}
X_2:sex — D_2: {man, woman}
X_3:occupation — D_3: {worker, employee, manager}

$A=(X_1=French)\wedge(X_2=woman)\wedge(X_3=manager)$
$B=(X_1=American)\wedge(X_2=woman)\wedge(X_3=manager)$
$C=(X_1=Belgian)\wedge(X_2=woman)\wedge(X_3=worker)$

We have $Gen(A,B)=(X_2=woman)\wedge(X_3=manager)$

Problem formalization

We have $\Omega = E \cup CE$ where E is the set of examples and CE the counter-examples. The problem is to find *good* assertions which recognize E and not CE. Each object $\omega=(d_1,...,d_p)$ is related to the assertion $A_\omega=\wedge\{ (Xj=dj) \mid j\in J\}$. The system $S_E= \{A_\omega \mid \omega\in E\}$ recognizes the examples and not the counter-examples. We want to find an assertion set more general than S_E which recognizes the examples and not the counter-examples.

Algorithm description

a) Initial properties :

The systematic research of all the assertions is a combinational problem. To reduce the complexity, we look for an assertion set according to discriminating and covering criteria. More precisely the assertions must satisfy the three following conditions:

1/ An assertion A has to be composed of l_{min} terms at least and l_{max} terms at most :

$l_{min} < length(A) < l_{max}$

Numbers l_{min} and l_{max} are integers between 2 and p.

2/ An assertion A has to be α-discriminating. It means that the number of counter-examples recognized by A must be inferior to α :

$card(\omega_A \cap CE) < \alpha$

If we want the assertion A not to recognize any counter-examples, we must take $\alpha=1$.

3/ An assertion A must recognize β examples at least :

$\beta < card(\omega_A \cap E)$

This condition allows the elimination of assertions which reflect too particular observations and provides security for a certain robustness of results.

b) The GENREG algorithm :

The algorithm, called GENREG, can be described as follows:

The generalized assertions are computed from assertions related to couples of examples. Assertions which do not satisfy condition 1) and condition 2), are eliminated. The whole process is repeated on these first generated assertions, and so on, until stabilization. We then take out of the rule base, those particular assertions that do not respect condition 3). We have a finite number of examples, the generated assertions lengths become smaller and smaller then the algorithm converges.

Let S be an assertion set. The '*generalized* ' calculation is based on the following resemblance measure, defined on S:

$A,B\in S$ $s(A,B) = 0$ if A and B have no common term
$s(A,B)=m$ if m is the length of Gen(A,B)

So s(A,B) is the number of the common terms of A and B.

The algorithm works out Gen(S) set calculation by including Gen(A,B) in Gen(S), if the length of Gen(A,B) verifies the previous condition.

c) Quality criteria for an assertion set :

The consequence of the third condition is that for some examples of E, it might be impossible to find, in the rule base, assertions which generalize them. Let E_S be the example set recognized by S. The percentage ρ_{wc} of well classified objects that is to say the percentage of examples recognized by S, is a criteria for measuring the quality of the recognition potential of an assertion set S:

ρ_{wc}= card(E_S) x 100 /card(E)

In the same way, let CES be the counter-example set recognized by S . The discrimination quality of the system S is measured by the percentage of badly classified objects ρ_{wc}:

ρ_{bc}= card(CE_S) x 100 /card(E)

If $\alpha = 1$, assertions do not recognize any counter-example and $\rho_{bc} = 0$.

d) Algorithm optimality :

We can prove that the proposed algorithm is optimal from the following point of view : it finds the most covering assertion(s) at the imposed conditions.

e) Algorithm complexity :

Firstly, one must calculate n(n-1)/2 similarity measures to obtain the generalized assertions. Algorithm complexity is at least n^2. Nevertheless, it is possible by appropriate methods to reduce it. If it is possible to find a partition $P=(P_1,...,P_K)$ of Ω such as : $\forall\ \omega_i \in Pi$ and $\forall\ \omega_j \in Pj$ with $i \neq j$ and $s(\omega_i,\omega_j) > l_{min}$ then it is useless to calculate similarity measures for objects which belong to distinct classes of the partition . If n_k denotes class k cardinality, the algorithm complexity becomes: $\sum_k n^2_k << n^2$. Such a partition can be realized by a descending connexity algorithm related to the single link method. For more details, see [Ralambondrainy, 1987] .

Application to Data Analysis structures :

The previous algorithm may be used to interpret the partition obtained by clustering methods applied to qualitative or binary variables. If the variables are quantitative, they must be transformed into binary variables. The algorithm GENREG can give help in describing factorial axis obtained by Multiple Correspondence Analysis (Principal Components Analysis applied to qualitative data) [Lebart and al., 1984] . For instance, the application of Multiple Correspondence Analysis, on data concerning students consumption of drugs,beer,alcohol,... gives the following result that the first axis discriminates "high consumers" against "weak consumers". We can precise the concept of "high consumers" by giving rules which describe them. For this purpose, we create the examples set as the students which are significantly correlated to the first axis. The other students are the counter-examples. We select the modalities, which have significantly contributed to the first axis. So we have a set of examples and counter-examples, characterized by binary variables, related to the concept of "high consumers", then the algorithm can be applied.

Generalization to multivalued and hierarchical variables

Classical D.A. techniques deal with data matrices where variables have a single value and for a given object, every variable must be defined. The inductive learning algorithm GENREG can be extended to multivalued variables, extra knowledge on variables is accepted and every variable needs not to be defined for the objects.

For instance, the variable X_1: "nationality" , of the example above, can be considered as a multivalued and hierarchical variable if we have set that European={ French,Belgian}. The hierarchy is as follows:

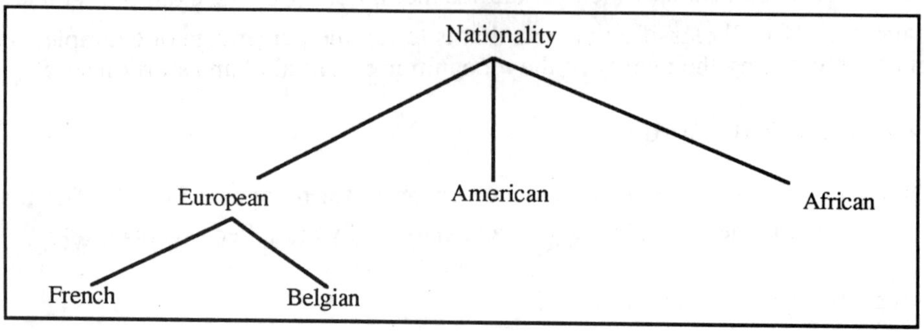

Example :

$A=(X_1=\text{European})\wedge(X_2=\text{woman})\wedge(X_3=\text{manager})$
$B=(X_1=\text{French})\wedge(X_2=\text{woman})\wedge(X_3=\text{manager})$
$C=(X_1=\text{Belgian})\wedge(X_2=\text{woman})\wedge(X_3=\text{worker})$

The hierarchy structure of the variable nationality gives the following implications :

French $\Rightarrow$ European and Belgian $\Rightarrow$ European.

So the generalization of the assertions B and C can be calculated :

$\text{Gen}(B,C)=(X_1=\text{European})\wedge(X_2=\text{woman})$ and $\text{Gen}(A,B,C)=(X_1=\text{European})\wedge(X_2=\text{woman})$

An example of application : edibility of mushrooms

The program GENREG was applied to generate rules which recognize mushrooms. We have a set of 630 observations on mushrooms characterized by 18 qualitative attributes :

(Hat color, foot color, flesh color, leaf color, leaf form, spore color, hat form, hat size, foot form, ring ?, splinter ?, volve ?, cortine ?, foot revetment, foot texture, does the flesh exude liquid ?, edibility, harvest season)

Each attribute has a set of possible modalities. For instance, for the attribute "hat color", the set of modalities is :

(white and whitish, red, black, pale to dark green, pale to dark yellow, pale to dark brown, pale to dark orange, pale to dark grey, pale to dark violet, pale to dark rose)

The mushrooms are then grouped into 7 classes with respect to their edibility :

23 : *fatal mushrooms*
80 : *heavily toxic mushrooms*
108 : *slightly toxic*
58 : *very barely edibility*
35 : *barely edibility*
209 : *good tasting*
87 : *excellent tasting*

From the data, the GENREG method gives 56 rules which may be used to recognize the mushrooms edibility :

Fatal mushrooms	are recognized by 3 rules : length min=14, length max=16
Heavily toxic mushrooms	are recognized by 6 rules : length min=9, length max=16
Lightly toxic mushrooms	are recognized by 5 rules : length min=6, length max=9
Very barely edibility mushrooms	are recognized by 8 rules: length min=5, length max=10
Barely edibility mushrooms	are recognized by 6 rules : length min=7, length max=15
Good tasting mushrooms	are recognized by 15 rules : length min=5, length max=14
Excellent tasting mushrooms	are recognized by 12 rules : length min=7, length max=15

An type of rule:

If	*foot color*	*= white*
and	*leaf color*	*= white*
and	*foot form*	*= bulbous*
and	*volve*	*= no*
and	*ring*	*= yes*
and	*splinter*	*= yes*
and	*cortine*	*= no*
and	*harvest season*	*= Spring/Summer/Autumn*

then ***excellent tasting***

5. CONCLUSION

We have shown in the present article, the interest of data analysis for statistical requests to relational data bases. The relational model is well adapted to data analysis because the common objects manipulated by statistical methods are easily represented by relations and therefore that relational D.B.M.S. 's are very useful for data analysis applications. The data analysis point of view enlarges the field of requests performed on a relational base. The PEPIN-SICLA prototype is operational on the Multics system of INRIA -Rocquencourt (France).

We have presented two applications of A.I. in statistics.The first one dealt with expert system in statistics. We think that expert systems such as CLAVECIN, which provides statistical and practical assistance, is more efficient than classical helps due to the system's ability to justify its reasoning, recommendations and choices. The system is operational on Multics and also on micro computer such as IBM-PC. The second application dealt with the learning algorithm GENREG. We noticed that A.I. techniques give importance to the understandability and to finding comprehensible concepts. D.A. techniques emphasize numerical efficiency. The method proposed combines the two points of view and it is a good tool for generating rules from data. This work is a first step to enlarging the application field of D.A. methods in analyzing more complex data than the classical data matrix.

For these applications, we have shown that it is a better approach to combine specialized systems in data management, data analysis and in artificial intelligence than to extend functionalities of a specific system. But this work would not have been possible without the collaboration of different teams in data management, data analysis and in artificial intelligence and the availability of the PEPIN, SICLA and K_X systems.

REFERENCES

Bouchet, P., Chesnais, A., Feuvre, J.M., Jomier,G., and Kurinckx, A., (1981), "Databases for microcomputers: the PEPIN approach".
ACM SIGMOD/SIGSMALL Newsletter, Vol 7, ≠2, Oct. 1981.

Carlsen, F., and Heuch, I., (1986), "Expert system utilizing standard statistical packages". Compstat 1986, Vienna: Physica-Verlag.

Chambers, JM, (1981), "Some thoughts on expert software." Proc. 13th Symp. Interf. Comp. Sc. Statist., New York , Springer 36-40.

Darius, P.L., (1986),"Building expert systems with the help of existing statistical software". Compstat 1986, Vienna: Physica-Verlag.

Demonchaux, E., Quinqueton, J., and Ralambondrainy, H.(1985), "CLAVECIN, un système expert en Analyse des Données". Rapport de recherche num. 431, Juillet 1985.

Demonchaux, E., Zinger J.L., (1987), "Knowledge Expert : Le Manuel de référence". Rapport Technique SEMA-METRA, Septembre 1987.

Gale, W. A.. (1985), "Artificial Intelligence and statistics". Editor Addison-Wesley Reading, Massachussetts.

Ho Tu Bao (1987), "Systèmes experts et Analyse des Données". Thèse de l'Université Paris 6, Janvier 1987.

Kobayashi, Y., and Ikeda, H., (1981), "Additional facilities of a conventional D.B.M.S. to support interactive statistical analysis".
1st LBL Workshop on SDBMS, Dec. 1981.

Lebart, L., and Morineau, A., and Warwick, K.M. J.P.,(1984), "Multivariate descriptive statistical analysis". NewYork: Wiley

Lefons, E., Silverstri, A., and Tangorra, F., (1983), "An analytic approach to statitical data bases". VLDB, Firenze, Italy, Nov.1983.

Michalski, R., Stepp R., (1983) " Learning from observations: Conceptual Classification". In *Machine Learning*, Michalski, R., Carbonell, J., and Mitchell, T. editor Palo Alto, Calif: Tioga Publishing Company.

Ozsoyoglu, G., and Ozsoyoglu, ZM., "Statistical database query langages". IEEE, Soft. Eng. Vol SE-11, no 10, Oct. 85.

Pregibon,D., and Gale W.A., (1984) "REX : an expert system for regression analysis". Compstat 1984, Vienna: Physica-Verlag.

Ralambondrainy, H., (1983), "Le système SICLA". 3rd, International Symposium on Data Analysis, Versailles 1983.

Ralambondrainy, H., (1987) "Génération de règles à partir de structures d'Analyse des Données". Research Report INRIA (to appear)

RULE BASED CLASSIFICATION OF IUE SPECTRA

R. Rampazzo (1)
A. Heck (2)
F. Murtagh (1)
R. Albrecht (1)

(1) Space Telescope–European Coordinating Facility, European Southern Observatory, Karl–Schwarzschild–Str. 2, D–8046 Garching, FR Germany. (Affiliated to Astrophysics Div., Space Science Dept., European Space Agency.)

(2) Centre de Données de Strasbourg, Observatoire Astronomique, 11 rue de l'Université, F–67000 Strasbourg, France.

1. INTRODUCTION

Classification of celestial objects is one of the central themes in astronomy and artificial intelligence techniques have already demonstrated their suitability for this purpose (Bernat and McGraw, 1986; Thonnat, Granger and Berthod, 1985). The increasing availability of stellar spectra in wavelength ranges other than the visual creates new problems for classifiers. In fact, the MK classification scheme that is defined on the visual range cannot, for example, be immediately extended to wavelength regions for which it is not calibrated, although a classical approach, similar to the MK one, may be used (Jaschek and Jaschek, 1984, 1987). This, together with the increasing quality and quantity of data available forces the astronomer to attempt various computer–aided automatic classification approaches (Kurtz, 1983).

The low resolution ultraviolet stellar spectra obtained with IUE are fairly homogeneous. Although they do not have the characteristics of a survey, it has nonetheless been possible to characterise standard stars independently of their visual classification (see e.g. Heck, 1987).

The ultraviolet classification of "normal" stars has already been defined (Heck et al., 1984) and that of "peculiar" stars has been started, but is not yet completed because of the even bigger lack of correspondence between the visual and the UV spectral ranges (Heck et al., 1986).

We present here the backbone of a Rule Based Classifier, the aim of which is to reproduce automatically the IUE classification, starting from a wavelength calibrated spectrum. The grid of the classification rules are directly deduced from a well classified set of stars on the basis of 29 parameters. Of these, 28 are a measure of line intensities and one, the asymmetry coefficient, a parameter that describes the continua and on which a reddening correction has already (empirically) been performed. Their definition is similar to that adopted in Heck et al. (1986).

The system (Fig. 1) — now partly implemented — analyzes the spectrum by building an initial command language procedure (in ESO's MIDAS image processing system) which roughly measures the features of the spectrum (e.g. the asymmetry coefficient). A table of values is passed to the expert system (ES) which uses low level rules to obtain general–type information (e.g. that the spectrum under investigation is a late type star). From these rules the ES gives information on the lines that have not been found, but potentially are present; it then builds a new command procedure to instruct the host system to find and measure these line intensities. A more precise classification is derived and a comparison with standard stars in the knowledge base (KB) gives an indication as to when to interrupt this iterative improvement.

Each operation on the spectrum is stored, together with the table of parameter values extracted from the original spectrum, in one unit working as a *blackboard* (a common data–structure which different parts of the expert system write into: see Noordzij, 1987). The blackboard reports on the actual status of the classification.

In the case in which no match with a standard spectrum is found the unit (i.e. the spectrum under investigation) will become part of the KB, defining, itself, a new class.

The advantages of using a rule based approach based on a development tool (KEE in our

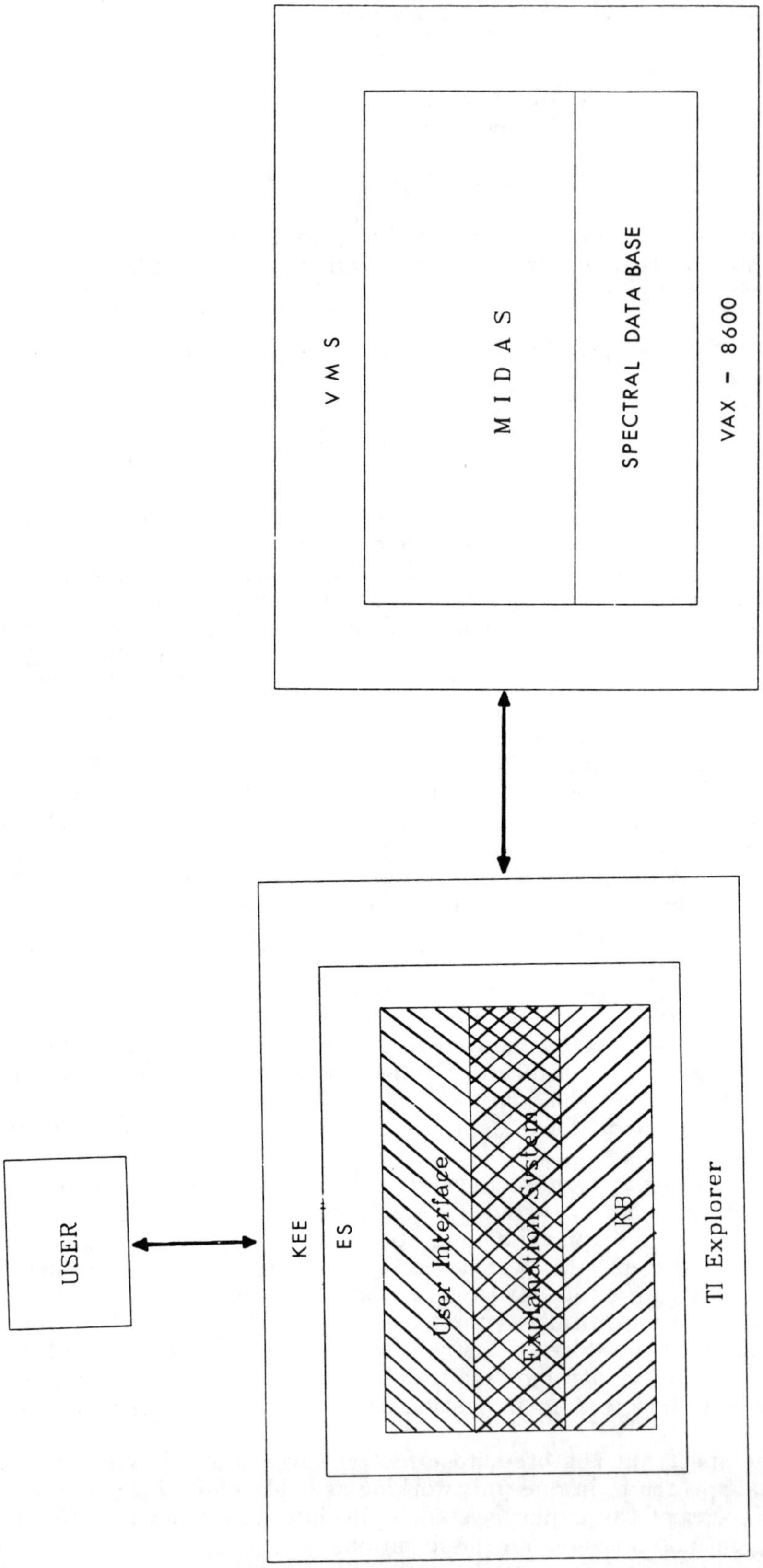

Fig. 1: Schematic structure of the ES and interaction with the host. The philosophy is that all operations involving computations have to be performed using well established and tested procedures in the MIDAS command language, with only the “reasoning” and classification being demanded of the ES. This latter, other than building batch procedures to analyse spectra, controls I/O and gives explanations to the user about the criteria used in the classification process.

case: see below) are that it permits:

(1) ease of programming and debugging;

(2) transportability (we started to develop the system on a Sun 3/160, we transported it to a Symbolics 3620 and then to a Texas Instrument Explorer);

(3) ease of control through various window processes, ease of plotting results, and of structuring the knowledge base;

(4) it allows the possibility of enlarging the knowledge base to classify objects in a different spectral range or even completely different objects; and

(5) it permits a visiting user to quickly work without prior knowledge of the image processing system, offering to him a complex batch procedure which is potentially modifiable by experts in subsequent implementations.

2. PROTOTYPE SYSTEM STRUCTURE

The basic parts of the system structure are the **Knowledge Base**, the **Explanation Subsystem** and the **User Interface**. The tool providing the shell in which the system is built is the KEE 3 *Knowledge Engineering Environment* produced by IntelliCorp (cf. general articles describing systems built with KEE: Fikes and Kehler, 1985, or Faught, 1986; or comparative appraisals of systems such as Wall et al., 1985, Gilmore et al, 1986, or Cross, 1986.

2.1 Knowledge Base (KB).

The knowledge base is composed of:

(a) standard stars for each class which the system tries to classify;
(b) the rules used for the classification;
(c) data and rules associated with the classification, but not directly connected with the spectral parameters extracted from the spectrum. This information may be, for instance, the colour (U–B) etc. which will be useful in the classification;
(d) new classes with their parameters specified by the interactive classification (e.g. class of peculiar stars).

The IF-THEN-DO rule structure in the KEE shell permits easily to build and, if necessary, modify the classification grid. The KEE backward chaining mechanism controls the sequence of operation of the rules (i.e. the system attempts to derive answers that are not available in the knowledge base, using rules and facts in this latter; if necessary information may be asked of the user during the reasoning process).

2.2 Explanation Subsystem.

This is clearly connected with the knowledge inside the the KB and the User Interface. It contains:

(a) information on the spectral features used to classify each spectral type;
(b) the lines expected, even if not used in the classification procedure, for each spectral type. This will be used in the iterative classification procedure, when the spectrum does not find an appropriate classification;
(c) explanation for the use of criteria other than the parameters of the spectrum for the classification;
(d) the criteria used when a new class is identified (particular lines, etc.);
(e) the lines that we expect to see with different signal-to-noise ratios.

2.3 User Interface.

The User Interface is mouse–oriented: the user may carry out all major operations by "clicking" on the relevant windows.

The lower level User Interface is

(a) the *KEE typescript window* through which the user exchanges information with the system, answers questions, and gets reponse;

(b) the *KEE output window* in which KB modifications may be followed graphically;

(c) a *KEE Lisp Listener window* which may be useful for the experienced user (but whose use is not necessary).

Upper level User Interface features include:

(1) **control panels** that perform the following:

(a) initialize the procedure;
(b) give help to start the procedure;
(c) start the classification;
(d) build MIDAS procedures; and
(e) comunicate with the host controller, which communicates with the MIDAS environment.

(2) a **selection panel** to select the spectrum to be classified from among those in the KB.

(3) an **explanation panel** which permits the user to follow step by step the processes carried out by the ES (e.g. the lines used for classification, etc.). This is the display for the explanation subsytem.

3. CURRENT DEVELOPMENT STATUS

An initial hierarchy of rules which starts identifying gross spectral classification, starting from the values of the asymmetry coefficient (see fig. 1), and achieving a detailed classification by working on the line intensities, has been implemented and tested on a set of 229 well classified spectra (Heck et al., 1984). In particular the rules to classify a set of normal O, B, A, F, and G supergiant, giant and d stars have been developed (fig. 2).

The procedure for extraction of the seeds to build the rules, currently made using pairwise plots of the lines in a manner similar to that shown in fig. 2 will in the future be carried out automatically.

Along the lines previously sketched out, a primitive User Interface (selection panel, control panels, I/O with the MIDAS environment) has been built. More flexible communication with the MIDAS environment and with the IDM database machine is necessary and is foreseen for the future.

4. CURRENT OBJECTIVES

As already stated, classification is a discipline widely used and shared by various branches of astronomy. The objects classified and the techniques used may vary, but the general process is very similar. A classifier ought therefore to be a general and reusable structure, well separated from the specific problem or technique used. Obtaining data or parameters, in fact, can more easily and efficiently be done through traditional procedural programming, especially using image processing packages already available to the astronomical community.

The experience being acquired in building this kind of ES relates to:

(1) building and controlling data reduction of large amounts of data through ES (all the data obtained will be reduced in the same way; Johnston, 1987);

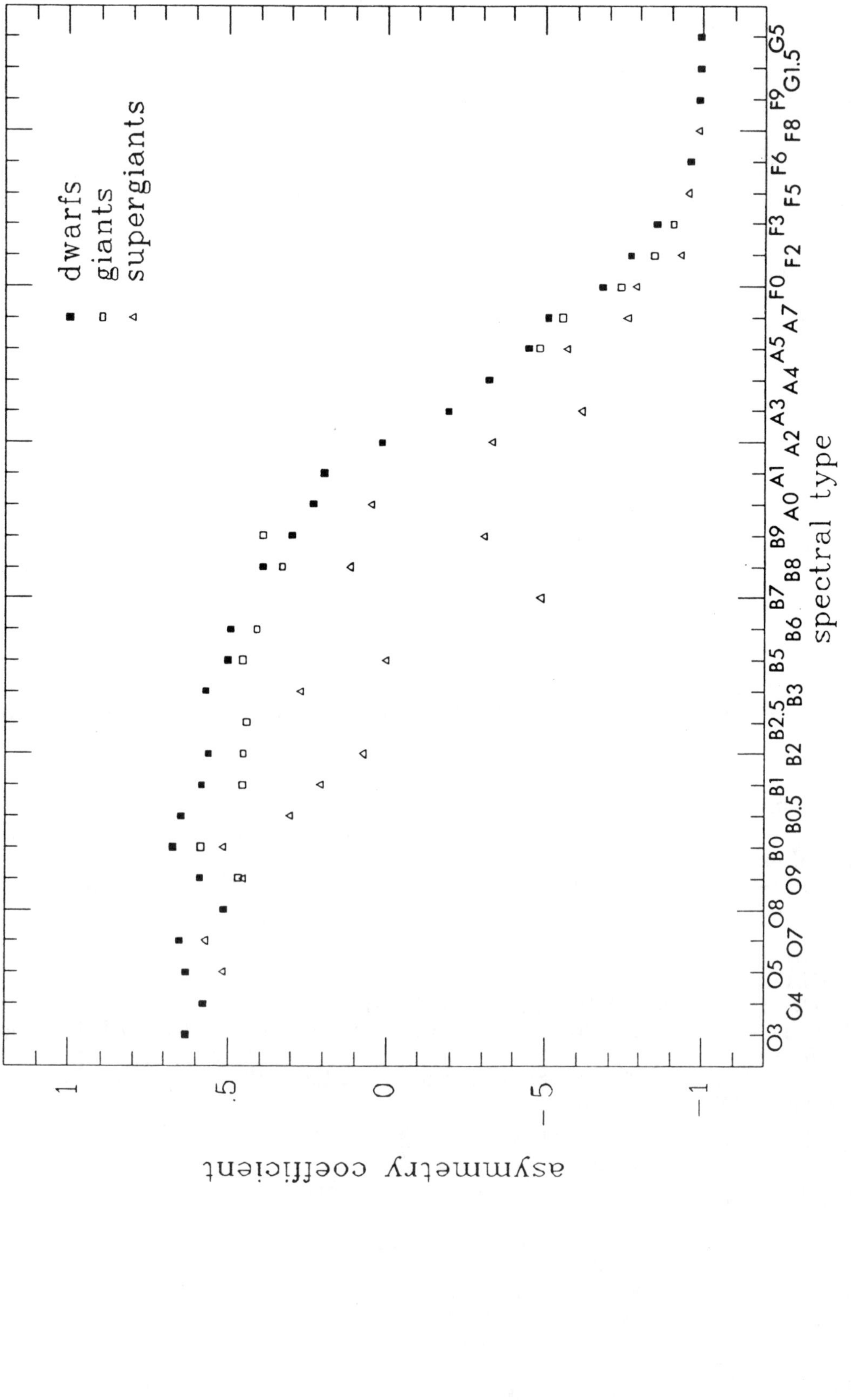

Fig. 2: The trend of the asymmetry coefficient as a function of stars' spectral types. This coefficient, which is a measure of the shape of the continuum, proves to be useful in building the grid of rules, in particular low level rules, that give the first hint of the spectral type.

(2) learning, where we seek to recognize similar structures in large data sets, thereby to analyse the data and to re–classify them on a clearer and more consistent basis;

(3) generating an ES structure that may reused in classifying different objects, and not only stars.

This will be useful when the interval of spectral range of celestial objects runs from the far ultraviolet to the far infrared, thereby enlarging considerably the possible set of peculiarities pertaining to an "optically normal" star.

Acknowledgements

Discussions and advice from Mark Johnston and Hans–Martin Adorf are appreciated.

REFERENCES

Bernat, A.P. and McGraw, J.T. 1986: in *Instrumentation in Astronomy VI*, ed. D. Crawford, SPIE, 89.
Cross, G.R. 1986: *Optical Engineering*, **25**, 436.
Faught, W.S. 1986: *Computer*, **19**, 17.
Fikes, R. and Kehler, T. 1985: *Communications of the ACM*, **28**, 904.
Gilmore, J.F., Howard, C. and Pulaski, K. 1986: in *SPIE Vol. 657 Applications of Artificial Intelligence IV*, 194.
Heck, A. 1987: in *Exploring the Universe with the IUE Satellite*, ed. Y. Kondo *et al.*, D. Reidel Publ. Co., Dordrecht, 121.
Heck, A., Egret, D., Jaschek, M., and Jaschek, C. 1984: *IUE Low Resolution Spectra Reference Atlas, ESA SP–1052*
Heck, A., Egret, D., Nobelis, Ph. and Turlot, J.C. 1986a: *Astrophysics and Space Science*, **120**, 223.
Heck, A., Egret, D., Hassall, B.J.M., Jaschek, C., Jaschek, M. and Talavera, A. 1986b: in *New Insights in Astrophysics — Eight Years of UV Astronomy with IUE*, ESA–SP 263, ed. E. Rolfe, 661.
Jaschek, C., Jaschek, M. 1984: in *The MK Process and Stellar Classification*, ed. R.F. Garrison, Dunlap. Obs., 290.
Jaschek, C. and Jaschek, M. 1987: *The Classification of Stars*, Cambridge Univ. Press.
Johnston, M. 1987: in *Space Applications of Artificial Intelligence and Robotics*, NASA, Goddard Space Flight Center.
Kurtz, M.J. 1983: in *the MK Process and Stellar Classification*, ed. R.F. Garrison, Davis Dunlap Observatory.
Noordzij, J. 1987: *Systems International*, March 1987, 57.
Thonnat, M., Granger, C. and Berthod, M. 1985: *Computer Vision and Pattern Recognition*, IEEE, 206.
Wall, R.S., Apon, A.W., Beal, J., Gately, M.T. and Oren, L.G. 1985: *Data and Knowledge Engineering*, **1**, 279.

AUTOMATED GALAXY RECOGNITION

Barry Rappaport, Computing Research Laboratory and Department of Astronomy, New Mexico State University, Box 3CRL, Las Cruces, New Mexico, 88003/USA

Kurt Anderson, Department of Astronomy, Box 4500, New Mexico State University, Las Cruces, New Mexico, 88003/USA

Abstract

Previous approaches to automated image processing have used both deterministic and non-deterministic techniques. These have not used any form of conceptual learning nor have they employed artificial intelligence techniques. Addition of such techniques to the task of image processing may significantly enhance the efficiencies and accuracies of the recognition and classification processes. In our application, the objects to be recognized and classified are galaxies.

Introduction

Classification presupposes recognition. The first step is to distinguish galaxies from stars and extended objects such as Planetary Nebulae and H II regions. Automated galaxy recognition has been done using classical pattern recognition techniques by various investigators. These recognition techniques can be categorized as deterministic or non-deterministic in nature. An example of a non-deterministic approach is the Bayesian classifier; this uses a statistical model for the probabilities of a pattern class to form a suitable classifier. Sebok (1979) successfully employed a Bayesian classifier to distinguish between non-stellar and stellar images on digitized photographic plates. Jarvis and Tyson (1981) used a deterministic classifier with their FOCAS (Faint Object Classification and Analysis System); extremely faint non-stellar objects could be identified. Fourier analysis is a third pattern classification technique which has not been fully exploited in the area of galaxy recognition.

The problems with most traditional classifiers is that they require the images' calibration and classification to have been taken under the same conditions using the same equipment. One of the primary goals of automated galaxy classification should be a classifier that can classify images taken on different equipment. It is hoped that a classifier developed from Artificial Intelligence (AI) techniques can be adapted to handle images taken from various sources with little or no initial massaging. The availability of an automated galaxy classification system will be important for many reasons. In particular, Space Telescope and other deep sky instruments will need such systems to rapidly identify and classify large number of images.

Artificial Intelligence' techniques fall into two categories. These are expert systems and various adaptive learning algorithms such as connectionism. Among the latter, the connectionist algorithm which utilizes backpropagation may be best suited for the task of pattern classification. It will be considered later in this paper. Finally, it should be noted that the dividing line between "expert systems" and "adaptive learning algorithms" is fuzzy at best.

Non-Deterministic Classification

One of the most useful non-deterministic classifiers is that based upon Bayesian decision theory. It employs a statistical model of the probabilities of pattern class (stellar or non-stellar, for example) to form the classifier. Each image is compared to this statistical model via the classifier and class membership probabilities are computed.

We might, for example, want to find $P(S|x)$, the probability that an object is a star given a set of measurements x. If we have some objects known to be stars, we can empirically determine $P(x|S)$, the probability of getting measurements x given that the object is a star. This is measurable from a photographic plate or other image recorder. Using Bayes' theorem, we can relate this to $P(S|x)$:

$$P(S \mid x) = \frac{P(x \mid S)P(S)}{P(x)}$$

where $P(S)$ is the a priori probability that the object is a star and $P(x)$ is the corresponding probability of getting measurements x.

If a classifier is needed to distinguish stellar and non-stellar images, a ratio of the probability of getting measurements x given the object is non-stellar, $P(x|\bar{S})$, and the probability of getting measurements x given the object is stellar, $P(x|S)$ can be used:

$$R = \frac{P(S \mid x)}{P(\bar{S} \mid x)} = \frac{P(x \mid S)P(S)}{P(x \mid \bar{S})P(\bar{S})}$$

where P(S) is the a priori probability that the object is non-stellar. The ratio $P(S)/P(\bar{S})$ can either be experimentally determined or can be left as an adjustable parameter. The problem then reduces to the modeling of $P(x|S)/P(x|\bar{S})$ from a "training set" of images of known type. The object is then classified as stellar if the computer value of R is greater than unity (for known $P(S)/P(\bar{S})$) or some other preset value.

Deterministic Classification

In FOCAS, Tyson and Jarvis (1981) use a deterministic classification scheme to separate and classify very faint images. Each image defines a seven-component feature vector. This vector contains information on the overall size, brightness, and intensity distribution. Clustering of the image vectors in 7-D space defines hyperellipsoids, multi-dimensional elliptical surfaces. The classification system is based upon the relationship of vector positions to the hyperellipsoids which have been created with information from a classification training set. With a good choice of classification parameters (vector components), the hyperellipsoids corresponding to the various classification types will be well-separated in the space.

Fourier Analysis

Fourier transforms (or at least a knowledge of the image intensity distributions) can be used in both deterministic and non-deterministic classification schemes. But there has not been developed any recognition method that is primarily based on Fourier transforms and Fourier analysis.

Images of galaxies and other non-stellar objects differ from stellar images in size and brightness patterns; they often display "structure" in the sense of non-monotonic intensity distributions (bars, rings, spiral arms, etc). Fourier transforms would show appreciable power at the corresponding scales. It may well be advantageous to use Fourier transforms for separation and classification of these objects. Indeed, Fourier transforms have already been used to separate and classify such objects as arrowheads and pottery shards. (Phillips, et al, 1987) The techniques should prove equally useful in the morphological classification of galaxies and other extended objects.

Expert Systems

An expert system is a computer program which simulates (and hopefully stimulates!) the thinking of the expert in the solving of a particular problem. Classically, the expert system uses a set of rules which are formulated by the expert and coded into the program. These rules can be thought of as the strategy that the expert uses to solve a problem. A simple example would be to define a fuzzy image as that of a non-stellar object. An expert-based quantification of "fuzzy," presumably based upon the sizes of stellar images, would be required.

Expert systems are now evolving from ruled-based systems to systems which act as "conductors." The expert system controls the data flow to and from the subprograms which actually do the problem solving. This may lead to hybrid AI programs - programs which are adaptive learning systems with an expert system framework.

Adaptive Learning Systems

Adaptive learning systems are computer programs which are able to change their own data flow based on input training sets; they can learn by example. The adaptive learning system which shows the most promise in the area of pattern classification is that of parallel connectionism with backpropagation.

Connectionism is a collection of the models of algorithms which control the flow of a neural network (directed graph). The network contains input and output nodes. The input nodes are the observational quantities. The output nodes are decision points within the problem solving process. For example, if the problem is to separate images into stellar and non-stellar categories, the input nodes could be the overall image size, brightness, and intensity distribution; while the output nodes could be "stellar" or "non-stellar." The connections between the nodes are weighted so that when certain input nodes are activated, the appropriate output node is turned on. The weighting of the connections is controlled by a backpropagation algorithm. This algorithm manipulates the weights in accordance with inputs and outputs which have been established from training sets.

Conclusions

There will be many problems in trying to create expert systems or adaptive learning algorithms that can handle the complexities and subtleties of galactic recognition and classifaction problems. One of the major problems to be addressed is the representation of the image data for use in the AI systems. This is the key when dealing with expert systems or adaptive learning algorithms.

The work which has been undertaken at NMSU is initially geared towards the task of deciding which AI technique or combination of techniques is best suited for the process of recognition and classification. Our analysis will include a "figure of merit" parameter based upon: cost (computer time, required database, data precision, data acquisition cost, etc.) versus accuracy of classification. An actual classifier will be built following this analysis.

Acknowledgements

We wish to thank M. Coombs, E. Dietrich, J. Jordan, and R. Hartley for their help and assistance with this work.

References

Barr, A. and Feigenbaum, E. 1981. **The Handbook of Artificial Intelligence.** Stanford: Heuristech Press.

Dietterich, T. and Michalski, R. 1981. "Inductive Learning of Structural Descriptions," **Artificial Intelligence, 16**, 257-294.

Duda, R. and Hart, P. 1973. **Pattern Classification and Scene Analysis.** New York: John Wiley.

Jarvis, J.F. and Tyson, J.A. 1981. "FOCAS: Faint Object Classification and Analysis System," **Astron.J., 86**, 476-495.

Jones, W.P. and Hoskins, J. 1987. "Back-Propagation," **BYTE 12**, 155-162.

Nilsson, N. 1980. **Principles of Artificial Intelligence.** Palo Alto: Tioga.

Phillips, K., Lopez, P., Gonzales, N., and Graham, N. 1987. "Local Shape Analysis of Archeological Artifacts by Computer: A Feasibility Study," **Computing Research Laboratory Report** MCCS-87-103.

Rich, E. 1983. **Artificial Intelligence.** New York: McGraw-Hill.

Sebok, W. 1979. "Optimal Classification of Images into Stars or Galaxies - A Bayesian Approach," **Astron. J. 84**, 1526-1536.

Smolensky, P. 1987. "On the Proper Treatment of Connectionism." (Pre-print).

Tyson, J.A. and Jarvis, J.F. 1979. "Evolution of Galaxies: Automated faint object counts to 24th magnitude." **Ap.J. (Letters)**, L153-156.

AUTONOMOUS STAR IDENTIFICATION FOR SPACECRAFT ATTITUDE CONTROL

Barry Rappaport, Computing Research Laboratory and Department of Astronomy, New Mexico State University, Box 3CRL, Las Cruces, New Mexico, 88003/USA

Ted Dunning, Computing Research Laboratory, New Mexico State University, Box 3CRL, Las Cruces, New Mexico, 88003/USA

Jay Jordan, Department of Electrical Engineering, New Mexico State University, Box 30, Las Cruces, New Mexico, 88003/USA

Keith Phillips, Computing Research Laboratory and Department of Mathematics, New Mexico State University, Box 3CRL, Las Cruces, New Mexico, 88003/USA

Richard Stanton, Jet Propulsion Laboratory, 4800 Oak Grove Drive, Pasadena, California, 91109/USA

Abstract

For many years, NASA and various scientists from several countries have been conducting basic scientific research using sounding rockets to carry instruments into the upper atmosphere and nearby outer space. These rocket payloads, originally built with surplus military hardware, have evolved over the last 20 years into remarkably able spacecraft with gas jet pointing systems capable of accuracies in the neighborhood of 10 arc sec. This sounding rocket payload technology has recently been transferred to a new generation of spacecraft, designed to be carried into near space via the Space Shuttle and to be retrieved on completion of a mission lasting days instead of minutes for a typical sounding rocket payload. Along with this new generation space research craft, called SPARTAN, has come a need for an advanced attitude control and guidance system capable of recognizing a number of bright constellations and of tracking specified stars in each constellation.

Present attitude control systems require months of flight testing and experiment programming. The spacecraft must be released at a precise time and position. Any slipping of launch date requires new flight control software. An attitude control system with the capability of recognizing certain constellations would significantly reduce the time to set up an experiment, resulting in faster turn around for experimenters with greater scientific yield at lower cost.

A star tracker with the ability to track stars with extreme accuracy has been developed by JPL for NASA, it is called ASTROS I (Advanced Star and Target Reference Optical Scanner). This star tracker has the ability to acquire and track stars with an accuracy of 0.1 arc sec. This paper will explain how ASTROS I's capability is being enhanced as to allow it to not only track but recognize its pointing position without human intervention, using pattern recognition techniques developed at NMSU's Computing Research Laboratory. This will enable the ASTROS I star tracker to not only achieve its goals pertaining to the planetary and Mariner Mark II missions, but also to serve as a replacement for the current generation of star trackers used in other space and military spacecraft such as the SPARTAN series.

Introduction

The problem of autonomously identifying star fields onboard a spacecraft has received increasing attention in recent years. If it were possible to point in an arbitrary direction on the celestial sphere, automatically identify the field stars in that direction, and calculate the pointing vector and field orientation, a wide variety of space projects would benefit significantly. This result is now achievable with the advent of the new generation of CCD-based star trackers (JPL ASTROS series) and large memory flight computers. The purpose of this research is to demonstrate that previously developed field identification algorithms, coded on a flight-scale computer, can be linked to a breadboard ASTROS star tracker to calculate pointing vectors in real time.

The Computing Research Laboratory (CRL) and the Physical Science Laboratory (PSL) of New Mexico State University have a significant interest in star recognition capability. Involved in sounding rocket technology since the 1950's, PSL has participated in the recent transfer to a new generation of spacecraft designed to be carried into space via the shuttle and retrieved on compleiton of a mission lasting days instead of minutes. Along with this new generation of research spacecraft, called SPARTAN, has come a need for an advanced attitude control and guidance system capable of recognizing star patterns and tracking specific stars in each pattern. One of the major research areas in CRL is pattern recognition and computer vision.

Likewise, JPL spacecraft would benefit from this capability in several ways. First as a means of providing continuous attitude updates, a simple form of star pattern recognition is already implemented for Galileo. If expanded for use on a three axis stabilized spacecraft, star recognition would not only provide attitude updates, but also a means for recovering from attitude anomalies such as occasionally experienced by Mariner spacecraft. Finally, for telescope pointing applications, the automatic calculation of pointing vectors could be a boon to the operation of shuttle- or Space Station-based observatories. For example, much of the Astro-1 Mission Managers' concern over a backup in the event that the Instrument Pointing System fails to locate scientific targets would be alleviated if this capability were implemented.

The Method and the Algorithm

The underlying assumption related to stellar recognition is that the stars in the celestial sphere form unique patterns. A simple graph-searching algorithm was implemented in the initial study to find stellar positions using some of these patterns.

In this algorithm, two different databases are used. The first is the Yale Bright Star Catalog (YBS) database. It contains the position and magnitude (brightness) of all the YBS stars. The stellar positions are delineated in terms of right ascension and declination while the stellar magnitude is actual visual stellar magnitude rounded to the nearest half-magnitude. The second database contains the information obtained from the image to be searched, i.e. the image seen by the satellite star tracker. It contains stellar positions in terms of an x and y position and a calculated brightness. The brightness is calculated using a simple pixel area number to magnitude definition.

A pattern tree is formed from the image. The nodes of the tree represent the stars and are labeled with magnitudes. The brightest star in the image is the root node. The edges are all pairs in which one node is the root; the edges are labeled with the distance between the pair of stars forming the edge. In the initial study, there are eight such pairs.

After the tree is completed from the image, the bright star database is searched for all of the bright star pairs that have similar magnitudes and distance separations to the initial pair of search tree stars. These pairs are stored. The search continues through the bright star database with the next pair of stars in the search tree. In order for a bright star pair to be stored, one of the components must have been previously stored.

If only one match is found, the algorithm gives the position of the brightest star in the image; if no matches are found, this is an error condition. In either case, the search is stopped.

Results of the Initial Study

The algorithm has been tested on two different sample maps generated by the astronomical mapping program. Neither map contained overlapping images. In both cases, the algorithm correctly found the position of the brightest star within the field image.

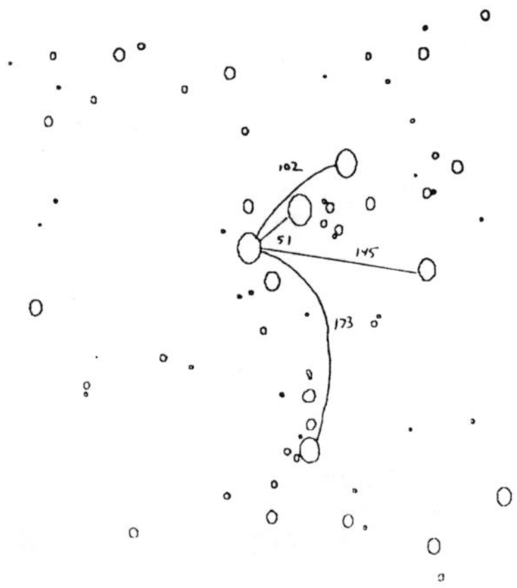

Fig. 1. Orion Region
(scale: 6cm = 11.5°)

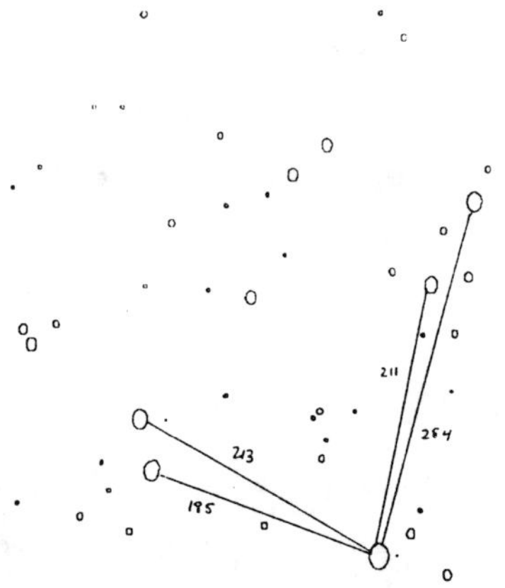

Fig. 2. Auriga Region
(scale: 6cm = 11.5°)

The first case was for an area in the constellation Orion. This area contained four bright stars: two second magnitude stars, one 2.5 magnitude star, and one third magnitude star. The algorithm only needed to use the distance between the two second magnitude stars to find the correct area of the sky. The first search cycle found only one match - the correct one. (Figure 1)

The second case was for an area in the constellation Auriga. This area contained three bright stars: one third magnitude star, and two fourth magnitude stars. The algorithm took three search cycles to correctly find the image area. (Figures 2 and 3)

The initial study proved the feasibility of using pattern recognition algorithms to increase the intelligence of satellite star trackers. It can be anticipated that a new generation of intelligent star trackers based on the methods used will be capable of tracking and pointing at specified targets without human intervention.

Orion

mag	# of stars
1.5	0
2.0	2
2.5	1
3.0	1
3.5	1
4.0	2
4.5	2
5.0	9
5.5	5
6.0	18
6.5	18
7.0	8
7.5	0

mag	star #	mag	star #	distance
2.0	34	2.0	45	51
2.0	34	2.5	51	102
2.0	34	3.0	12	173
2.0	34	3.5	33	145
2.0	34	4.0	7	292
2.0	34	4.0	32	37
2.0	34	4.5	1	291
2.0	34	4.5	29	180

Cycle	Matches
1	1

ra = 5 hours 36 minutes
dec = -1 degrees 12 minutes

Auriga

mag	# of stars
1.5	0
2.0	0
2.5	0
3.0	1
3.5	0
4.0	3
4.5	1
5.0	4
5.5	4
6.0	14
6.5	16
7.0	5
7.5	1

mag	star #	mag	star #	distance
3.0	2	4.0	10	185
3.0	2	4.0	16	213
3.0	2	4.0	37	284
3.0	2	4.5	29	211
3.0	2	5.0	21	312
3.0	2	5.0	26	221
3.0	2	5.0	39	300
3.0	2	5.0	42	319

Cycle	Matches
1	16
2	6
3	1

ra = 3 hours 42 minutes
dec = 47 degrees 47 minutes

Fig. 3. Pair Matches and Results

Further Development

Further development will be in correcting problems with the labeled tree pattern recognition algorithm and modifying it for improved recognition, intensive testing of real data, and investigation of other recognition methods.

The current algorithm is intolerant of incorrect magnitude determination. This problem can probably be corrected by increasing sensitivity to allow for slight magnitude variations, not just broad categories. It will also be necessary to bypass a number of bright objects within the field which correspond to non-stellar objects such as planets, asteroids, and galaxies. Another problem is the determination of the magnitudes of overlapping objects such as double and multiple star systems. Methods must be developed to recognize these objects and determine their magnitudes. One method to be tried will use Fourier analysis techniques.

In the current implementation, distances on planar projections are used. As the image (or satellite) is rotated, the planar map projections change. A new projection system will be used in which images are projected onto the surface of a unit sphere instead of a plane. Pattern matches will be found using the angle between pairs instead of the distance between the stars. The orientation of the search image can be found by using a least squares fit of search image and the database. Verification of the position of the image center can then be accomplished by image subtraction of the search image and an image produced from the database.

The method used in the initial study will actually work no matter which star is chosen to be the lead matching star; it need not be the brightest in the image. This in itself is part of the true beauty of this pattern recognition algorithm. Of course, if there are #n stars in the pattern this requires #n patterns in the database. The system will be tested and tuned on real data obtained on several deep space missions and available from NASA.

A second recognition method, based on Fourier transforms, will also be investigated. In either the planar or unit sphere projection method, a unique pattern is determined by a given collection of stars (constellations). Viewed from a particular position, this pattern can be coded and analyzed using either a finite Fourier transform or the continuous Fourier transform of a polygon determined by the stars. Pattern recognition methods based on Fourier transforms are being used at CRL for a variety of identification projects. An attempt will be made to adapt these methods to constellation recognition.

Conclusion

In conclusion, this work is just in its early stages and it can be anticipated that there will be many interesting results obtained and it is hoped that eventually knowledge gained from this research will be added to the future generation of star trackers.

APPLYING ARTIFICIAL INTELLIGENCE TO ASTRONOMICAL DATABASES
A SURVEY OF APPLICABLE TECHNOLOGY

Donald A. Rosenthal
NASA Ames Research Center
Moffett Field, CA. U.S.A.

ABSTRACT

Artificial Intelligence Technology has begun the transition from laboratory curiosity to serious software engineering alternative. Artificial Intelligence, which tries to model intelligent behavior normally exhibited by humans, has many subfields of research including expert systems, machine learning, natural language understanding, automatic programming, and automatic theorem proving. This paper surveys several emerging technologies which are relevant to astronomical database issues such as interface technology, internal database representation, and intelligent data reduction aids. Among the technologies discussed are natural language understanding, frame and object representations, planning, pattern analysis, machine learning and the nascent study of simulated neural nets. It is becoming clear from AI research and commercial AI technology that these techniques will become increasingly important for astronomical research, and in particular, for applications with large databases.

INTRODUCTION

Database interface technology refers to the interaction between the database and the user. Most present database users are familiar with at least one of the standard query languages such as SQL. The investigation of natural language understanding has provided tools which allow less restrictive and syntactically rigid interactions with database retrieval systems. There have been several commercially viable products constructed utilizing these ideas, and in the domain of astronomy, a highly extensible language-like interface has been constructed for use with HST observing proposals. Interfaces which allows fuzzy, goal-directed, queries have also been developed. In the future the hope is to be able to circumvent the command language problem altogether by utilizing pictorial and graphically object-oriented techniques. There have also been some recent results in developing truly associative memories which are relevant to this approach.

There are several standard internal representations utilized by database management systems. Among these are the hierarchical, network, and relational models of database organization

[Ullman]. The field of artificial intelligence has produced a rich set of powerful data representations. Current work is attempting to adapt structures such as frames and objects to databases.

Database analysis is potentially the most fruitful application of artificial intelligence to astronomical databases. As the volume of data generated by new astronomical facilities increases, the need for automated analysis will likewise increase. There are attractive analysis applications from the two AI fields of planning and machine learning.

The technique of refinement of skeletal plans, applied to molecular biology in the 1970's [Friedland], is relevant to many scientific domains. A prototype which aids astronomical reduction by the application of this planning methodology has proved the feasibility of automating data reduction at a very high level. A Bayesian classifier, developed to investigate untutored machine learning, was applied to the IRAS Low Resolution Spectra database with surprising results. The classifier (named "AUTOCLASS"), which requires no user provided a priori classification information, has since been used to analyze the contents of other astronomical databases. This classifier may well provide the basis for systems designed to categorize the contents of newly generated data archives, where the complete range of object types making up the catalog is not known in advance.

A short review paper cannot properly cover all these topics in appropriate detail. Therefore, major concepts and promising future work will be described and appropriate publications cited for further investigation by the reader.

INTERFACE TECHNOLOGY

Natural Language Interfaces

The intent of a database interaction language that emulates a natural language is twofold. The first objective is to move away from rigidly structured command languages. The second is to allow a more intuitive, almost conversational interface to the database, obviating the need for memorization of commands that may seem obscure to the user. If the user of an astronomical database wishes to know how many pulsars in the catalog have periods in a given range, it is much more direct to simply ask, "How many pulsars have periods greater than .5 seconds and less than .75?", than to learn SQL and construct the appropriate query. In general, these natural language interfaces are restricted to a quite narrowly focussed domain of discourse, but within that domain they can be very powerful. Solution of the overall natural language understanding problem, however, is still a long-range research goal.

An interesting example of a such an interface has been built at the Space Telescope Science Institute [CohenW86b]. Its function is to aid in the process of review and selection of proposals to use the Hubble Space Telescope. This system (named TACOS, for Telescope Allocation Committee Operations Support) provides an intuitive, robust and extensible command interface to its users. Its capabilities include standard DBMS functions such as:

- Answering queries about data
- Generating and printing reports
- Modifying or adding information

The query system is indeed very English-like. The following examples, taken from the TACOS User's Guide [CohenW86a] should illustrate the potential of natural language interfaces to astronomical databases. The TACOS prompt ("**Tacos>**") is shown at each user initiated action.

- Simple queries:

Tacos> For all proposals, display type, pi_last_name, time_on_target, scores, and title
(Type, pi_last_name, and time_on_target are columns of the database table.)
Tacos> Display average real_time, total real_time, and cumulative real_time for all proposals

- Report generation:

Tacos> output exposure_time - dark_time in 10 columns and pi_last_name in 25 columns

- Command completion: TACOS infers a correct command from a partial input by supplying appropriate implied clauses. For instance, if the user has requested:

Tacos> for all proposals, type pi_last_name and title
and next simply:
Tacos> type exposure_time
the exposure time for all proposals will be output. It assumes the intent is to continue working on the last specified set of proposals until a new set is specified. The specification of the proposal sets can be arbitrarily complex: "for proposals 10, 20-50, and 100-last." If the command following the last two illustrations were:
Tacos> for the proposals with time_on_target < 30, or Pi_last_name = Galileo except proposal 20
the pi_last_name and title would be displayed for this new range of data objects

- Reasoning about commands: The system checks for consistency between verbs and explicit or implicit objects when given a command. For example, "limit" is a command to set an upper limit for numerical values such as exposure time. If limit is the verb used in an operation, and attributes to be limited are not specified, the system assumes the user intended to limit the attributes for which this action would be sensible, that is, only the numerical attributes. TACOS would not attempt to apply "limit" to string valued attributes such as the proposer's name.

TACOS is not restricted to a given vocabulary, the user can define new verbs, nouns, and adjectives. For example:

Tacos> define "summarize" as "display pi_last_name and title for"
This would allow the user to now specify:
Tacos> summarize proposals 3-10.

Tacos> define "summary" as "pi_last_name, title in 30 columns"
Tacos> display summary of proposals 3-10
Tacos> display scores and summary

Tacos> define "regular_viewing_time" as "exposure_time" - "dark_time"
Tacos> display summary and regular_viewing_time

The define command can also be used in the sense of asking for a definition. Given the actions above, the following might occur:
Tacos> define "summarize"
Definition: "summarize" means "display pi_last_name and title for"

Define also allows the user to build up collections of data objects:
Tacos> define "real_time_proposals" as " all proposals with real_time != 0"
Tacos> define "real_time_proposals" as "real_time_proposals except proposal 2"
Tacos> define "real_time_proposals" as "real_time_proposals and proposals 100-last"
Tacos> define " real_time_proposals"
Definition: "real_time_proposals" means "proposals 100-last and all proposals with real_time !=0 not in proposal 2"

Finally, TACOS allows the user to define adjectives:
Tacos> define "short" as "with time_on_target < 25"
Tacos> define "complexity" as "if real_time !=0 or dark_time > time_on_target/3 then 'high', else 'low'"
define "complex" as "with complexity = 'high'"

TACOS operates on standard relational databases, and levies no requirements on the data, other than that it be organized in tables. The next topic, goal directed interfaces allows "fuzzy" queries, and achieves this through adding structure to the database relations

Goal Directed Interfaces

Database queries can be specific requests, such as, "display scores and summary for all proposals", or they can define a target qualification, and query for data which is close to this target [Motro]. Let us return to the pulsar period example. Suppose the intent of the query, "How many pulsars have periods greater than .5 seconds and less than .75?" was really to find out the number of pulsars with medium length periods. To do this, a query had to be constructed with fixed quantitative boundaries. If there was an entry for a pulsar with period .75001 seconds, the query as written would not find this entry, even though it would probably be of interest to the user. A more striking example would be the case where this pulsar, with a period slightly outside the exact range specified, were the *only* pulsar in the database with a period close to that range. In that case, the user would find no medium length period pulsars in the database.

The technique of answering questions with information that is *close* to what is being sought utilizes similarity measures. Such capabilities are achievable at the cost of adding semantic information to the database.

This additional semantic information is termed "distance". The distance between two data objects is a function of the distances between their respective attributes. The notion of distance for a numeric attribute such as the pulsar period is obvious. The distance between symbolic attributes, such as galaxy types, may be stored in the database in a relation specifically utilized for this purpose. There are several forms that such a relation might take. For example, the relation may have instances of every pair of attribute values with the distance as a third attribute. Alternatively, distance may be specified by a relation that provides an ordering of all possible values of an attribute. Distance might also be calculated by a function, with appropriate pointers to functions stored in the database.

Distances between attributes may not all be equivalent. For example, morphological similarity might be specified as being more important than galactic position, magnitude, etc. The relative importance of each attribute must therefore be given a weighting factor. For some comparisons, certain attributes may have no significance whatsoever, and can be given a weight of 0. In addition, attributes will likely have differing ranges of values, possibly differing by many orders of magnitude (e.g. recession velocity versus magnitude). Thus attributes may be assigned associated scaling factors, so that they can be adjusted into values of comparable significance. Finally, similarity levels, or degrees of closeness, (known as "neighborhood radii") must be defined. These are the numerical limits of closeness of the scaled distances, which determine whether two entities can be considered similar.

Although these queries are still based on seemingly arbitrary numerical thresholds, there is a subtle difference between this method and simply giving numerical ranges to queries. These similarity measures are based on multiple attributes. Thus, two objects which are quite different in many ways, but are very similar in one important respect might be retrieved in the same query as objects that are only slightly different from each other for each attribute. In addition, given neighborhood radii, a single query might generate sets of data objects, each at specified levels of similarity.

Pictorial Interfaces

So far, this paper has dealt solely with issues relating to reduced or catalogued data; data that can be representation as a relational tuple. However, unreduced data, such as image or spectral line archives, comprise a substantial amount of information which is of little use if it is not easily accessible or retrievable. One cannot construct friendly natural language queries about data whose features have yet to be extracted and placed in relational tuples.

How might such archives be utilized? There are at least two standard operations that can be imagined:

• Given a known feature (particular grism spectra, emission feature, etc) one would like to retrieve all examples of that feature from a database.

• Given a known object (star, galaxy, spectral line) or combination (star field) one would like to retrieve the particular image(s) on which they appear.

The input might be data from another image archive, for example a CCD image generated by a different detector, edited so that only the objects of interest remained. It may in fact be a freehand rendering of an emission line drawn with a paint program at a workstation.

A system to use such inputs as keys for retrieval of data, would have to take into account imprecision in the input (for example, if it is hand drawn) as well as transformations of the image in scale, translation, and rotation. While the author knows of no such systems implemented to date, there are promising technologies which may soon afford these capabilities. Two approaches will be presented and discussed.

Bayesian Classification

Bayesian probability is based on calculating a posterior probability (conclusions) given a prior probability (evidence) and conditionals. Specifically, Bayes' rule is:

$$P(A|B) = \frac{P(A)\ P(B|A)}{P(B)}$$

The probability of A, given B, is the probability of A times the probability of B given A, divided by the probability of B.

A Bayesian based classification system has been developed as a testbed for research into untutored machine learning at NASA's Ames Research Center [Cheeseman]. This system has been utilized to analyze astronomical databases by separating the data into categories by mutual similarity. This system (known as AUTOCLASS) is of tremendous potential benefit as an analysis tool that can be used to classify data sets. AUTOCLASS can operate on numeric, symbolic, or mixed data, and its role as an analysis tool will be discussed in depth in the section on data analysis. The important point in the context of intelligent interfaces, however, is that the same technique may be used to correlate a new data object (input from the users) with the class it most resembles, in order to retrieve those objects which the input resembles.

When AUTOCLASS is run on a database, it generates a mean and standard deviation for the real-valued attributes of each class. For a database of spectral lines, this would amount to an average curve shape for each class. Given an additional data object as input, AUTOCLASS can easily determine which classes that object resembles, if any. AUTOCLASS can also return a parameter corresponding to the quality of the resemblance.

An even more intriguing aspect of AUTOCLASS as database retrieval tool is the fact that AUTOCLASS can be instructed to ignore certain attributes of the objects it is processing. For example, there may be a subtle feature that is shared to a degree amongst classes, but which is not significant enough to affect the overall classifications. However, the user may wish to reclassify the data with respect to this feature (i.e., "retrieve" all objects that show a very weak particular absorption feature). This can be accomplished by having the classifier ignore all but the attributes (wavelength bins in this case) where this feature appears, reprocessing the data from this new viewpoint.

Associative Memories

A database user would like the system to be able to recognize and retrieve an object given only a partial description. For tabular data, this is accomplished using relational tables. For image databases, this would mean retrieving an image from the archive given only a subimage, potentially an inaccurate subimage—the second scenario described above. This capability would be based on image pattern recognition technology and the ability to associate input data with data in the archive, without requiring a precise match.

Such matching or associating is an important feature of biological memory. The brain appears to operate by generating connections between large number of neurons, where the combination of enabling and disabling inputs from neighbors determines whether a given neuron fires, passing its new state to its downstream neighbors. This collection of connections between extremely simple processing elements seems to be the basis of encoded information, as opposed to storage in separate memory elements. Such "connectionist" patterns have now been emulated both in hardware and software to produce associative memories

One connectionist approach is the type of simulated neural network now being studied by researchers in the field of machine learning [Rosenberg, CohenN]. Nodes, connections, and weights of the connections are represented as matrices within the program, where weights of connections between nodes is found, for example, by using node IDs to index into an array. Weights can be positive (excitatory) or negative (inhibitory). These systems are "trained" by presenting them with input/output pairs, where inputs and outputs have been reduced to numerical vectors. The main areas of research are in development of automatic methods for reorganization, i.e. adjustment of internal weights and connections, until the system does a reasonable job of generating expected outputs from given inputs across the entire set of training data. Two of the most intriguing aspects of these systems is that they are indeed self organizing, and are also able to generalize. That is, they have not just encoded a look-up table of the input/output pairs. When given previously unseen inputs, they can produce reasonable outputs.

For example, after having been taught a set of partial of family trees, by describing the relations between individuals, a query can be made about a relationship that was not part of the initial training data. When asked, "who is the sister of John?", it will return names of females in John's family who are in the same generation, even if the specific examples of John's sister were not previously presented to the system [Hinton].

An important advance relevant to the retrieval of archived images has been made by producing an optical simulation of neural nets [Abu-Mostafa]. This system has demonstrated the capability of recognizing a stored image from a partial input image. The two major components which comprise this system are:

- a two dimensional array of optical switching elements—the neurons, which can be interconnected via light beams

- a hologram which contains the actual interconnection information.

This system has the capability of storing an immense number of neurons and interconnections. The investigators report that a one cubic centimeter hologram can specify one trillion connections, meaning that it could specify all possible connections between one million nodes. The actual system being used presently simulates approximately 10,000 neurons.

The neurons in this system are "fired" whenever the combined intensity of light striking them from all specified inputs is greater than a given threshold. Recognition is produced by projecting an image onto the two dimensional array of switching devices. The archive images, stored in holograms, interact with the reflected input via an arrangement of lenses, mirrors, and an array of pinholes. The resulting array of points of light is then fed back into the system by reflecting it off the switching array, reinforcing the matching process. The best match between the input and stored images is determined by the brightest image issuing from the holograms.

INTERNAL REPRESENTATIONS

AI researchers have incorporated the notions of frame-like organizations of data originally discussed by philosophers and psychologists. Frames are structures that store the attributes or properties of entities via named "slots". The attributes can include relationships to other slots. Hierarchical relationships are typical (for example, the frame for M51 might have "galaxy" as the value of its "is-a" slot. Likewise the frame for galaxy might have "M51" in the list of values for its "instances" slot.) Slots and their values can be inherited from related frames, so that a hierarchy of objects can be created without the need for duplication of representations. Objects are frames which have slots not only for their attributes, but also for the operations (known as attached procedures or methods) that can be performed on the objects. Object-oriented databases can support complex objects, multiple connectivity, and methods that deal with, among other things, maintaining the consistency of the database [Nguyen], and utilization of the data such as its calibration, presentation, display, or update.

For example, to continue with the pulsar example, the user may wish to query the database in such a way that, not the number, but the pulse profiles of all pulsars with medium length periods are displayed on the workstation screen. If this were a query on a database of objects, the query might involve locating the appropriate pulsar identifiers, linking to another database that contains the pulse profiles, and invoking the associated normalization and display routines for each profile. The advantage of such an organization becomes clear when it is realized that this representation allows multiple, inhomogeneous databases to be conveniently used in concert.

For example, suppose that there is not one, but multiple archives of pulsar profiles, each generated by different receiver and antenna configurations, possibly at different frequencies, potentially even at different observatories. The original query would locate those pulsars which match the query parameters, and for each, would examine the appropriate slot to find the link into the proper database containing its profile. The associated display method would send a message to that data object. On receipt of this message, the object would set about displaying itself. This operation might entail invoking its associated calibration and normalization methods prior to display. As the profiles generated by the various experimental setups would probably have differing numbers of datapoints, a sophisticated display method might involve interpolation or extrapolation routines, to fit into a display size specified by the user

This of course, achieves no functionality not available with other representation schemes. But it does achieve a cleaner, more conceptually direct system architecture. Instead of storing, at the system level, enormous lookup tables that associate every possible operation with every possible data object type, the pointers to those operations (methods) are stored with the objects themselves. When new data types or objects are added, or old ones are removed, no system level changes need be made. Additionally, as these representations allow object hierarchies (analogous to taxonomic trees) methods can also be inherited. Thus, a new data object which is a refinement of an existing type, need only contain method slots for those methods which are new and unique to its type, or which are meant to override inherited methods.

The object representation is also a powerful aid to maintaining database consistency. For example, there may be operations which must be performed when an object is deleted or updated (to remove or update links, for example). If a pulse profile is removed from an archive, the links to that profile from other databases must also be removed. That cleanup operation can be stored with the object itself as one in a set of database maintenance methods. Methods can also be used to implement database security mechanisms. For example, object based database systems have been built which present method menus to the user, with menu selections determined by the privileges of the user [Maier].

DATA ANALYSIS

AUTOCLASS

The AUTOCLASS system has already been introduced as a potentially useful technique for pictorial database interfaces. It has already been used successfully to categorize spectral line data, among other things. Its main attributes are:

- based on Bayes' rule
- needs no input other than the data itself
- provides a measure of the goodness of fit

The second point is worth discussing in detail. AUTOCLASS needs no a priori classification information. That is to say, given the data, and only the data, it will attempt to produce statistically reasonable classes. The classification that it does generate is assured to be unbiased by any features not in the data itself. This allows AUTOCLASS, in effect, to discover what is in the data; an important capability, especially for new or incompletely studied databases. Many clustering techniques in use today require the number of classes as input. Their job is then to assign each data object to classes based on mutual similarity. Performing classification in this manner can produce meaningless classes if the similarity is based on "noise"—the user never knows if the resulting classes are meaningful or not.

AUTOCLASS operates on relational tables, or any data that can be set up as a vector of values. It assumes independence of each of the values within a class, and operates by comparing the same attribute (position in the vector) across all data objects in the dataset. Values can be symbolic or numeric. The method of generation of prior probabilities is not constrained—for numeric data it is typical to use the mean value of an attribute across all data objects. An important feature of AUTOCLASS is that the system can find the null classification. In other words, when given random data as input, it generates a single class only. That is, it finds that all data objects are equally related to each other.

The first astronomical database that AUTOCLASS was given to process was the IRAS Low Resolution Spectra database. The spectral bins make up the attribute vector, with the spectra normalized by integrated power. This database was an attractive astronomical database because, as red-shifted spectra were not expected, all curves could be compared bin by bin.

The results of this analysis were quite intriguing. When given the entire LRS, AUTOCLASS found about 70 classes. One of the results that is immediately obvious is that spectra are separated out by data quality. Generally, for each main class, AUTOCLASS has a number of representative classes, distinguished by signal to noise ratio. Other results were more subtle. For example, two classes with seemingly very similar spectra were separated out. Even when superimposed, only slight differences are distinguished—essentially a slight shift of the peak of the main feature. When the sky positions of the constituents of the classes are plotted, however, there is a marked difference in their distribution, one class being essentially wholly within the plane of the galaxy. AUTOCLASS appears to be especially sensitive to shifts of features, probably due to the assumption of independence of the data attributes. As AUTOCLASS does not make any comparison across different bins, similar curves that are shifted relative to each other may in fact no longer resemble each other in AUTOCLASS. In fact, AUTOCLASS would give the same results if the order of the spectral bins was randomized.

AUTOCLASS can perform some interesting subclassifications by performing the procedure recursively on its subclasses. For example, given the full dataset, one class of about 250 high quality black body curves was found. When given just that one class to examine alone, it found 25 subclasses. In the first instance, it was reasonable that all the black body spectra more closely resembled each other than any other objects in the catalog—they were being compared to a curve that was an average of all spectra in the database. However, in the second case, given the average black body curve as a prior probability, features of the individual spectra of this type became significant. In fact, when each subclass was plotted with the Rayleigh-Jeans tail removed, distinct features in the residual curves became apparent.

Data Analysis Assistant

A prototype of an analysis system which determines a plan for data reduction has been built at the Space Telescope European Coordinating Facility in Munich. This data analysis assistant project is an astronomical data reduction system which applies the technique of "refinement of skeletal plans" to HST data. Specifically, the system interacts with the user to determine the analysis needs. In skeletal planning, a generic plan for the particular situation is developed or retrieved from a library of plans and refined down to the level of primitive tasks. In this case, after plan refinement is completed, the user is consulted, and an executable procedure is generated which utilizes the specific analysis packages available on that computer. A test-case involving CCD detector calibrations and two data analysis systems, SDAS/IRAF [Shames] and MIDAS [Crane], is presently running. In the prototype, plans are generated using data descriptors, rather than inferring this needed information from the actual data.

This prototype was built to investigate planning technology in a response to several important problems encountered in reducing astronomical data. Specifically, there exist a wide range of analysis systems, each with its own approach, capabilities, and command language to be learned. The specific contributions of this project included:

- Development of a representation for knowledge about data, instruments, and analysis techniques
- Demonstration of the capability to develop language independent analysis plans
- Demonstration of the capability to convert these plans into executable procedures in the user's choice of target languages

Plan generation is accomplished by use of a four step skeletal plan for CCD calibration:

1. extraction of a subimage of data to be calibrated
2. bias subtraction
3. dark current correction
4. division by a flat field

Plan generation includes the determination of which dark and flat field images (if any) correspond to each data image. Refinement consists of replacing the steps of the skeletal plan with appropriate operators (termed "tasks" in this system). Operators may be "compound" or "primitive". Compound operators are further refined until the entire plan can be expressed entirely in terms of primitive operators. (Primitive commands may actually be further refined during the translation step, if these operators must be implemented by a series of commands in the target system's command language). During the refinement process, redundant operators are recognized and "merged" into a single task. Merging is implemented at a higher priority than expansion, in order to remove redundancies as early as possible, avoiding unnecessary computation.

An interesting implementation detail of this system is the use of object-oriented data representation. For example, each primitive task has a "format" method which performs the actual command generation. This method takes as input the user's choice for target analysis system. Given this approach, the format method becomes the only place in the entire system where command language dependent knowledge need be stored.

The success of this prototype demonstrates that these planning techniques are relevant and useful to the domain of astronomy. Hopefully, this planner will be used as a model for future analysis systems, so that high level analysis tools will begin to become available to the astronomical community.

SUMMARY

This paper has illustrated three areas where Artificial Intelligence Technology may be of use in the domain of astronomical databases: intelligent interfaces, internal representations, and data analysis.

Within the area of intelligent interfaces, the example of a natural language query system developed for use by the Time Allocation Committee of the Hubble Space Telescope was presented. The capability of fuzzy queries was illustrated by the technique of goal directed queries, using similarity measures. Finally, two technologies which might lead to the development of pictorial interfaces were presented: one based on Bayesian probabilities, the other on associative memories.

Frame structures, and their property of inheritance were discussed. A specialization of frames, termed "objects", which store procedural information in "method" slots were described. The utility of objects, especially with regards to collections of inhomogeneous data sets, was discussed.

Two of the many potential applications of AI to data analysis were presented. The first was a system based on Bayesian probabilities, which has the capability of discovering classes of data within a database, without any information other than the data itself. The second was a prototype which has demonstrated the feasibility of automatically generating plans for the reduction of astronomical data, refining those plans, and translating them into executable procedures which actually implement the data reduction in a choice of analysis systems.

With these examples in mind, it is clear that AI provides great opportunities to those working with large astronomical databases.

ACKNOWLEDGEMENTS

The author wishes to acknowledge the valuable contributions made to this paper by P. Cheeseman, S. Colombano, P. Friedland, J. Kelly, M. Sims, and M. Zweben, all of whom read and commented on earlier drafts.

Bibliography

[CohenN] Cohen, N., Abrams, T., Harley, S., Tabor, L., Sejnowski, T., *Skill Learning and Repetition Priming in Symmetry Detection: Parallel Studies of Human Subjects and Connectionist Models,* Proceedings of the Eigth Annual Conference of The Cognitive Society, (August 1986).

[Cheeseman] Cheeseman, P., Stutz, J., Self, M., *Automatic Generation of Optimal Classes,* Nasa Ames Research Center, Internal Document , (June 1987)

[CohenW86a] Cohen, W., *TACOS User's Manual,* STScI internal document;(3 July 1986)

[CohenW86b] Cohen, W. *ST Proposal Entry Processor Design Manual, TAC Tool Section;,* STScI internal document (July 1987)

[Crane85] Crane, P. Banse, K. Grosbøl, P., Ounnas, C., and Ponz, D., *MIDAS,* in Data Analysis in Astronmy, ed. di Gesu´ et al. (Plenum Press: 1985)

[Friedland] Friedland, P., Iwasaki, Y., *The Concept and Implementation of Skeletal Plans.* Journal of Automated Reasoning 1,2 (Spring, 1985), 161-208.

[Hinton] Hinton, G. , *Learning Distributed Representations of Concepts,* Proceedings of the Eigth Annual Conference of The Cognitive Society, (August, 1986)

[Johnston] Johnston, M., D., *An Expert System Approach to Astronomical Data Analysis.,* Proceedings of the 1987 Goddard Conference on Space Applications of Artificial Intelligence and Robotics, (May 1987)

[Maier] Maier, D., Nordquist, p., Grossman, M., *Displaying Database Objects,* Proceedings of the First Internal Conference on Expert Database Systems, (April 1986)

[Motro] Motro, A., *Supporting Goal Queries in Relational Databases*, Proceedings of the First Internal Conference on Expert Database Systems, (April 1986)

[Nguyen] Nguyen, G., *Object Prototypes and Database Samples for Expert Database Systems*, Proceedings of the First International Conference on Expert Database Systems, (April,1986)

[Rosenberg] Rosenberg, C., Sejnowski, T., *The Spacing Effect on Nettalk, A Massively-Parallel Network,* Proceedings of the Eigth Annual Conference of The Cognitive Society, (August, 1986)

[Shames] Shames, P., Tody, D., *A User's Guide to the IRAF Command Language*, Space Telescope Science Institute SDAS/IRAF Workshop (December 1986)

[Ullman] Ullman, J., D., Principles of Database Systems, (Computer Science Press: 1982)

The Generic Applications and Monitors Environment

Dieter Teuber
Astronomical Institute
Westfaelische Wilhelms-Universitaet
Wilhelm-Klemm-Strasse 10, FRG

Keywords: GAME, generic environment, expert system, Muenster Redshift Project (MRSP)

Astrophysics from large, heterogeneous and distributed databases encounters the problems of homogenizing the input data and storing the results in a well defined way using a database system. The evaluation of millions of objects contained in the database requires the introduction of an expert system. To meet the requirements arising from both premises a support system named Generic Applications and Monitors Environment (GAME) was created. Its properties are parametrized and GAME itself is organized like a database system.

Before presenting some details of GAME, the scientific objectives will be introduced. GAME supports the Muenster Redshift Project (MRSP, Horstmann et al., this conference). The goal of MRSP is to search for three-dimensional structures in the univers up to high redshifts. Direct Schmidtplates from the ESO/SRC-survey and objective prism plates are digitized with our improved microdensitometer PDS 2020 GM, using a sample width of 15 µm. Each plate yields approximatly one gigabyte of data. The data are reduced leading to hundreds of thousands up to millions of object images, depending on the sky area. These images have to be corrected, processed and classified by an expert system. Fig. 1 illustrates from left to right how the task is transfered to less abstract levels until the computer hardware is reached. The vertical direction is assigned to the course of the enquiry towards more abstract levels.

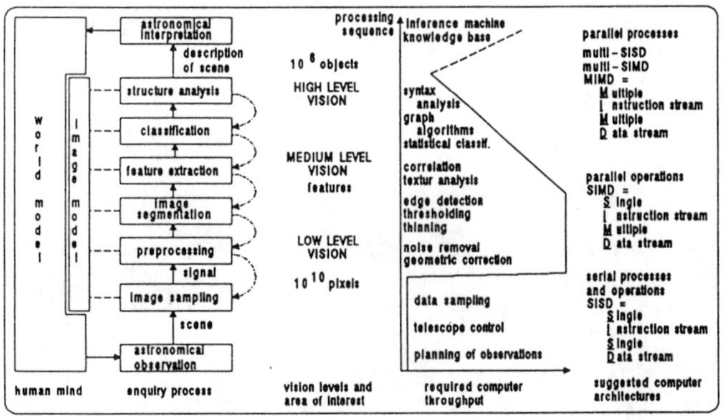

Fig. 1: How to take the MRSP expert knowledge to the computer hardware

INTERNAL STRUCTURE

GAME forms the basis for the MRSP expert system. Some principles of the GAME concept are outlined by Teuber (1985). The practical aspects concerning data reduction shall be discussed here. Fig. 2 shows the internal structure of GAME and its interfaces to the outside world. The functional units of GAME are named administrators (ADM). Each ADM possesses its own data structure optimized for the specific needs of its operations.

The most complex unit is the administrator of monitors. Based on definition files stored on disk, the monitor administrator generates the actual monitors. A GAME monitor interfaces to the user via a communication method (language, menu), supervises the processes initiated by the user and communicates results and status information to the user. The functional details of each monitor can be varied over a wide range by setting up the definition file. Thus the monitor ADM possesses some properties of an user interface management system (UIMS, Green 1985). The monitor ADM is supported by the interaction administrator which links the application programme to the monitor ADM.

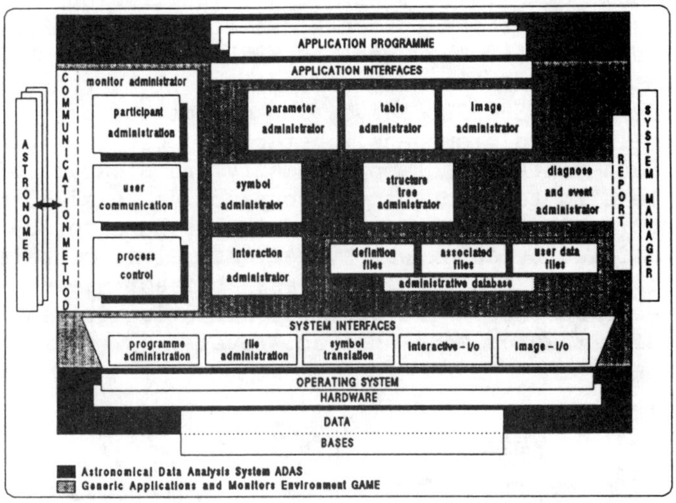

Fig. 2: GAME interiors and interfaces

The symbol administrator serves as a synchronizing mechanism (in the sense of computer science this is also a monitor) of the internal activities of GAME. To the application programme it is mostly visible through the LOGIN and LOGOUT interfaces. The discovery of errors within GAME is assisted by the administrator of diagnoses. It communicates to the system operator by dumping messages to the output device. With revision 3.0 the diagnose ADM will be expanded into an event server. For every user call GAME provides a standard performance. The user can modify or suppress these activities by adding or removing events from an event list.

Data storage is organized by structure trees (not to be confused with tree structures). A structure tree is made recursively of nodes which themselves are structure trees, i.e., each node contains in itself a structure tree. A branch of the tree terminates with a leaf containing user (scientific) data together with a description. In many aspects structure trees resemble a hierarchical data base system. Fig. 3a shows an example of a structure tree. The user has to specify only the path to find a value, he does not need to know about the storage structures which have to be passed. The path may be either a key to a certain location or a directive to search for a number of instances. It is apparent from the scheme of the administrator in Fig. 3b that the menu like character of the administrator programme makes it easy to add new structures to the system. A "path jump" facility enables the redirection of a user specified path into another (part of a) structure tree.

Image display and graphics interfaces are available, but at present not yet fully integrated into GAME. They will meet existing standards as IDI (Terret 1986) or AGL (Fini 1986).

The bottom layer towards the operating system consists of operating system interfaces to resolve operating system dependencies. They perform inter-task control, intertask communication, file administration, operating system/GAME symbol translation and data transfer.

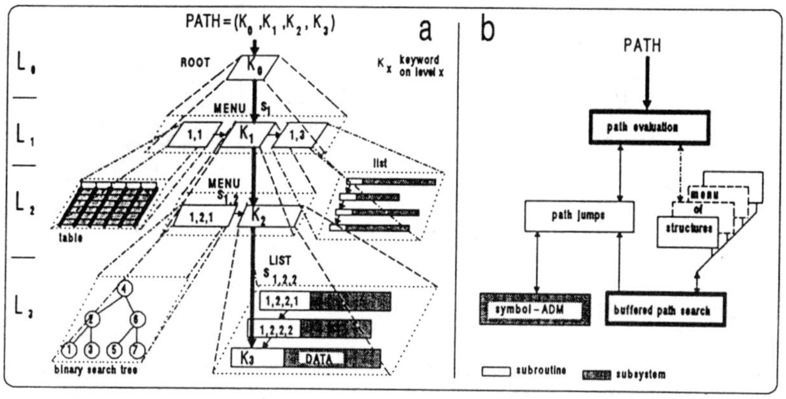

Fig.3: a) a structure tree
b) scheme of the administrator of structure trees

FORMATS OF SCIENTIFIC DATA

The global relations between the GAME data structures concerning user data handling are schematically presented in Fig. 4. When GAME supports an expert system instructions to the system and messages from the system are also considered as data (procedural data). An instruction for the data reduction system (DRS) can be generated and a message from the DRS can be interpreted by the expert system. The parts inside the dotted boundary show the monitor which copes with procedural data. The monitor can function either as an interactive user interface or as user port and DRS interface to the expert system. Using the appropriate interface calls the application programmes may store and retrieve data formatted as image, table or parameter.

An image is a compound format (named ADATA) which uses a header made of parameters and tables to describe the data in the datacube. Several images may be packed into one file and interrelated by a structure tree. In its simplest form a structure tree may serve as a list-structured or a hierarchical directory. It may be extended to form a kind of hierarchically organized database. Images written in FITS format according to Wells et al. (1979) may processed on-line, i.e., they need not to be preprocessed by a FITS-reader.

Parameters are the simplest data units. A parameter consists of a name which is the key (when referenced within a structure) to a storage area which contains a control and a data segment. The control segment holds the description of the structure in which the values are arranged and the format of the coding. The structural description permits to handle the parameter as a single value, array, stack, ring buffer and/or menu.

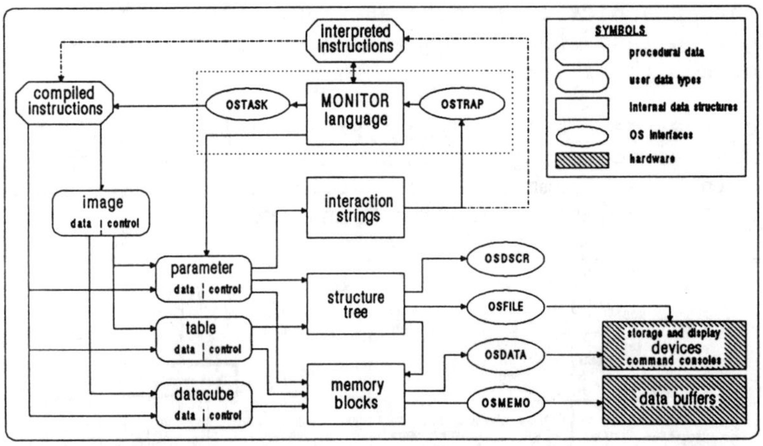

Fig. 4: Data access paths of GAME (Rev. 3.0)

INTERFACING APPLICATION PROGRAMMES TO GAME

As in all VOS like systems, application programmes communicate to GAME via calls to interface routines. These interfaces must be carefully designed to meet the requirements at the time of definition, but should also be developpable with little expense for the applications programmer.

An interface must be as simple as possible, but also cover as many incidents as possible. The first case implies a short argument list, while the latter demands a complex one. Another problem is the management of interactive user communication which arises in the case of missing parameters. As in other data analysis systems the solution is found in offering high and low level interfaces. High level interfaces allows a conveniant formulation of operations. The specifications must be interpreted by the system at run time which often results in a degradation of performance.

GAME high level interfaces have been designed to inherit generic properties. Their default performance can be adapted at run time by the application programme setting up an event list. High level interface calls can also be compiled automatically into low level calls by a source code preprocessor. In GAME the versatile functions of high level interfaces are driven by a command string. The command string is translated into a parameters for the low level interface. GAME definitions are such that the precompiled code is still FORTRAN and in agreement with the user level definition of GAME. Other systems as IRAF (TODY 1987) have gone one step further and use a proper language instead of FORTRAN.

Application programmes are introduced into GAME via the preprocessor. Besides the transcription to low level interfaces, the preprocessor reduces the amount of task code by extracting from the source code monitor scripts for setting defaults and for the interaction with the user. These scripts are stored in the associated information files of programmes or programme packages. GAME updates concerning application programmes are as far as possible performed by the GAME software update processor.

APPLICATION OF GAME AND FUTURE ENHANCEMENTS

GAME, as its name implies, is open to non-GAME data formats. Images in FITS format may be processed on-line (as mentioned above). Catalogues of various formats may be processed using a corresponding definition file. Once the file is created all GAME users are able to access the respective catalogue.

Running a programme without DRS monitor, i.e., with operating system support only, may be tideous, but in the interactive mode the use of a monitor is a matter of convenience. The monitor becomes a necessity as user port as one steps towards an expert system. The definition (file) of such a monitor may well be elaborated using an AI language such as LISP.

An view of the projected application area of GAME is given in Fig. 5 .

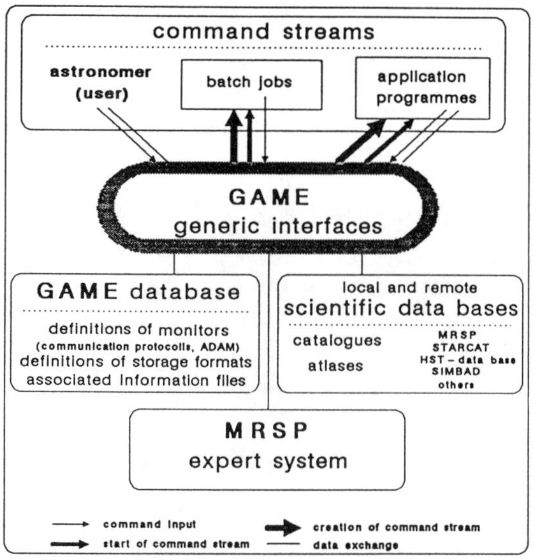

Fig. 5: GAME application area

GAME has been developped on a Perkin Elmer OS32 based computer. It is scheduled to be ported to a UNIX based computer system in 1988. The above text refers to GAME version 3.0 which will become operational in November 1987. Due to shortage of programmers tables are currently only available as part of an image header.

Acknowledgements are sent to R. Budell who was instrumental in implementing the ADATA storage format.

References

Fini, L. (1986), ASTRONET Graphic Library Reference Manual, Vers. 2.1

Green, M. (1985) in: Workshop on User Interface Management Systems, Proceedings, ed. G.E. Pfaff. Springer, Berlin, p. 9

Horstmann, H., Schuecker, P., Gericke, V. (1987), these proceedings.

Terret, D., Shames, M. (1986), An Image Display Interface for Astronomy (DRAFT), Version 0.4. STScI, Baltimore

Teuber, D. (1985) in: Proceedings of the International workshop on Data Analysis in Astronomy, eds. V. DiGesu, L. Scarsi, P. Crane, J.H. Friedman, S. Levialdi. Plenum Press, New York, p. 235

Tody, D. (1987), The IRAF Data reduction and Analysis System, in IRAF System Handbook 3A, NOAO

Wells, D.C., Greisen, E.W., Harten, R.H. (1981), Astron. Astrophys. Suppl. Ser. 44, p. 195

Databases: Current Trends

THE INCA DATABASE FOR THE PREPARATION OF THE HIPPARCOS MISSION

Arenou F. & Morin D.
Observatoire de Meudon, F-92195 Meudon Principal Cédex

ABSTRACT : The INCA database, including all stars proposed for Hipparcos observation, has been created as a subset of SIMBAD database. It has now evolved independently and contains about 214,000 stars (with 5,000 stars not yet in SIMBAD). The data contained in INCA are daily improved with the help of specific softwares.

1. WHY THE INCA DATABASE

The Hipparcos project, included in the scientific programme of the European Space Agency (E.S.A) in 1980, aims at the very accurate measurement of the positions, parallaxes, and proper motions of about 110,000 preselected stars, with a precision of about 0.002" on positions and parallaxes, and of 0.002" per year for proper motions.

The INCA Consortium, leaded by C.Turon, is carrying out the preparation of the INput CAtalogue, which will contain all the stars to be observed by the Hipparcos satellite.

From the very beginning of this work, in early 1982, the support of the Centre de Données de Strasbourg (C.D.S.), and the intensive use of the SIMBAD database turned out to become essential in order to sort out redundancies among the 600,000 stars submitted by more than 200 proposers of the international astronomical community.

At that time, SIMBAD clearly was the only set of identifications and stellar data which could be interrogated using any identification, thus allowing to clarify most of the redundancies (see Fig 1). Cross-identifications were done using the Strasbourg-Cronenbourg Univac computer from Meudon Observatory.

The completeness of SIMBAD for all the stars up to about 9th magnitude allowed the construction of a basic list of bright stars (survey) accounting for half of the Input Catalogue (see Turon, Gómez & Crifo, this volume).

The first release of the catalogue (1983) contained 160,000 stars in a sequential file. Including the survey stars and stars newly identified, and eliminating redundancies, the INCA database now contains about 214,000 stars. The necessity of interrogating, of making statistics, and of updating very frequently the catalogue naturally implied, early 1985, the creation of a specific database, named INCA (acronym for INput CAtalogue), suited to our task, with the same structure and basic data as SIMBAD, thus allowing the use of all the softwares developed for SIMBAD, which have proved to be very efficient.

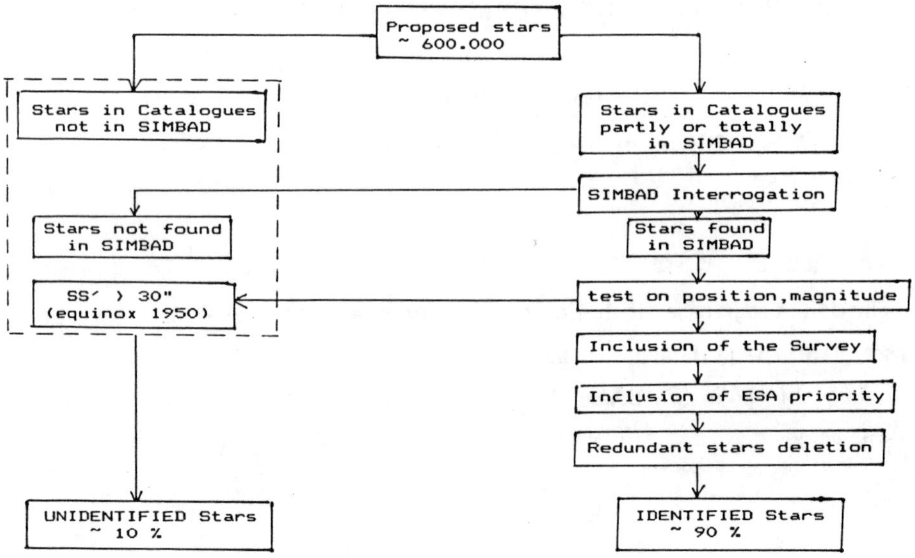

Figure 1

Situation before the INCA Database creation.

2. A SPECIFIC DATABASE

Once created, the INCA database was then developed in an autonomous way, though in close collaboration with the C.D.S. team (see Fig 2). The INCA database includes additional stars, different entries and new identifiers and data :

2.1 New stars

The INCA database now contains about 214,000 stars : 209,000 stars from SIMBAD and 5,000 additional stars which were not yet in SIMBAD.

2.2 Different entries

Due to the specificity of the Hipparcos satellite, stars in double and multiple systems received a special treatment according to the following rule :

1 entry if $\rho < 10''$ (AB) and 2 entries if $\rho \geq 10''$ (A+B), where ρ is the separation between two components A and B.

2.3 Identifiers

In addition to identifiers already in SIMBAD, some other identifiers (see Fig 3) were required to take into account :

- by which proposal each star was demanded ("*I*" identifier)
- the location of a star in 0.81 square degrees cells (size of the field of view of the satellite), to study the sky distribution ("*IBIL*" identifier) ;
- the Lausanne photometric identifier to facilitate the introduction of new photometric data provided by the Lausanne team ("*LID*" identifier) ;
- information about components of double and multiple systems given by the Working Group on Double & Multiple Stars, Observatoire Royal de Belgique ("*CCD2*" & "*CCDM*" identifiers).

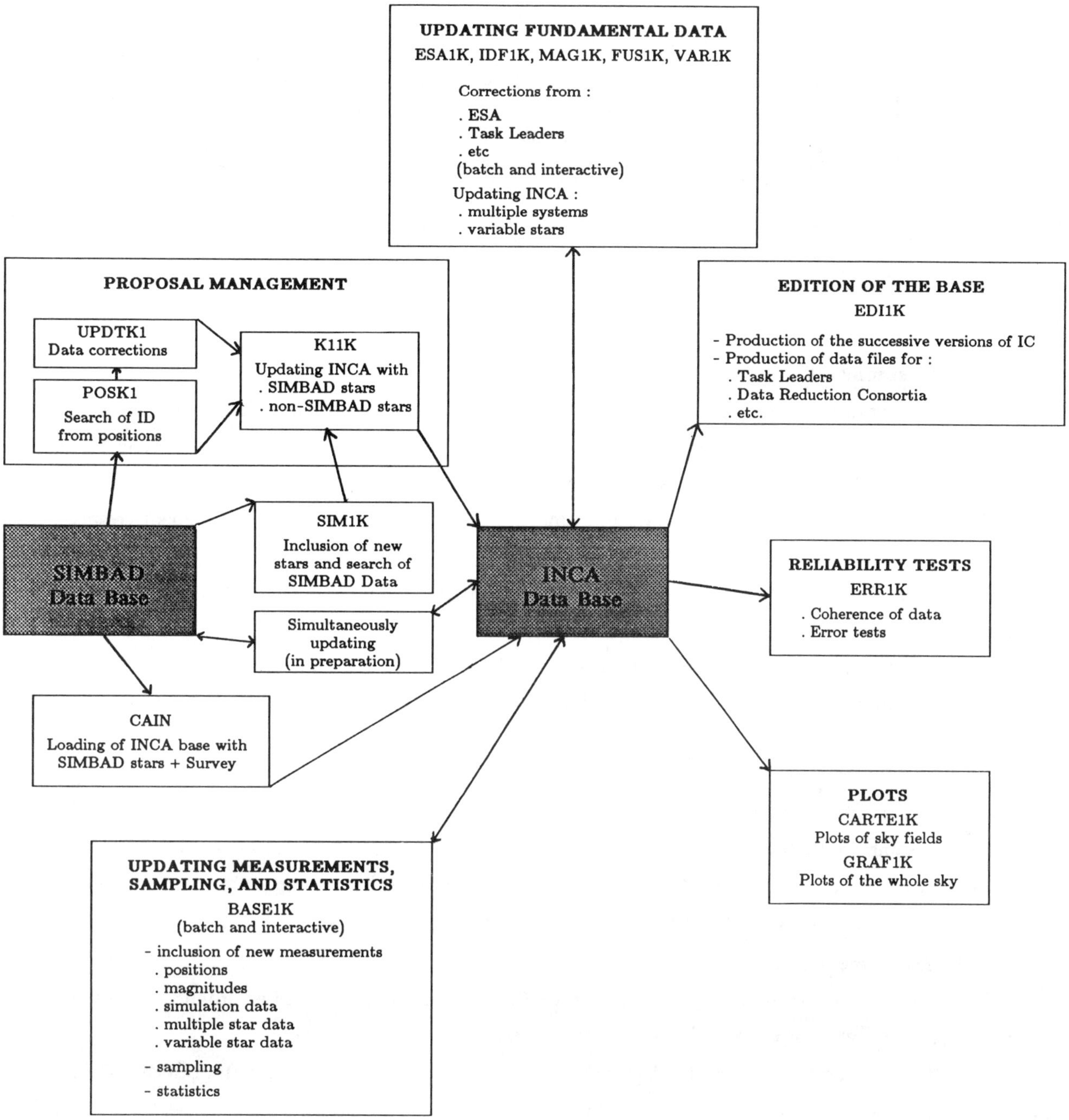

Figure 2
Input Catalogue preparation : New softwares.

INCA Identifier

Identifier name	Proposal number	Running Nr in the proposal	ESA priority	Proposer's priority	error code
I	54	2182	4R	0	0

IBIL Identifier

Identifier name	Cell number in galactic latitude : IB	Cell number in galactic longitude : IL	Discriminating number of stars in the cell
IBIL	-78	124	1

Figure 3
"I" and "IBIL" identifiers.

2.4 Measurements

New measurements, continuously improved, obtained by ground-based observations and measurements or compiled in the literature by other working groups of the INCA Consortium are introduced in several catalogues. The situation in October 1987 is :

- "*pos*" and "*pm*" containing positions and proper motions : 181,000 and 189,000 measurements respectively ;
- "*pH*" and "*vH*" for photometry about normal and variable stars : 86,000 and 12,000 measurements respectively ;
- "*sH*" for parameters concerning numerical mission simulation : 1,600,000 measurements ;
- "*CCDM*" containing data about double and multiple stars.

3. NEW SOFTWARES

Apart from the softwares written by the C.D.S. staff, many new softwares operating on the INCA database and including a total of 36,000 PL/I statements and 6,000 FORTRAN statements have been written for 3 years, to solve different problems specifically related to the elaboration of the IC, though these softwares are compatible with SIMBAD structure.

3.1 Management of proposals

This software allows the exploration of SIMBAD, within a 10 arcmin radius around each of the 50,000 stars which were proposed using identifiers not recognized by SIMBAD.

Another software introduces the data concerning the newly recognized stars from SIMBAD into INCA .

3.2 Updating the INCA database

New identifiers, fundamental data and measurements are updated following the recommendations of ESA, INCA task-leaders, working groups coordinators, and proposers. This is performed using a C.D.S. SIMBAD software as well as new softwares described in Fig 2. Another software allows corrections of the entries in double & multiple systems.

3.3 Sampling and making statistics

In order to analyse the sky distribution, the effects of the different simultations and to detect redundancies, softwares were designed to perform samples and statistics using different astrometric, photometric, and spectroscopic parameters, ESA priority, etc.

3.4 Plots

Graphic softwares were written, to quickly figure out the local and global distribution of the considered stars. Figure 4 illustrates the aspect of the sky distribution of the stars proposed by proposal #53 ; Figure 5 displays a zone of 4 square degrees around the 502th star of proposal #53.

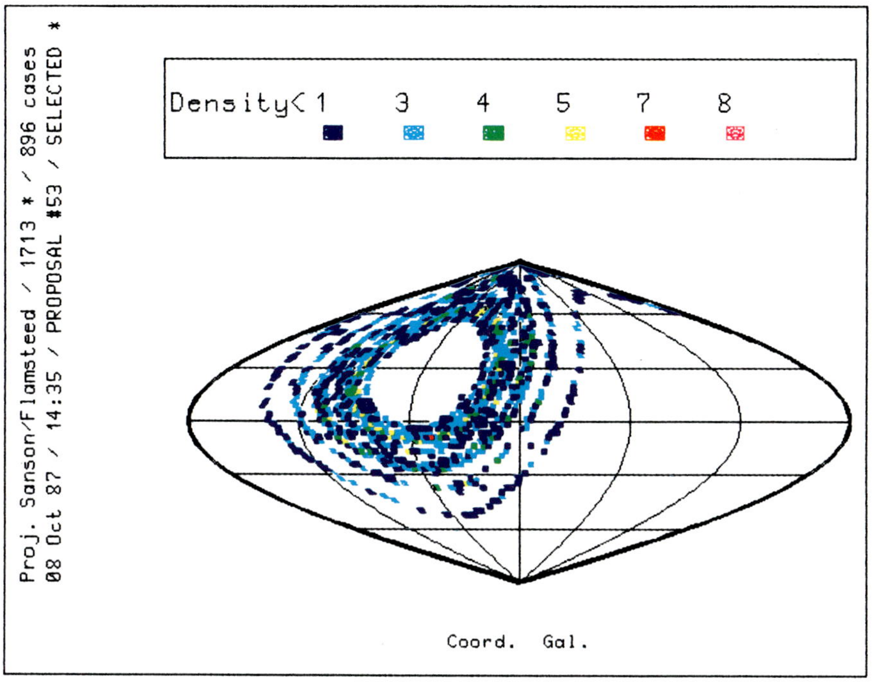

Figure 4

Figure 5

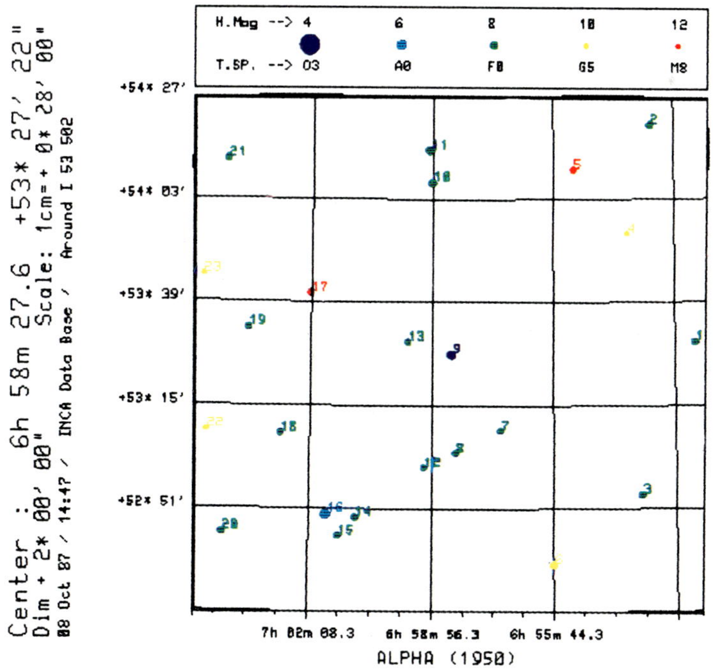

3.5 Edition of the INCA database

One important software allows the edition of the database for the different INCA working groups. Several editions have been necessary, each one containing improved data, for the simulation group, for ground-based observers, for ESA, for Data Reduction Consortia, etc.

3.6 Reliability tests

In order to improve the reliability of the Input Catalogue, a software has been designed to detect possible errors on astrometric, photometric and spectroscopic data, errors on cross-identifications and redundancies ; for each star, the consistency of all the data is tested : within each kind of data, between different types of data (e.g. colour index vs spectral type), and between identifiers and data.

4. TECHNICAL ASPECTS

4.1 Staff

Apart from the staff of the other working groups who provides the newly compiled or observed data for the preparation of the Input Catalogue, 3 scientists, 2 software engineers, and 4 technicians are specifically involved in the updating of the INCA database at the Observatoire de Meudon.

4.2 Internal structure of the database

The structure is the same as that of SIMBAD :

- 1 main file, containing all the data concerning one star : fundamental data, list of identifiers and measurements ; each field be variable in length ;
- 3 identifiers index files in order to allow access to an object from any of its identifiers ;
- 1 coordinate boxes data file in order to allow a quick access to stars when interrogating by their coordinates ;
- the bibliographical texts data file is the SIMBAD one ; it contains the texts of the references of the measurements.

4.3 Content of the database (October 1987)

stars : 214,000 ;

identifiers : 2,125,000 (547,000 "*I*" identifiers) ;

measurements : 3,166,000 ;

The INCA database contains 153 MBytes :

- 47 MBytes for fundamental data, system information and free bytes reserved for future use ;
- 63 MBytes for identifiers and measurements (90% proper to INCA) ;
- 43 MBytes for the index files ;

On average, each star is read 120 times/year and modified 12 times/year (see Fig 6)

4.4 Performances

The performances of the INCA database are similar to those of SIMBAD (Wenger, 1985) :

- interrogation by identifier : instantaneous answer in dialogue mode, 1h30m for the whole database in batch ;
- interrogation by coordinates : 2 sec in dialogue mode ;
- sampling : less than 5 min for a sample requiring the reading of the whole database.

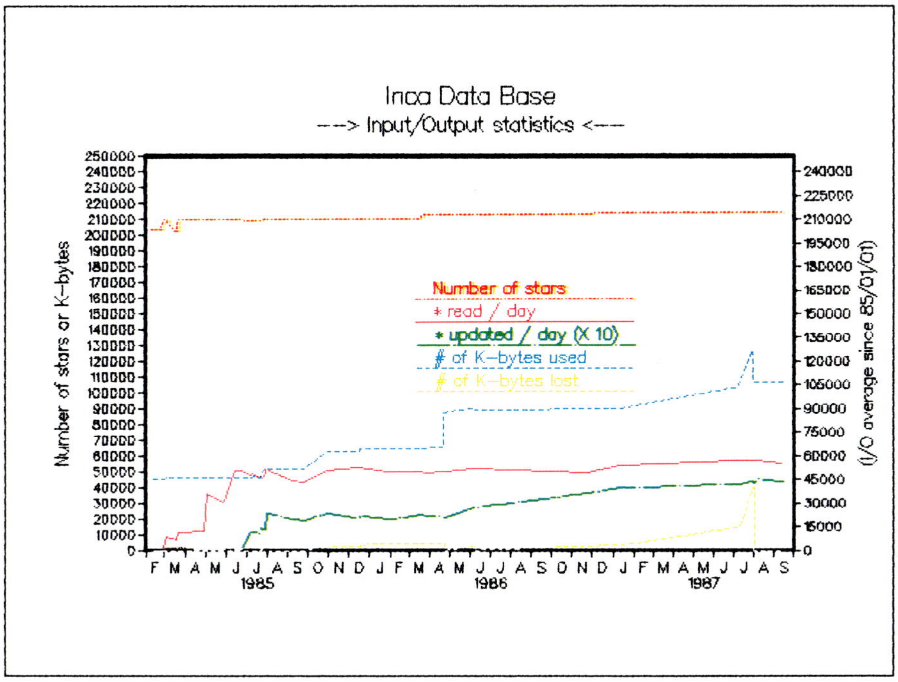

Figure 6

REFERENCES

- "Hipparcos : scientific uses of the INCA database", Turon C., Gómez A., Crifo F., this volume.
- "The INCA database, sub base of Simbad", Morin D., Arenou F., 1985, INCA Colloquium "Scientific aspects of the Input Catalogue preparation", Aussois, pub ESA SP-324, Perryman & Turon eds.
- "Construction of the Input Catalogue", Gómez A., Crifo F., Morin D., Arenou F., 1985, INCA Colloquium, Aussois, op.cit.
- "Presentation of the astronomical database SIMBAD", Wenger M., 1985, INCA Colloquium, Aussois, op.cit.

OBJECT NOMENCLATURE IN THE IUE PROJECT

M. Barylak
ESA IUE Observatory (VILSPA)
Apartado 54 065
28080 Madrid, SPAIN

I. ABSTRACT

In cooperation with the Centre de Donnees Astronomiques de Strasbourg (CDS) the IUE project has undertaken the task to homogenize the object identifications in the observing log.

For 30000 images (out of ~50000) the observed object could be uniquely identified. For the cases of no or doubtful identification and to ensure a homogeneous nomenclature in the IUE project a working group has been established. The current status is presented.

II. INTRODUCTION

With the on-line availability of the IUE database for interactive query sessions, many people have discovered and criticized that the use of on object identifier or even object coordinates as search key will not provide them with a complete list of IUE observations of this object. They innocently have entered the domain of increasing confusion in the IUE (and generally the astronomical) nomenclature. Nowadays object are being explored in many different wavelength ranges (Gamma ray, X-ray, UV, visible, IR, radio, etc.), each representing a highly specialized field of research and employing different object designations. Hence it comes as no surprise that an object may appear under a variety of names in the IUE observing log depending on the astronomical discipline of the observer. This circumstance is bound to reappear in any other astronomical project to come. The need for an astronomical nomenclature hierarchy will increase with time.

III. PROPOSALS ON HOMOGENEOUS OBJECT IDENTIFIERS

In 1984 the Long Range Planning Commitee recommended to the IUE project to consider adopting the IAU hierarchy for the "homogeneous object identifications" in the IUE data banks. Unfortunately, a formal action could not be taken because a complete hierarchy for all astronomical objects has not been agreed upon by the International Astronomical Union (IAU). But considering the fact that the IUE satellite primarily provides spectral data, it was proposed that the identification of stars should follow the one given in primary spectroscopic catalogs (Warren 1985).

Hopes that the next IAU General Assembly in New Delhi (Nov. 1985) would provide some advice on how to proceed in solving the nomenclature problem were not fulfilled. Hence a new proposal was presented to the IUE Three Agency Coordination meeting (Barylak and Wamsteker 1986) following the recommendations of Bull. Inform. CDS no. 18 (1980) and taking into account the "First Dictionary of the Nomenclature of Celestial Objects" (Fernandez et al. 1983). This proposal also considered the format (layout and alignment) of the homogeneous object id. field in the IUE database. In cooperation with CDS (Heck and Egret 1986), one of the world-recognized authorities in astronomical nomenclature, the venture was started to standardize the object identifications and to provide proper coordinates (epoch 1950.0) as the coordinates in the IUE log suffer numerous mistakes and round-off errors.

IV. RESULTS

The first result of the homogenization of the object identifiers in the IUE data bank is summarized in the following table. Three "catalogs" were invented for sake of simplicity:

- IUE: which contains all IUE related images such as WAVECAL, NULL, CALUV, T FLOOD, UV FLOOD, etc.
- ZZ : the CDS invented "Z"olar "Z"ystem catalog
- AOO: the catalog of "A"ll "O"ther "O"bjects which holds currently unidentified objects or with designations not commonly used.

Taking SIMBAD (Set of Identifications, Measurements and Bibliography of Astronomical Data, Egret and Wenger, 1987) as reference, it came as a very positive surprise that for 30000 images (out of 46648) the object observed could be uniquely identified. For 6699 images the identification was doubtful and further investigation is needed. Finally, for 9949 images no identification was possible, but around one third of them are IUE specific images.

For the cases of no or doubtful identification and to ensure a homogeneous nomenclature in the IUE project a working group has been established. This working group welcomes any suggestions, comments or advice to improve the current situation of the IUE nomenclature.

First results of homogenization attempt of IUE object identifiers

Catalog	Nr.entries		Catalog	Nr.entries	
ABCG	117		MRK	430	
"AOO"	3994		NGC	2390	
BD	2219		NOVA	338	(novae)
CD	195		PG	344	
COM	350	(comets)	PK	133	
CPD	156		QSO	519	
ESO	110		SK	305	
GD	105		SN	137	(supernovae)
HD	26662		V*	2126	
IC	128		WD	245	
"IUE"	3262	(IUE specific)	X	188	
LMC	241		"ZZ"	1954	(solar system)

V. CONCLUSION

The current process of introducing homogeneous object identifications in the IUE data banks together with accurate coordinates (which will NOT substitute the ones in the observing logs) is a critical task in view of the creation of the "Final Archive" of the IUE project. Once an homogenization of the object identifications has been achieved an adoption to any given nomenclature hierarchy should be easy. The user should put considerable thoughts into the possibility of finding (and tracing) his object(s) of interest in the huge variety of astronomical databases by just a single name. The moment to voice requests, suggestions and comments has come and is now, as new projects (ESIS, SPAN, ...) will make use of and/or provide access to the IUE archive.

VI. REFERENCES

Barylak M., Wamsteker W.: 1986, IUE Three Agency Coordination Meeting, ESTEC, June 25-27, 1986.

Egret D., Wenger M.: 1987, SIMBAD: present and future, see these proceedings

Fernandez A., Lortet M.-C., Spite F.: 1983, Astron. Astrophys. Suppl. Vol. 52, no.4.

Heck A., Egret D.: 1986, private communications

Warren W.: 1985, IUE Three Agency Coordination Meeting, Vilspa, April 1985.

THE UNIFORM LOW DISPERSION ARCHIVE OF THE INTERNATIONAL ULTRAVIOLET EXPLORER SATELLITE

M. Barylak, J.R. Munoz, C. Driessen
ESA IUE Observatory (VILSPA)
Apartado 54 065
28080 Madrid, SPAIN

I. ABSTRACT

The Uniform Low Dispersion Archive (ULDA), as developed at the ESA IUE observatory and accepted by the three agencies (NASA, ESA, SERC) as a further product of the International Ultraviolet Explorer (IUE) satellite, is presented.

This low dispersion absolutely calibrated spectral archive of the IUE satellite will facilitate the access to the IUE archive and hence allow a better utilization of these data together with the possibility to retain standard quality control assurance by the IUE project.

Together with the ULDA a software support package (USSP) is being provided which will allow any scientist to help himself interactively to search for, to select and to downlink low dispersion spectra to his own computer for subsequent data reductions.

II. INTRODUCTION

Over the last five years an average of 360 images per month was requested from the VILSPA IUE database (Sastre 1987) . In the United States archived IUE spectral files are available from the Colorado Regional Data Analysis Facility, the National Space Science Data Center (NSSDC) and the Goddard Space Flight Center (GSFC). The total average of IUE images requested from these centers is 828 images per month (Thompson 1987). In view of this large demand for archival data from the IUE database efforts were made to facilitate the access to released IUE images. As a first result the generation of ULDA (Munoz 1984) has been decided.

The ULDA will be generated on magnetic tapes, but the final purpose is to have this archive on a fast access device (ie. disk), together with a software support package which allows to manipulate the data interactively. The ULDA is a compacted form of the low resolution spectra in the IUE image archive as it is made of the Merged Low Resolution spectra only.

III. CORRECTIONS TO IUESIPS

The IUE Spectral Image Processing System (IUESIPS), which provides the standard output products of the IUE satellite, has undergone a continuous evolution, enhancing the quality of the data and removing various software deficiencies and errors. The changes to IUESIPS are summarized in the NASA IUE Newsletter #25 (Sep.1984) The following table shows the mayor milestones in the IUESIPS history and define the type of corrections which has to be applied:

IUESIPS CHRONOLOGIES

```
Beginning              VILSPA                 GSFC
of IUE        ======================================
              |    V     |          |    V     |
              |    e     |          |    e     |
              |    r     |          |    r     |
              |    y     |          |    y     |
              |          |          |          |
              |    O     |          |    O     |
              |    l     |          |    l     |
              |    d     |          |    d     |
              |          |          +----+-----+   -->  22May78
              |   S/W    |          |    |  B  |
14Jun78  <--  +----+-----+          |    |  a  |
              |    |  B  |          |    |  d  |
              |    |  a  |          |    |     |
              |    |  d  |          |    |  I  |
              | O  |     |          | O  |  T  |
              |    |  I  |          |    |  F  |
              | l  |  T  |          | l  +-----+   -->  07Jul79
              |    |  F  |          |          |
07Aug79  <--  | d  +-----+          | d        |
              |          |          |          |
              |   S/W    |          |          |
              |          |          +----------+   -->  03Nov80
10Mar81  <--  +----------+          |          |
              |    N     |          |    N     |
              |    e     |          |    e     |
              |    w     |          |    w     |
              |          |          |          |
              |   S/W    |          |   S/W    |
                  ...                   ...
```

ITF = Intensity Transfer Function S/W = Software (IUESIPS)

III.1 FLUX CORRECTION

This error arises in images processed during the Very Old software period (all cameras). The correction is applied to the merged spectra by multiplying the extracted Flux Number (FN) by a constant, depending on the camera:

LWR: New FN = Old FN * 1.78

SWP: New FN = Old FN * 1.83

III.2 WAVELENGTH CORRECTION

This anomaly is due to the usage of bi-weekly dispersion constant calibrations in low dispersion for the SWP camera. It occurred during the Very Old Software and Old Software Bad ITF periods.

The correction is performed by comparing the original dispersion constants with an adopted set of mean dispersion constants values, considered as the more accurate for the given period :

$$\text{Corrected Lambda} = \text{Old Lambda} * m + d$$

with

$$m = (B_2 * B_2' + A_2 * A_2') / (B_2^2 + A_2^2)$$

$$d = [B_2 * (B_1' - B_1) + A_2 * (A_1' - A_1)] / (B_2^2 + A_2^2)$$

Primed values are the old dispersion constants (ie. the ones originally used to process the image) and un-primed values are the set of adopted mean dispersion constants, ie.:

	LWR	SWP
A_1	-298.22	981.37
A_2	0.30242	-0.46657
B_1	-266.66	-263.68
B_2	0.22577	0.37618

$$\text{Sample} = A_1 + A_2 * \text{Lambda}$$

$$\text{Line} = B_1 + B_2 * \text{Lambda}$$

III.3 BAD ITF CORRECTION

This error was caused by the use of the SWP ITF (Intensity Transfer Function) with incorrect 20% exposure levels. It applies to the Old Software Bad ITF period and only to the SWP camera.

The method used to correct this error is the so called SNAFU correction, described thoroughly by Cassatella et al. (1980). This method requires the Line-by-Line spectrum. If a spectrum is affected by both wavelength and ITF errors, the wavelength correction has to be applied first.

III.4 ABSOLUTE CALIBRATION CORRECTION

This deficiency originated from the non-existence of a definitive Inverse Sensitivity Function. It applies to all images processed during the periods of both Very Old and Old Software. The algorithm applied is a spline interpolation of the NET spectrum using the currently defined Inverse Sensitivity Function.

III.5 IUE EXPOSURE TIMES

To achieve an absolute calibrated spectrum the observed flux has to be divided by the exposure time. This exposure time is filled in during real time operations and is recorded in the IUE raw image header label at a specific position. But in many cases this time does not stand for the exact duration of the exposure, especially in double aperture images, trailed spectra, very short exposures, modifications to the label, etc.

Hence the exposure times have been taken from the IUE database where they are closely controlled and monitored by the IUE observatories (GSFC and VILSPA). These more reliable exposure times have been compared with the ones given in the image header label. And in the case of discrepancy, the IUE observatory staff of VILSPA has looked up the right information from the hand-written observatory logs.

III.6 MISSING ENTRIES IN RECORD ZERO AND LABEL CORRECTIONS

This deficiency arises from the non defined Record Zero entries and from the fact that some of them are incorrect or missing. With the New S/W implementation, these errors have been corrected. This correction applies only to the Very Old and Old software periods. The corrected entries are:

- Target coordinates
- Dispersion constants
- Aperture code
- Exposure time
- Camera and Image number

IV. ULDA FINAL FORMAT

The ULDA is kept on magnetic tapes in FITS format (Wells et al. 1981, Munoz 1987). Each absolutely calibrated spectrum represents one FITS file with the following structure of four FITS blocks (2880 bytes) :

Block #1	Header for Data
Block #2	Data
Block #3	Header for IMAGE Extension
Block #4	Data Quality Flags (Epsilons)

V. THE ASTRONOMICAL CONTENTS OF ULDA VERSION 1.0

The following table should provide some hints on the astronomical contents of ULDA version 1.0 (ie. all low dispersion spectra from the beginning of the project till the end of 1983):

Class	Description	nr.
00 - 08	Objects of the Solar System	2 382
10 - 11	W C and WN stars	934
12 - 19	O type stars	2 801
20 - 29	B type stars	4 454
30 - 39	A type stars	1 971
40 - 49	late type stars (F, G, K, M)	3 636
40 - 53	Variable stars	976
54 - 56	Dwarf novae, Novae, Supernovae	1 247
57 - 59	Symbiotic stars, T Tauri, X-ray	1 575
60 - 66	Shell stars, pulsars, nova-like, etc.	796
70 - 76	Planetary nebula, HII regions, SNR, etc.	2 010
80 - 90	Galaxies, globular clusters, quasars, etc.	2 757
98 - 99	wavecals, null and flat fields	1 079
	Total number of spectra :	26 618

("Class" is the object classification scheme used by the IUE project - see any ESA IUE Newsletter or any Merged Log microfiche).

VI. ULDA SOFTWARE SUPPORT PACKAGE (USSP)

A system is being developed jointly for the expectant European astronomical community which will allow any scientist to help himself interactively to low dispersion spectra taken by IUE. Amongst other things, one will be able to search for and select those ULDA spectra which one wants to downlink to one's own computer for subsequent manipulation.

The user will normally access the system via intercomputer links (modem connections, Packet Switching Data Networks (PSDN X.25), etc.), hence the time required to actually get one's hands on the data should be reduced considerably.

VI.1 BRIEF SYSTEM OVERVIEW

The system's software (Driessen 1987, Driessen et al. 1987) is distributed over three types of computing center, namely:

a) Principal center - responsible for generating the ULDA and for a variety of house keeping actions. This will be VILSPA in the first instance.

b) National centers - one center per country serving the country's astronomical community. By having many national centers we get the following benefits:

- Data transmission costs are kept to a minimum by avoiding international telephone charges.
- The system can be allowed to grow to embrace a large number of countries without degrading an individual country's mean response time. Furthermore the software at a national center can be tailored to meet the country's inter-computer network system and the country's perceived needs.
- By spreading the load total capital, development and maintenance costs are shared.

c) Individual institutes - where the end users lurk. These require software to transform the data transmitted from a national center into a form palatable to the user's image processing system.

VI.2 EASE OF USE

Considerable thought (Driessen 1986) has gone into trying to make the system as easy to use as possible and, perhaps more important, easy to pick up. This is all the more crucial since we are talking of what is basically a retrieval system, hence will probably be used infrequently by any one individual, which in turn implies one will have plenty of time between uses of the USSP to forget how to drive it. The basic strategy adopted is to prompt the user what he can do next at every stage in his session. HELP's will be provided but will be small in size, the idea being that any reasonably intelligent user will be able to follow his nose.

VII. CONCLUSION

The ULDA-USSP will facilitate the access to data in the IUE image archive and therefore allow a better utilization of these in a way which will allow the IUE project to be able to retain its standard quality control assurance on its products. The ULDA version 1.0 is complete until December 1983 and regular updates are expected during the lifetime of the Project. At the risk of laboring a point the ULDA and its software support package will come close to putting a large number of spectra on-line to a multiplicity of users and as such is surely a precursor to systems which will dramatically increase the availability of astronomical data from a variety of observatories in the future.

VIII. REFERENCES

Cassatella A., Holm A., Ponz D., Schiffer F.H.: 1980, "A corr. algorithm for Bad ITF Low Dispersion SWP Spectra", NASA IUE NL. #8.

Driessen C.: 1986, "USSP User Requirements Document", Software Descr. SD/9019-01/CD/861212, IGCS/VILSPA.

Driessen C.: 1987, "USSP Software Requirements Document", SD/9020-01/CD/870316, IGCS/VILSPA.

Driessen C., Murray J., Pasian F., Russo G.: 1987, "USSP Architectural Design Document", SD/9022-00/CD/870220, IGCS/VILSPA.

Munoz J.R.: 1984, "IUE low dispersion absolutely calibrated spectral archive (ULDA)", SD/4013-01/JRM/840328, IGCS/ VILSPA.

Munoz J.R.: 1987, "IUE FITS: FINAL", Three-Agency Coordination Meeting, June 1987, ESA IUE Observatory, Madrid.

Sastre C.: 1987, "VILSPA Use of Archive", Three-Agency Coordination Meeting, June 1987, ESA IUE Observatory, Madrid.

Thompson R.: 1987, "Requests of Archival Data in the US", Three-Agency Coordination Meeting, June 1987, VILSPA.

Wells D.C., Greisen E.W., Harten R.H.: 1981, Astron. Astrophys. Suppl. 44, 363.

A DATABASE FOR THE MAGELLANIC CLOUD CLUSTERS

R K Bhatia and H T MacGillivray
Royal Observatory, Blackford Hill, Edinburgh EH9 3HJ Scotland

SUMMARY

A technique has been developed for automatically identifying clusters of stars in the Magellanic Clouds from objective measurements made with the COSMOS plate-scanning machine. The technique is being used to provide an objective, homogenous and statistically complete database of clusters in and around the clouds which will enable their detailed properties and relationship with the properties of their parent galaxies to be examined.

1. INTRODUCTION

The study of the Magellanic Clouds (Large Cloud – LMC; Small Cloud – SMC) provides valuable insight into a very wide spectrum of astrophysical problems. An important component of the clouds is the rich cluster systems: the LMC is estimated to have around 5000 clusters (Hodge 1980) and the SMC around 2000 (Hodge 1986). At present very limited information is available for these clusters. To begin with, no complete survey has been done for identifying them. Even for the ~1700 clusters already identified in the LMC, basic data such as accurate positions, magnitudes, colours, star-counts, shapes and orientations are known only for about 200 of the most prominent ones. Any conclusions drawn from such scanty data is bound to suffer from severe selection effects. It hardly need be emphasized that the time is ripe for an attempt to fill in this serious gap.

It becomes clear at the outset that identifying and obtaining data for each individual cluster (as has been done in the past) is a prohibitively time-consuming and daunting task. For example, a visual search for clusters in and around the LMC would require hundreds of person-hours (remembering that a total of 300 sq. deg. are to be covered!); and given the subjective nature of such a search, there would always be an element of doubt. Moreover, even if such a search were to be successfully conducted, how would, for example, other data like the positions, ellipticities and orientations be obtained?

This is where the powerful combination of the UK Schmidt telescope (UKST) and the automatic high-speed plate scanning machine COSMOS comes into the picture. Briefly, COSMOS scans a plate and outputs all the data onto a magnetic tape (for details see MacGillivray and Stobie 1984). Among these data are the celestial coordinates of all the objects identified by the COSMOS software, their ellipticities, orientations, instrumental magnitudes (COSMAG), etc. The software used is the "crowded-field" package developed at ROE which separates out component images in a multiple system and provides also profile information on each component. The only problem then becomes to develop a technique based on objective criteria for identifying the clusters: the rest of the data follows. It is as simple as that! The main purpose of this paper is to describe the technique we have developed for identifying clusters using the data provided by COSMOS. We shall also discuss how the other data are obtained.

2. CLUSTER IDENTIFICATION TECHNIQUE

In order to determine a technique for identifying clusters, scans on a small 6 sq. deg. field of the UKST sky survey plate (J1953) of the ESO/SERC sky survey field number 57 were made. The area selected was a strip to the north-east of the bar of the LMC in the direction due east. The back-ground density of the stars changed in the strip towards the east, enabling us to test our method in very crowded regions also. The data from COSMOS consisted of several hundred thousand images down to B~22.5. In order to reduce this number down to a more manageable number of cluster candidates, any object picked up by the COSMOS software which had a "multiplicity" of four or more was initially classified as a cluster. A list of such objects was prepared. Their xy positions were then used to make an overlay for visually identifying the objects on a film copy of J1953 with a hand-held magnifier. From this visual inspection, a number of objects in the list were found to be "spurious": these were galaxies, nebulosities and in particular bright stars with haloes and diffraction spikes. Another method had to be devised to identify and reject these contaminants.

Further rejection was done by defining a two-parameter space from the COSMOS data. The first parameter used the magnitude histogram of the images constituting the object: Δm was defined to be the difference between the brightest and second brightest image. The second parameter p used the intensity profile information at four different points, avoiding the peak intensity as this was likely to suffer from machine noise in the saturated cores of the bright images. A plot of Δm vs p was used to subsequently remove most of the contaminants.

Fig. 1 (a) shows the (Δm,p) plot for the bright objects in the sample (COSMAG $<$ -7, equivalent to about B~15). For the brighter clusters the separation is p-dependent: a vertical cut-off in p at 0.26 neatly removes most of the contaminants. For the fainter objects, the separation is Δm dependent: as Fig. 1 (b) shows, a horizontal cut-off in Δm at 2 separates the clusters. The criterion used is evidently magnitude dependent.

This method of defining a "candidate" sample of clusters and subsequently cleaning up the sample is quite efficient. For example, the initial list for clusters brighter than COSMAG =-5 (B~17), prepared on the basis of multiplicity, contained more than 800 objects. After applying the (Δm,p) criterion, the number was reduced to 300, out of which 205 finally turned out to be clusters. In total, we detected 287 clusters (down to COSMAG ~-4.5 (B~17.5)), which more than triples the number of known certain clusters in that area. This is consistent with the findings of Hodge (1980).

Unfortunately, mainly due to the crowded background, there is still an element of subjectivity in the last step (after applying the above criteria). Out of the 287 clusters identified, 205 were classified as certain, 42 as probable and 32 as uncertain. Most of the uncertain clusters were small and faint and situated in regions with a crowded back-ground. We are at present investigating methods for eliminating human intervention at the last step. However, it should be emphasized that this can still be regarded as tidying up an initial sample selected from the use of objective criteria.

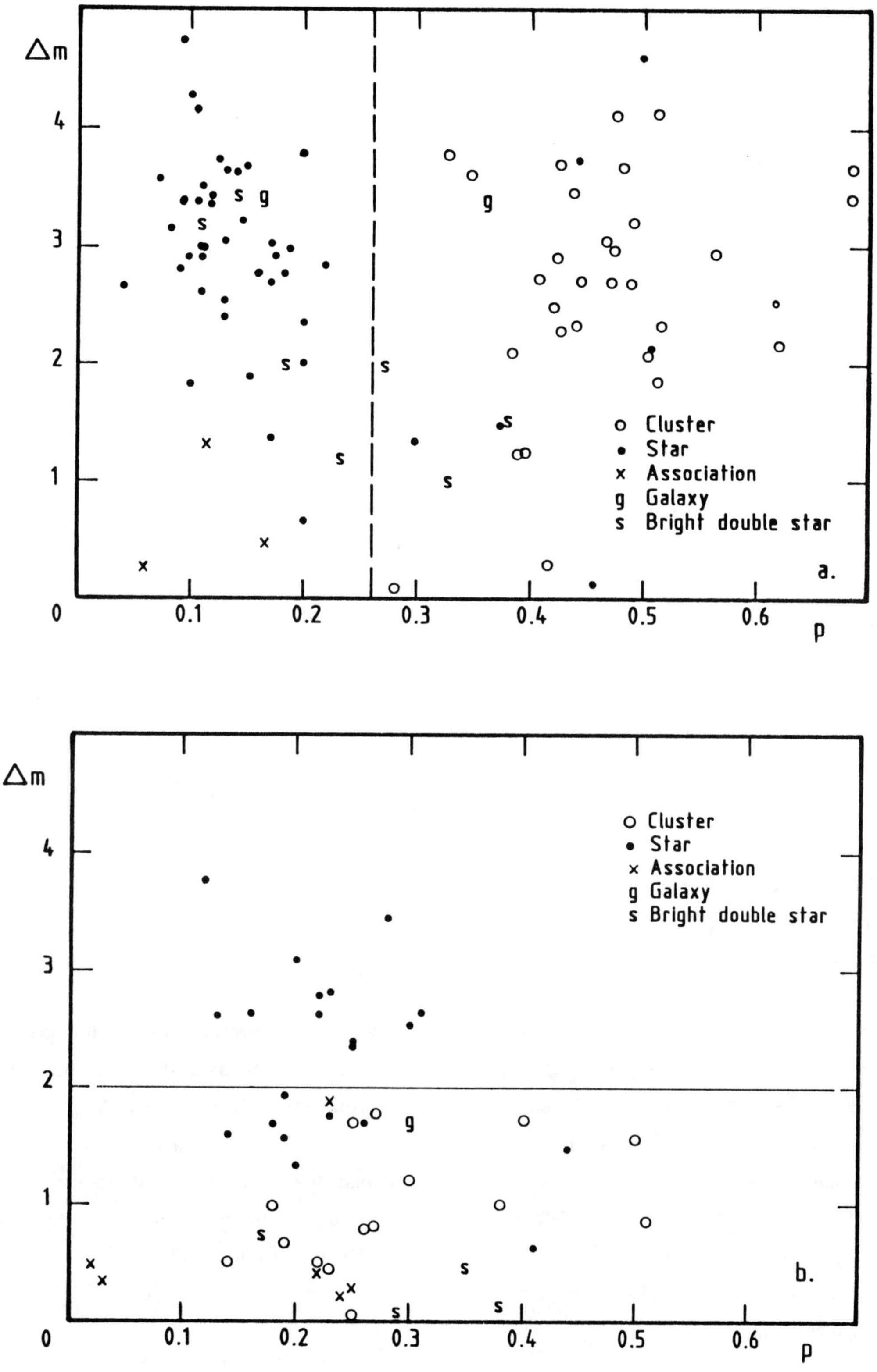

Figure 1: *The quantity* Δm *versus p for (a) the bright cluster candidates, and (b) the faint cluster candidates. See text for details.*

3. OTHER INFORMATION OBTAINED

a: Magnitudes and colours:

We have demonstrated elsewhere that it is possible to calibrate, using CCD observations, the instrumental magnitudes for these clusters obtained from COSMOS (Bhatia, Cannon and MacGillivray 1987). The rms scatter in B is ~0.15 and in R~0.11, which compares favourably with the results obtained from photo-electric photometry. We have also shown that reliable B−R colours can be obtained from these calibrations. This colour seems to be a good indicator of age (Mateo, private communication). Again, the power of the technique becomes evident: through this calibration it is possible to obtain the colours and hence the ages for several hundred clusters in a single field.

b: Projected ellipticities and orientations:

Previous studies of ellipticities of individual MC clusters have been done by Geisler and Hodge (1980) from micro-densitometer tracings, by Frenk and Fall (1982) from eye-estimates, by Geyer et al (1983) from the Agfa-Contourfilm technique and by Kontizas et al (1985) from microdensitometer tracings. Our technique is based on a moment analysis and gives both the ellipticity and orientation of the cluster. An analysis of the ellipticities of the LMC and SMC clusters will be presented elsewhere. That our technique gives reliable results can be seen from the following: for the SMC cluster NGC 121, we get e = 0.294. This compares well with the value of 0.298 obtained by Geyer et al (1983) and 0.25 by Kontizas et al (1985). Again, once the clusters are identified, obtaining their ellipticities and orientations becomes a simple matter.

c: Star-counts and diameters:

Star-counts around the clusters (except, of course for the very smallest ones) can be used for obtaining important parameters like the core and tidal radius (using King's models). Moreover, it is also possible to obtain the ellipticities and the diameters. It is a simple matter to use the COSMOS data for obtaining completely objective star-counts in annuli around the clusters from their centres outwards. Fig 2. shows the stellar density as a function of radius for a cluster in our sample. The star-counts well away from the cluster-center help define the back-ground, and we deduce a diameter of 10 pc. For comparison, the COSMOS moment-analysis method gives 9 pc. The same cluster had been assigned a diameter of 5.5 pc by Hodge (1980) from eye-estimates. We suggest that this discrepancy is caused by the tendency of the eye to be biased by the few bright stars in the cluster center, resulting in a lower value for the diameter. This again demonstrates the value of a survey based on objective, machine-produced data.

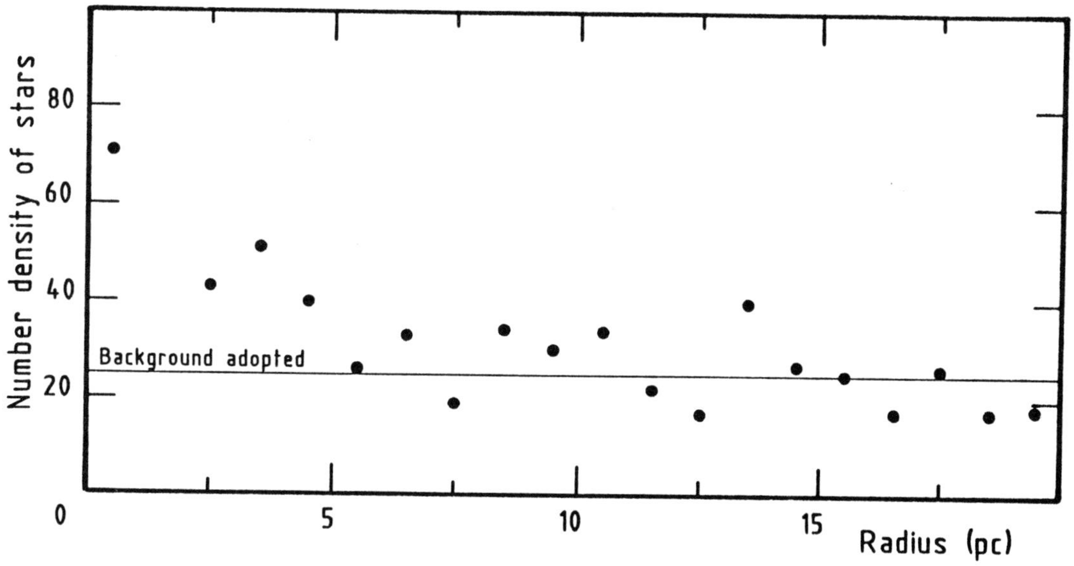

Figure 2: *The stellar density as a function of radius for one of the clusters in our sample. The radius we deduce (5pc) is approximately double that given by Hodge.*

4. CONCLUSIONS

We have shown that the powerful combination of UKST plates and the COSMOS high-speed machine can give an enormous amount of information for the clusters in the MCs in an efficient and reliable manner. We have developed a technique for automatic cluster identification and successfully applied it to a 6 sq deg. area in field 57 in the LMC, which has more than tripled the number of known clusters (Bhatia and MacGillivray 1987). Once the clusters are identified, other information such as ellipticities, orientations, magnitudes and colours can be obtained in a straight-forward manner. B and R plates covering the areas for both the Clouds have already been scanned. We expect to be able to produce a complete and reliable catalogue giving all the basic data for the clusters, in the near future. This will form a valuable data-base for the analysis of the cluster system of the Magellanic Clouds.

REFERENCES

Bhatia, R.K., Cannon, R.D. and MacGillivray, H.T. (1987) in proceedings of the ESO work-shop on "Stellar Evolution and Dynamics in the outer halo of the Galaxy" (in press).

Bhatia, R.K. and MacGillivray, H.T. (1987) submitted to Astron. J.

Frenk, C.S. and Fall, S.M. (1982) Mon. Not. R. Astr. Soc. 199, 565.

Geisler, D. and Hodge, P.W. (1980) Astrophys. J. 242, 66.

Geyer, E.H., Hopp, U. and Nelles, B. (1983) Astron. Astrophys. 125, 35.

Hodge, P.W. (1980) Astron. J. 85, 423.

Hodge, P.W. (1986) P.A.S.P. 98, 1113.

Kontizas, E., Dialetis, D., Prokakis, T. and Kontizas, M. (1985) Astron. Astrophys. 146, 293.

MacGillivray, H.T. and Stobie, R.S. (1984) Vistas Astron. 27, 433.

ASTRONOMICAL DATA ANALYSIS FROM REMOTE SITES

Supriya Chakrabarti, Carl Dobson, George Kaplan, Herman Marshall,
Michael Lampton, Roger Malina, Stuart Bowyer
Space Sciences Laboratory, University of California, Berkeley, CA 94720, U.S.A.

Jeff Star
Remote Sensing Research Unit, University of California, Santa Barbara, CA 93106, U.S.A.

Abstract

Given the progress in the communication technology, it is expected that during the space station era the mode of instrument operation and data analysis will be dramatically different from those at present. A consortium of several universities and NASA centers are evaluating various aspects of design and operation of scientific instruments and data analysis over various computer networks from a remote site. Such a scheme has officially been termed *telescience*. We will report on the development of methodologies for *teledesign, teleoperation* and *teleanalysis* and the verification of these concepts using the *Extreme Ultraviolet Explorer (EUVE)*, a satellite payload scheduled for launch in 1991. The EUVE telescopes will be operated remotely from the EUVE Science Operation Center (SOC) located at the University of California, Berkeley. Guest observers will remotely access the EUVE spectrometer database located at the SOC.

Distributed data processing is an integral part of telescience. We will describe our experience with the *Browse system,* currently being developed at the University of California at Santa Barbara through a grant from NASA for remote sensing applications. We will discuss the suitability for its adoption for astronomy applications. Browse allows the examination of a subset of the data to determine if the data set merits further investigation. The examination can be as simple as looking for a specific data element based on its location, date of observation, quality indicator, spectral coverage etc. It also allows the viewing of data in various modes depending upon the available resources at the user's end (e.g., graphics terminal vs. dumb terminal), level of data compression applied, required display format, and transmission of the data over a network to a local graphics display station.

Introduction

With the availability of sensitive array detectors the data rates for the modern astronomical instruments have grown tremendously over the past few years. Even with the availability of high speed computers, the processing of the voluminous data these instruments generate is tackled only one experiment at a time. During the space station era, we anticipate a large number of collaborative, multi instrument experiments conducted from geographically distributed sites which will easily exceed the database of typical experiments conducted today. Furthermore, since most of these experiments will be conducted by astronomers from different institutions, a new mode of science operation will need to be employed, which will further complicate the data analysis.

It is likely that the database generated by these instruments will be maintained at several sites. This distribution of the database implies a need for efficient data transfer between the sites for analysis purpose. ''Electronic browse'' of the available data set will help minimize the required data transfer. Furthermore, it is obvious that each site will have certain unique expertise, analysis tools etc. which need to be shared between the collaborating scientists.

The Office of Science and Application (OSSA) of NASA has recognized these needs and has set up a set of testbed activities to evaluate the feasibility of remote operation of instruments and correlative data analysis from the experimenters' home institution. This new mode of conducting experiments and data analysis has been termed *telescience.* An important goal of these activities is the identification of critical issues for the design of the Science Applications Information System (SAIS) and the Space Station Information System (SSIS). We are involved in several of these activities at the University of California at the Berkeley (UCB) and Santa Barbara (UCSB) campuses. The Berkeley efforts use various aspects of an Extreme Ultraviolet (EUV) satellite mission, while the Santa Barbara work is focussed primarily on remote sensing data. In this paper we describe our joint efforts to explore the possibility of adapting a distributed database system developed at UCSB for use with UCB's EUV satellite data.

The Extreme Ultraviolet Explorer (EUVE)

The EUVE instrument consists of three scanning telescopes, which will perform an all-sky survey in the extreme ultraviolet (80-800 Å), and a deep survey/spectrometer telescope that will collect imaging and spectroscopic data along a narrow band near the ecliptic (Bowyer et al., 1986). Raw data will consist of lists of time-tagged photon detections, which are then remapped to sky positions using spacecraft attitude data. At the Space Sciences Laboratory (SSL) of UCB a network of Sun workstations will be used for data analysis. This is the same network that is currently in use for development of the analysis software, as well as for instrument design, test, and calibration.

A major goal of the Telescience initiative is for experimenters to be able to operate their (possibly space-based) instruments from their home institutions as well as analyze the data obtained from one or more experiments. With the increasing number of collaborative research efforts involving widely separated sites, it will also be important to develop methodologies for these institutions to cooperate on the design of the experimental instruments, and data analysis schemes. These methodologies, *teleoperation* and *teledesign,* are the focus of three Telescience testbed activities involving the EUVE project at UCB.

In the area of teleoperation, an experiment is underway to simulate the operation of the EUVE science instruments from a remote site using one of the existing TCP/IP networks. The instrument are represented in these tests by a "Kiwi" (a "flightless bird") consisting of breadboard electronics and other prototype hardware. It is connected to a single-board computer in essentially the same configuration as will be used with the real instruments during orbital operations. This workstation receives telemetry and sends commands to the onboard computer of the instrument. Software to relay instrument commands over a network using the TCP/IP protocols has been written to allow a second workstation to be used as the command interface. Testing will proceed in steps over progressively longer distances. Initial tests, using two workstations on the SSL ethernet, have already been performed. Subsequent tests will be performed with the second workstation at Stanford University, using the Bay Area Regional Research Network (BARRNET), which connects several research sites in the San Francisco area over high speed (T1) links.

In the final system the two workstations will be on opposite coasts, with the ground support workstation at GSFC and the SOC workstation at UCB-SSL. The actual link between the SOC and GSFC could be either a dedicated line or a connection through an existing network such as ARPAnet or NSI. A network connection has certain advantages over a dedicated line, but new concerns over command validation arise due to the large number of users who have access to the networks. The goal of the teleoperation experiments is to help answer these concerns as well as others about whether the network can supply the bandwidth necessary for control of the EUVE instruments.

In the teledesign area, we are designing a scheme which allows software development by a Software Control System (SCS), that allows programmers and users to coexist peacefully. That is, versions of many of the analysis software modules are already in use for instrument testing and calibration, at the same time as more advanced versions of the same modules are under development. A "promotion" scheme provides a controlled mechanism for installing revised software modules without causing unpleasant surprises for the users.

In order to encourage the sharing of software resources, all software and associated files are globally accessible from anywhere in the SSL computer system. This is not accomplished with a central computer or file server. Instead, each workstation has its own disk, with files being accessible from any of the other workstations, using Sun Microsystems' Network File System (NFS). From the perspective of teledesign, it is natural to extend the locally shared software environment to be used between institutions on a wide area network such as the ARPAnet. As part of the Telescience Testbed Program, the SCS is being extended to provide such a service, but replacing the NFS file-sharing facility with a "borrow" mechanism that propagates updated versions of modified files at daily intervals.

The Browse System

As a part of the telescience testbed activity an electronic browse capability is being developed at the Santa Barbara campus of University of California. The Browse testbed utilizes a database consisting of images and texts with the goal of allowing the user to preview the database from remote locations. The primary database used by the Santa Barbara group is remote sensing data from Landsat and other sources. The key goal of this testbed is to demonstrate the viability of a distributed database system using local database and specialized processing software with users accessing it through wide area computer networks. Preliminary results detailing experience with the current Browse implementation is outlined elsewhere (Star et al., 1987a).

The Browse project is aimed at defining methodologies for managing large volumes of data expected from future missions (estimated at a terabit per day for the Earth Observing System (EOS), for example, Arvidson et al., 1986). With such large databases, the ability to electronically browse the database before retrieval is essential. The present estimate of the number of user requiring such browse capability ranges from 1500 to 15,000 (Star et al., 1987a). It is reasonable to expect that the familiarity of these users with the database will vary widely. Therefore a need exists for several levels of user interface to the database.

In order to explore these issues a prototype system consisting of an image and other spatial data and descriptive information about the data (metadata) has been created which serves as one node of the distributed image database system. The present database runs on a Digital Equipment Corporation microVAX II minicomputer running the VMS operating system. The database management software being used is RIM, a relational database management system (DBMS) created by the Boeing Aerospace Corporation. An IBM PC/AT microcomputer is used as a remote user workstation with a telephone line using standard communication protocol as the communication network.

In this experiment the microVAX is being used as a dedicated database computer. A set of utilities has been written for use on the remote user workstation which includes communication software, graphics software for defining the geographic area of interest as well as image display software which allows the user to view subset images of selected data sets (Star et al., 1987b). These utilities, in conjunction with the database software at the host node, allow the user to locate and preview potentially useful data sets for a specific application. For example, a geographic region can be described using the interactive graphics software on a map displayed on the user's workstation. After the area of interest is chosen, available data sets for that region can be selected based upon several constraints such as date, sensor type, cloud cover and other scene-specific criteria.

As shown in Figure 1, each node of the database consists of three major components, the *directory,* the *catalog,* and the *user* databases. The directory is a database containing information regarding the nodes in the distributed database and a description of the contents at each node. It is resident in each node of the network and contains one entry for each node in the network. The entries describe the nodes with a set of attributes such as, the number of data objects available, the geographic and scientific emphasis of the node, and the data format.

The catalog, which is unique to each node, consists of detailed attributes of the data, which can be used by users to select appropriate images. These attributes include *the form of the data* such as film, printout, digital etc.; *the description of the data* such as date and time of acquisition, platform, sensor, spectral coverage etc.; *quality indicators* such as cloud cover, sun angle; and *any processing history* which may have been performed on the data.

The user database contains information regarding local users, such as their field of specialization. Since the current implementation is based on remote sensing applications, it contains the user's scientific and geographic specialization. It also contains information regarding system usage, thereby allowing the users to obtain information regarding other scientists involved in related research.

In order to evaluate the performance of this strawman system a group of scientists has been selected to use it. Since no network links are currently in place, a toll free telephone number has been established to allow the test users to access the system at UCSB. Through this link the users can query the database, download the required image data and then perform graphic manipulations at their local workstations.

A typical session with present system will involve specifying a set of attributes through a menu driven procedure. The geographic area of interest is next identified interactively by drawing a window on a map with a pointing device. The user is then prompted for more specific information. For example, a climatologist interested in the land surface energy balance research in the Santa Barbara area might enter the following attributes:

Database of interest:	*CATALOG*
Platform:	*ALL*
Sensor:	*ALL*
Begin Date:	*June 1, 1979*
End Date:	*Sept 1, 1979*

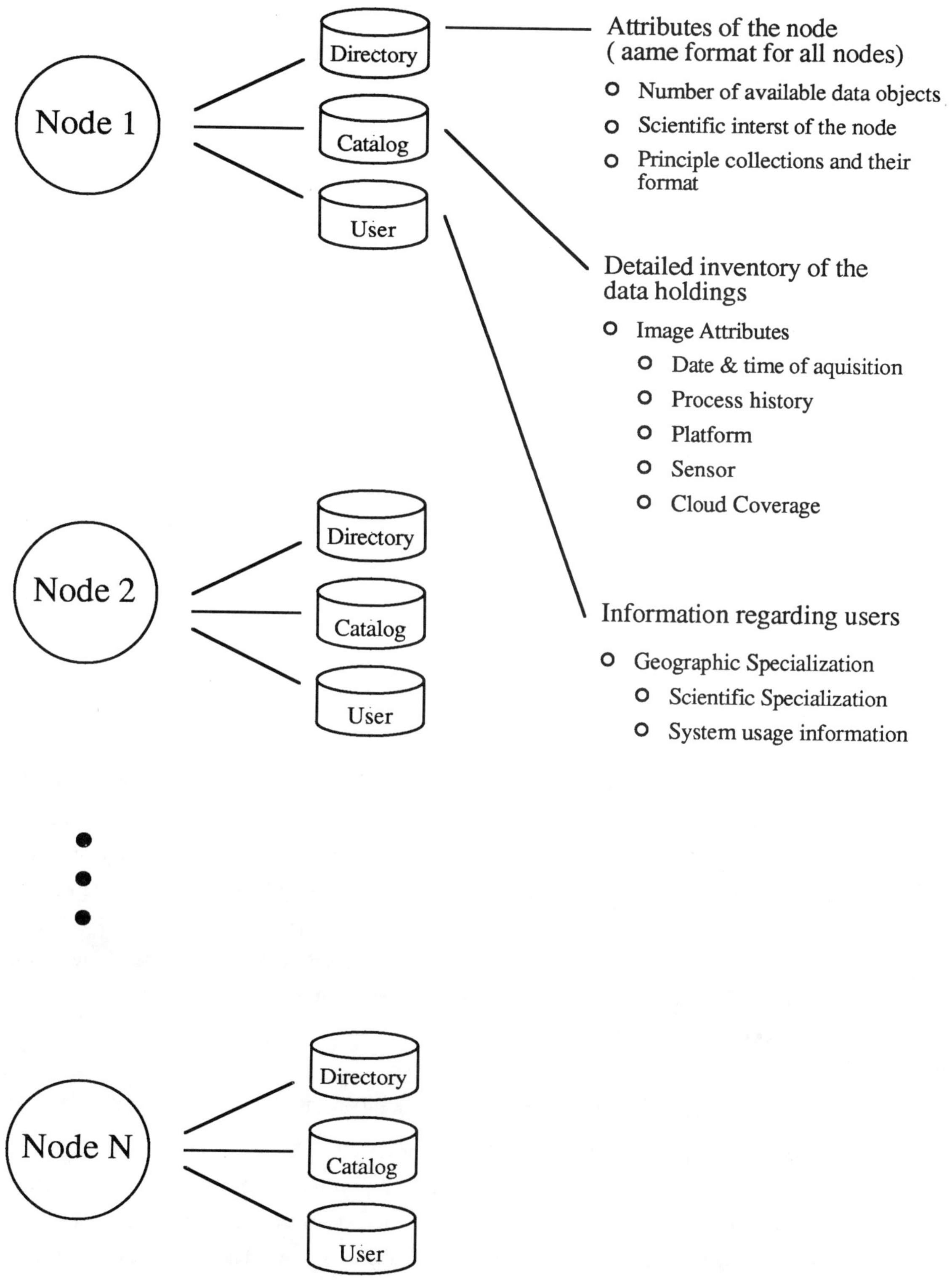

Figure 1. A schematic of the Browse distributed database management system.

to which the system might respond with information of the following nature:

Browse Database: 2 items found

1. Heat Capacity Mapping Mission,
   July 3, 1979
   3 bands on line, 600m spatial resolution
   scene id # A-A0062-16032-2

2. Landsat-3,
   August 15, 1979
   4 bands on line, 600m spatial resolution
   scene id # 8246517502500

For more information please enter the scene id or ESC to escape:

At this point the user may obtain more detailed information such as the sensor specifications, resolution, number of lines, sun angle, etc. by entering the scene id. This is how the user ''browses'' through the image database and identify potentially useful data sets. When an appropriate image is found, the user exits the database to download the image data at the remote workstation. The image display software can then be used to manipulate the actual image data.

The Adaptation of Browse to EUVE

One of the important components of the EUVE program is the Guest Observer (GO) program for spectroscopic observations with the Deep Survey/Spectrometer instrument. The current plans for the GO data analysis include, a local database and distributed scientific users having various level of access to the database. These components are all similar to those in the Browse system. Furthermore, the EUVE data makes heavy use of images of astronomical objects, another similarity with the Browse system. Finally, the spectroscopic data are obtained by imaging spectrometers, and so they can be treated like images. For all these reasons we are exploring the possibility of incorporating the Browse system or some of its key components in the EUVE GO data analysis system. The following section describes the GO data handling plans and its similarities with the Browse scheme. Since the EUVE GO plans are in their infancy and the Browse is not available for external users, we describe only our current plans for the GO data analysis.

The EUVE Guest Observer Data Analysis Plan

Data processing for GO's will proceed in much the same manner as with sky survey data (see Marshall *et al.*, these proceedings). At the discretion of the guest observer, the off-line mode of operations will require permanent files that are generated in the production processing. Binned photon and exposure maps are generated that may be deconvolved for some studies (e.g., extended sources and crowded fields), while detailed photon and exposure data are saved in "pigeon hole" files for studies of individual sources. It is expected that the latter data set will be more commonly used. The collection of detailed data into pigeon holes (as applied to spectra) is governed by the list of sources detected in the deep survey instrument which provides a direct image of the target region. These data will be similar in format to those in the Browse database. As with the survey data, sources may be found which were not expected or which may be brighter than expected and thereby contribute to noise in the spectrum. Thus, off-line analyses are required to collect data for sources that were not in the original catalog. This off-line processing should run in a fashion similar to that of the scanner data (Marshall et al., Figure 4). Data for newly discovered sources will be collected from prior observations for future analysis.

The main differences between the handling of survey and spectroscopic data are in the treatment after collection into a permanent database. Figure 2 shows a preliminary scheme for handling GO data. Generally, all data are handled by the GO interface software which allows data and software transfers via phone lines, networks, or physical media (e.g., optical disks for large requests or printout for users without access to computer networks). Browse has schemes that account for the different levels of access to the database; even different types of user environments (e.g., dumb terminal vs. graphics workstation interfaces) are also included in their interface. We will examine these schemes before selecting the GO interface.

The GO Interface selects the applicable portion of the database to minimize file transfer time and should warn the remote user if the requested data are not in the current database. In such a case, a special request is generated that requires action by the database manager, which controls the archive data. Requests for unavailable data are handled is handled by generating a set of new coordinates for guiding the selector program via the database manager. This process insures that the raw data are managed properly by an on-site operator of the database managing program. Unlike Browse, we plan to allow remote logins but access to the production network is restricted for security. Temporary files may be created by running remote processes and then transferred to the GO's home institution.

The GO database plan calls for a file called *history*. It contains general orientation information, instrument housekeeping information and other parameters such as sun angle, positions of the moon and other planets, etc. There are also plans for storing intermediate results such as exposure information, photon maps, data quality indicators, etc. This database is therefore essentially the same as catalog in the Browse system, which contains information that describes the quality of the data and other attributes.

It is expected that a number of specialized I/O and data analysis packages will be written specifically for use in the IRAF data analysis environment, an image analysis package developed by the National Optical Astronomy Observatories (NOAO). These packages will be available for remote execution or transfer to the GO's home institution for installation in IRAF. This is similar in concept to the software Browse will provide to its remote users for manipulation of images. The packages supplied for EUVE data analysis will run on any hardware configuration to which IRAF has been ported. Using IRAF's potential networking functions, one may also invoke this package at the remote site without transferring the applicable files.

We also anticipate several simultaneous observations with EUVE by other instruments (ground- or space-based). These GO nodes will contain special data not in the EUVE database as well as specific scientific interest and expertise. These characteristics are also incorporated in Browse and are included in the directory and user databases. We anticipate that the EUVE GO database management plan will include similar features.

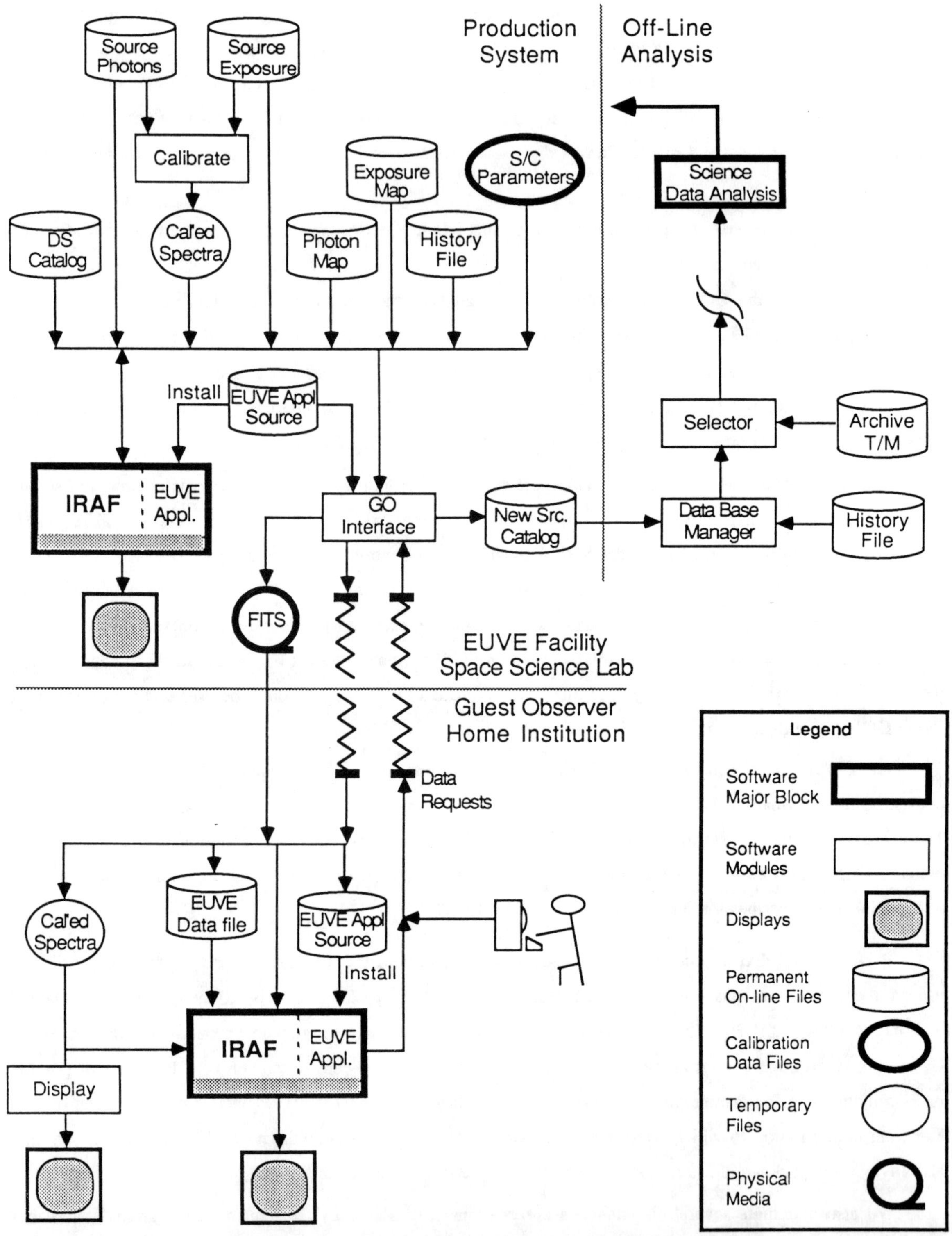

Figure 2. A schematic of the EUVE guest observer data management plan.

Unfortunately Browse, as it is presently implemented, uses a VMS® operating system. All of the EUVE software is built around a network of Sun workstations running Unix®. Our SCS scheme, described earlier makes heavy use of UNIX facilities. Therefore, we will not be able to incorporate Browse readily. We will, however, include several key concepts of Browse.

Acknowledgements

We thank Mark Friedl, William Marchant and Danial Quinlan for helpful discussions. We also thank Aileen Corelli for editorial assistance. This work has been performed under GSFC contract NAS5-29298, NASA subcontract NASW-4234, and the University Space Research Association subcontract 800-68 to the University of California, Berkeley.

References

S. Bowyer, "The Extreme Ultraviolet Explorer and the local interstellar medium", *Adv. Space Res.*, **6**, 153, 1986.

H. L. Marshall, C. A. Dobson, R. F. Malina, and S. Bowyer, "Plans for the extreme ultraviolet explorer data base", *this proceedings.*

J. L. Star, M. A. Friedl, D. M. Stoms, and J. E. Estes, "Browse capability for spatial databases", To appear in the proceedings of *GIS-87, San Francisco, the 2nd annual international conference, October 1987a.*

J. L. Star, D. M. Stoms, M. A. Friedl, and J. E. Estes, "Browsing for suitable input data" to appear in the proceedings of the *International Geographic Information System Symposium,* Crystal City, Arlington, Virginia, November, 1987b.

MAINTENANCE OF A DATA BASE

Pascal DUBOIS
CDS, Observatoire de Strasbourg

A. Introduction

To make possible that a database survive, one has to consider the maintenance as being of prime importance. This consists mainly in two aspects: updating and cleaning. This might apply not only to the data but also to the software. The work implied by these operations is frequently grossly underestimated. Most of the comments presented below arise from the experience gained with SIMBAD, the database of the "Centre de Données de Strasbourg" (Egret and Wenger, 1987) and try to show the real dimension of these operations.

B. Updating

The implementation of data in a database could be made in two different modes: at demand or by batch.

1. Demand mode

In the case of SIMBAD, this mode is used for the introduction of the bibliography (Carpuat et al. 1984) and correction of a small amount of data. It has the advantage to benefit of immediate tests with preexisting data and then to reduce the probability of introducing errors. Nearly every day papers containing errors or imprecisions in the identification of objects appear in the literature. The checks can be made immediately before introduction in the database. This shows the necessity to give always at least two types of identification: for instance two names, or a name and position, or a photographic reproduction of the field plus position, etc... It should be noted that objects with exotic denominations or imprecise data have a small priority to be introduced in the database and then the acces to the object and to the paper in which it is used, will be lost.

2. Batch mode

In case of a great amount of data, the batch mode is preferable. Some tests can also be performed, but all tests cannot be imagined before introduction of the data. Then the introduction of a new catalogue is also the main source of errors in a database.

The introduction of a catalogue consists in two main phases. In a first step it is necessary to check the consistency of the denominations used in the catalogue with those of SIMBAD. It is not trivial if , for instance, the identification is given by a letter like B which signification could be BD, Batten, Bailey, Barnard, Bracesi, ... The integration is also not immediate when the identification is given only by a number coming from different catalogues, the format of the number giving the name of the catalogue. As an example, CPD, COD, BD, HD, ... numbers could be found in the same column. After this preparation the errors detected in comparing the new catalogue and SIMBAD have to be corrected. Generally the cross-identifications have also to be carried out. Typically, about 20 % of a catalogue has to be verified and 1% of objects give difficulties. This means, for a catalogue of 100000 objects, that 20000 objects (identification or data or both) must be verified. The operations of implementation take a few days for a catalogue of 1000 objects and about six months, for one person, for 100000 objects. These operations take more time if there are many data of the same type in the database allowing many tests and then detection of more errors. On the other hand, it takes only a few days to introduce 50000 new data for which no other comparison data are available.

C. Cleaning

The cleaning operation implies the correction of errors which can be detected either systematically or at random.

1. Systematic detection

For a systematic detection, some tests have to be imagined and applied to the data. For example missing numbers in a catalogue could be detected. This is easy in a catalogue with a running numbering. This is a little more difficult when the catalogue is ordered by zones and could be rather complicated in some other cases.

2. Random detection

For a detection at random the first way, in the case of SIMBAD, is through introduction of bibliographical references. The second important contribution is the detection of errors by the users.

D. Conclusion

A database is a living object. The result of an interrogation is generally different from a similar interrogation made previously because of additions and corrections brought meanwhile.

It exists a kind of "natural selection". When data, and especially the identifications of an object, are imprecise they cannot be introduced in the database. The reference to the publication is consequently not made. This implies the necessity to give always at least two types of identifications for an object.

The volume of this "plumbing", for a database of 1 million objects, needs the efforts of several astronomers working full time.

References:

- Carpuat C., Damge M., Dubois P., Kirchner S., Lagorce A., Laloe S., Ochsenbein F. and Wagner M.J. : 1984, CDS Information Bull.26,58

- Egret D. and Wenger M.: 1987, this conference.

SIMBAD : SURFACE DISTRIBUTIONS OF SOME BASIC ASTRONOMICAL DATA

Daniel Egret and André Heck
CDS, Observatoire astronomique, Strasbourg, France

Abstract

Distributions on the celestial sphere are presented for some basic astronomical data provided by SIMBAD, the CDS astronomical database.

Introduction

SIMBAD, the database of the Centre de Données de Strasbourg (CDS) is the largest set of basic astronomical data in the world (see, e.g.: Heck and Egret, 1987a; Egret and Wenger, 1987). It is regularly updated and, then, represents pretty well the status of the observational knowledge on astronomical objects at the present date.

We illustrate here the surface distributions of fundamental data, updating the work of Egret et al. (1976) carried out more than ten years ago, with a database containing, at this date, only 430 000 stars and no galaxies.

Stars and Galaxies in SIMBAD

Figures 1 to 3 display respectively the surface distributions of all the stars, galaxies and non-stellar galactic objects contained in SIMBAD, and for which coordinates are available. There is still a number of objects for which the coordinates are not available in machine-readable catalogues: a major part comes from the Durchmusterungen which were not machine-readable until very recently (Warren, 1987); clusters for which only maps are given in the literature are another case.

The figures are drawn in galactic coordinates using a Samson/Flamsteed projection. Longitude 0° is on the left of the figure, 180° at the center and 360° on the right. The *pixel* in the drawings is an elementary field of size $2.5° \times 2.5°$. Colors show the number of objects per elementary area, on a linear scale given in the captions. This density is to be divided by 6.42 to get the number of objects per square degree. Note that a white area indicates a complete lack of objects in the corresponding field: this is, for instance, the case for galaxies towards the galactic equator.

Features easily recognizable are, towards the galactic plane, Cygnus ($l \approx 80°$), Perseus ($l \approx 120°$), Orion ($l \approx 200°$) and Carina ($l \approx 290°$), and, in the Southern hemisphere, the Magellanic Clouds (LMC: $l \approx 280°$, $b \approx -35°$; SMC: $l \approx 300°$, $b \approx -45°$).

Objects with bibliography

The amount of information available for a given object may vary from a single mention in a catalogue or an astronomical paper to hundreds of bibliographical references and dozens of observational measurements (Ochsenbein and Dubois, 1986).

Figures 4 and 5 display a subset of Figures 1 and 2 respectively: namely those objects for which bibliographical references are available. The difference between the numbers of objects in Fig. 1-2 and 4-5 comes from the fact that the latter ones do not include objects appearing only in large astronomical catalogues and surveys. Papers containing lists of more than 1 000 objects are considered as catalogues. The bibliography is exhaustive from 1950 onwards for stars, and from 1983 onwards for the other objects.

Examples of basic data: UBV and uvby photometry

The two following figures display, for the stars in SIMBAD, the density of objects for which the following observational data are available.

Figure 6 gathers 50 536 stars with photoelectric photometry in the UBV system (Mermilliod, 1987), while Figure 7 presents 12 858 stars with uvby photometry (Hauck and Mermilliod, 1985).

SIMBAD and space data

An example of the intersection of space data and SIMBAD is given in Fig. 8 displaying the surface distribution of the targets observed by the IUE satellite for which identifications and coordinates are available in SIMBAD. This figure has been drawn from the homogenized IUE log prepared for ESA by the CDS (Heck and Egret, 1987). Here again, the Magellanic Clouds clearly appear as a maximum of density, together with several regions in the Galactic plane.

A plot of the same kind for the Input Catalogue of the HIPPARCOS space astrometric mission is presented by Turon et al. (1987).

Conclusion

We illustrated here some surface distributions of basic astronomical data offered by SIMBAD. They are representative of the general status of available data. In a more comprehensive paper in preparation, we shall consider other types of data and discuss biases and incompletenesses that should be taken into account for astronomical studies.

Acknowledgement: The drawings were made using the software developed by Frederic Arenou for the INCA consortium (Morin and Arenou, 1987).

References

Egret, D., Heck, A. and Ochsenbein, F.: 1976, Astron. Astrophys. Suppl. Ser. **26**, 65

Egret, D. and Wenger, M.: 1987, this conference

Hauck, B. and Mermilliod, M.: 1985, Astron. Astrophys. Suppl. Ser. **60**, 61

Heck, A. and Egret, D.: 1987a, Messenger, **48**, 22

Heck, A. and Egret, D.: 1987b, Homogenized log of IUE observations, CDS special report (in press)

Mermilliod, J.-Cl.: 1987, UBV Photoelectric Catalogue (1986), I. Original Data, Astron. Astrophys. Suppl. Ser. (in press)

Morin, D. and Arenou, F.: 1987, this conference

Ochsenbein, F. and Dubois, P.: 1986, Bull. Inform. CDS **31**, 137

Turon, C., Gómez, A., Crifo, F.: 1987, this conference

Warren, W. H. Jr.: 1987, Bull. Inform. CDS **32**, 67

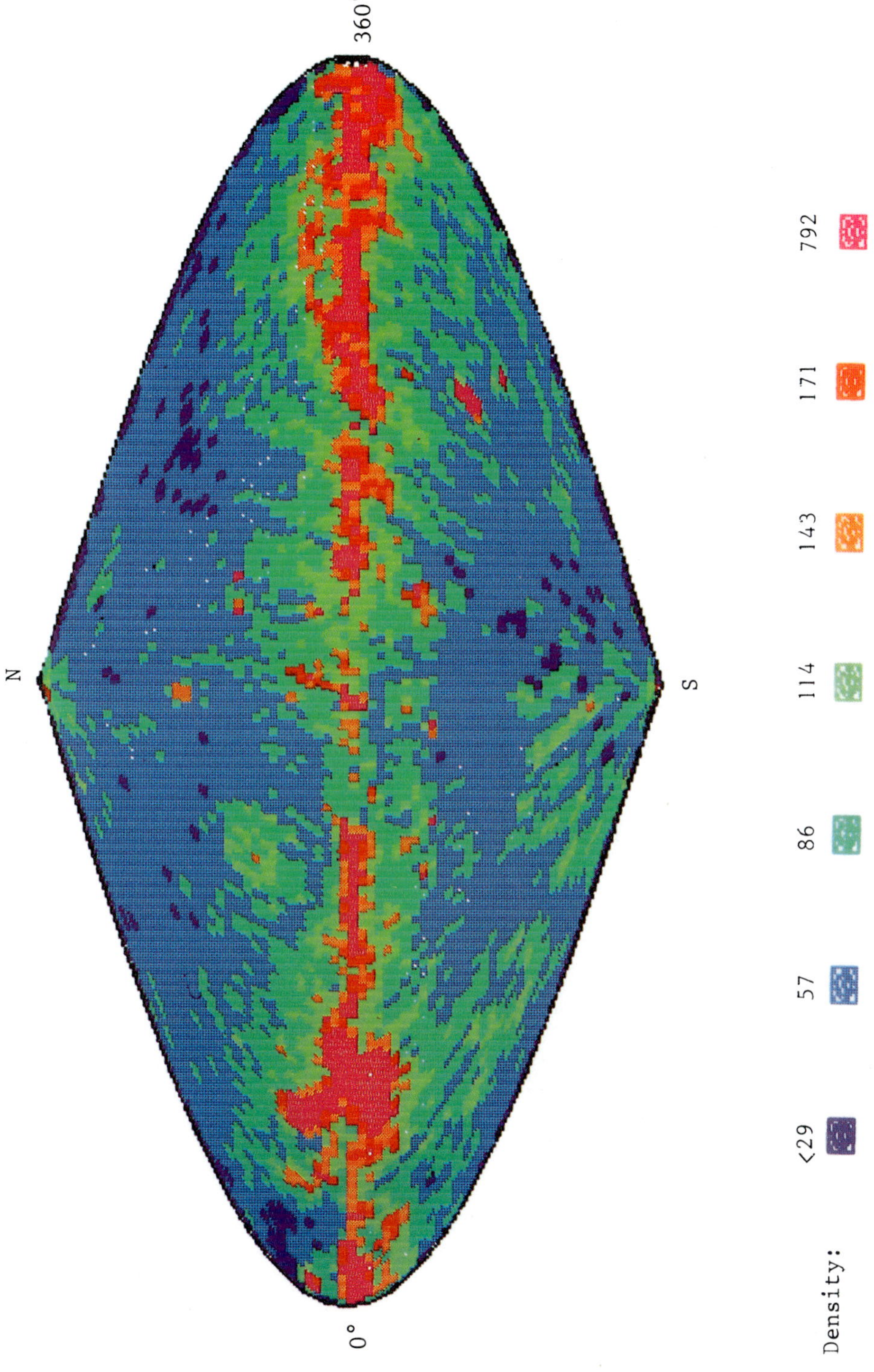

Fig. 1 - Surface distribution of the 531,945 stars contained in SIMBAD. (Refer to the text for the definition of the density).

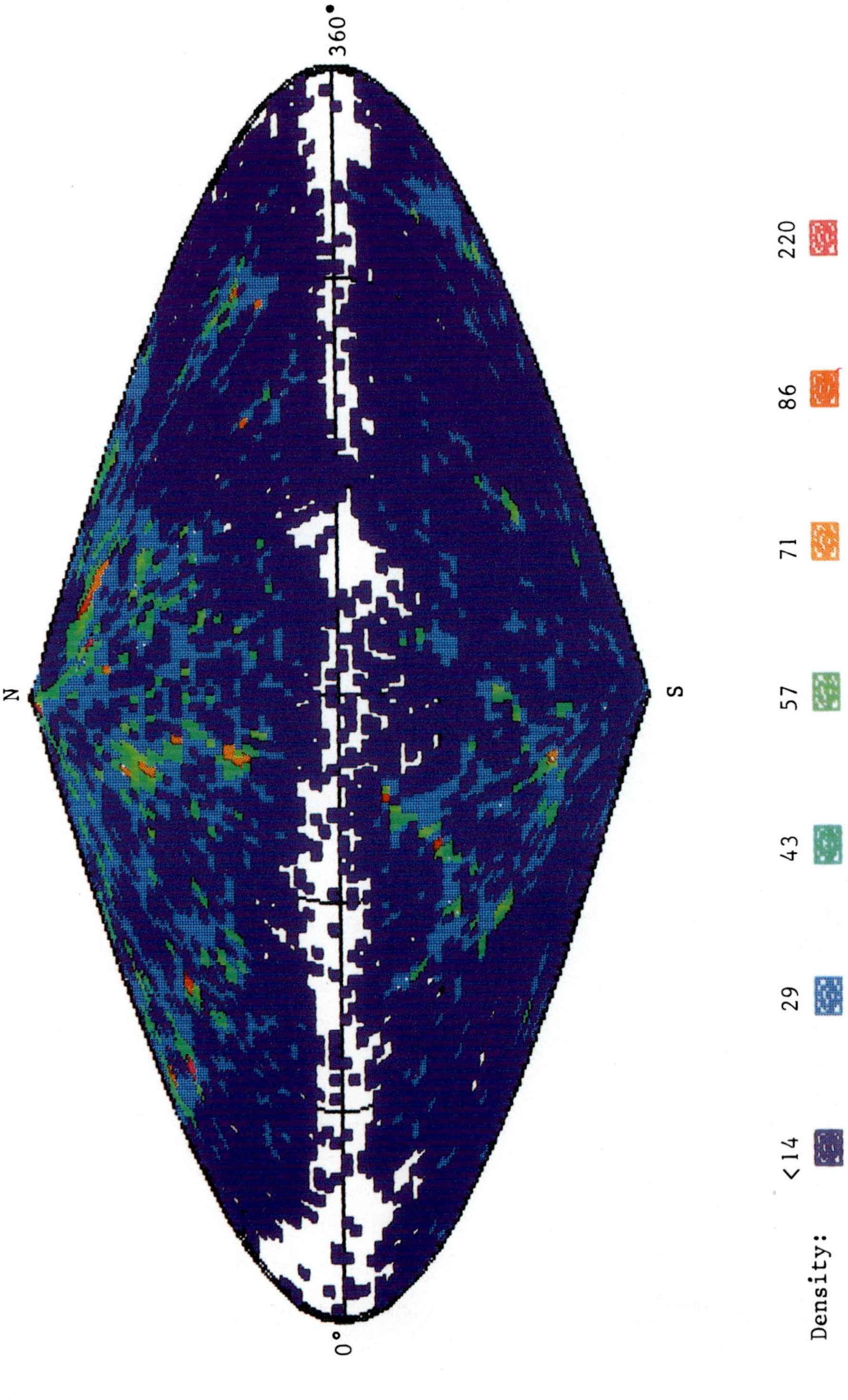

Fig. 2 - Surface distribution of the 67,164 galaxies contained in SIMBAD.
(Refer to the text for the definition of the density).

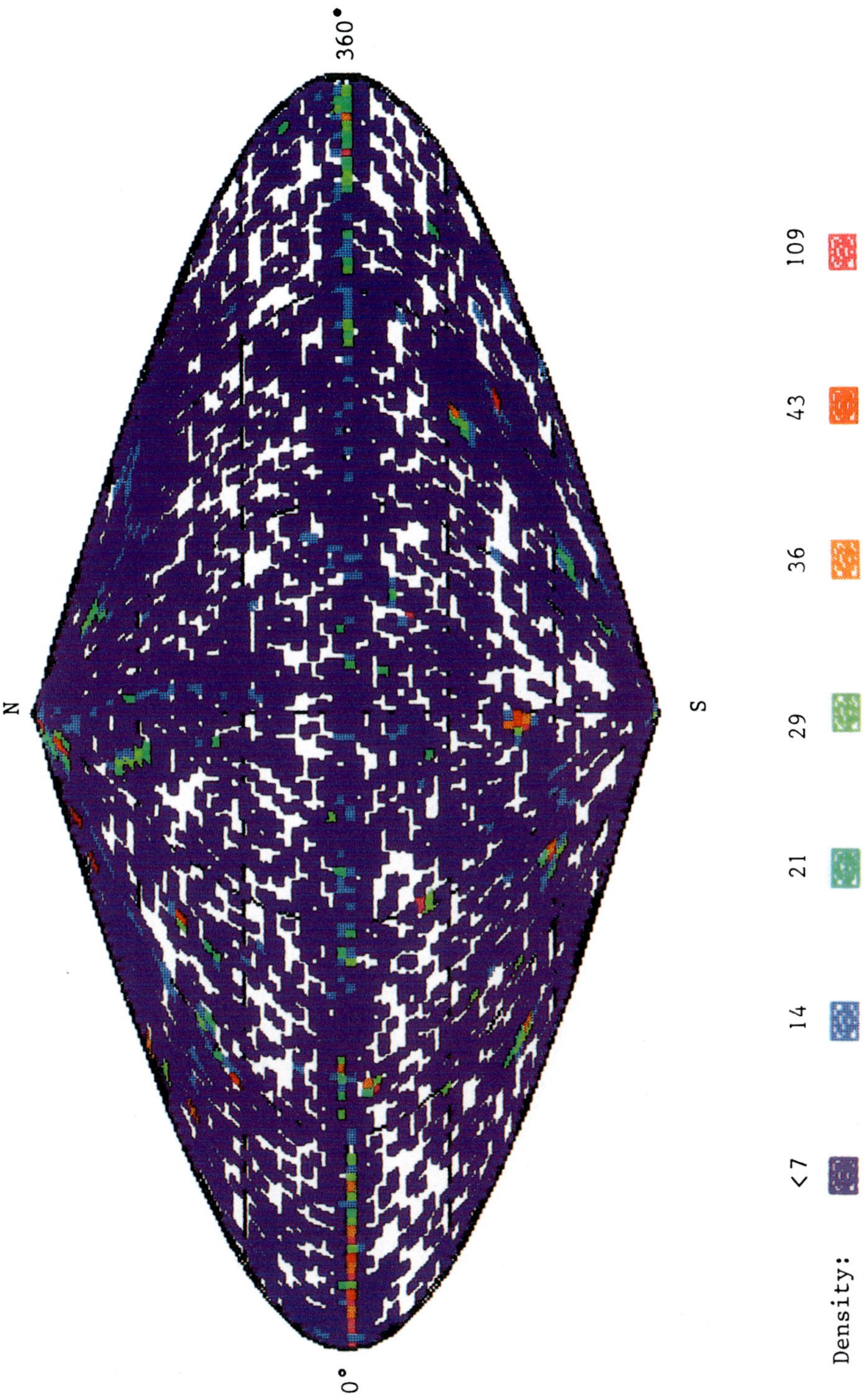

Fig. 3 - Surface distribution of the 15,180 non-stellar galactic objects (planetary nebulae, HII regions, globular clsuters, etc.) contained in SIMBAD.
(Refer to the text for the definition of the density).

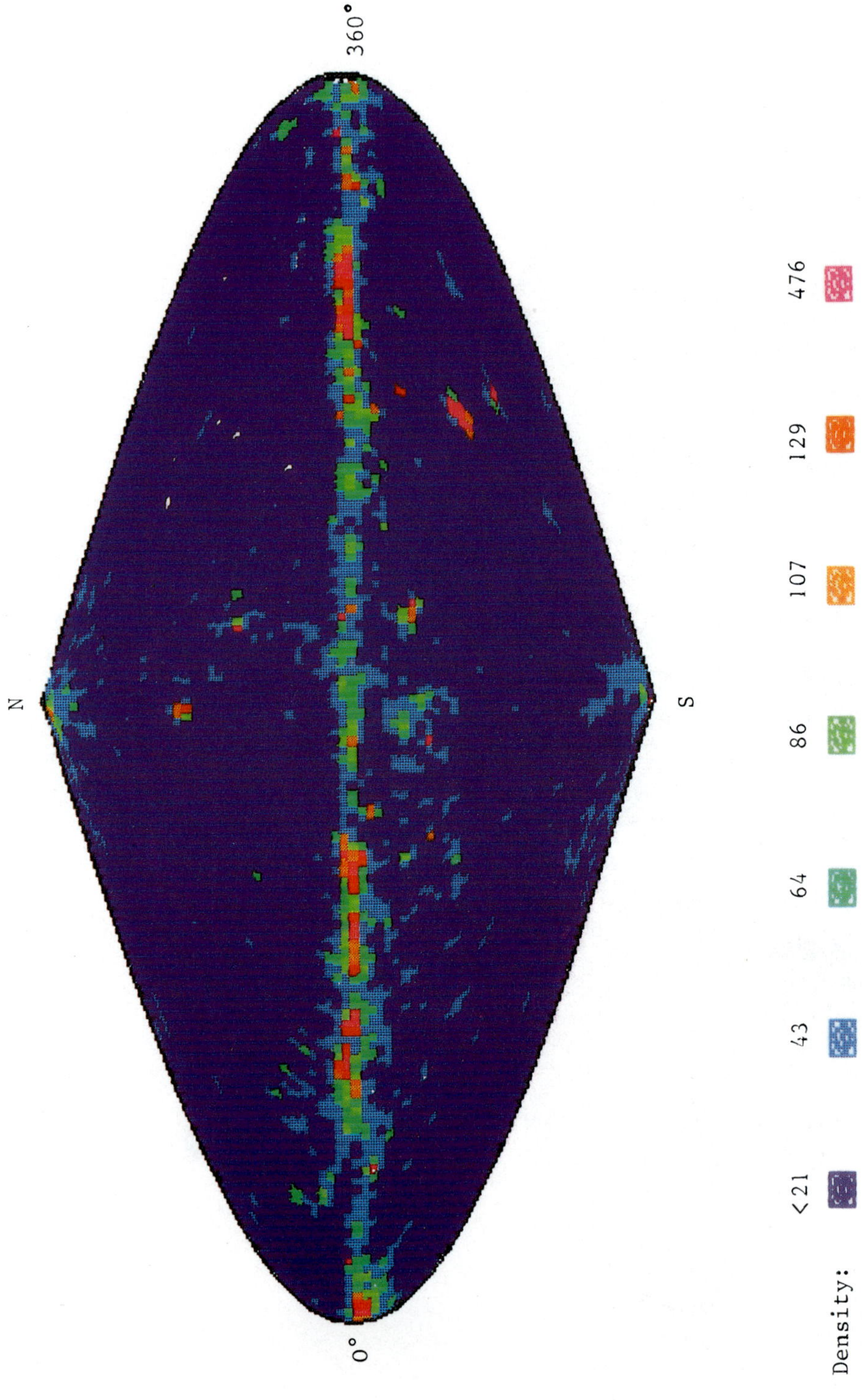

Fig. 4 - Surface distribution of the 112,523 stars with bibliography in SIMBAD. (Refer to the text for the definition of the density).

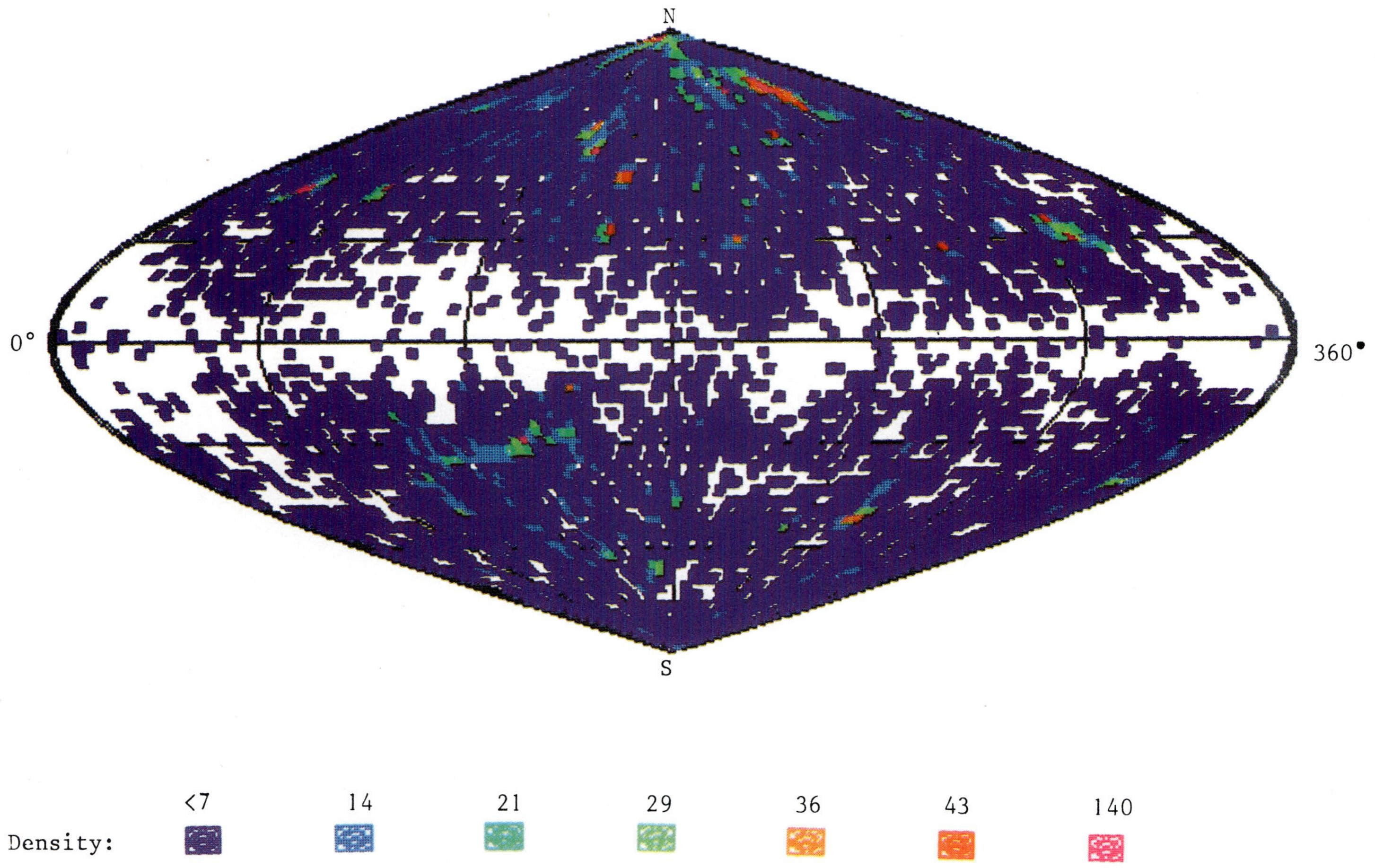

Fig. 5 - Surface distribution of the 15,691 galaxies with bibliography in SIMBAD.
(Refer to the text for the definition of the density).

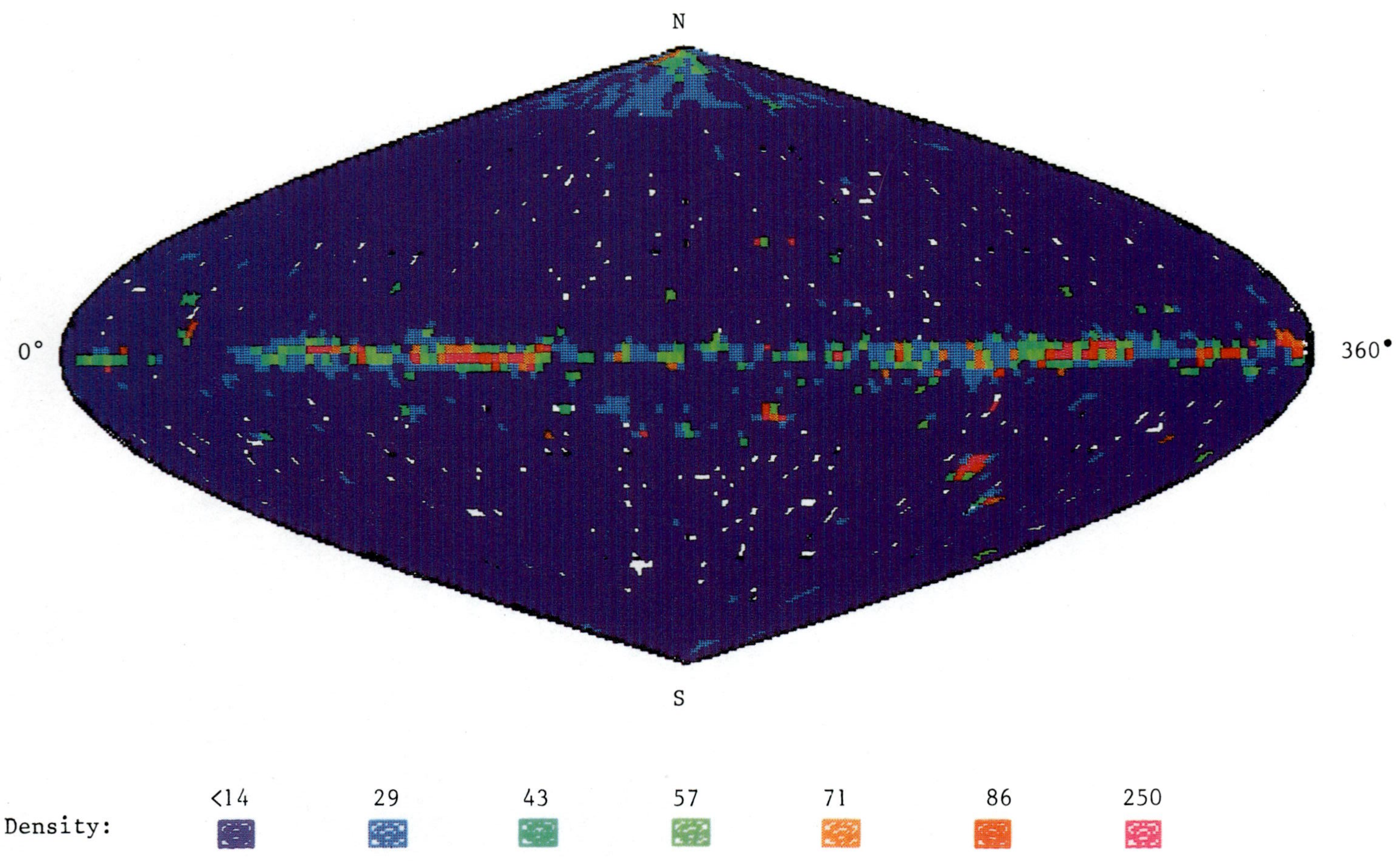

Fig. 6 - Surface distribution of the 50,536 stars with UBV photometry in SIMBAD. (Refer to the text for the definition of the density).

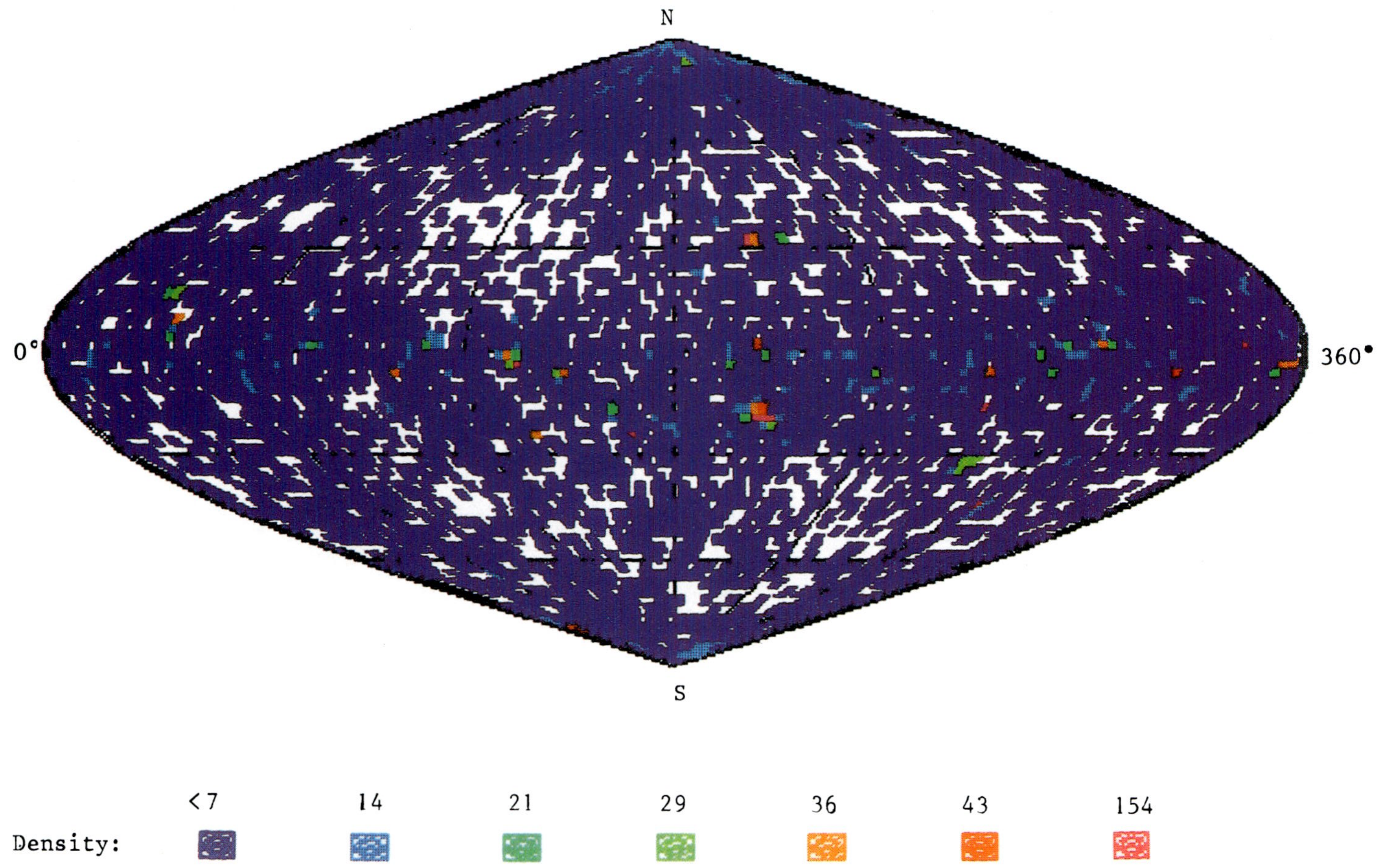

Fig. 7 - Surface distribution of the 12,858 stars with uvby photometry in SIMBAD. (Refer to the text for the definition of the density).

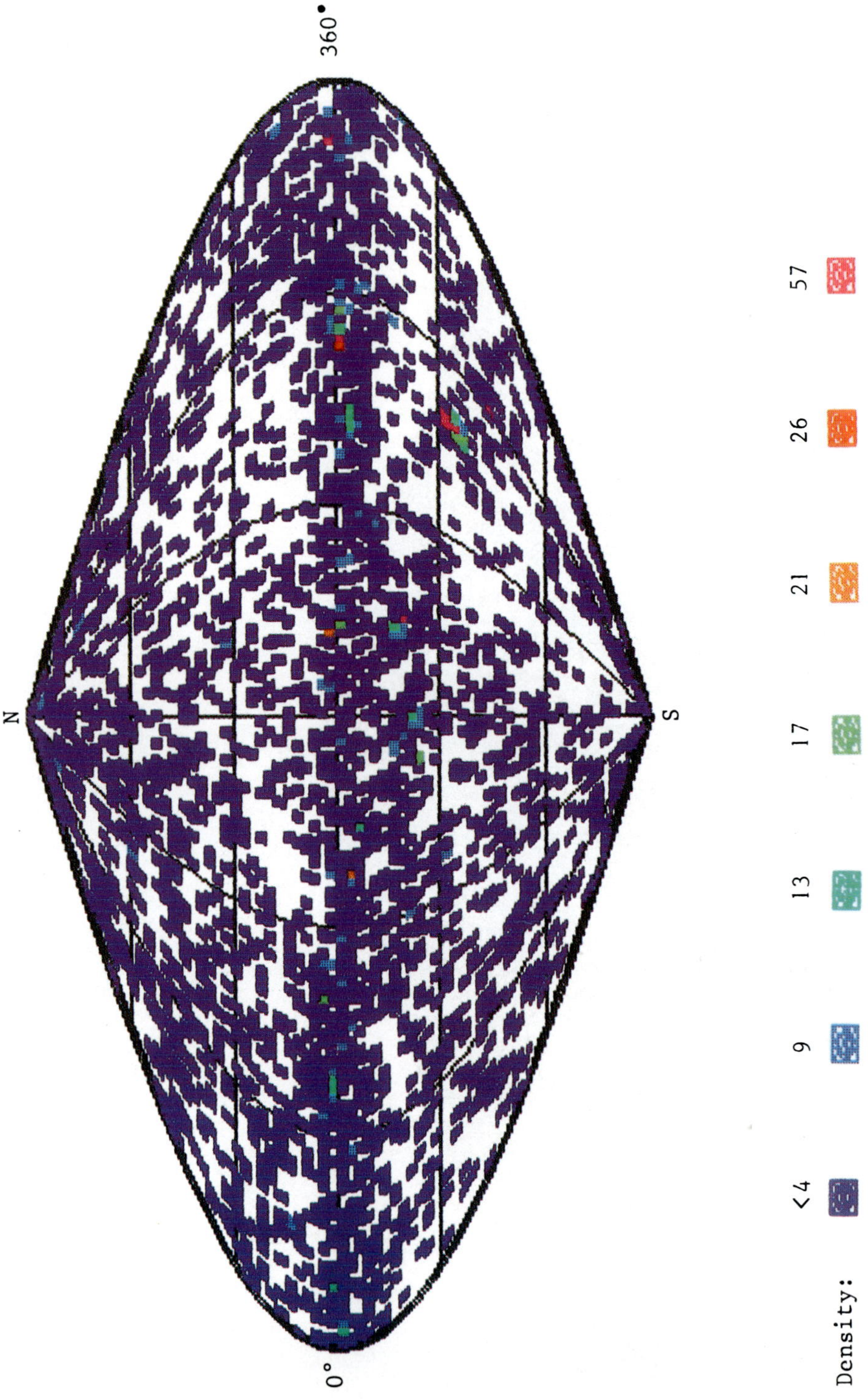

Fig. 8 - Surface distribution of the 4,480 IUE targets in SIMBAD.
(Refer to the text for the definition of the density).

CROSS-REFERENCING THE GSC WITH SIMBAD DATABASE

Daniel Egret[1] and Brian McLean[2]
[1]CDS, Observatoire Astronomique, Strasbourg, France
[2]Space Telescope Science Institute*, Baltimore, Maryland, U.S.A

Abstract

The TYCHO project (measurement of B and V magnitudes, and positions, for 500 000 stars brighter than about 11.0 mag., using the Hipparcos star mapper) will use the Tycho Input Catalogue (TIC) as a finding list for the data reduction (Egret and Høg, 1987).

The TIC is to be derived from the Guide Star Catalogue (GSC) of the Hubble Space Telescope (described by Jenkner, 1987), with additional data from the SIMBAD database of the CDS (Egret and Wenger, 1987).

The cross-matching of GSC and SIMBAD is now entering into an operational phase, with the final cataloguing of GSC plates. Some preliminary results, based on a wide test region covering different galactic latitudes are presented here.

1. The TYCHO Input Catalogue

The TYCHO Input Catalogue (TIC) will contain some two million stars and be complete to approximately 12.5 mag. Only a subset (those stars really brighter than the detection limit) will be finally measured in the (B, V) TYCHO system.

The primary source for elaborating the TIC is the *Guide Star Catalogue* complete to 15th magnitude towards the poles and up to 13th in the galactic plane. Presently being compiled at the STScI, it will cover the whole sky.

The main reason of cross-matching the GSC with SIMBAD is to complete (and improve) the bright end of the GSC: the brightest stars cannot be properly measured on the photographic plates, while their positional and photometric parameters are in general accurately known in SIMBAD. We also want to flag variable stars and high proper motion stars for which the parameters given in the GSC may have changed at the time of the mission. We will finally cross-identify lists or catalogues of astrophysical interest (photometric and astrometric standards, Input Catalogue for the Hipparcos main mission, etc.).

This work is undertaken as a collaboration between the *Space Telescope Science Institute* (the staff of the Guide Star Selection System will deliver a subset of their Guide Star Catalogue in advance of publication) and the *Centre de Données de Strasbourg (CDS)* (where the cross-matching with SIMBAD will be undertaken).

2. The cross-matching of GSC and SIMBAD

A subset of the Guide Star Catalogue is made available, in advance of publication, for the TYCHO project. A sequential file, containing the necessary data and cross-identifications is

---

* Operated by the Association of Universities for Research in Astronomy, Inc., under contract to the National Aeronautics and Space Administration.

extracted from SIMBAD. The cross-matching is based on an automatic comparison of the position, magnitude, and classification (*stellar* or *non-stellar*) of the objects from both files (GSC and SIMBAD).

In the course of the task, the following improvements are to be incorporated into the Catalogue:

- when the coordinates stored in SIMBAD are more accurate than the GSC one, then the SIMBAD position will replace the GSC one;
- the same thing will happen with the magnitudes;
- when high proper motions are found, the star is flagged and a position at epoch 1990 is computed;
- other flags related to known characteristics of the star are also included;

The relative accuracies of SIMBAD and GSC data are discussed as a function of the magnitudes and position on the plate of the stars as well as of their estimated errors.

An error box is computed for each star, using individual estimates of the errors on position and magnitude. For each cross-identification, a *reliability* index is computed by comparing the differences of positions and magnitudes (SIMBAD-GSC) to the estimated errors given in each file.

Ambiguous cases are treated through an iterative process. The most difficult cases will imply an interactive procedure using a graphic display of the area. More details on this application are given in a technical report by Didelon and Egret (1987).

3. The test region

The test region is an area in the Northern hemisphere (the scanned plates come from the Palomar Quick V survey) between 12h 16 and 17h 18 in right ascension, 22°30' and 30° in declination. For this test we have included stars slightly fainter than will be necessary for the operational phase (up to 13.5 mag.).

In the area defined above, 38 403 stars brighter than 13.5 are catalogued in the GSC. 4 356 additional redundant entries, due to overlap between plates, have been identified during the production of the GSC, and are not considered here. The estimated errors on the position given in the GSC have a mean value of 0.27 arcsec. For the V magnitude, the mean value is 0.39 mag.

In the same area, 5 438 objects are found in SIMBAD (including 194 known variable stars, 380 entries in multiple systems, 273 stars with annual proper motion larger than 0.2 arcsec., etc.).

1 461 objects have a position given to the arc minute, and 26 a position even more badly known. The histogramme of magnitudes for both files is given in the Figures 1-2.

4. Results

For the test area, the results of the automatic procedure are the following:

- 3 420 cross-identifications of one SIMBAD star with one GSC entry are found. For 595 of those objects the position or magnitude difference is significantly larger than expected from the estimated errors (more than 3 times the estimated error). In the contrary a very reliable identification is found for 1752 stars (differences smaller than the estimated errors).

- 1 734 other SIMBAD stars have no certain counterpart in the GSC file. Among them, approximately 330 stars brighter than V=9.0 mag. are objects missing in the GSC (images too large on the plates) and will be added in the TIC. The other ones are faint objects (720), galaxies (30), or secondary components in multiple systems (90). One third consists in objects for which the errors on position are larger than expected, making difficult a reliable identification to a unique GSC entry.
- The 284 remaining stars lie on the boundaries of the region and will have to be cross-matched with contiguous areas.
- 793 stars brighter than 10th magnitude in the GSC are unknown in SIMBAD; these entries will be checked in order to find possible misclassifications in the GSC. This gives also a very important information about the completeness of SIMBAD in different zones of the sky.

Conclusions

We have to compare two very different files: the GSC is a homogeneous survey from a rather uniform set of photographic plates, while SIMBAD, as a complex database, is a compilation of many heterogeneous catalogues from various sources. The result of the task will be a considerable improvement of both GSC and SIMBAD.

For the TYCHO project, the cross-matching will certainly not be 100 percent achieved because of the tight schedule (launch of the satellite in April 1989). We are now discussing the idea of working with a *clean* subset of SIMBAD, including, beside the bright stars, only those stars with reasonably well-known parameters, which are also the most useful for TYCHO (and the GSC).

The complete integration of the GSC into SIMBAD appears more as a long-term project.

Acknowledgement

The cross-matching software has been written by P. Didelon, through a contract of CNES and INSU. D.E. acknowledges the support of INSU (ATP télescope Spatial) and CNRS (GRECO Astrométrie Spatiale) for this work. Computations were made on the IBM 3081K of the Centre de Calcul de Strasbourg-Cronenbourg.

References

Didelon, P. and Egret, D.: 1987, CDS Inform. Bull. **32**, 27

Egret, D. and Høg, E.: 1987, *in* IAU Symp. 133, *Mapping the Sky*, in press

Egret, D. and Wenger, M.: 1987, this conference

Jenkner, H.: 1987, this conference

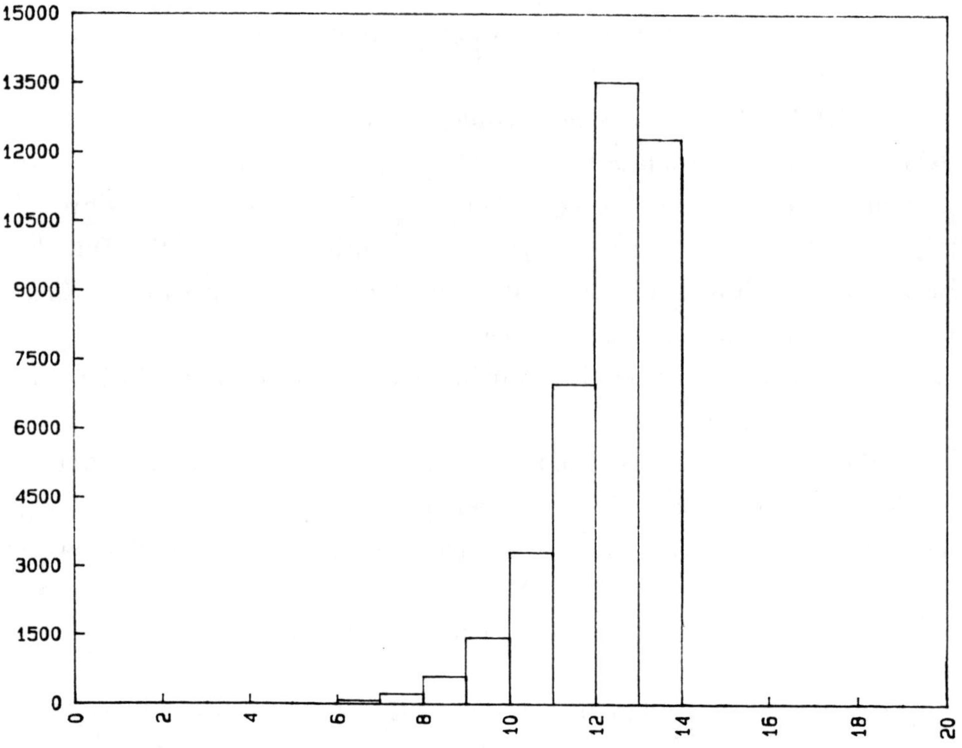

Figure 1: Histogramme of the magnitudes of GSC stars

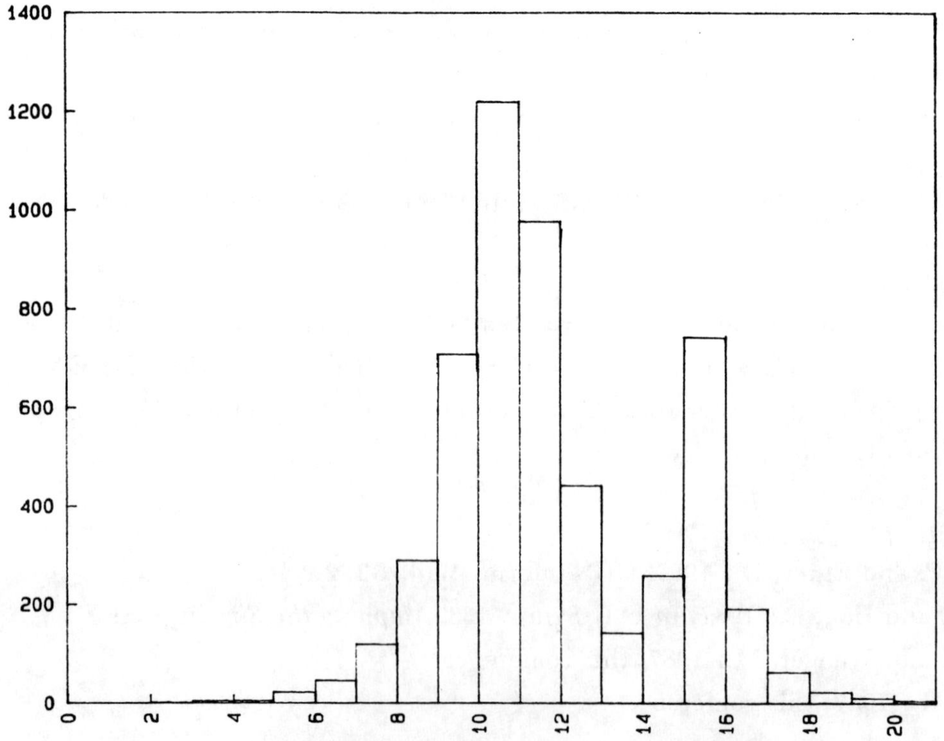

Figure 2: Histogramme of the magnitudes of SIMBAD objects

SIMBAD - PRESENT STATUS AND FUTURE

Daniel Egret and Marc Wenger
CDS, Observatoire astronomique, Strasbourg, France

Abstract

SIMBAD is the astronomical data base produced and updated on a continuing basis by the Strasbourg astronomical Data Centre (CDS), at the Observatoire de Strasbourg (France). S.I.M.B.A.D. is the abbreviation for **S**et of **I**dentifications, **M**easurements, and **B**ibliography for **A**stronomical **D**ata. We describe here the present status of the database, and the future developments.

I. The present status

Stars and Galaxies

SIMBAD is the largest set of basic astronomical data in the world. Originally created for stellar data, SIMBAD now includes galaxies and galactic objects outside the solar system.

It contains:

- 700,000 objects, including
- 600,000 stars, and
- 100,000 non-stellar objects (mostly galaxies)
- for which 2,500,000 identifications
- and more than 1,000,000 measurements are provided on line.

These figures will be soon out-of-date with the planned inclusion of the Guide Star Catalogue of the Hubble Space Telescope (of the order of 20 million objects).

The bibliographic index contains references to stars from 1950 onwards, and to galaxies and all other objects outside the solar system from 1983 onwards. Presently there are more than 740,000 references taken from 42,000 papers published in the 90 most important astronomical periodical publications.

Retrieving data: three query modes

The data, measurements, and bibliography in the SIMBAD database can be retrieved:

- through any object designation (about 400 different types)
- or by object coordinates (equatorial or ecliptic, at any equinox, or galactic). In this latter mode, one may request to get all objects within a circle of given dimensions around a given position.
- in the sampling mode, where criteria can be specified on parameters such as position, magnitude, existence of various types of data, etc. More details are to be found in Egret (1986) or in the SIMBAD user's guide (Wenger and Ochsenbein, 1985).

How to gain access to SIMBAD ?

Apart from a simple terminal, the only requirement for accessing SIMBAD is obtaining an account number from CDS which will be used for invoicing. SIMBAD is accessible through the public packet switching data networks: TRANSPAC in France, DATEX-P in Germany, TELENET in the U.S., etc. SIMBAD may also be accessed by direct telephone call, or, in France, through TELETEL.

Astronomers without access to a data network can also mail their requests to the data centre which will then return a printout with the corresponding data. In the same way, copies of individual catalogues (from a list of more than 500) can be obtained on magnetic tapes. Data transfer is also possible via the major electronic-mail networks (EARN, BITNET, USENET, ARPANET). Note that these E-mail networks, which do not allow remote log-in, are not suitable for an interactive access to SIMBAD.

More than 100 connected Institutions

To the present date (12 Oct 1987) there are 117 observatories or astronomical institutes having an account number on SIMBAD. The figures are rapidly increasing.

Stations with access to SIMBAD include : the IUE Observatory of ESA in Madrid, the STScI in Baltimore, the NASA Goddard Space Flight Center, the ST-ECF in Garching, through its host, the European Southern Observatory, the STARLINK node at RAL, Caltech (Pasadena, CA), the Center for Astrophysics (Cambridge, MA), the CFHT in Kamuela (HI), the Anglo-Australian Observatory in Epping, the South African Astronomical Observatory in Capetown, etc. (see map in Heck and Egret, 1987a).

Updating SIMBAD: a very large cooperative effort

SIMBAD is continuously growing and kept up-to-date, not only by the Strasbourg CDS staff, but also by many cooperating persons in other institutions. This includes a complete survey of the literature in order to give an immediate access to the current *bibliography* of each object, and the compilation of published observations by experts in the field, for the *measurements* section.

The bibliography is updated on a daily basis, thanks to the enthusiastic efforts of the librarians of the IAP (Institut d'Astrophysique de Paris) and Observatoire de Paris, in collaboration with the CDS (see e.g. Carpuat et al., 1984). Catalogues of measurements (original data and mean values when available), produced by specialists, are integrated into SIMBAD by the CDS staff. Only published data are included, and a reference to the source is given.

Collaboration with institutes having specialization in specific fields is then a natural consequence. This is particularly the case for: Bordeaux, Meudon and Paris (bibliography), Geneva and Lausanne (photometry), Heidelberg (astrometry), Lyon (galaxies), Marseille (radial velocities).

Some catalogues, prepared at CDS itself and available as CDS Special Publications, are made in fields where the Strasbourg personnel has specific qualifications. Thus the Catalogue of Stellar Groups lists some 30,000 stars according to their spectral peculiarities.

SIMBAD and the space programs

On a much larger scale, CDS takes also an active part in space projects like HIPPARCOS and TYCHO which are heavily dependent on SIMBAD for the preparation of their respective input catalogues.

Thus, the INCA database of the HIPPARCOS Input Catalogue consortium is organized as a subset of SIMBAD, sharing the same computer resources (Morin and Arenou, 1987).

As a collaboration between the Hubble Space Telescope GSSS team and the CDS, the TYCHO Input Catalogue will result of a cross-matching of SIMBAD with a subset of the Guide Star Catalogue (the stars brighter than thirteenth). This work will also be used to complete and improve the bright part of the GSC (Egret and McLean, 1987).

CDS acts also as the european disseminator of the IRAS observational material and has been requested by ESA to homogenize the IUE log of observations (Heck and Egret, 1987b).

Publications and Documentation

A very complete User's Guide for SIMBAD is available on request (last update: August 1985) either in French or in English. A new issue is planned for the end of 1987, together with a new release of SIMBAD including some improvements of the user's interface and new facilities for file transfer.

CDS stays in touch with its users and other interested persons by a six-monthly Information Bulletin. Apart from keeping readers updated on CDS services and the latest developments, it contains also general papers and news about other data centres' activities.

II. The future: a new release

Since the time SIMBAD was built, many things have evolved concerning astronomical data base content and design: data volumes have increased, a wide range of computers are now accessible with increased power and lower prices, networking becomes common and fast.

These considerations drive us to re-design SIMBAD, taking in account all these trends (Wenger, 1987). Several aspects of database design will be regarded for this new version.

Data base management system (DBMS)

The information unit in SIMBAD is the astronomical object. Such an object can be described by one identifier and its coordinates. It can also consist of 30 identifiers, 20 measurement types, each type containing from one to fifty measurements, and 600 bibliographical references. Up to now, no DBMS is able to fit the requirements concerning such variability in the amount of data for different objects.

Although it is theoretically possible to structure SIMBAD using the relational data base model, it would need to split an object into so many tables and rows that gathering all informations for a well studied object would need several hundred input/outputs.

So appears the necessity of creating a specific DBMS for SIMBAD, in order to keep its orientation towards astronomical object and also its performances.

Machine independance - portability

Designing a large software system today needs to take especially care of its portability. The portability of a software can be defined by its ability to run on different non compatible machines, with as less adaptation work as possible.

Such portability is now absolutely required to preserve the enormous amount of work needed for writing a software. Such a software can commonly be in use during ten years and even more. On the other hand, computer technology advances so quickly that it will have to run on different machines in its life. Also a software can be run on different kinds of machines, crossing technologies: mainframe computers, work-stations and microcomputers. These kinds of machines are rarely compatible !

To achieve this portability, we will use for the next SIMBAD version, a now well-tried solution: write the whole software in C language and do not use machine and compiler dependent features. A good way to reach this is to avoid reading any hardware, operation system or compiler specific documentation for the system you are using.

Object-oriented programming as structure

To preserve software investment, it is also necessary to be able to easily upgrade software, by adding to it new features and adapting the existing one to new requirements. Such software modifications are facilitated by a well structured software, each function being as independent as possible from any other one.

Building a structured software will be achieved in SIMBAD by using the concept of object-oriented programming as described in Goldberg and Robson (1983).

An OBJECT is an abstraction made of the encapsulation of the data needed to describe it and the functions, here called METHODS, devoted to the manipulation of these data. Objects having the same properties are grouped in CLASSES. An object is manipulated through MESSAGE TRANSMISSION. Such a message consist of the receiver object, the method to activate and eventually a list of parameters.

This concept of objects allows simple software design (by specifying each object), leads to modularity (objects interact only through message transmission), ensures data integrity (they are concealed in the object definition) and facilitates interfaces with other softwares (objects are manipulated by a single and simple concept: the message transmission).

To stay performant, and also portable, object management in SIMBAD will be implemented through a tool of standard C language functions.

Managing large volumes of data

Although SIMBAD can be considered as a large data base with its 700,000 astronomical objects and 180 Mbytes, it will reach its conceptual limitations with the arrival of new data gathered by satellites or automated ground-based detectors.

These data will not only consist of new measurements to add to the presently known objects, but also of a huge amount of new objects: the delivery of the Guide Star Catalogue for the Hubble Space Telescope will increase from one to twenty millions the number of objects in an astronomical data base like SIMBAD. Also the storage of images will require capacity for large data volumes.

Classical file access methods are rapidly out of breath when file sizes have to be counted in gigabytes. So we define our own file management system, wich will be a four layers software:

- SIMBAD oriented functions
- ACCESS METHODS consisting of classical access methods (sequential,direct,indexed files) working on virtual files.
- VIRTUAL FILES made of sequential numbered blocks. These blocks are addressed by a way involving only the number of the block, independently of its length. This will allow to address easily 4 Gigablocks with a four bytes address. Length of the blocks will evolve with central memory capabilities (4 Mbits chips are already available).
- PHYSICAL FILES will at last support the virtual blocks through a mapping function. To ensure portability of the whole system, these physical files will be UNIX-like files, available by each C compiler.

Towards a European Astronomy network

Presently, SIMBAD can be accessed from all over the world through Packet Switching Networks (X25). To enlarge access possibilities, we plan to integrate SIMBAD in other networks, facilitating by this way access to new users.

CDS is playing a rôle in the various discussions which should lead in the near future to the setting up of a European astronomical data network. SIMBAD is also involved since the first stages in the European Space Information System (ESIS) project (Albrecht et al., 1987).

Ideally SIMBAD could be connected through such a network with the observing logs of the main ground and space observatories. A user, coonected to a database will have access to the information contained in each other data base. Moreover, he will query all these data bases through the same user interface.

User Interface/Astronomical Query Langage

More and more astronomers are using more and more databases: it becomes necessary to offer, as far as possible, a uniform query language for all these bases. This implies a collaboration with the other institutions elaborating data bases. The ST-ECF at Garching engaged a process in this way by creating a "Data Base Working Group" conducting a reflexion on these topics. Products like Termwindows (ESO/IPG, 1987) and Proteus (ESO/ST-ECF, 1987) developed at ESO, in the User Interface domain are already available.

CDS is strongly involved in these projects and will adopt the standardisation rules when they issue.

References

Albrecht, M., Russo, G. and Richmond, A.: 1987, ESIS Pilot Project, User Requirement Document - 1st draft, ESA/ESRIN

Carpuat, C., Damgé, M., Dubois, P., Kirchner, S., Lagorce, A., Laloë, S., Ochsenbein, F. and Wagner, M.-J.: 1984, CDS Inform. Bull. **26**, 83

Egret, D.: 1986, CDS Inform. Bull. **30**, 25

Egret, D. and McLean, B.: 1987, this conference

ESO/IPG: 1987, TERMWINDOWS, A Terminal Independent Screen Formatting and Window Package, ESO, Garching

ESO/ST-ECF: 1987, PROTEUS, A Generalised User Interface Construction Kit, ESO and ST-ECF, Garching

Goldberg, A. and Robson, D.: 1983, Smalltalk-80 - The language and its implementation, Addison-Wesley Publishing Company

Heck, A. and Egret, D.: 1987a, Messenger, **48**, 22

Heck, A. and Egret, D.: 1987b, Homogenized log of IUE observations, CDS Special Report (in press)

Morin, D. and Arenou, F.: 1987, this conference

Wenger, M.: 1987, SIMBAD: reconversion or new release, CDS Inform. Bull. 32, 23

Wenger, M. and Ochsenbein, F.: 1985, SIMBAD User's Guide, CDS, 86pp.

A FEW CONSIDERATIONS ON PHOTOMETRIC DATABASES

A. Heck[1] & J. Manfroid[2]

[1] C. D. S., Observatoire Astronomique, Strasbourg, France
[2] Département d'Astrophysique, Université de Liège, Belgium
* Maître de Recherches au Fonds National de la Recherche Scientifique

Abstract

Problems related to the homogenization of photometric data are commented, in the perspective of existing and potential photometric data banks.

The need for keeping individual photometric measurements of programme and standard objects accessible is stressed. A complete description of each instrumental system should also be archived.

Introduction

The construction of databases of photometric indices is certainly very appealing to the astronomers who could easily retrieve the information about as many objects as they would need.

Ideally, all observations should be brought back to a standard system through a homogenization procedure and averaged values should be readily available. Practical contingencies prevent however the achievement of a fully reliable database of averaged values for most photometric systems.

With the development of an ever better instrumentation, an ever larger proportion of stars is found to be variable or "micro-variable". Hence, this is another reason to take into account the archiving into databases of individual measurements in addition to lists of averaged values. In the case of the Johnson, Geneva or Stroemgren systems, this would mean millions of records.

For all photometric databases the most stringent requirement concerns consistency and this will be discussed hereafter.

To homogenize or not to homogenize?

Large compilations of averaged photometric data are available for different systems as:

- Geneva (Rufener, 1981),
- UBV (Nicolet, 1978),
- UBVRI (Lanz, 1986),
- UPXYZVTS (North, 1984),
- uvby (Hauck & Mermilliod, 1985),
- VBLUW (Python, 1979).

Ochsenbein (1974) has also carried out an important work of homogenizing the V data available in SIMBAD, the CDS database (Egret & Heck, 1987; Egret & Wenger, 1987; Heck & Egret, 1987). The Geneva, UBV and uvby compilations are also available in SIMBAD. Details are generally given by the authors on the homogenizing procedures, but they are essentially based on systematic differences and personal weights linked with the observers.

Some difficulties of homogenizing photometric data have been mentioned in Hauck (1973) and Nicolet & Hauck (1977). In a previous paper (Manfroid & Heck, 1984), we cautionned against the use of averaged values - even weighted - in the catalogues of so-called homogenized photometric data. We even strongly restricted the validity of such catalogues since it appeared that many stars observed with different realizations of a photometric system could not be brought back to the standard system (see Manfroid, 1984 for an illustration of the impact of different Stroemgren systems).

Theoretical arguments (Young, 1974; Manfroid, 1985), extensive numerical experiments (Manfroid, 1984, 1985) and observations conducted mainly in the uvby system (Manfroid & Sterken, 1987; Sterken & Manfroid, 1987) have confirmed this conclusion. It appears that transformation equations between the standard and the natural systems can be found only for some stars which are specially well suited to the particular photometric system.

For the uvby, these well-behaved stars are the main-sequence A- to G-type stars with $0<b-y<0.4$. But even for that restricted group, it is sometimes necessary to use several sets of linear transformations according to the color of the star. When the color transformations are applied to other stars, the results are system dependent, and errors as high as several tenths of magnitude can arise.

It is significant that the use of transformation equations involving a larger number of parameters (such as the more complete equations described in Manfroid & Heck, 1983, 1984) does not change significantly the validity domains. In some instances, this could even lead to larger errors when extrapolating. These algorithms are better within the validity regions but not beyond. This confirms that analytical equations are unable to give for all stars a correct mapping between the different systems.

It is not justified to average values with such systematic errors in order to homogenize a catalog. The final result would be inconsistent. On the other hand, it is probable, depending on the skill of the observers and the quality of the instrumentations, that each set of original results is internally consistent although in a non-standard system. All homogenization procedure can only degrade that coherency.

Caveats

Data considered for inclusion in a photometric catalog or database should first be checked for their degree of standardization. Hence each potential contributor should make available not only his/her final data, but also:

- a complete description of the instrumental system (passbands, response curves, etc.),
- a list of the measurements of standard stars and of well-known stars covering as broad a range of spectral characteristics as possible.

This will allow to determine carefully the subsets of stars admitting a color transformation. The photometric indices (averaged values as well as individual measurements, for variable stars for instance) of stars belonging to this nucleus can and should be published in an homogenized form in the standard system.

The non-transformable stars should be treated differently. Meaningful averaged data could be produced only within each natural system. If some transformation is applied it should be stated explicitly. In all cases the values of the standard stars should be available together with the adopted transformation equations, so that a re-evaluation of the standardization method can be done.

In various instances, different lists of standard values, obtained by different authors with different instruments are available. Systematic differences between those lists are another source of error. Hence the standard values used by each observer should be known, so that this effect can be removed.

Present situation

Among the available important photoelectric catalogues, only the Geneva catalogue can be considered as being homogeneous. The consistency of the treatment of the data and the stability of the instrumental system ensure a unique homogeneity making many of the above requirements unnecessary or automatically satisfied.

On the other hand, the Johnson and Stroemgren averaged or "homogenized" catalogues which are accessible for instance through the SIMBAD database fall short from those desiderata. The simultaneous availability of the UBV and uvby original photometric indices listed in Mermilliod (1987) and Hauck & Mermilliod (1985) respectively goes in the sense of our recommendations, although the situation cannot be fully satisfactory because of the frequent lack of complementary information in the corresponding original papers.

A minimum step to be taken in any photometric database would be to make always available the original data properly referenced as it is the case in SIMBAD. Averaged or homogenized values of photometric indices would be fully reliable only in the above-mentioned conditions are satisfied.

Homogenizing observations made by amateur astronomers is even a more perilous exercise. A few comments can be found in Heck & Manfroid (1985) and in Heck et al. (1985) in the context of compilations of observations relative to the variable stars RR Tel and V348 Sgr respectively.

References

Egret, D., Heck, A. 1987, SIMBAD: Surface distributions of some basic astronomical data, this conference

Egret, D., Wenger, M. 1987, SIMBAD: Present status and future, this conference

Hauck, B. 1973, Photometric data in machine-readable form, in Spectral Classification and Multicolour Photometry, eds. Ch. Fehrenbach & B.E. Westerlund, D. Reidel Publ. Co., Dordrecht, 285-287

Hauck, B., Mermilliod, M. 1985, uvbyß photoelectric photometric catalogue (magnetic tape), Astron. Astrophys. Suppl. 60, 61-62

Heck, A., Egret, D. 1987, SIMBAD, the CDS database, Messenger, 48, 22-24

Heck, A., Houziaux, L., Manfroid, J., Jones, D.H.P., Andrews, P.J. 1985, Photometric variations of the irregular variable V348 Sgr, Astron. Astrophys. Suppl. 61, 375-385

Heck, A., Manfroid, J. 1985, Variations in the variability of RR Tel, Astron. Astrophys. 142, 341-345

Lanz, T. 1986, Photoelectric photometric catalogue in the Johnson UBVRI system, Astron. Astrophys. Suppl. 65, 195-197

Manfroid, J. 1984, Stability and accuracy of the uvby system, Astron. Astrophys. 141, 101-104

Manfroid, J. 1985, Traitement numérique des données photométriques, D. Sc. Thesis, Univ. Liège, 198p.

Manfroid, J., Heck, A. 1983, A generalized algorithm for efficient photometric reductions, Astron. Astrophys. 120, 302-306

Manfroid, J., Heck, A. 1984, On the accuracy of the reduction of the uvby photometry, Astron. Astrophys. 132, 110-114

Manfroid, J., Heck, A. 1985, On the homogenization of photometric data in Calibration of Fundamental Stellar Quantities, ed. D. S. Hayes, L. E. Pasinetti & A. G. D. Philip, D. Reidel Publ. Co., Dordrecht, 565-567

Manfroid, J., Sterken, C. 1987, Instrumental effects and the Stroemgren photometric system, Astron. Astrophys. Suppl., in press

Mermilliod, J. C. 1987, UBV photoelectric catalogue (1986), I. Original data, Astron. Astrophys. Suppl., in press

Nicolet, B. 1978, Catalogue of homogeneous data in the UBV photoelectric photometric system, Astron. Astrophys. Suppl. 34, 1-49

Nicolet, B., Hauck, B. 1977, Critical evaluation of photometric data, in Compilation, Critical Evaluation, and Distribution of Stellar Data, eds. C. Jaschek & G. A. Wilkins, D. Reidel Publ. Co., Dordrecht, 121-124

North, P. 1984, An up-to-date catalogue of the Vilnius UPXYZVTS photometry, Inform. Bull. Strasbourg Stellar Data Center, 27, 133-134

Ochsenbein, F. 1974, On the relationship between the apparent magnitudes given in several catalogues and the UBV system, Astron. Astrophys. Suppl. 15, 215-252

Python, M. 1979, Photoelectric photometric catalogue (magnetic tape), Astron. Astrophys. Suppl. 38, 463-464

Rufener, F. 1981, Third catalogue of stars measured in the Geneva Observatory photometric system, Astron. Astrophys. Suppl. 45, 207-366

Sterken, C., Manfroid, J. 1987, Quasi-consistent filter passbands and the homogeneity of photometric indices, in Observational Astrophysics with High Precision Data, 27th Liège International Symposium, in press

Young, A. T. 1974, Observational technique and data reduction, in Methods of Experimental Physics, ed. N. Carleton, Academic Press, New York, 12A, 123-192

A NEW EXTRAGALACTIC DATABASE

George Helou and Barry Madore
California Institute of Technology
Pasadena, CA 91125
U.S.A.

ABSTRACT We describe a proposed "National Extragalactic Database" that will make available data from extragalactic and other catalogs, as well as data and references from published articles dealing with all extragalactic objects. The goal is to remain current in the literature, and work toward completeness in past literature. All data that can be summarized in a few numbers will be included, and those that cannot will be referenced with adequate description attached. The Database will also provide references to observing logs and data archives kept by observatories or other facilities.

The Database will reside at the Infrared Processing and Analysis Center at Caltech, will be made remotely accessible, and will provide services similar to those offered by SIMBAD. The basic capabilities will be: 1) literature searches, and summaries of published data for objects specified by name; 2) positional searches for all objects in the vicinity of a specified location on the sky; and 3) keyed searches for all objects whose observed parameters satisfy a given set of conditions.

I. INTRODUCTION

We describe here a proposed Database containing all data available on extragalactic objects, and accessible to the astronomical community; this would be a service parallel and complementary to the stellar data service offered by SIMBAD (Set of Identifications, Measurements and Bibliography for Astronomical Data). Our colleagues on this proposal are: C. J. Lonsdale, W. Rice, B. T. Soifer, and C. A. Beichman.

The impetus for the Extragalactic Database derives from the need for a central archive accumulating all published extragalactic data, organized for fast retrieval and flexibility, and accessible to all astronomers. This need has been increasing as more surveys (redshift surveys, radio emission, *IRAS*, *etc.*) have collected larger data-sets, and can only increase as more astronomers take a multi-spectral approach to astrophysical problems, where *all* the data on an object are essential.

Furthermore, the proposed Database is easily manageable with today's technology. The total amount of summary data on galaxies available today is less than the volume of data contained in the *IRAS* Point Source Catalog. This would make a manageable database that could be assembled at reasonable cost and effort.

The longer we delay setting up the Database, the larger the initial effort becomes. To minimize the cost of establishing the Database, we rely on the environment of the *Infrared Processing and Analysis Center* (IPAC), which brings to the job of creating the Database expertise in three essential areas, namely, astronomical databases, extragalactic compilations, and national user-oriented facilities. We also rely on collaboration with the *Centre de Données Stellaires* at the *Observatoire Astronomique de Strasbourg* (the keepers of SIMBAD), with whom we have a preliminary understanding concerning exchange of data and expertise.

II. DATABASE CONTENTS

The National Extragalactic Database will make available a wealth of data on galaxies, quasars, and clusters of galaxies, as well as on objects that are possibly extragalactic but have not been certified as such on an individual basis (e.g. double-lobed radio sources). For each object, the Database will contain all published measurements that can be summarized in a few numbers, such as positions, diameters, inclinations and position angles, magnitudes and fluxes at all available wavelengths (X-ray, ultraviolet, optical, near and far infrared and radio), morphological types, recession velocities, distance estimates, and group membership. Initially, the Database will not contain dimensional data such as rotation curves, spectra, maps and images. However, it will contain a complete and current listing of all references pertaining to each object, with an indication about their data content.

The basic philosophy is to include all data as published, without any effort to homogenize them, adjust them to a standard scale, or "correct" them. An effort will be made however to eliminate redundant entries (e.g. data from a catalog reproduced in a paper), and to provide as many qualifiers as necessary to help the first order interpretation of the data, most notably the exact frequency, position, and date of the observation, instrument used and beam size, assumed distance to the object, or assumed cosmological constants affecting the interpretation. Published data that have been modified (i.e. adjusted to another scale, corrected for various effects, *etc.*) will be included with an appropriate flag to indicate they are not new measurements. In gray areas the tendency will be to include rather than reject data, since completeness is an important goal.

A cornerstone of this effort (as with SIMBAD or any major catalog) will be the establishment of a table of identifications linking together the various names the object has been given. Unlike our approach to the data themselves, we anticipate in this area making decisions based on detailed case-by-case examination. Since it is necessary to exercise scientific judgement in this area, the system will have to accommodate some identifications which will remain tentative for lack of conclusive evidence or because of positional uncertainties.

Such "positional associations" could arise between galaxies named in different catalogs, or sources detected at different wavelengths, e.g. an infrared source and a quasar.

We envisage tapping two main sources of data: first, published data (§II.A), including major catalogs as well as articles in past and current literature, with which we will keep up on an continuing basis; and second, target lists and indices to data archives at observatories or data centers (§II.B). Other sources of data (such as original unpublished data) might be eventually considered. In §II.C, we estimate the size and expected growth rate of the Database.

II.A. Published Data

Published catalogs and papers containing original measurements will provide the main ingredient in the Database. The minimum requirement will be that all data published since 1983 be included, as well as all substantial data sets published previous to that year. 1983 is the year in which SIMBAD started entering extragalactic references into their files. Another intermediate completeness date to consider is 1974, the last year for which the Second Reference Catalog of Bright Galaxies has compiled bibliographic references. Continuous up-dating will keep the Database complete in terms of current publications. The source for current references will be the systematic literature scan carried out at the *Institut d'Astrophysique de Paris*, which also feeds the SIMBAD reference files.

An essential part of the Database will be the major catalogs of galaxies and quasars, a sampling of which appears in Table 1, with an approximate count of entries. These catalogs have a substantial number of objects in common, but the various names are in sufficiently common use that the Database should be able to recognize all of them as legitimate names for the same galaxy. Apart from the main data entries in the catalogs, the Database will also incorporate the invaluable "notes", especially those in the RC2 and the UGC.

The other class of catalogs to be included in the Database will be those resulting from surveys at wavelengths other than optical, and where a reasonable fraction of the sources are extragalactic. Obvious candidates are the various *IRAS* catalogs, radio survey catalogs away from the Galactic Plane, and X-ray catalogs. For example, the *IRAS* Faint Source Survey will produce about 100,000 new extragalactic sources, at least half of which will correspond to galaxies which are not presently cataloged, and about 5% of which will correspond to objects that are not visible on the Palomar Sky Survey Plates.

Another class of quantities provided by the Database will be quantities that are functions only of position on the sky, such as the redshift correction to go from one frame of reference to another, or various estimates of Galactic extinction.

Table 1. SOME MAJOR EXTRAGALACTIC CATALOGS

Name of Catalog	Abbreviation	No(objects)
Morphological Catalogue of Galaxies	MCG	30,000
Catalog of Galaxies and Clusters of Galaxies	CGCG	28,000
ESO/Uppsala Survey of the ESO (B) Atlas	ESOUP	16,000
Uppsala General Catalog of Galaxies	UGC	13,000
Catalog of Galaxies and Quasars Observed by IRAS	CGQ	11,000
Catalogue of Peculiar Galaxies and Associations	CPGA	6,800
Southern Galaxy Catalogue (U Texas)	SGC	5,500
Second Reference Catalog of Bright Galaxies	RC2	4,400
A New Optical Catalog of Quasi-Stellar Objects	HB	3,700
The Catalogue of Quasars and Active Nuclei	VCV	3,500
Virgo Cluster Catalog	VCC	2,100

II.B. Archival References

An essential service to Database users will be in the form of references to data collected but not necessarily published yet, or to raw data available for re-processing. This service will make use of lists of observed targets kept by observatories, and summaries of data archives kept by data centers. Target lists are presently maintained by some ground-based observatories (e.g. the Very Large Array of the National Radio Observatory), as well as all space-borne ones. The most significant target list and data archive in the foreseeable future will be generated by the Hubble Space Telescope. In the case of instruments whose fields of view are substantially extended (e.g. the *IRAS* Additional Observations), the archival referencing will also point to objects that were not targets themselves, but were within the field at the time of an observation.

II.C. Size Estimate

Almost 100,000 galaxies will appear in the data base immediately, together with a comparable number of sources that may be *a priori* suspected of being extragalactic (e.g. *IRAS*, radio, X-ray sources away from the Galactic plane). For the majority of these, little more than positions and possibly magnitudes or diameters are known. About twenty thousand objects will have entries in the major systematic survey catalogs (e.g. UGC or ESO/Uppsala). Beyond that, about a thousand objects will have well over a dozen measurements associated with each of them, and a hundred or so galaxies and quasars will have in excess of a hundred references each. Based on these numbers, we estimate that 500 megabytes of data would be input at start-up if we meet the nominal goal of completeness in all published data. The volume of data will still be more than half that much if we only meet the minimum goal of including all data published since 1983, and major contributions before then. We estimate the volume of data to double every five years, assuming that we archive only summary quantities.

III. USER'S VIEW OF DATABASE

Users will perceive the Database as an environment to which they can connect remotely through a simple procedure, and within which they can query for information, submit more involved queries for processing in batch mode, or ask for files to be transmitted back and forth. The system will be user-oriented, soliciting input in conversational or menu-driven modes, and offering on-line documentation. We describe services provided by defining three major modes of interaction between user and Database, ordered in priority: Enquiry about objects specified by name (§III.A), object selection by specified requirements (§III.B), and user feedback (§III.C). A major long-term goal will be to add the capability to handle storage and dissemination of one- or two-dimensional data. The proposed approach to the data is to be all-inclusive rather than selective. Thus all published data will be carried as published, and it will be the user's responsibility to discriminate between "acceptable" and "spurious" data.

III.A. Enquiry about Objects Specified by Name

In the fundamental interrogation mode, the user will provide an object's name and obtain any or all of the following items:

i) a list of all alternate names;
ii) a summary of basic data, e.g. size, magnitude, redshift, morphology;
iii) summaries of other data available;
iv) a list of bibliographic references.

In this mode of enquiry, the user will have a choice between an interactive run where the data are provided for one object at a time, and a batch job on a list of objects which is constructed via direct input, or transmitted to the Database as a file. Eventually, a language will be developed in which to express the search request for each object separately, so the files can contain a different request for each object, rather than a blanket request for all objects on the list. Under items (iii) and (iv) above, listings can be limited by selecting on attributes such as data type or date of publication. Output will consist of printed information, and optionally, of machine-readable files. Output from an interactive session would be transmitted directly via the connection to the user, whereas output from a batch job would be collected off-line and mailed to the user.

III.B. Object Selection by Specified Requirements

The second major use of the Database will be to request a listing of all extragalactic objects meeting certain criteria, including:

i) position, thus allowing vicinity searches out to a given radius;
ii) availability of certain data, e.g. CO measurement, or IUE spectra;
iii) restrictions on data values, e.g. redshift between 0.09 and 0.11;
iv) restrictions on origin of data, e.g. sizes only from RC2;
v) exclusions, e.g. exclude UGC galaxies.

In addition to the full-sky option, this kind of search capability will also be available on a list of objects provided by the user. Initially, these searches will all be restricted to the batch mode. Eventually it should be possible to have these searches made as part of an interactive session. In any case, it will be possible for the user to inspect the results of a search, and use the output file as input for a new run, either selecting by specified requirements again, or making a blanket inquiry for detailed data or references (§III.A) on the objects selected.

III.C. User Feed-Back

The third mode of interaction will be in the form of information flowing from a user to the Database. Besides a general message utility, the user will be able, when in the "selected object" mode, to enter comments specifically related to that object, such as corrections, notes, warnings, or information on improved identifications or new discoveries. Such information might be eventually made part of the Database if it is intended to be public by the user, and if it is clearly useful to other users.

IV. SCHEDULE

The proposal desrcibing the Database was submitted to the National Aeronautics and Space Administration in June 1987, and is still under review as of this writing in October 1987. We expect that about one year after we start work on it, the Database will be open to remote access by astronomers, even though data entry will still be taking place. Access software will achieve its final configuration about six months later. The start-up phase will be concluded about one year later, at which point all activity will concentrate on keeping the Database current rather than adding data from the literature of previous years.

GALACTIC HI SURVEY DATA – A DATABASE PROBLEM

L.A. Higgs and P.E. Dewdney

Dominion Radio Astrophysical Observatory
National Research Council of Canada
Penticton, B.C., Canada

ABSTRACT

The synthesis radio telescope of the Dominion Radio Astrophysical Observatory is starting a survey of galactic hydrogen emission in the 21-cm line in the portion of the galactic plane above $\delta = 30°$. Initially, observations of 2°-diameter fields will be made at 24 field centres, spaced at intervals of 5° in galactic longitude, and the survey will be incompletely sampled. Each observation will yield a data cube of 256 × 256 × 128 pixels (RA × DEC × velocity). This will provide a database having approximately 1.5-arcmin spatial resolution and 2.6 km/s velocity resolution which will be very useful for statistical studies of the interstellar medium. Although this database, even if observing time is found later to make it a fully sampled survey (about 75 field centres), is only in the category of a moderate-size database ($\approx$ 10,000 small images), it presents problems of accessibility and analysis common to larger databases. Solutions to these problems are being sought.

INTRODUCTION

The primary instrument at the Dominion Radio Astrophysical Observatory (DRAO) is an aperture-synthesis radio telescope providing east-west baselines up to 600 metres at frequencies of 1420 and 408 MHz (Roger et al., 1973; Veidt et al., 1985). It consists at present of four 9-metre dishes and, after 35 days of observation, provides 130 images of the region of sky being observed. Two are radio-continuum images, at 1420 and 408 MHz, with spatial resolutions of about 1 arcmin and 3.5 arcmin respectively. The field of view is approximately 2° at the higher frequency and 8° at 408 MHz. The other 128 images are at closely spaced frequencies covering the 21-cm line of neutral hydrogen, in a bandwidth that can range from 0.125 to 4 MHz.

Three additional antennas are being added to the telescope and, within two years, the observing time required for a full synthesis should drop to 12 days. The additional array elements will permit the implementation of self-calibration techniques which should greatly increase the dynamic range of the instrument.

The synthesis telescope is operated as a National Facility for Canadian and visiting astronomers and is used for research in which its large field of view and sensitivity to low-surface-brightness objects are useful. As a "background project" at the Observatory, the synthesis telescope is being used to survey the northern galactic plane in a systematic fashion. At 408

MHz, the survey will completely cover the portion of the galactic plane visible to the synthesis telescope.

THE GALACTIC PLANE SURVEY

The portion of the galactic plane at declinations above 30° (resolution of an east-west array is rapidly degraded at lower declinations) can be covered completely ($|b| < 3°$) at 408 MHz by 24 field centres spaced at 5° of galactic longitude. The survey is expected to provide maps of low brightness emission nebulosity in the plane as well as a catalog of some 10000 unresolved sources of radio emission. The simultaneous 1420 MHz observations will not, of course, cover the galactic plane completely. Nevertheless, the incompletely sampled 21-cm data will still be valuable in providing an unbiased statistical sample of the structure and dynamics of atomic hydrogen (HI) gas in the galactic plane. So far, about three of the 24 fields have been observed and data reduction is in progress.

A sample of the 21-cm data for the field centred at $\ell = 140°$, $b = 0°$, is presented in Figure 1. The individual maps in the figure have a frequency resolution of 5 km/s, and the spatial resolution is 1.5 arcmin. Low-spatial frequency components missing from the synthesis telescope data have been derived from observations of the same field with the DRAO 26-metre antenna. Intensities are given in terms of brightness temperature. For each map, an average temperature has been subtracted. Thus the maps present deviations from the average HI emission.

DATABASE CONSIDERATIONS

For effective use of this HI database when the survey is completed, the information must be made easily available to the user, and software and hardware for the examination and analysis of the database must be provided. HI data cubes are unusual in that they are almost "irreducible", i.e. the features appearing in the two-dimensional images are parts of three- dimensional, connected structures, sometimes immersed in a smoothly varying three-dimensional background. Although the analysis of three-dimensional data structures is an area of active study, it is currently difficult to automatically isolate these features completely so that they can, for instance, be catalogued. Furthermore, visual appreciation and comprehension of diffuse three-dimensional structures is difficult for human observers, and computer-aided presentation is essential to the understanding of HI data. Rapid three-dimensional coordinate transformation is required as well as linear and non-linear spatial filtering. Since, most display devices are two-dimensional, projection of three dimensions into to two dimensions is particularly important. For example, longitude-velocity maps are often crucial for dynamical interpretation of HI features.

HI data are often difficult to understand in isolation from other astrophysical information, and comparison with models and other data is particularly important. A good example of a database with which comparisons are often useful is the four-wavelength data cube of

IRAS observations. They resemble HI observations in the sense that they contain many extended structures (essentially three-dimensional in the data cube). A future example could be a database of observations of the C^+ line at 157μ. Such a database would resemble the HI database quite closely, and comparison will be very important in understanding the thermodynamics of the interstellar medium. In the DRAO HI database, the removal of velocity structure resulting from galactic rotation, based on models, will be necessary in deriving kinematics of the interstellar gas.

The quantity of data in one DRAO HI observation (128 maps of 256 × 256 pixels) is not great, amounting to about 34 Mbytes if floating-point representation is used, but practical difficulties arise in the efficient examination of the data on affordable machines. Software packages for the manipulation of such a data cube already exist at DRAO and other institutions (e.g. the MADR package at DRAO, and the AIPS software supported by the National Radio Astronomy Observatory), but it is generally cumbersome and slow. The rapid manipulation of data needed for efficent analysis requires fairly large machines and probably that all the data be in random access memory. A cheaper alternative to continually recalculating transformations or projections of the data in a large machine might be to permanently store common transformations of the data cube as separate databases, i.e. multiple copies of the data in various forms. An optical disk system could easily accommodate the storage requirements for a reasonable number of transformations, and the CPU power required to examine the data would be greatly reduced. Users could access the data on small, affordable workstations. Of course, not all transformations could be forseen, and occasional resort to a brute force approach would be neccessary. Even in this situation, one might hope that the provision of a number of "starting points" would be helpful.

Even if multiple transformations or projections of the data- base were available, there are still severe limitations in the area of hardware display devices. The information in any one plane of the data cube can only be interpreted by comparison with the information in neighbouring planes. Thus one requires hardware devices that allow the rapid visual presentation of the various data planes to the researcher. Current image devices generally lack the storage capacity for quickly handling this amount of data and various institutions have developed special-purpose hardware towards this end. Examples of this are the video-disk hardware in the Groningen GIPSY image system and the image storage unit developed at NRAO. The arrival of new laser disk technology (erasable disks) on the commercial market will open new possibilities and developments in this direction are being monitored. The presentation of the data in a form suitable for the printed page poses further problems. The confused jumble of data presented in Figure 1 takes on more discernable structural forms when presented in a visual gray-scale display.

New software approaches to the identification of HI features in data cubes are also required. HI maps are usually noisy, and some of the most interesting features are faint. New algorithms are needed which can "find" connected features in this environment. Unless the whole data cube is in random access memory, straight-forward algorithms are exceedingly slow even in the absence of noise. One possible but limited approach would be to integrate along all possible "view lines" through the cube, hoping that some view lines would coincide with one- and two-dimensional features in the data, i.e. string-like and sheet-like structures. Even this approach would probably require a supercomputer. But the existence of a fast algorithm for doing this kind of task would open new avenues for the identification of the structure of interstellar clouds.

CONCLUSIONS

The survey of the northern galactic plane now underway at DRAO will provide 24 data cubes of 21-cm line information that will be of importance in statistical studies of the structure of the interstellar medium. However, the efficient examination of even this small amount of data poses problems using the existing hardware and software resources of DRAO. It is hoped that before the survey is completed (in about 5 years time), new image-analysis systems, optimized for data cubes, will be available.

REFERENCES

Roger, R.S., Costain, C.H., Lacey, J.D., Landecker, T.L., and Bowers, F.K.: 1973, *Proc. I.E.E.E.* **61**, 1270.

Veidt, B.G., Landecker, T.L., Vaneldik, J.F., Dewdney, P.E., and Routledge, D.: 1985, *Radio Science* **20**, 1118.

Figure 1. – Sample survey data for the first field observed. Contours are of brightness temperature. The contour and shading interval is 10K and the first white shading level is at 15K. An "average" brightness temperature has been subtracted from each map. The number at the top left of each map gives the LSR velocity in km/s.

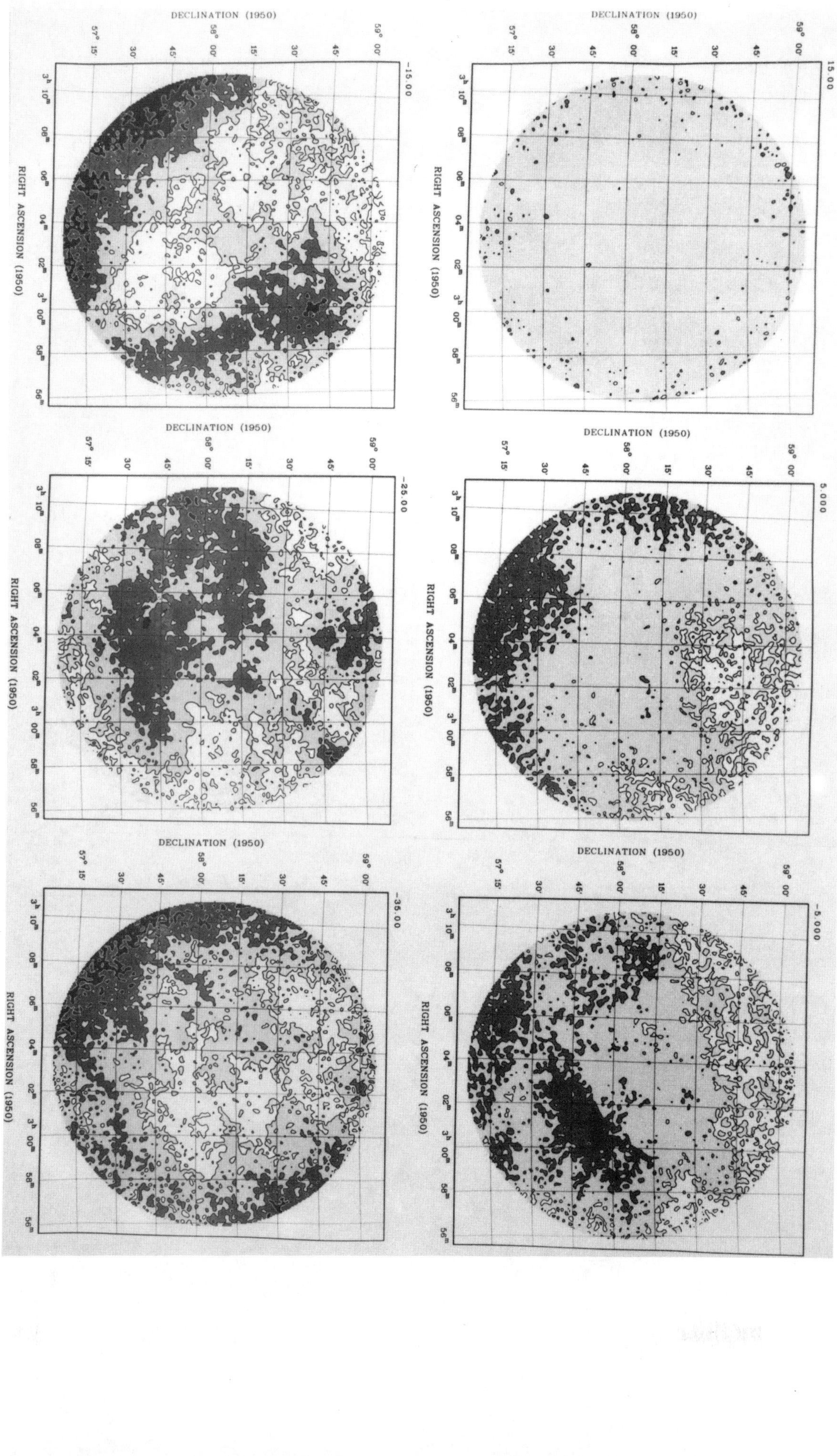
DECLINATION (1950)
RIGHT ASCENSION (1950)
-15.00
15.00
-25.00
5.000
-35.00
-5.000

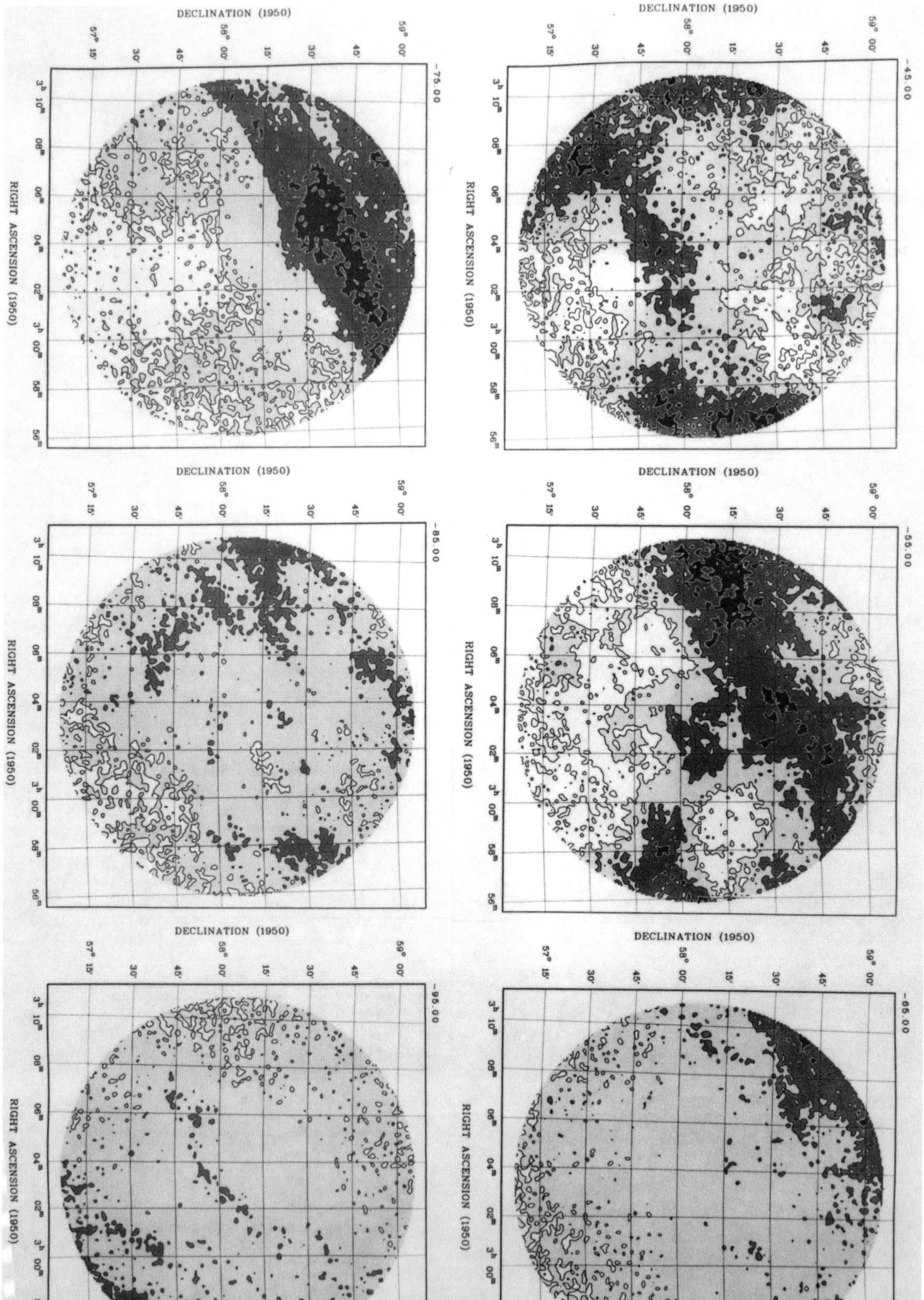
DECLINATION (1950)
RIGHT ASCENSION (1950)
-45.00
-55.00
-65.00
-75.00
-85.00
-95.00

VOCABULARY AND GENERAL REMARKS

Carlos Jaschek
CDS, Observatoire Astronomique, Strasbourg, France

Our aim is to present the definition of some terms we shall use frequently during this meeting. I think it useful to define terms so that we agree on what we are talking; the history of sciences is full of terms which meant different things to different peoples so that long lasting misunderstandings occured.

Let us start with the word *data*. We define data as parameters which can be derived from observations by a convenient model or theory. This implies that data are based upon observations, although the connection may be remote. The average density of matter in a sphere of one Megaparsec radius around the sun is an essential observational datum for cosmology, but it is not clearly related to a given observation. The flux at 12μ is also a datum; here one can pinpoint its origin to a series of measurements obtained by an observer with a given detector. Please observe that it data do not follow from *observations* without model or theory. This relates each datum to a well-defined epoch in the history of science but means also that data perish with the model, a fate not shared by observations.

(Astronomical) observations are defined as the detection of a signal carried out by somebody at a given place and time, with a certain instrument and for a certain purpose. Please notice that in this definition we have already a clear difference with observations in physics: to observe the velocity of light we do not need to know where and when the observations were carried out -we expect the result to be independent of these parameters. In astronomy we must keep in memory all old observations, because they are the only thing left to us. We may regret that Chinese astronomers did not provide us with a photoelectric lightcurve of the supernova of 1054, but the visual observation they made are the only observations of the lightcurve of this object.

Observations are sometimes called raw data. Notice that if one uses this terminology, one must also include with the (astronomical) observation all engineering parameters relating to the telescope and the detector. Assume that I observe the position of a star on the sky. I specify its altitude and its azimuth, the time, date place, temperature and atmospheric pressure (for the refraction correction) as well as engineering data on the instrument. Each item may consist of several readings (azimuth read three times, temperature twice, etc.) but we call all this raw data. Its preservation is essential if one wishes to re-use the data for instance with an improved model.

If one makes one step upwards in complication one finds the calibrated (or edited) data. These are refined raw data, with the different readings averaged and discordant readings rejected, the auxiliary data removed (temperature and pressure used to calculate refraction) and so on. This step is more or less complex depending upon the complexity of the intervening technology. The number of informations to be stored can be diminished sharply when passing from raw to calibrated data, a fact which is of obvious interest for data banks. Please observe that calibrated data never substitute entirely raw data -this is a hard question for data archives, which has to be decided for each kind of data differently. The usual answer that one can drop "raw" data when the reduction process is well understood, is a good recipe, but when are reduction procedures well understood? Perhaps after fifty years, like in photoelectric photometry, but here again one may have surprises.

After calibrated data one passes then to reduced data, expressed in physical or astronomical units. For instance from my observation of a star position, I end up with right ascension and declination at a given instant. This, in a certain way is the end result of the observation. Again here we have an obvious compression of the volume of information, with regard to calibrated data (and even more respect to raw data).

The storage of raw and/or calibrated data is called an archive. The storage of reduced data is called a data bank. A data bank may be defined as a set of coherent information collected for further use. For instance my personal observing file is a data bank. If a data bank is prepared for further use by computers and prepared in a computer readable way, we talk of a data base. Thus data bases are the logic evolution of data banks; the conversion became possible through computer technology.

A data bank needs to have a clearly defined purpose; the data must be ordered and the data must be explained. When we convert the data bank into a data base, we must make provisions concerning the security, the integrity, the resilience, and the standardisation. All these points are important but also very technical and I shall not discuss them further. Let me insist on the other side a bit more on differences between data bases. We can distinguish broadly two types: the "compilation" approach and the "critical" approach. In the first approach, the authors collects all data of a given type and lists them; the user is thus obliged to make himself a collation of the data. In the second approach the authors makes a critical evaluation of the data and replaces -say the existing ten UBV photometric values observed for a given star- by a single set of UBV data for this object. It is clear that both approaches are useful for different public. Being a spectrocopist I would mistrust any "recommended" spectroscopic value provided by somebody -I would like to see myself the individual data. But if I need a proper motion I trust whatever a good astrometrist proposes me, because I lack the expertise to deal with six different values of the proper motion. Since we all are usually specialists in only one field, both types of data banks have their own usefulness and their own limitation. It goes without saying that the quality of a data base depends critically on the quality of the compiler -a good specialist may produce a good data base, but a trustworthy data base without a good specialist behind does not exist.

On top of data bases we have the data centers. A data center is an institution devoted to the collection, critical evaluation and distribution of data. This definition excludes explicitely "data production": the accent is on data handling. The reason is that a data center is neither an observing facility (like IRAS), nor has the competence to produce a critical appraisal of all kinds of data. For instance at Strasbourg we have no staff member who is photometrist; therefore all critical evaluations of photometry are done at Lausanne- Geneva by the group of Prof. Hauck. This is an important feature of data centers, ignored by many peoples who think naively that NASA or CDS can produce something which the specialists in that area could not do. This is clearly similar to assuming that librarians can produce better books than what they have in their library! (Although this may be true in isolated cases, it is wrong in general). The existing data centers are a few: NASA Goddard in the USA, Astrosoviet in the USSR, Kanazawa at Japan, Potsdam in the GDR and of course the CDS. Data bases on the other hand are much more numerous; surely in astronomy one has about a hundred. An updated list of data bases is being published in the Bulletin of the CDS.

A further difference between data bases and data centers is that the latter are always public, ie open to all astronomers, whereas many data bases are closed, ie just accessible to the producer and a group of colleagues. This is so because as we shall see in this meeting, a data base is a research tool and as such its producers think most often in putting it to full use for their own research, rather than to open it to the public -in which case even his scientific competitors may use it. Since the establishment of a data base implies a large effort, such a viewpoint reasonable. One can however argue that if the establishment of a data base is expensive in time, manpower and money, sponsoring agencies should insist that the data base -after some reasonable time- be opened to the community. The main reason is that in a time of decreasing resources for sciences, one avoids much duplication of work this way. It would be easy to quote examples! Another reason for opening existing data bases to the public is that all data bases are piece-work in the sense that they organize only the knowledge in a very small area of science. Their real importance comes out only when the different informations can be confronted -ie through integration with other data. A data base on the photometry of galaxies is a valuable enterprise by itself, but obviously it is much more useful to be able to combine these photometric data with others, like radial velocity, classification, etc.

How can this "opening" of the individual data bases be done most easily? I think the only answer is through existing data centers. So for instance the producer of a data base (say on radial velocities of galaxies) should make it available in the form of a catalogue on tape to one of the existing data centers. Since all of them are linked through exchange agreements, it is indifferent to whom the information is given. In this way both the author of the data base keeps the intellectual property of the work and the data can be integrated with others. Data centers are also usually better equipped for distributing information than data bases, as any user will find out very soon.

The evolution of data centers led naturally to the integrated data centers (IDC). This term designates a data center in which all existing informations on an astronomical object are integrated. For instance for a star all known photometric information is listed, as well as its radial velocity, binary star characteristics, spectral type, etc. and also the references to all research in which this object was discussed. For the moment, the only existing IDC is the

SIMBAD data base of the CDS, but as we shall see in the US exist plans for an extragalactic IDS.

It is clear that an IDC can only be created from a combination of existing data bases, each one produced by a specialist, and this illustrates nicely the other sense of the term "integrated" in the IDC: what one integrates are not only the data, but also the expertise of all colleagues who elaborated the data bases upon which the IDC is based.

Let me turn next to my last topic, namely data archives. I shall call data archive an ordered collection of documented raw and/or calibrated data, collected at a certain observatory and organized in a way to permit permanent storage and retrieval. Although most archives keep only raw and/or calibrated data, one may also keep reduced data. What distinguishes then a data archive from a data base? The fact that an archive contains data obtained at one establishment or facility, whereas a data base contains data without restriction of place. So a data base on galaxies shall contain data obtained by different astronomers with a variety of instruments at a variety of places. The ESO archives on the contrary will contain data obtained at ESO telescopes, and the IRAS archives data from just one satellite and so on. Both types of data collections have their own merits, both are much needed by the community, but it is plain that they are different. Let me illustrate this with some examples.

a) Identification.

In an archive, a set of data is attached to the object the observer noted in the log book. This sounds fine until one realizes that observers usually do not care how to denote an object -they use for a star a HD number, a name, the Durchmusterung number and so on. There are even rumors in the air that some observers use obscure names to impossibilitate colleagues to know what they observed but this may be a tale. This inevitably produces a certain disorder. As an example of what that means, let me quote that for a single log book of one space experiment it took six months of work of one astronomer at the CDS to produce a cleaned version, ie a log book which provided one single name and a single position for each object observed.

On the contrary in a data base and an IDC correct object identification (and cross-identification of different names) is basic, right from the start, because otherwise one can never be sure that the same object does not appear under two different names.

In some archives this problem was attacked by comparing the position of a source observed with the position in a catalogue of a certain type of objects. So for instance the IRAS team compared their source catalogue with more than twenty catalogues of stars, galaxies, etc. As a result this may end up with an object having two different identifiers (which moreover may be identical). Whereas this is fine for the IRAS archive, because it points out that the object may be identified with X_1or X_2,in a data base, or even worse, in an IDC you cannot do that; if you want to intercompare all information available on an object it is impossible to add as a note "This object may (or may not have) the following infrared fluxes".

b) Limits.

A second characteristics is that an archive is essentially a closed data collection: it starts when the telescope and receiver was put into operation and ends when either one is shut down. For instance when the IUE satellite stop working, the IUE archive will be closed. But a data base, being a combination of many observational data, is practically open and the same happens to an IDC. You might say that when techniques change, certain kind of data do not arrive any more, like visual estimates of brightness or photographic magnitudes. If you run an IDC you would be surprised how wrong you are, because both techniques, yield nowadays several 10^4-10^5 data per year!

c) Inclusion.

A third characteristics is that a data archive is not including all kinds of objects, or all types of information. ESO telescopes for instance cannot observe all stars of the northern sky, nor do ESO archives contain observations of stellar magnetic fields -and conversely IRAS did not observe the ultraviolet, not IUE the infrared. On the contrary IDC's collect all kinds of data, over all the sky.

It seems to me that a recognition of their different nature would help to put things in the right perspective and avoid trying to set up data archives as if they were IDC. We all can only gain in a fruitful cooperation, but lose a great deal of effort by duplicating things or re-inventing the wheel.

SPECIALIZED DATA BASES

Mercedes Jaschek and Carlos Jaschek
CDS, Observatoire Astronomique, Strasbourg, France

Summary

The paper discuss data bases on classification (spectral classifications or galaxy classifications). Because of the complexities of both the observational material and the classification procedure, the combination of independent estimates of the classification of an object presents considerable difficulties. The analysis shows that the only feasible solutions is a 'critical" catalogue based upon the decision of an expert.

Data bases are basically of three different types, which will be called (Jaschek 1984):

1 - observational (containing the observations of the author or of his group)

2 - bibliographic (containing a listing of all data found in the literature)

3 - critical (providing a reduction or an average of the data, so that one single ('best") value is given for each object).

Each data base is adressed to a certain audience; the natural audience for DB's of types 1 and 2 are the specialists in the field. DB of type 3 appeal more to the non-specialist who finds himself helpless before the variety (and disagreement) of existing data. Quite naturally therefore specialists tend to have a very critical attitude on type 3 DB's because they feel they de not need them. But since obviously each one of us is a specialist in a small field, we all are happy to use DB's of type 3 at some stage of our own research for data outside our own field. In what follows we shall discuss in particular one type of data bases, namely those based upon estimates rather than upon measurements. A specific case is that of spectroscopic data bases, a field in which both authors have worked some time. However our considerations apply equally well to the data bases on classifications of galaxies (Hubble (1926), Holmberg (1958), de Vaucouleurs (1959), van den Bergh (1960)).

Broadly speaking under spectroscopic data we find:

a) spectral classification

b) descriptions of spectral peculiarities

c) quantitative data like equivalent widths, abundances, spectrophotometric data.

The last class of data are similar to other kinds of quantitative data and their handling can be done on a similar basis as for instance photometric catalogues -see the poster paper by Manfroid and Heck. We shall therefore not discuss them here, and concentrate upon data of type a). Data of type b) follow the same pattern of data of type a).

The most important characteristics of spectroscopic data is that they are the result of an estimation, not of a measure. As a consequence they are described by discontinuous parameters, usually expressed in alphanumeric code.

Whereas the discontinuous character presents no major difficulties for further treatment, the fact that spectroscopic data are estimates rather than measurements produces some difficulties. Let us start with some basic considerations. A spectral classification system is defined by a set of n standard stars, whose types are settled by definition. The classification of a star is then the identification of this star with one out of the n standards. Ideally there should be one standard for each type defined in that peculiar classification system; for the MK system, which is a bidimensional scheme with 44 spectral types and eight luminosity classes, one should have at least 352 standards, one for each intersection. In fact not all intersections are defined by a standard star and one must thus interpolate between nearby standards for classifying. The mesh is thus incomplete, and on top of this some of the standards have been changed over the years, so that the mesh of 1956 is not exactly the same as in 1987. Stars classified some time ago may thus have been classified slightly differently over the years.

Let us forget for a moment these difficulties and assume that we do have a complete and invariable mesh of standards. Clearly the precision of the classification is closely related to the distance between neighboring cells. A classifier working on an B2 star can only decide that it be either B1, B2 or B3 -by definition the system has no B1.5 stars.Classification should thus be accurate within ± 1 tenth of a spectral class and analogously within ± 1 luminosity class. In fact an intercomparison of different classifiers has shown this to be true. (Jaschek and Jaschek 1965, 1973) When one intercompares the lists of different classifiers, one must also take into consideration two more facts. The first is that any classification system is bound to a certain technique — slit spectrograms on spectrographic plates for instance. Usually such a configuration is described with the help of some parameters like wavelength range covered, height of the spectrum and resolving power. For the MK system typically this corresponds to $\lambda\lambda$ 370–480 nm, 10^{-3} m and 0.16 nm but these figures are only indicative because the system has shifted over the years to better resolution and a wavelength range more extended into the ultraviolet. If one uses a different instrumentation — for instance objective-prism spectra, or higher or lower resolution — one is strictly speaking outside the MK system. Provided the instrumental parameters are not too different of those of the MK system one may try to put oneself on the MK system. However carefully this is done it is bound to produce complications and additional scatter somewhere. Please observe that this happens in all cases when one 'ties in" estimates.

The second fact summarizes what one can call loosely the 'quality" of the classifier or the classification. It implies that there are 'good" classifiers and 'less good" ones; the difference may depend on the eye sight, practice, carefulness and experience. Usually it is impossible to be precise on this, but the fact is that each classifier has its own scale of weighting (dimly!) the work of colleagues.

With the two facts in mind, what to do then when one has for one object several discrepant classifications? One easy solution is to list everything — ie. produce a 'bibliographic" catalogue and let the user decide what to do. One of us produced such a catalogue (C. Jaschek et al

1964), which was used by many colleagues. It was heavily criticized by specialists because of the fact that 'everything" was included; they felt that one should omit 'poor" classifications because their inclusion produced the impression that the MK system is worse than it is. Not feeling like God Almighthy, I have always refused to reject lists because of subjective criteria. A more sensible criticism came from non-spectroscopists who wondered very much what to do with the several types listed.

As a solution, some non-specialists have proposed to use the average of the different spectral types given for a star, in order to produce a 'best" type. It is easy to show that this is incorrect — the average of A2V, A3V and A3V cannot be A2.67 because this does not exist : we only have A2V and A3V. In other cases, the MK system does not carry all decimal groups. So the average of G5III, G8III and G8III is G7III, but G7III has no standards in the MK system. Besides these two difficulties we have the more serious one to know if all classifications published can be considered to be truly independent. If not, obviously the whole procedure is worthless. A more reasonable answer to the problem can be a 'catalogue of selected spectral types", in which a spectroscopist makes a choice of what seems to him the 'best" classifications. His choice must be guided by a consideration of the different factors mentioned above. Such a catalogue was produced by M. Jaschek (1978). It was welcomed by non-specialists and run into criticism from specialists who objected the way the selection was made. Such a criticism is however unfounded because all classifications (besides the one chosen) are easily obtainable from the CDS; if somebody wants to use other selection criteria he has all existing data available to him.

One problems which occurs repeatedly in the process of selecting the 'best" value, is what to do when one has two different sources, neither one fulfilling exactly the MK specifications. To give one example consider the case that we have to decide between a source 'A" of a classifications based on 20 $A.mm^{-1}$ slit spectrograms and another source ('B") based upon 108 $A.mm^{-1}$ objective prism material. Since the MK system is (now) attached to 80 $A.mm^{-1}$ slit spectra, it is clear that neither sources obey the specifications. In favor of the material 'A" one may say that since it is based upon better resolution this should be favored over 'B". But if in the source 'B" the author did classify a very large number of spectra, one can still make a case for source 'B", since the classifier has a unique experience and the classifications provided are surely affected by smaller errors. As can be seen the solution is not easy; the most one can say is that it depends also on the use one wants to make of the classification.

In view of such difficulties one could be tempted to give up entirely this selection based upon personal appreciations and to look for 'objective" methods of selecting the 'best" classification. The most direct way is through the use of colors. Once one has decided that the object is normal (ie. has no spectral particularities) its color should relate closely to the spectral type. A comparison of different classifications with (unreddened) colors usually permits to eliminate discordant classifications. The difficulty is that only a few specialized photometric systems permit a separation of luminosity classes; this makes that one can check easily the spectral type, but not the luminosity class. And of course the procedure is unapplicable when the object has a peculiar spectrum. Despite these difficulties, such tests are specially useful to eliminate outright blunders, originating for instance through wrong identification of stars.

Another, more indirect check upon the internal consistency of the classifications produced by an author of a long list of classifications is an examination of the frequencies of the different classifications. Assume that one considers the groups of stars classified as B2V, B3V, B2III and so on. If one examines the frequency of stars in each group, one would expect intuitively a smooth relation between the frequencies along or across the main sequence. If at any point a discontinuity appears, we may suspect that some systematic error came in — for instance some authors classify more stars as class IV than others. Of course such tests can only be applied if the total number of objects is very large.

From these considerations one sees that 'objective" tests are only helpful in a few cases, but cannot solve the general problem. We are thus back to the 'critical" catalogues compiled by specialists and this is the most important conclusion of this presentation. In areas where one has to combine subjective estimates, one cannot use the common statistical procedures of weighting and averaging. The best one can do is to trust the judgement of a specialist to select what are the 'best" classifications, and this situation shall probably last upon the day when the estimates are replaced by measurements, as it is done in automated classification. Let it be stressed that such a critical compilation should be accompanied by a collection of all available data, so that in cases of conflicting opinions one may go back to the original source.

As a final comment let me add that many topics mentioned here are discussed in detail in a forthcoming book *Data in Astronomy* to be published by the Cambridge University Press in 1988. A complete list of specialized data bases is also included.

References

de Vaucouleurs G. (1959) Handbuch der Physik 53, 275

Holmberg E. (1958) Lund Medd. 136

Hubble E. (1926) Astrophys. J. 64, 321

Jaschek C., Conde H., de Sierra A. C. (1964) Publ. La Plata Ser. Astr 28, part 2

Jaschek C. and Jaschek M. (1965) in IAU Symp. 24, ed. Loden K, Loden L. O. and Sinnerstad U., Reidel Publ. Co.

Jaschek C. and Jaschek M. (1973) in IAU Symp. 50, ed. Fehrenbach C. and Westerlund B. E., Reidel Publ. Co.

Jaschek C. (1984) Quart. J. Roy. Astr. Soc. 25, 259

Jaschek M. (1978) CDS Inform. Bull. 15, 121

Van den Bergh S. (1960) Astrophys. J. 131, 215

THE CALIBRATION DATABASE SYSTEM: CONCEPTS AND CONTENTS

Helmut Jenkner 1)
Space Telescope Science Institute 2)
Baltimore, Maryland, USA

Abstract: Several parts of the ground system in support of Hubble Space Telescope operation are involved in instrument calibration activities. As an integral part, databases of instrument data and astronomical reference data are being established.

I. INTRODUCTION

In general, a calibration is considered to be the measurement of any aspect of the signature which a scientific instrument (including the telescope itself) imposes on the observed data, the removal of such signatures from the data, and the transformation of science data into physically meaningful units.

In the context of Hubble Space Telescope (HST) operation, the following subsystems of the Science Operations Ground System (SOGS) are involved in calibration activities:

- o the Routine Science Data Processing (RSDP) as part of the Post-Observation Data Processing System (PODPS), and

- o the Calibration Database System (CDBS).

In addition, databases of instrument-related and astronomical reference (standard) data are required to perform calibration and monitor instrument performance.

II. ROUTINE SCIENCE DATA PROCESSING

The Post-Observation Data Processing System (PODPS) is the portion of the HST ground system which provides the software to automatically (i.e., in a non-interactive or pipeline mode) manage the receipt, editing, calibration, and archiving of data from HST. In addition, it generates certain types of output products from the calibrated data. The Routine Science Data Processing (RSDP) is the portion of PODPS which performs this automatic processing.

The RSDP data processing function edits the science data to insert fill data to take the place of any missing data, then evaluates the data for acceptable quality. Once the science data are accepted, data processing converts them to a generic format. The RSDP calibration function then processes it, using a

---

1) On assignment from the Astrophysics Division, Space Science Department, European Space Agency.

2) Operated by the Association of Universities for Research in Astronomy, Inc., under contract to the National Aeronautics and Space Administration.

standard instrument-, mode-, and format-specific calibration algorithm. RSDP also receives Astrometry Science and Engineering data from the HST Astrometry and Engineering Data Processing (AEDP) system; these are simply received, cataloged, and archived.

Edited and calibrated science data, astrometry science data, engineering data, and definitive orbit data are placed in the on-line archives as the routine processing of the data finishes.

III. CALIBRATION DATABASE SYSTEM

The Calibration Database System (CDBS) transforms raw and reduced observations taken by each of the HST Science Instruments (SIs) to provide data needed for calibration of science data in the pipeline processing (RSDP) and for the interactive analysis of the science data as performed by the general observer (using the IRAF/SDAS system). In addition, CDBS generates data describing past and present status of the instruments, data needed for commanding the instruments, and data on the instrumental and telescope performance. The description of the instrumental behavior includes trend analysis, a fact which underscores CDBS's continuing observatory responsibility.

Conceptually, CDBS can be thought of as consisting of two parts:

- o the instrument software, and
- o the database (system) software.

The database software stores the parameters, pointers to parameters, parameter generation history, and moves parameters to the RSDP calibration pipeline. It provides utilities which allow users to peruse the database to determine parameter generation histories, list parameters obtained under specified instrument configurations, obtain the file specifications of parameters contained in the SOGS archive, and other data query/storage functions.

The instrument software contains specific reduction and analysis routines for HST science instruments, specifically:

- o Wide Field/Planetary Camera (WFPC),
- o Faint Object Camera (FOC),
- o Faint Object Spectrograph (FOS),
- o High-Resolution Spectrograph (HRS), and
- o High-Speed Photometer (HSP).

The analysis for Fine Guidance Sensor (FGS) science data is performed with the Astrometry and Engineering Data Processing (AEDP) system.

In addition, the instrument software provides applications to duplicate the calibration processing in RSDP, in order to allow re-calibration as well as testing of algorithms or new calibration parameters.

While the operational CDBS and its databases are installed in the ground system environment, the instrument software functions and (read-only) access to the databases are made available to the general user.

IV. DATABASES

The databases associated with CDBS contain the following types of data:

- o Engineering parameters,
- o Instrument data, and
- o Astronomical reference data.

Engineering parameters give detailed information about the status and conditions of the instruments and the telescope itself.

Instrument data, such as flat fields or dead-time corrections, are required to populate RSDP. In the pre-launch era, these data are provided by the Instrument Development Teams, who will also derive updates at the end of the Science Verification period (i.e., after the first six months of HST operation). For the remaining parts of the operational life of HST, these data will be maintained by STScI. In addition, passband information on the passbands of all optically active components (telescope, filters, detectors, etc.) will be available.

In order to provide a comprehensive set of astronomical standard sources, a combined effort of outside contractors in the astronomical community and of STScI scientists is being undertaken. Standard sources are required for the following areas:

- o UV Spectrophotometry (programs led by Bohlin, Holberg, Savage; 71 targets).
- o Optical Photometry and Spectrophotometry (Baum, Courvoisier, Hesser, Landolt, Lub, Oke, Stone, Tapia, Turnshek, Walker; more than 280 targets or fields).
- o Wavelength Calibration (Hobbs, Duncan; 25 targets).
- o Astrometric Calibration (Duncombe, Franz, van Altena; 62 targets and fields).
- o Polarimetric Calibration (Schmidt, Tapia; 55 targets).
- o Flat Fields (this area does not contain actual target data, but rather provides the concepts of how to obtain flat fields from orbit).

Thus, a database of standard data for about 500 specific targets and fields will be available in CDBS.

In the area of spectrophotometry, these data will be used to derive uniform spectra covering the entire HST wavelength region, i.e., the range from 1200 to 12000 Angstroms. In addition, a "grid" of absolute model flux distributions for a variety of sources (stellar as well as non-stellar) will be derived and stored in CDBS.

V. CALIBRATION CONCEPTS

To derive updates to calibration parameters used in RSDP or provided for general users, data of standard sources observed with HST are compared with data predicted for this specific source and instrument configuration (mode) based on existing calibration information.

In order to achieve this goal, a "passband generator" is used to combine the various component passbands relevant for the specific instrument mode, making use of a mathematical representation of the various instrument ("instrument graph"). The resulting mode passband is then combined with the spectrum of the standard source from the database, yielding the predicted data.

The comparison of the observed and predicted data then results in updates to the instrument parameters (passbands), but also to the target spectra. In order to achieve higher calibration accuracy and reliability, data from various modes and instruments may be combined to isolate the effects of individual components or standard data items.

THE GUIDE STAR CATALOG: STRUCTURE, PUBLICATION, AND FUTURE PLANS

Helmut Jenkner 1), Barry M. Lasker, and Brian J. McLean
Space Telescope Science Institute 2)
Baltimore, Maryland, USA

Abstract: The current structure and content of the Guide Star Catalog (GSC), as well as future enhancements in this area are described. An overview of the forthcoming publications is given, both with regard to scientific papers and electronic media.

I. INTRODUCTION

The Hubble Space Telescope (HST) will use off-axis guide stars to achieve its pointing performance. The selection of these guide stars uses a catalog specifically constructed for this task, the Space Telescope Guide Star Catalog (GSC), which consists of about 20 million stars and objects in the range of 9th to 16th magnitude.

The catalog is based on microdensitometer scans of the UK SERC J survey in the south and a 1982 epoch "Quick V" survey made with the 48-inch Schmidt on Palomar Mountain. The hardware aspects, image processing and calibration algorithms, precision estimates, and GSC properties have been discussed in recent publications and conference proceedings (Jenkner 1983; Jenkner 1984; Kinsey et al. 1984; Lasker 1984a and 1984b; Lasker and Kinsey 1984; Lasker et al. 1986; Russell 1986; Russell and Williams 1986; Jenkner et al. 1987; Lasker et al. 1987; Lasker, Sturch, et al. 1987; Russell et al. 1987).

II. STRUCTURE AND CONTENT

Access to the GSC has to be accommodated by identification as well as by celestial area. Together with the expected volume of data, this led to the formulation of a structure which would allow fast access, but would not require inordinate amounts of storage space nor specific software components.

The GSC is organized into 10000 regions; each region therefore holds a few thousand individual object entries. In order to account for the varying population of the catalog as a function of galactic latitude, the region size is not fixed, but varies, so that the number of objects per region is nearly constant.

Each object in the GSC carries a 10-digit number as identification. The first five digits encode the region number; the last five specify the number of the star within the particular region. Since only a few thousand objects are contained in each region (leaving some room for future expansion), this

---

1) On assignment from the Astrophysics Division, Space Science Department, European Space Agency.

2) Operated by the Association of Universities for Research in Astronomy, Inc., under contract to the National Aeronautics and Space Administration.

identification scheme is noncontiguous, i.e., there are many 10-digit numbers which do not correspond to an entry in the catalog.

Although this nomenclature as such does not encode coordinates, the region structure of the GSC would provide a hidden coordinate-encoding if the region boundaries were fixed; assuming coordinate updates, this would lead to the well-known problems of coordinate-related nomenclature techniques. Therefore, an additional feature has been adopted for the GSC regions: they are not (primarily) defined by their celestial coordinates, but by their content. In other words, once an object has become a member of a region, it remains part of it regardless of position updates. Therefore, the maximum and minimum coordinates of regions may overlap slightly, once updating activity has occurred.

Each region is thus characterized by its center (reference) position and extent, as well as by the number of objects it contains. (Back-pointers to the individual plates from which objects appear in the region as well as several organizational items are available in addition.) For each entry the following information is contained in the GSC:

- o Identification (GSC Number).
- o Position, in the form of offsets (xi, eta) to the region reference position.
- o Position standard deviation (one per position).
- o Magnitude band descriptor.
- o Magnitude and magnitude standard deviation.
- o Object classification (star versus non-star) and quality.
- o Source plate (or catalog) identification.
- o Various flags.

Note also that object data derived from different plates (i.e., for objects appearing on more than one plate) are stored separately, but with the same identification.

III. PUBLICATION

The publication of the Guide Star Catalog is scheduled for 1988; it will consist of a descriptive part in a number of papers in the astronomical literature, and the bulk of the data will be published in electronically readable format. The papers will contain detailed descriptions of all aspects related to the GSC, specifically:

- o Background information about the GSC and the Guide Star Selection System in the context of the Space Telescope mission.
- o Astronomical data (plates, catalogs).
- o Software system used for catalog construction.

- o Hardware system used for plate scanning.
- o Image processing algorithms.
- o Calibration (photometry, astrometry).
- o Catalog structure.
- o Description of data formats and access software.

In addition, these publications will contain detailed error analyses and extensive statistical information on all aspects of the GSC, the processing, and the various reference materials used in the derivation of the data.

The catalog data themselves, as well as the statistical information mentioned above, plus basic access software and user information in the form of text, will be made available in computer-readable form.

The volume for the entire set of data is expected to be in the range of one to two Gigabytes. Therefore, distribution on magnetic tape would require up to 20 rolls of tape (even at high density and large blocking); the associated effort with regard to tape duplication is prohibitive for mass distribution, and may only happen in exceptional cases. Currently, compact optical disks (CDs) seem to be the medium of choice; the data would require two or three platters; both the manufacturing and distribution process are comparatively simple, and readers for compact optical disks are available at low cost. However, details of volume and file structure need to be addressed before publication.

At ST ScI, the Guide Star Astrometric Support Program (GASP, McLean et al. 1988) has been modified to use a sample compact optical disk containing portions of the GSC as well as image information, and has thus demonstrated the feasibility of this approach.

IV. ENHANCEMENTS AND PLANS

At the time of this conference, the Guide Star Catalog is complete to more than 90%. The number of objects in the complete catalog will be somewhat larger than 20 million.

Plans for enhancements in the image processing and calibration areas, as well as with regard to the incorporation of new surveys, have been described in the previous parts of this series. From the organizational (data base) point of view, we anticipate an enhanced catalog structure for the future, which will, however, retain the basic GSC nomenclature and region scheme. This new structure is based on the need to:

- o preserve raw data (image features) obtained directly from the underlying plates;
- o perform re-calibration of selected plates without affecting the rest of the catalog;
- o provide corrections in an orderly and trackable way; and
- o provide cross-references to other catalogs.

To achieve this goal, it is expected that the main part of the catalog will contain object names, image features, and other information, directly derived from the plates, but without any additional calibration. A second part will contain these constructs, e.g., plate solutions and photometric transformations. Addenda and error corrections will be carried in a third part. Cross-references to other catalogs (as well as the catalog data themselves) will also be added to the new structure. The various pieces of information related to one object would be combined by a software module - the catalog server - upon demand, i.e., upon a request by a user for data from the catalog. This enhancement would allow the efficient incorporation of new survey data, and also of other catalog data as they become available in the community.

REFERENCES

Jenkner, H.: 1983, "Astronomical and Statistical Algorithms Used in the Space Telescope Guide Star Selection System", in: Proc. Statistical Methods in Astronomy, ESA SP-201, 65.

Jenkner, H.: 1984, "Principles of Astronomical Catalog and Data Base Construction and their Application towards the Space Telescope Guide Star Selection System", in: Proc. International Course on Data Handling in Astronomy and Astrophysics, ed. B. Hauck and G. Sedmak, Memorie Societa Astronomica Italiana.

Jenkner, H., Russell, J. L., Lasker, B. M.: 1987, "The Guide Star Catalog. III. Structure and Publication, Status and Plans", in Proc. of IAU Colloquium 133 "Mapping the Sky", Reidel Publ. Corp. (in press).

Kinsey, J. H., Denman, R. J., Evzerov, A.: 1984, "Laser Interferometer Enhanced Positional Precision of a Scanning Microdensitometer for Astronomical Plate Scanning", Rev. Sci. Instr., 55, 1960.

Lasker, B. M.: 1984a, "Schmidt Astronomy and the Space Telescope", in: Astronomy with Schmidt-Type Telescopes (Proc. IAU Colloquium No. 78), ed. M. Capaccioli.

Lasker, B. M.: 1984b, "The Guide Star Selection System for the Space Telescope: Long-Term Activities", in: Science from Measuring Machines (Proc. Royal Greenwich Obs. Workshop), ed. I. N. Reid and P. C. Hewett.

Lasker, B. M., Damashek, M., Garnavich, P. M.: 1986, "A Shaped Aperture for Two-Dimensional Microdensitometry", Publ. A.S.P., 98, 125.

Lasker, B. M., Kinsey, J. H.: 1984, "PDS Sampling", Bull. AAS, 16, 498.

Lasker, B. M., Jenkner, H., Russell, J. L.: 1987: "The Guide Star Catalog. I. Overview, History, and Prospective", in Proc. of IAU Colloquium 133 "Mapping the Sky", Reidel Publ. Corp. (in press).

Lasker, B. M., Sturch, C. R., et al.: 1987, "The Guide Star Photometric Catalog", Ap. J. Suppl. (in press).

McLean, B. M., et al.: 1988, "The Guide Star Astrometric Support Package," (in preparation).

Russell, J. L.: 1986, "The Guide Star Selection System and the Guide Star Catalog for Space Telescope", in: Astrometric Techniques (Proc. IAU Colloquium No. 109), ed. H. Eichhorn and R. J. Leacock, p. 721.

Russell, J. L., Williams, C. A.: 1986, "Testing Schmidt Plates for Astrometric Purposes," in: Astrometric Techniques (Proc. IAU Colloquium No. 109), ed. H. Eichhorn and R. J. Leacock, p. 721.

Russell, J. L., Lasker, B. M., Jenkner, H.: "The Guide Star Catalog. II. Astrometric and Photometric Algorithms and Precision", in Proc. of IAU Colloquium 133 "Mapping the Sky", Reidel Publ. Corp. (in press).

THE GLOBAL OSCILLATION NETWORK GROUP (GONG) HELIOSEISMIC DATA REDUCTION and ANALYSIS SYSTEM

James R. Kennedy
James A. Pintar
National Solar Observatory
National Optical Astronomy Observatories†
P.O. Box 26732
Tucson, Arizona 85726 U.S.A.

HELIOSEISMOLOGY

Studies of the solar interior have long been limited by the fact that, expect for neutrinos, virtually all of our knowledge is based on observations of conditions in the Sun's outermost layers. Models based on this information have been unsuccessful in simultaneously resolving the problems of the primordial helium abundance and the apparent deficiency of neutrinos. *Helioseismology*, the use of acoustical oscillations to study the Sun, holds great promise as a tool for making significant advancements of our knowledge of the solar interior and, by extension, stellar interiors, in general.

The Sun is continuously engaged in a myriad of acoustical or seismic oscillations. Known as the "five-minute oscillation", this effect appears at the surface as essentially radial oscillations with periods of from 2.5 to about 11 minutes which produce Doppler shifts on solar spectral lines. These waves represent free standing oscillations trapped in spherical-shell cavities. They are confined on the outside by reflection from the steep density gradient of the photospheric surface, and on the inside by refraction imposed by the inwardly increasing sound speed. The exact temporal oscillation frequency depends critically on conditions within a given wave's cavity (see for instance, Christensen-Dalsgaard, *et al.*, 1985; Deubner and Gough, 1984).

Though the overall surface pattern at first seems chaotic, it is actually due to the superposition of a large number of different normal modes. The surface patterns for the various modes are described by the spherical harmonic functions. For a given frequency, the longer the horizontal wavelength of a mode the more deeply that mode penetrates into the solar interior. The longest modes reach into the solar core itself and their acoustical line widths suggest that they have lifetimes of weeks or even months.

Helioseismic waves can serve as *probes* which may be used to generate accurate pictures of the solar interior. The inner refraction depth of a given mode is determined by the sound speed and thus, the temperature, at that point. By combining the information from a great many modes, it is possible to infer the temperature of the solar interior as a function of depth using inversion methods in some ways similar to those used for some geoseismic studies. Alternately, the parameter space in solar models can be explored to attempt to converge predicted and measured oscillation frequencies.

---

† Operated by the Association of Universities for Research in Astronomy, Inc., under contract with the National Science Foundation.

THE GONG NETWORK

Operationally, helioseismic studies require very accurate frequencies of the resonant modes. Precise measurement of these frequencies by a single station is obstructed by the data gaps imposed by nighttime interruptions and weather. Diurnal gaps contaminate the time series with a frequency of one/day. Its harmonics form sidelobes which are superposed onto every "real" signal to produce a complex mix of real and false spectral features. Unfortunately, the spacing of the sidelobes imitates those of many real features making it essentially impossible to differentiate between them (Hill and Newkirk, 1985). The obvious solution is to obtain a virtually *unbroken* time series of data free from *periodic* interruptions. With only random gaps, observing the Sun about 90% of the time appears to be sufficient for an essentially perfect reconstruction of the acoustical spectrum. Since the resonant periods extend down to about 150 seconds, samples should be obtained every 75 seconds or less. As some of the modes are very long-lived, observations on a scale of years is required to fully resolve them.

The Global Oscillation Network Group (GONG) will achieve these goals with a network of six imaging helioseismic observing instruments located nominally at 60° increments of longitude around the world (Harvey *et al.*, 1987). The GONG project, being managed by the National Solar Observatory (NSO) of the National Optical Astronomy Observatories, will develop and operate both the network and a central data analysis center as a facility for the international solar physics community. Presently the community has formed seven science teams to handle the development and execution of the scientific analyses. These teams are actively engaged in a wide variety of related activities.

The GONG network itself will consist of small, automated, observing instruments. A full-scale prototype field system is currently under development and should be operational in mid-1988. The complete network is expected to go into operation in late 1990, and will run for a minimum of three years.

DATA ACQUISITION

The surface oscillation modes number in the millions. The separation of the waves into individual modes requires imaged Doppler measurements to characterize the structure of the surface patterns. A 256x256 format will be employed to resolve horizontal surface wave numbers up to about 250. Unfortunately, as the observed patterns represent the sum of millions of different mode combinations, untangling them is a significant computational challenge.

Individual solar modes may exhibit velocities of 20 cm/s or less, while the sum of all the modes at a given point is only a few hundred m/s. To be limited by the Sun's "random" surface motions, the GONG instruments must be capable of making imaged velocity measurements with a precision of less than one m/s. The instrument of choice is based on the the Michelson interferometer and is called a "Fourier Tachometer" (Beckers and Brown, 1978). It has the effect of sampling the entire disk image at three different points in a spectrum line profile in successive frames. Subsequent arithmetic on the respective pixels in each of these three frames produces a velocity image, or Dopplergram, of the Sun. The resulting image is then stored on a rotary-head magnetic tape cartridge, along with an intensity image and other information, for subsequent reduction at the central data reduction and analysis facility.

DATA REDUCTION

It is clear that the real challenges in the area of GONG data reduction and analysis are presented by two factors: a monumental volume of data and a long sequence of complex computing tasks. Each station in the network will produce at least 200 Mbytes of data every day. The whole network will generate over a Gbyte per day, seven days a week, for three years. Over this time the total accumulation data from the field network will be about 1.5 Tbytes. Including the processed data products, the total archive will exceed 4.3 Tbytes.

In the reduction process , each of the three individual 64K pixel frames of each station must be corrected for a variety of instrumental, photometric, geometric, and known motion effects. Commonly, as many as three adjacent GONG stations will observe simultaneously for periods of up to several hours. These data must be merged into a single stream of "best" frames and, thus, quality measures for each and every frame must be computed on a continuing basis to make this selection. Furthermore, to be merged properly, the pixels from each of the different stations must match cartographically. Thus, optical and atmospheric distortions must be measured and used to re-map every pixel of each station back into a "standard image frame".

Once these preliminary tasks are complete, several more computationally intensive reductions will be performed. The solar velocity images are decomposed into their spherical harmonic coefficients. The time series of these coefficients is further processed by way of a Fourier transform to the desired frequency spectra. The principle processing bottleneck in imaged helioseismology is the spherical harmonic decomposition. With currently available algorithms, this process alone requires a throughput rate of more than 1 MFLOP/s, 24 hours a day, to keep up with the incoming data. When added to the other reduction processes the total rate reaches nearly 2.5 MFLOP/s for continuous operation or 5 MFLOP/s in a 12-hour day. With a reasonable allowance for scientific analysis, equipment outages, and so forth a minimum sustained throughput rate of 7 MFLOP/s is expected to be required.

In all, the GONG pipeline processes will produce several product data sets in a format suitable for scientific analysis. It is anticipated that standard product sets will include, at least:

1. Time series of Velocity and Intensity Images,
2. Time series of Spherical Harmonic Coefficients,
3. Power Spectra,
4. Tables of Mode Frequencies,
5. Velocity Time Averages,
6. Magnetograms,

DATA ARCHIVING

Scientific analyses will proceed from the archived data sets. The raw and reduced data sets will be archived in two basic forms, depending on the degree of ready access expected. Data sets which may be accessed very infrequently, such as the various raw intensity images, will be archived on inexpensive 2 Gbyte rotary-head tape cassettes. As presently conceived, more frequently accessed data, such as the reduced product sets, will be stored on 2 Gbyte 12 inch "write-once read-many" (WORM) optical disks.

ANNUAL GONG DATA ARCHIVES

Data Product	Archive Size (bits)	Medium
Raw Images	4.0×10^{12}	Rotary Head Tape
Corrected Images	4.0×10^{12}	Rotary Head Tape
Merged Images	1.3×10^{12}	Rotary Head Tape
Intensity Images	4.4×10^{11}	Optical Disk, Off-line
Velocity Images (Dopplergrams)	4.4×10^{11}	Optical Disk, Off-line
Modulation Images	4.4×10^{11}	Optical Disk, Off-line
Spherical Harmonic Coefficients	2.2×10^{11}	Optical Disk, Off-line
Power Spectra (daily, monthly, to date)	6.6×10^{11}	Optical Disk, On-line
Modes Frequency Tables	7.7×10^{7}	Optical Disk, On-line
Velocity Time Averages (1-hour)	9.2×10^{9}	Optical Disk, On-line
Magnetograms (1/hour)	9.2×10^{9}	Optical Disk, On-line
Other Products	2.0×10^{11}	As Appropriate
Total	**11.7×10^{12}**	

By employing a combination of tapes and optical disks, both medium costs and physical storage space can be minimized. The use of tapes and disks with the same storage capacity allows for user-transparent, compatible data structures. Data archived on tape can be easily "promoted" to optical disk, essentially with a one-to-one mapping. Those data stored on optical disk will be divided into two categories: on-line and off-line. On-line data will be stored in an optical disk "jukebox" system for ready access. Systems are currently available with capacities exceeding 200 Gbytes.

To enhance compatibility with other astronomical data systems, the GONG archives will probably employ the optical disk format which was developed by the Space Telescope Science Institute (STScI) for the Hubble Space Telescope Data Management Facility (DMF). This scheme stores data in the widely used FITS format using software which has already been developed by STScI (McGlynn *et al.*, 1985).

The STScI DMF software system includes a file server which acts as the system interface to the optical disk system. It uses magnetic disks as staging buffers between the optical disks and the rest of the computer system. Thus, the file server conceals the peculiarities of the optical disk system by making it appear like a conventional magnetic disk to the user.

Since all archived disks are cataloged in the file system, whether or not they are on-line, the file server also transparently handles the mounting of optical disks that are not currently in the jukebox. The STScI system will be adapted to treat the off-line rotary head tapes in a similarly transparent fashion. Thus, a user may access any data set in the archive through identical I/O calls, irrespective of the storage medium or its state of readiness. The file server also will handle other traditional tasks, such as copying from optical disk to tape and translating between standard data formats.

USER ACCESS

Most scientific users will access the GONG product data sets with the same software interface: the GONG Reduction and Analysis Software Package (GRASP). This package will provide a database management facility, including a catalog, browse, and display capability, together with a data piping system and attached analysis tools library. This will allow the user to quickly and efficiently browse, view, and select a block of data from a given product data set and submit this block to any of a number of standard analysis routines provided in the package. In addition, "hooks" will be provided to allow the user to develop custom analysis codes and easily integrate them into the library along with the standard tools.

Access to GONG data will be provided in a highly portable software environment suitable for several levels of distributed processing. The GRASP system will be developed within the Image Reduction and Analysis Facility (IRAF) processing environment. The user interface will not only be highly portable, but will have software compatibility with many other IRAF-based astronomical data analysis systems, such as those employed by STScI, ROSAT, and more than 150 other sites. Any of these systems could import GONG data and GRASP analysis tools virtually effortlessly.

The principle purpose of a portable user interface will be to support and encourage the use of networking and distributed processing. As presently conceived, NSO will provide direct access to the GONG data to scientific users on-site, as visitors, and remotely via network connections to the GONG Data Management and Analysis Center in Tucson. In addition, subsets of the GONG database will be available at one or more of the NSF supercomputer centers for local or remote access. NOAO is a member of the San Diego Supercomputer Center Consortium (SDSC) and is currently tied to it by a 56 Kbps satellite link. It is expected that SDSC will be the node for distribution of data subsets to the other supercomputer centers.

It should also be noted that the basic data will be maintained in a standard FITS format which can be exported on tape to any facility equipped to read FITS tapes. Furthermore, by maintaining optical disk format compatibility with the HST effort, data can be exported on this medium to any facility equipped to read HST disks. Thus, there can be broad access to the data quite independently of the GONG GRASP system, or its IRAF host environment . Users who might choose to use another astronomical or image processing software environment will free to do so.

HARDWARE REQUIREMENTS

The GONG DMAC will provide for the processing, archiving, and support facilities for a helioseismic database exceeding 35×10^{12} bits. The intake of data and some archiving responsibilities will be handled by a pre-processor of the SUN-4 class. Main pipeline processing will be carried out on a 10 MFLOP/s mini-supercomputer in the Alliant FX/8-8 or Multiflow 7/200 category. This main processor will also be responsible for affecting pipeline output archiving and scientific analyses. The database management will be supported by a dedicated catalog processor in the SUN-3 class.

Raw field data and corrected images will be generated and archived on inexpensive 8mm rotary-head data tape cassette systems, such as the 2 Gbyte Exabyte EXB-8200 system. The merged image stream and all subsequent generations of data products will be stored on 2 Gbyte 12 inch WORM optical disks. Those data which will be maintained on-line will be handled by optical disk jukeboxes, such as the 120 platter, 240 Gbyte, Cygnet system. In all, GONG will have about 8 tape cassette drives, 500 Mbyte magnetic disk drives with a total capacity of 10 Gbytes, four individual disk optical drives, two jukeboxes, as well as archival space for 1,750 8mm tapes (3.5 Tbytes) and 460 optical disks (0.9 Tbytes).

SUMMARY

The GONG project is actively developing a data management facility to support an international collaboration in helioseismology. This facility will become operational concurrently with the data acquisition network in late 1990. The system will include pipeline reduction, data archiving, and user data analysis systems. The pipeline reduction system will process some 1.2 Gbytes of incoming data every day. The raw and reduced data will be archived on a mixture of media including rotary-head tape cartridges and WORM optical disks. The total archive over the project three-year observing run will exceed 4 Tbytes. The user interface will be written in the IRAF environment and will include basic database management features, as well as standard analysis tools. This interface will be portable and extensive use of distributed processing and networks is anticipated. The system hardware will center on a "mini-supercomputer" capable of sustaining in excess of 7 MFLOP/s. It will be serviced by a dedicated catalog processor and an I/O preprocessor to maximize the throughput efficiency of the system.

REFERENCES

Beckers, J.M., Brown, T.M.: 1978, *Future Solar Optical Observations: Needs and Constraints*, (eds. G. Godoli, G. Noci, and A. Righini; Firence), 189.

Christensen-Dalsgaard, J., Gough, D., Toomre, J.: 1985, *Science* **229**, 923.

Deubner, F.L., Gough, D.O.,: 1984, *Ann. Rev. Astron. Astrophys.* **22**, 593.

Harvey, J.W., Kennedy, J.R., Leibacher, J.W.: 1987, *Sky & Telescope* **74**, 470.

Hill, F., Newkirk, Jr., G.,: 1985, *Solar Phys.* **95**, 201.

McGlynn, T., Russo, G., Richmond, A., Shames, P., Ochsenbein, F.: 1985, *Data Management Facility- Design Guide*, STScI, Baltimore, Maryland.

Design of the ST Data Management Facility Catalog

Stephen H. Lubow
Space Telescope Science Institute and Johns Hopkins University

Abstract

The Space Telescope (ST) Data Management Facility (DMF) is an interim data archive for ST data. The DMF catalog consists of summary information about ST observations and associated data. The catalog contains an observation index which is used to provide information about successfully proposed, uninitiated observations, as well as information about completed observations. In the latter case, the observation index includes information sufficient to locate the actual data on optical disk. The catalog will also contain information about calibration and engineering parameters. Potential catalog users include archival researchers, ST proposers and users, instrument scientists, and engineers.

Several design issues and decisions are discussed. These include the use of a commercial relational database system, the choice among existing relational systems, use of standard interfaces, the structure of the observation index, and the flow of data into the catalog.

I Introduction

The Data Management Facility (DMF) for the Hubble Space Telescope consists of three parts. They are the user interface (Richmond 1985), the file handler which manages the archived data that resides on optical disk or tape (McGlynn et al. 1985), and the online catalog that describes the archived data (Lubow 1986). Typically, the user interface will directly interact with the online catalog and will forward data requests to the file handler for the optical disk archive.

This paper describes the design of the catalog portion of the DMF.

II Requirements

The DMF is the main source of the archive data for user processing. The hardware/software system should provide reasonably short access times (seconds) in a multi-user environment (> 15 simultaneous users). Furthermore, it should permit rapid access for predefined queries and yet permit some *ad hoc* querying capabilities.

The DMF catalog needs to contain sufficient descriptive information to characterize archived data sets. Calibration data should also be provided online to permit researchers to check on the calibration parameters used for processing observations (Botway and Lezon 1987). Engineering data should also be made available online (Schneider 1987), since this data includes instrument performance information, pointing information, and other relevant data.

The DMF is designed to impose only a minimal load on other STScI computer systems. Requests to keep the catalog current should not burden the ST ground support system (SOGS) or the proposal entry system (PEPSI), which are both sources of data to the DMF catalog. The DMF will act as a server system, accepting data from SOGS and answering archive requests from SOGS and science users at both clustered systems and workstations.

III Choice of Database System

In the last decade, considerable progress has been made in the development of commercial relational database systems. Relational database systems provide great flexibility with *ad hoc* querying capabilities, unlike the earlier hierarchical and network systems (see Codd 1982). With relational systems, data is always available to the user in the form of simple tables which can be combined in complex ways at run-time. Unfortunately, this added power has historically carried an associated cost of slow querying speed (see Date 1986a). Recently, several commercial relational database products have emerged that provide good querying speed, although only a few provide good multiuser performance (see e.g., Cattell et al. 1985, Boral and DeWitt 1984). These systems provide powerful querying capabilities typically through a nonprocedural query language. That is, the query specifies what results are desired, rather than a procedural description of how to obtain the results (Date 1986b). The most popular query language is now SQL, which has become or is becoming an ANSI, ISO, and US government standard (ANSI 1986, Date 1987).

There now exist several powerful query optimization techniques that have greatly helped in individual query performance (Ullman 1982). Index and sometimes hash access methods are available for users to create fast access structures into the database. Query optimization is largely concerned with choosing the appropriate access structures and their order of application. Optimization is done automatically by the database system software, transparently to the user. Often optimization involves building intermediate tables and access structures. These optimization techniques are not exact and depend on certain assumptions and heuristics to obtain a good processing strategy. It is often hard for the system to estimate the size of intermediate tables in order to decide on the best query processing strategy. Such sizes can depend upon the exact data values used in the query qualifications. Statistical information of the distribution of data can help to overcome this problem (see Epstein 1987). We have found that very complex queries, involving more than 4 tables and several nestings of queries can be quite slow (many minutes to hours) on even the best relational systems.

To obtain good performance, database systems have essentially created their own operating system. Conventional, general purpose operating systems lack the characteristics desired for the database realm. For example, the UNIX file system is based on the assumption that most files are fairly small – a terrible assumption for databases. As a result, UNIX can suffer from file a lack of file localization with large files. (Some recent versions of UNIX have attempted to correct this problem through the use of cylinder groups (Quaterman *et al.* 1985).) Other issues such as database protection, recovery, and locking are not well-addressed by conventional operating systems.

Good multiuser (> 15 users) performance is available with only a few systems today. To perform well in a multiuser sense, means that the typical query response time scales linearly with the number of users. System overhead can introduce nonlinear behavior, such as thrashing, that limits the number of users. Some database systems suffer badly in a multiuser environment because there is insufficient sharing of code and data among users. Multiuser systems operate best in a server architecture where users run interface software on host machines that communicate with a server that is dedicated to database processing (Gray 1987). Host-server communications typically take place over a local area network. The host requests are multi-threaded on the server. To as large an extent as possible, the server will centralize the database code and data.

Frequently used queries can be precompiled and stored within the database system. Some systems, like SYBASE, permit the query processing plan, the optimized form of the query, to be saved. Others, like the IDM, only preparse such queries, but must recompute the processing plan with each invokation. Relational systems are also well-equipped to handle *ad hoc* queries, since the users' queries are not constrained by preexisting relationships among data items.

There is currently some debate about whether special purpose hardware systems (database machines) are really necessary to obtain good performance or whether special software systems, such as the special purpose operating systems, are sufficient. In principle, special purpose hardware can deliver better performance. However, general purpose hardware has the advantage of being lower in cost, more available, and being serviced by a large support organization. Special purpose hardware often will not be able to track the latest hardware advances, such as parallel processors and fast disk channels, as closely as general purpose hardware.

Our internal benchmarks, still in progress, indicate that some relational database products do meet our needs, both in terms of performance and querying capabilities. It appears that the IDM with a VAX host and probably the SYBASE system running on SUNs are both within our requirements. The IDM is based on special purpose hardware, while SYBASE runs on general purpose hardware.

Preliminary single user benchmarks indicate that SYBASE was about 40% faster than the IDM in terms of cpu time, and several times faster on I/O times, largely due to differences in disk hardware. We have devised a multiuser benchmark for the DMF, which we have run on the IDM system. We found that the IDM is able to handle 50 simultaneous users without excessive system overhead. A similar benchmark will be run soon on the SYBASE system, which claims to handle this number of users quite well.

IV Interfaces

All commercial database systems provide some tools for accessing a database system. Such tools typically include a report-writer for producing well formatted query output, a forms or menu based interface for displaying data, an interactive query facility, and a programming interface for obtaining results from database queries within a program. Although these tools are often very helpful and perhaps essential, their exact implementation is highly vendor dependent. Unfortunately, no general standards currently exist for these tools. As a result, such tools have the severe disadvantage of limiting the extent to which software can be shared between organizations that use different database systems. Large applications cannot be easily ported to other database systems without extensive reprogramming.

We have developed a method to cope with this programming interface issue. Currently, several types of vendor-supplied programming interfaces now exist. We have come to prefer an interface in which requests to the database system are forwarded from within an application as text strings through vendor supplied library calls. Unfortunately, these library calls are highly vendor-specific. To increase the portability of our database software, we have devised a programming interface, called the STDB interface (Lubow 1987a,b). The interface creates a layer of insulation from the vendor-specific calling sequences. The STDB interface also simplifies the calling sequence required by a programmer to access a database system. A similar idea has recently resulted in a several commercial products along these lines, such as DEC's DSRI.

The STDB interface is presently implemented on the IDM and SYBASE and supports compiled queries (stored commands), as well as text forms of queries.

V Catalog Structure

The catalog contains several groups of tables. The major groups include the calibration data tables, the engineering data tables, and the observation index tables. The calibration tables for the DMF are obtained from the calibration database system (CDBS) that resides on SOGS, the ST ground support system.

Engineering data should also be queryable on the DMF, although it is not feasible to keep all engineering data online in the database system, since more than 200 MB/yr is expected. This data is passed to the DMF from the SOGS system, where it is routinely used to produce instrument trend analysis, etc. On the DMF, the engineering data tables will contain only recent data from various instruments. Earlier engineering data can be brought into the DMF database from the optical disk archive, when needed. It is likely that *ad hoc* queries will need to be constructed to study unforeseen instrument problems.

The observation index tables provide a description of both successfully proposed, as well as completed observations. This information is organized by exposure, as listed in an ST observing proposal. Within the ST scheduling system of SOGS, proposed exposures become transformed in complicated ways for scheduling purposes. In fact, the data that is archived is actually labelled by its transformed id. We felt, however, that is was necessary for observers to view the database as organized along the lines of an ST proposal. The transformed state is not a natural organization for researchers to view the database. Schemes are currently being investigated to store within the SOGS database the mapping from the transformed id back into the original proposal exposure id. This mapping is generally many-to-one, since a single proposed exposure may be broken up for scheduling purposes.

Data in the observation index includes items such as target name and position, apertures, configuration, mode, spectral elements, PI, etc.

For completed exposures, the observation index also contains the information needed to locate the data in the archive. The observation index and optical disk archive are loosely coupled. File names in database tables are purely logical names that do not directly specify the actual disk directory locations of files in storage. The actual mapping from the logical file names used in the DMF database relations and to directory locations in the archive file system takes place in the file handler. This mapping provides a clean separation between the catalog and the data storage systems, and supports handling proprietary access to data.

VI Data Flow

The flow of data into the DMF required careful attention to the requirement that there be minimal impact on the other STScI computer systems. This requirement had to be balanced with the need to be able to maintain up to date information on the DMF.

a CALDB

CALDB data in the DMF is maintained by a combination of incremental and batch updating. Tables within CALDB were designed to each contain a tuple identifier field (tid). The tid is an integer that uniquely identifies tuples within each table. Incremental updates to the CALDB tables in the DMF are to be performed daily. To perform this updating, daily queries will be run on the DMF versions of the CALDB tables to obtain the maximum tuple number on each such table. These numbers are then used to run queries on the corresponding CALDB tables within SOGS to forward any new tuples back to the DMF. Additionally, we will occasionally perform a bulk copy of CALDB tables from the SOGS system into the DMF.

b Observation Index

The initial source of data for the observation index is the successful proposals. After proposal selection, a bulk copy of successful proposal data will occur from the proposal database (PEPSI) into the DMF. The DMF labels such data as being in the proposal stage only.

Within the SOGS system, completed data sets contain a header file that provides a description of the data. The DMF uses information from the header file to build and update in the observation index. Nearly all necessary data for the index can be obtained from the header files. This scheme imposes very minor overhead on the SOGS system.

The DMF then needs to verify that a proposed exposure exists for each completed one. If it does not, a warning will be issued to a log file that is checked daily. The DMF will do some additional checking, such as that the target position in the proposed and completed exposures are in good agreement.

VII Futures

Several issues may be investigated in the future. One such issue includes the use of a generalized interface between a user-interface and an application program that makes database requests. Such an interface would both increase application portability and allow a range of user interfaces to be used with minimal impact on the application code. The application then need not be concerned whether the user interface is based on menus, forms, voice, etc.

Remote access to this catalog would provide researchers with an opportunity to explore the contents of the ST archive from their home institutions. Another issue involves expanding the observation index to include more descriptive information about each exposure. For some instruments, object classifications might be stored in the DMF. There are many other issues yet to be explored, such as support of queries across different catalogs.

VIII Acknowledgements

I thank L. Hurt, T. McGlynn, A. Richmond, G. Russo, and P. Shames for many useful discussions.

REFERENCES

ANSI 1986, *Database Language SQL*, Document ANSI X3.135.

Botway, L. and Lezon, K. 1987, "Calibration Database Design", Space Telelscope Science Institute Internal Document.

Boral, H. and DeWitt, D.J. 1984, University of Wisconson-Madison, Dept. of Computer Science, Technical Report 155.

Cattell, R., Berger, R., Chambers, L. 1985, "Simple Database Benchmarks", Sun Microsystems Internal Report.

Codd, E.F. 1982, *CACM*,**25**, 2.

Date, C.J. 1986a, *Relational Database: Selected Writings*, Addison-Wesley, Reading, Mass., ch. 5, 6.

——— 1986b, *An Introduction To Database Systems*, vol. I, 4th edition, Addison-Wesley, Reading, Mass., ch. 6.

——— 1987, *A Guide To the Standard SQL*, Addison-Wesley, Reading, Mass.

Epstein, R. 1987, *Unix Review*, **5**, 5, p28.

Gray, J. 1987, *Unix Review*, **5**, 5, p42.

Lubow, S. 1986, "Design of the Data Management Facility Catalog", Space Telelscope Science Institute Internal Document.

——— 1987a, "Towards A General Database Programming Interface", Space Telelscope Science Institute Internal Document.

——— 1987b, "The STDB Interface", Space Telelscope Science Institute Internal Document.

McGlynn, T., Russo, G., Richmond, A., Shames, P., and Ochsenbein, F. 1985, "Data Management Facility", Space Telelscope Science Institute Internal Document.

Quarterman, J.S., Silberschatz, A., and Peterson, J.L. 1985, *ACM Comp Surv*, **17**, 4, p379.

Richmond, A. 1985, "Architecture of the User Interface", Space Telelscope/ECF Internal Document.

Schneider, W. 1987, "Science Instrument Performance Analysis Software", Space Telelscope Internal Document.

Ullman, J.D. 1982, *Principles of Database Systems*, 2nd edition, Computer Science Press, Rockville, MD, ch. 8.

THE LUND CATALOGUE OF OPEN CLUSTER DATA

Gösta Lyngå
Lund Observatory
Box 43, S-221 00 Lund, Sweden

ABSTRACT

The 5th edition of the Lund catalogue of open cluster data (Lyngå ,1987) is now being disseminated through the CDS and NASA data centres. A print-out is on display at this meeting.

Apart from updated information, this edition has a new approach to the presentation of the precision of the data. Weight classes have been assigned to the listed values of cluster distance, turn-off point and reddening. The methods of weight judgement follow certain set rules, so as to make the procedure as objective as possible. It is maintained that some kind of reliability classification is needed for the optimal use of any collection of heterogeneous astronomical data.

1. THE CATALOGUE CONTENT

In the Lund catalogue of open cluster data I have collected data from the literature on properties of open clusters as such rather than specific information about cluster stars. More than 1000 clusters in our galaxy are identified, i.e. they have at least an identification name and positional data. For most of them, angular diameters and some descriptional parameters are also available. Almost 500 clusters are studied well enough to have estimates of distance, age and effects of interstellar reddening. Detailed observations leading to determination of abundances are available for less than 100 clusters. Approximately the same number of open clusters have useful radial velocity determinations.

The catalogue consists of six files, all consisting of 7920 byte records and all written in ASCII code. It is designed for print-out or micro-fiche pages with 60 lines, each 132 characters. The contents of the files are as follows:

1. Introduction. A description of the format and content of the catalogue.

2. A formatted catalogue. This is mostly used for print-outs. Essential data for each cluster are collected in one column.

3. The reference list. An alphabetical list of references. Each record is 264 bytes. The record number equals the reference number in file 2 or file 6.

4. Alias names for the clusters. This turns out to be quite useful as some clusters are referred to by several different names in the literature.

5. A list of the clusters in longitude order.

6. The full data catalogue. This file is the comprehensive one. It is also the most suitable one for computer access; each cluster occupies 528 bytes, i.e. there are 15 clusters to a record.

2. ASPECTS OF INFORMATION USAGE

In this section it is assumed that the catalogue is accessed by a program and it is the properties of this program, which determine the procedures for utilising the data. Naturally, a print-out of the files can be used directly for a small amount of information.

a. *Reading data.* The most common use of the data set is retrieval of single data values for one cluster or another. Mostly, it is then important to have the literature reference for the listed data value. Each cluster can be found by its common denotation, by the IAU number or by a running number. The built-in tracers make this a fast procedure. The file of references is most conveniently handled by direct access.

b. *Editing data.* The purpose of editing may be to change a data value and the corresponding reference, to insert or delete a value, or to insert or delete a cluster entry. These tasks will in principle be carried out by the program working on the computer resident files. Only after detailed checking will the original files be altered. Whenever substantial changes have been made, i.e. every second year or so, new editions of the cluster catalogue are submitted to the CDS and NASA astronomical data centres.

c. *Plotting data.* By far the most common method of learning something new from data values in a catalogue is to plot various quantities. The cluster catalogue handling program will directly generate a plot file of two parameters; a third parameter may be used to control the symbol size (more about weighting methods below) and selection of data to be plotted can be made according to a fourth parameter.

d. *Analysis of the material.* The ultimate purpose of this data collection for clusters is to analyse and draw conclusions about the global properties of clusters and of the cluster system.

In addition to plotting, the program allows other statistical methods of analysis. The papers by Janes et al. (1988) and Lyngå and Palouš (1987) show a number of examples of actual use. The presence of a thick disk in the distribution of stellar clusters, a new determination of Oort's constant A and more accurate values of the galactic gradients of metallicity and age are some results that are based on the Lund catalogue of open cluster data.

3. HOMOGENEITY OF INFORMATION

Some of the best studied clusters have reliable data for all catalogue entries, while the majority of clusters have hardly any information apart from positions on the sky. Even when a certain entry is available, its precision can vary widely from cluster to cluster. This lack of homogeneity is the main problem in the present application, a problem that recurs in the use of most astronomical data collections. Unless this problem is dealt with in a purposeful manner, the data can obscure reality rather than help to penetrate it.

4. RELIABILITY CLASSIFICATION

In our efforts to study the characteristics of the cluster system of the Galaxy, we (Janes et al., 1988) have introduced a system of weight classification for determinations of distance, turn-off point and reddening. This has been applied to almost 500 clusters. The resulting weight classes are given in the new edition of the cluster catalogue. Their precise definition is given in the paper by Janes et al. Here, I will only give a brief description of the characteristics that have been used in generating the weight classes:

a. *Number of observed stars.* The weight assigned to a determination is proportional to the square root of the number of photographically observed stars. Photoelectric measurements get double weight.

b. *Scatter around the main sequence.* Whether the scatter of photometric data is due to imprecision in the observations or is caused by differential reddening, it is a source of error in the determination of distance or turn-off point. An exponential function is used, such that a dispersion of 0.3 in colour index gives a very low weight class and 0.05 gives a high weight class.

c. *Range of main sequence.* Distance and turn-off point determined for a cluster with a main sequence range of ten magnitudes get high weights, while zero weights are assigned if the observed main sequence is one magnitude or less.

d. Limiting absolute magnitude. If a cluster photometry only concerns stars more luminous than V=-4, the weight is very low, while the highest weight from this aspect is given to clusters observed down to V=+5 or fainter.

5. THE WEIGHTING APPROACH

Astronomical results based on the Lund catalogue of open cluster data are being published elsewhere by myself and other users. The purpose of the present paper is to emphasise the importance of weight assessment for heterogeneous data collections and to show how this can be achieved for the cluster catalogue. Similar methods should be considered for other astronomical data bases.

REFERENCES

Lyngå,G.: 1987, Catalogue of open cluster data, 5th edition, distributed as magnetic tape or microfiche by CDS or NASA data centres; CDS VII-092

Lyngå,G. and Palouš. J.: 1987, Astron. Astrophys. in press.

Janes, K.A., Tilley, C., Lyngå, G.: 1988, Astron. J., submitted

A SUPPORT ENVIRONMENT IN THE X-RAY ASTRONOMICAL DATA ANALYSIS: SOME PRELIMINARIES.

M.C. Maccarone [1], N.R. Robba [2], O. Di Rosa [2]

1 - Istituto di Fisica Cosmica e Applicazioni dell'Informatica, C.N.R., Palermo, Italy
2 - Istituto di Fisica, University of Palermo, Italy

Abstract.

General concepts of a relational database, with pointers to image and tables data, is here developed in the case of X-ray astronomical data analysis.

Various operators derived from the relational algebra have been implemented in the MIDAS system to build an easy and complete query language; this should be used together with the MIDAS features to define a decision support environment to help researcher along the complex X-data analysis procedure.

Introduction.

The large amount of data coming from X-ray astronomical observations show many and often complicated formats mainly depending on the instruments and satellite telemetry.

Nevertheless in the scientific analysis phase the collected data can be seen as belonging to the following three types: time series, energy spectra, images.

At the same time the analysis environment must provide researchers with access to data catalogues and archives, to user catalogue management, to calibration reference files, to algorithms and application modules, to output devices.

Then it would be very useful to include data, information and biblio-references in a relational database to be managed by a high level query language that supports user in the choice among various analysis methods and in the consistence control of the results.

Aim of this paper is to verify the possibility to build such a database for X-ray astronomical applications, within an already existing structure for data analysis, the MIDAS system.

In the following we will show how X-ray astronomical data are transformed from the raw tape format to the MIDAS ones and then in the database, what are the relational operators implemented to manage it, what is still to be done.

Some notes about the MIDAS system.

MIDAS system has been developed by the Image Processing Group of the European Southern Observatory in Munich to analyse astronomical data (mainly in optical field).

MIDAS consists of a control program (monitor) and an unrestricted set of application programs (both of general and specific use) built around a common data structure. The application programs interact via keywords which allow easy communication of data, parameters and global conditions.

MIDAS essentially supports three generic data structure: bulk data frames, tables and keywords; data in either format may be converted into any other.The efficient MIDAS table structure is moreover based on the data base relational model and on criteria of semantic integrity.

At the end, all user interaction and terminal output is logged by the monitor. This logfile may then be stored for a history of a MIDAS session and may also be used to replay the entire session later on (with or without updating via editing).

Data structure for X-ray astronomical application in MIDAS.

X-ray data are usually obtained as single photon counts, defined by position, arrival time and energy, also if, cause to the telemetry, they are often reduced by onboard computer before their transmission to the ground station (energy histograms, time profiles, images).

Let's suppose that a data analysis for the study of periodic pulsation or spectral features about a given source is required. In the analysis procedure it is possible to have many different paths to follow because of the various choices depending on formalizable and/or not-formalizable constraints. Therefore all these choices are not necessarily automatic and the human knowledge determines the path and controls the results consistence also via a comparison with previous analysis results and/or information coming from other sources (bibliographic references).

Then the database structure must take into account not only the possible different data types but also the information related to the satellite telemetry, to the on-board instruments, to the biblios.

In this sense the main part of the database has been identified in **G**eneral **I**nformation **T**ables **GITs** for the particular anlysis field, in which, for each observation, are defined identification, housekeeping data, observational modes, original telemetry, observational times, dead-time correction, energy resolution, exposure integral, auxiliary data, pointers to calibration and to scientific data files,...

The calibration and the scientific data have been stored in files **CALs** and **VALs**, whose type is depending on their intrinsic meaning: so, time series have been stored in MIDAS table format while the bulk data frame format was used for images and for energy spectra. (We remember that data in either format may be converted into any other in MIDAS). Each scientific data file has a descriptor containing user remarks or specific historical information.

Bibliographic references may reside in **BIBLIOs** tables related to the particular analysis to perform on the data (time variability or energy spectra analysis for example) and name of which is referred also in the **GITs**.

Data transfer from tape to MIDAS format is performed by using simple interactive procedures that fill each time a row of a **GIT** and its related files. The **GIT** rows may also be used as input parameter to get data from the physical archive and to transfer them in **VALs** or **CALs** files; in fact they do not necessarily reside in the working memory.

The relational database for MIDAS.

The more the representation of an application is rich and jointed in descriptive terms, the more the formulation of the procedure to obtain results is easy.

In the relational model the database is divided into two parts:

- the one constituted by the physical archives of the data,
- the other constituted by catalogues with the descriptions and the references to the data archives.

By enriching the catalogue content with accurate descriptions of the applicative context, it is possible to delegate the management system for a greater and greater number of controls and tasks, with consequent reduction of the complexity of the analysis programs.

Along this direction, data coming from X-ray astronomical observations have been transferred to MIDAS as described above.

To build the relational database management system devoted to X-ray astronomical applications, we thought it convenient to implement some basic operators of the relational algebra in MIDAS, in addition to the commands already existing in it to manage table data (creation, selection, sorting, merge, ...).

Some operators that usually work on homogeneous tables with same arity, have been here implemented in a more general sense, with the only limitation that labels of equal attributes (columns) in different tables must however be referred to data having the same meaning. In particular, union, difference and intersection of two tables may be summarized as follows (* indicates an undefined value):

UNION/TABLE

:A	:B
a	b
c	d

∪

:A	:C
a	e
g	f
a	e

→

:A	:B	:C
a	b	*
c	d	*
a	*	e
g	*	f

DIFFERENCE/TABLE

:A	:B	:C
a	b	c
b	d	e
f	a	a
g	h	h

−

:A	:B	:D
a	f	c
a	b	f
f	f	b
g	h	s

→

:A	:B
b	d
f	a

INTERSECTION/TABLE

:A	:B	:C
a	b	c
b	d	e
f	a	a
g	h	h

∩

:A	:B	:D
a	f	c
a	b	f
f	f	b
g	h	s

→

:A	:B
a	b
g	h

Projection operator allows selection of columns; it is not submitted to any limitation:

PROJECT/TABLE

:A	:B	:C
a	f	d
f	g	b
b	c	h
a	f	d

$\pi_{:B,:C}$ ⟹

:B	:C
f	d
g	b
c	h
f	d

Join operator has been implemented for the natural and teta options.

Natural-join may be applied only between tables with a common attribute:

NJOIN/TABLE

:A	:B
a	b
a	c
b	b
c	d

|X|

:B	:C
f	c
c	d
c	t
d	f

⟹

:A	:B	:C
a	c	d
a	c	t
c	d	f

Teta-join, that is join under a given condition between two attributes of the two tables, may be applied on tables without common attributes (different labels of columns):

TJOIN/TABLE

:A	:B
1	5
2	7
3	13

|X| :B<:C

:C	:D
10	2
2	2
15	3

⟹

:A	:B	:C	:D
1	5	10	2
1	5	15	3
2	7	10	2
2	7	15	2
3	13	15	3

Also the cartesian product, **PRODUCT/TABLE** command, is available only if the two tables have not common attributes; in this sense it corresponds to the teta-join without conditions on the values.

In addition to these, some very simple commands on tables have been implemented to a faster management and updating of the database, e.g.:

- **DELETE/ROW** to delete from a table one or more chosen rows,
- **COMPRESS/TABLE** to reduce the redundance of equal rows in a table.

Such a set of relational commands allows to create many and more complicated operations on tables and then to obtain a more direct interactive query-language for the user. For example, to a query like this:

"Perform the time series analysis of all observations (**GITtime** *) with pointing A when a particular detector B was active, conditions under which image analysis (* **GITimage** *and* **BIBLIOimage** *) was already performed"*

it is possible to answer via a combination of relational operators: selection on the **GITs**, intersection between **GITimage** and **BIBLIOimage**.

Final remarks.

We are now facing the definition of the effective query language and the **BIBLIOs** structure.

The use of this relational database in MIDAS, together with the general features of this system (keywords, modularity, logfile,...), is in progress to define a support environment to analyse X-ray astronomical data.

References.

- 'Data Analysis in Astronomy II', V. Di Gesù, L. Scarsi, P. Crane, J.H. Friedman, S.Levialdi, (Eds.), Plenum Press, 1986.
- C.C. Yang, 'Relational Databases', Prentice-Hall, 1986.
- 'MIDAS User's Guide', Image Processing Group, ESO Garching, 1987.

A DATABASE OF OBJECTS TO B~22 FROM COSMOS SCANS OF WIDE-ANGLE SCHMIDT PHOTOGRAPHS

H.T. MacGillivray,[1] R.J. Dodd[2] and S.M. Beard[1]

1. Royal Observatory, Blackford Hill, Edinburgh, EH9 3HJ, U.K.
2. Carter Observatory, P.O. Box 2909, Wellington, New Zealand.

SUMMARY

The COSMOS high-speed, automated, photographic-plate measuring machine is being used to produce a database of objects down to B=22 in the Southern sky from scans on glass copies of the SERC IIIaJ Southern sky survey. The database contains for each object its celestial coordinates, magnitude, size, shape, position angle and star/galaxy classification. The data are already being used for a number of large-scale astronomical projects. Plans for the future include digitisation of second epoch survey plates taken with both the UK and Palomar Schmidt Telescopes in order to allow a greater scope of investigations (e.g. utilising the colour and proper motion information) to be carried out.

1. INTRODUCTION

Photographic plates taken with modern wide-angle Schmidt cameras (such as the UK, ESO and Palomar Schmidts) still represent the most efficient means of recording information over large areas of sky in optical passbands. The wealth of information is enormous. On a single plate covering some 40 square degrees of sky, at 10 microns resolution (~0.67 seconds of arc), there are some 10^9 elements of information ("pixels"). Some 2000 such plates are required to cover the whole sky in a single passband (~2 x 10^{12} pixels), and generally astronomers require to work in more than one passband and with plates separated by time intervals. Already, the Palomar Schmidt telescope is embarking on a second epoch survey of the Northern sky in 3 passbands (J, R and I) thereby complementing the original 2 passband Northern sky survey, and the UK Schmidt (UKST) likewise is about to embark on a second epoch survey of the Southern sky. The resulting information content of the Northern and Southern sky surveys, when combined, is quite staggering!

The primary disadvantage of optical surveys of the sky on photographic emulsions is that the information is not readily available for quantitative study. However, by means of high-speed photographic-plate measuring machines it is possible to extract the information contained thereon and to make that information available for meaningful statistical analysis.

At the Royal Observatory Edinburgh, we are currently engaged in an undertaking to digitise the whole of the Southern sky at a relatively high resolution (<~ 1 arcsec) from scans made with the COSMOS automatic, high-speed, plate-measuring machine on glass copies of the Southern sky IIIaJ atlas. The data are already being used for a wide variety of large-scale astronomical projects. These scans are nearly complete and the result will be (when finalised) a database of objects down to B~22 over the whole of the Southern sky.

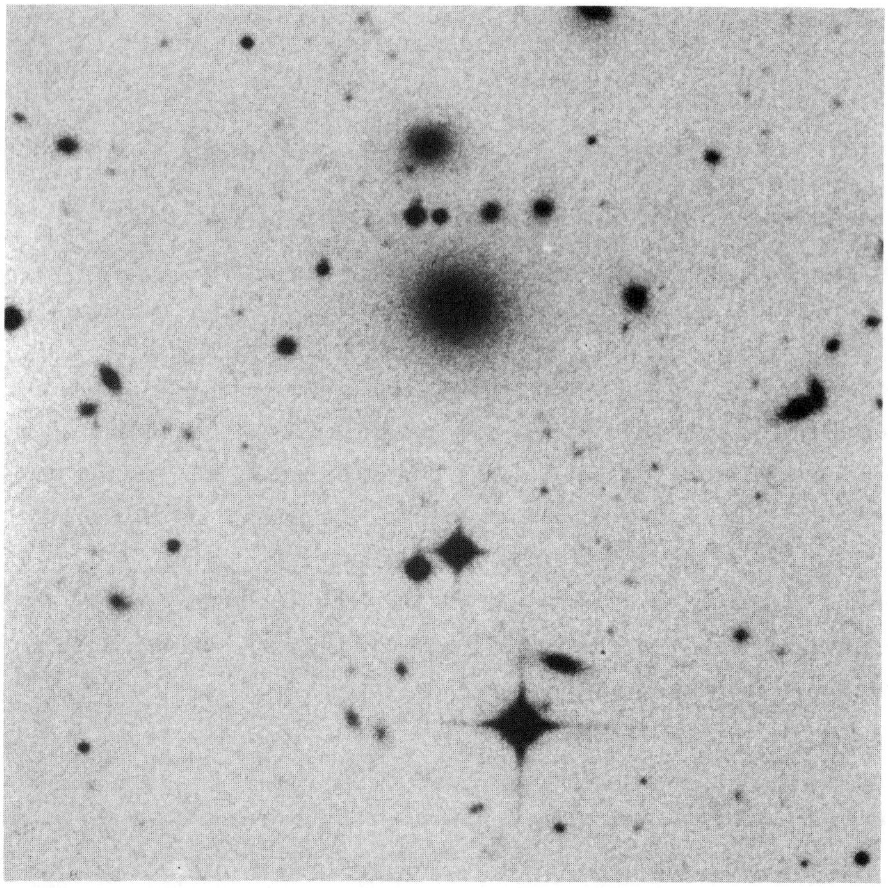

Figure 1: *Photograph showing a small region of sky approximately 5' x 5' from the southern sky survey field number 156.*

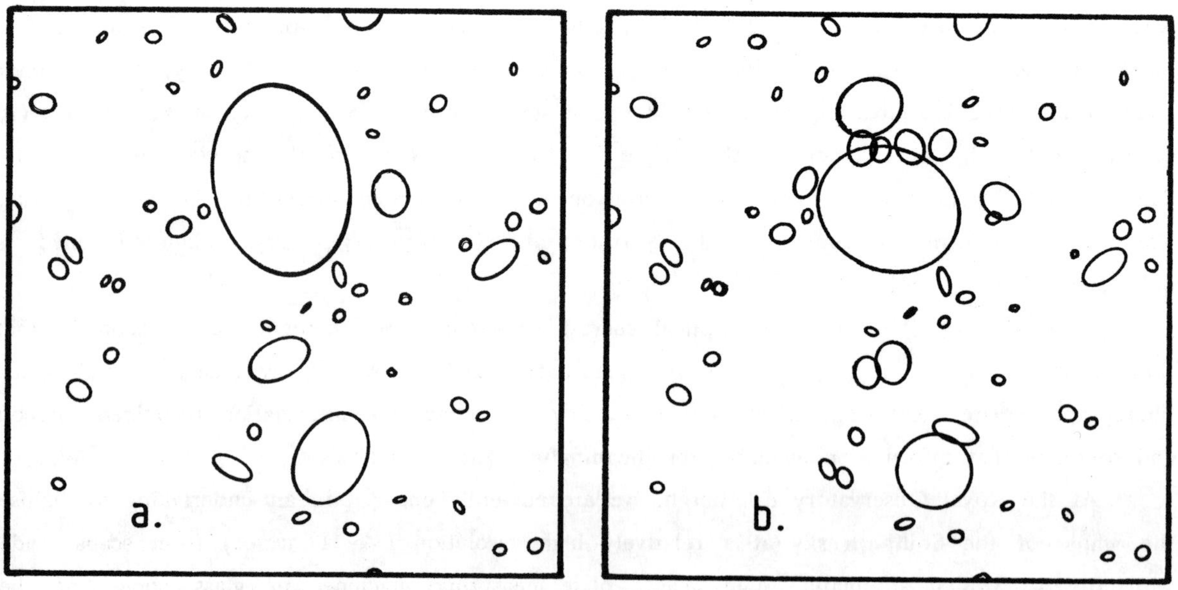

Figure 2: *The parameters obtained for the images detected down to B~22 from figure 1 after applying the COSMOS software Figure (a) shows the data without applying the "crowded-field" software, while figure (b) shows the results after applying it.*

2. THE COSMOS SYSTEM

Details of the COSMOS system have been given elsewhere (MacGillivray and Stobie 1984). Briefly, COSMOS is a photographic-plate digitising device, converting the photographic density at each point (or pixel) into a 14-bit transmission value (i.e. 16384 grey levels). The machine is operated principally in two modes:- a "Mapping Mode" (MM) where the transmission value at every pixel position on the plate is output to disk and recorded, and "Thresholded Mapping Mode" (TM) where only pixels above a certain predetermined fixed isophotal threshold level above the local sky background are recorded on disk. In both cases, off-line software exists to interpolate the sky background intensity value at each pixel position, thus allowing sky-background variations over the plate to be followed. Transmission to intensity conversion is carried out by means of spots or wedges of known relative intensities which are recorded on the plate during the exposure.

TM mode is the more frequently used of the two and represents a compression by a factor of ~10–20 over the data storage requirements of MM mode. At a resolution of 16 microns (i.e. pixel size), the scan of a complete Schmidt plate in MM takes 4 hours on COSMOS. In TM mode, however, a "prescan" of the plate at lower resolution is also performed in order to determine the sky background over the plate for image detection, and hence in this mode a complete scan of a Schmidt plate is carried out in ~6 hours. At this rate the latter mode COSMOS scans typically 3 Schmidt telescope plates each day.

Software exists to take data in either MM or TM modes in order to detect the images above threshold and to calculate a set of basic parameters such as image centroid, instrumental magnitude, profile, shape, orientation and celestial coordinates. Recently, the software has been greatly improved, enabling it to work in crowded regions of sky (where images may be "blended" i.e. their pixels overlap above threshold level). The improved software enables peaks in the pixel data to be detected and for the separate images to be reconstituted from the unambiguous profile information (see figures 1 and 2). An advantage of our approach is that we make no assumptions concerning for example a "standard" profile, and hence our software is quite generally applicable and can be applied to situations involving the overlapping of different types of object.

The computer configuration on COSMOS is shown in figure 3. The machine is controlled by means of a PDP 11/24 computer which sends data directly (by means of a high-speed data transfer link) to magnetic disks on the VAX 11/750 computer, which is the dedicated data-processing computer. The VAX computer has two optical disk drives (OSI Laserdrive 1200) which are used for large-capacity (1 Gbyte per side) data archival purposes. The possibility of using video cassettes for the purpose of large-capacity data-storage is also currently being explored.

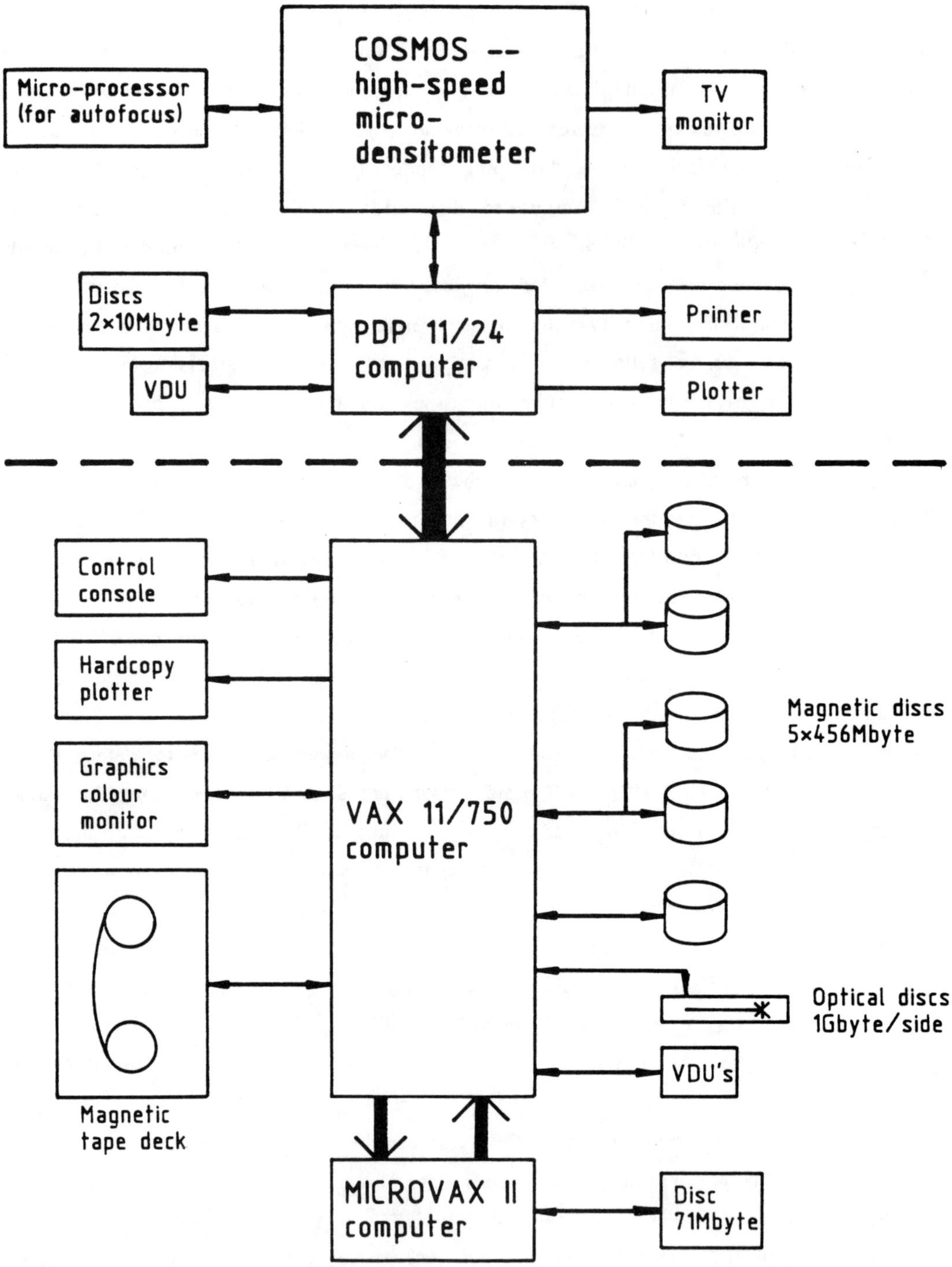

Figure 3: *The computer system on COSMOS.*

3. OBSERVATIONAL MATERIAL AND PROGRESS TO DATE

Our observational material consists of scans made with COSMOS in TM mode with a 16 micron pixel size on glass copy plates of the SERC IIIaJ Southern sky survey. After processing by the software described above, the final data for each plate consists of a catalogue of typically somewhere in excess of 150000 objects detected (and in many cases nearer some millions) down to B~22. Fainter objects are also present, but we are limiting the database of objects to B=22 for the sake of completeness. The raw TM scans and the final image catalogues are stored on optical disk for convenient access.

Although a catalogue of objects is easily produced by the COSMOS system for each plate scanned, some further processing of the data is necessary before it is accepted for the database:- "field effects" (i.e. image parameter shifts due to emulsion sensitivity variations, machine saturation, etc) are corrected for; contaminants such as satellite trails, "ghost" images of bright stars, etc. are removed; spiral galaxies containing HII regions etc. which are "deblended" by the software are reconstituted; separation of the images into stars and galaxies is carried out; etc. It is also emphasised that all of these processes are performed completely automatically, with no manual intervention, using the software we have developed.

Finally, calibration of the data is another very important consideration. We know from our previous work (e.g. MacGillivray and Stobie 1984) that magnitudes for galaxies from COSMOS follow a 45° slope (over a wide range of magnitudes - 10<B<22) when compared with other galaxy photometry (CCD, photoelectric and photographic). Providing magnitudes for galaxies from COSMOS is, therefore, very straightforward with only an accurate zero point being required. For stars, however, the relationship is more complicated and is not linear. CCD or photoelectric sequence stars are required to calibrate the COSMOS stellar photometry, and some deep sequences have already been obtained for this purpose. On the other hand, however, we have seen that for both stars and galaxies there is a very tight relationship between the magnitudes from overlapping plates, and hence calibration in every field is not required. Sequences can be carried from field to field in a leap-frog manner.

This project to provide a database of objects has to compete for time on COSMOS alongside the many other major projects being undertaken. Nonetheless significant progress has already been made and approximately 50% of the Southern sky fields (i.e. 450) have been scanned at the time of this meeting. The region having the most complete coverage (>95%) is the region with declinations South of −15 degrees and Galactic latitudes South of −10 degrees. It is estimated that scans for the whole of the Southern sky will be completed by the middle of 1988.

In essence, two databases of the sky are being produced. The first contains the basic parametrised (i.e. positions, magnitudes, etc.) information for each object and will be the database which will be used most frequently by astronomers. However, with the raw compressed TM mode data available (at very high storage density) it is also possible to reconstitute any part of the sky using software we have available and to display it as a 2-D image on, for example, a monitor screen in colour-coded form. It is felt that this second lower-level database could also potentially have valuable use for astronomers.

4. PROJECTS ALREADY BEING CARRIED OUT

The ultimate intention of the scans we have described above using COSMOS is to produce a large database (estimated to contain when complete some 100–200 millions) of objects in the Southern sky. However, already several projects are underway making use of the material. The most important of these are listed below.

a) Study of Galactic structure from counts of stars as a function of latitude and longitude.

b) Structure of the Magellanic Clouds and Magellanic Stream (e.g. Albers et al 1987a, 1987b)

c) Identification and analysis of Magellanic Cloud star clusters (Bhatia and MacGillivray 1987a, 1987b)

d) A search for halo star clusters and Local group dwarf galaxies (MacGillivray et al 1987).

e) Identification of point sources from the IRAS catalogue (Wolstencroft et al 1986; Leggett et al 1987).

f) The Edinburgh/ Durham Southern galaxy survey (Collins et al 1987).

g) A catalogue of rich galaxy clusters.

In addition to the major survey described above to scan the Southern sky in the IIIaJ passband, essentially for large-scale statistical analyses, COSMOS is also engaged in many other surveys from other types of plates. These other surveys involve scans of time-separated plates or plates in several passbands, but are generally confined at present to a smaller number of fields. The more important of these are:-

a) A search for brown dwarfs (Hawkins 1986b).

b) A search for variable objects (Hawkins 1986a).

c) The UKST-COSMOS intermediate Galactic latitude survey (Miller and Mitchell 1987).

d) The Edinburgh-Cape faint blue object survey (Stobie et al 1987).

e) Population structure of the SMC (Hatzidimitriou, in preparation).

In all of these cases, the data are being archived on optical disk so that the catalogues can eventually be made available to all interested astronomers.

5. FUTURE PLANS

Although the database of objects described above will be able to fulfill a large number of applications, the situation is never static. Improvements in COSMOS are being planned (in the areas of accuracy, speed and resolution) which, it is hoped, will be able to exploit fully the information content of photographic emulsions, in particular with regard to the availability of second epoch survey plates.

The second epoch Southern sky survey will shortly be underway with the UKST. The intention is to do this survey in the R passband rather than in J. Thus the second epoch survey will open the way for much further investigation than possible with the first, particularly with regard to colour as well as proper motion information. It is our intention to scan the plates for the second epoch Southern sky survey also, and to combine the data with the existing first epoch data. In this way, we hope to produce eventually a database containing not only the B magnitudes for objects down to B=22, but also the B–R colour and "displacement" in position. This displacement will be determined relative to the fixed reference frame of background galaxies, as opposed to a constantly changing reference frame defined by stars. Looking further ahead, a similar project for the second epoch Palomar sky survey would also be of extreme value.

The wealth of data available from the sky surveys, in digital form, will be enormous and will provide a powerful database for astronomers, enabling a wide range of astronomical investigations to be carried out.

REFERENCES

Albers, H., MacGillivray, H.T., Beard, S.M. and Chromey, F.R., 1987. Astron. Astrophys., 182, L8.

Albers, H., MacGillivray, H.T., Beard, S.M. and Chromey, F.R., 1987. in "Stellar Evolution and Dynamics in the Outer Halo of the Galaxy", In Press.

Bhatia, R.K. and MacGillivray, H.T., 1987. Astron. J., Submitted.

Bhatia, R.K. and MacGillivray, H.T., 1987b. in "Astronomy from Large Databases: Scientific Objectives and Methodological Approaches", These proceedings.

Collins, C.A., Heydon-Dumbleton, N., MacGillivray, H.T. and Shanks, T., 1987. in IAU Symposium No. 130, "Evolution of Large Scale Structures in the Universe", In press.

Hawkins, M.R.S.H., 1986a. Mon. Not. R. astr. Soc., 219, 417.

Hawkins, M.R.S.H., 1986b. Mon. Not. R. astr. Soc., 223, 845.

Leggett, S.K., Clowes, R.G., Wang, G., MacGillivray, H.T., Parker, Q.A., Savage, A. and Wolstencroft, R.D., 1987. in "Astronomy from Large Databases: Scientific Objectives and Methodological Approaches", These proceedings.

MacGillivray, H.T., Bhatia, R.K., Dodd, R.J. and Beard, S.M., 1987. in "Stellar Evolution and Dynamics in the Outer Halo of the Galaxy", In press.

MacGillivray, H.T. and Stobie, 1984. Vistas astron., 27, 433.

Miller, L. and Mitchell, P.S., 1987. In preparation.

Stobie, R.S., Morgan, D.H., Bhatia, R.K., Kilkenny, D. and O'Donoghue, D., 1987. in "Faint blue stars", IAU Colloquim No. 95, In press.

Wolstencroft, R.D., Savage, A., Clowes, R.G., MacGillivray, H.T., Leggett, S.K., and Kalafi, M., 1986. Mon. Not. R. astr. Soc., 223, 279.

PLANS FOR THE EXTREME ULTRAVIOLET EXPLORER DATA BASE

Herman L. Marshall, Carl A. Dobson, Roger F. Malina, and Stuart Bowyer
Space Sciences Laboratory, University of California, Berkeley, CA 94720, U.S.A.

ABSTRACT

We present an approach for storage and fast access to data that will be obtained by the *Extreme Ultraviolet Explorer (EUVE),* a satellite payload scheduled for launch in 1991. The EUVE telescopes will be operated remotely from the EUVE Science Operation Center (SOC) located at the University of California, Berkeley. The EUVE science payload consists of three scanning telescopes carrying out an all-sky survey in the 80-800Å spectral region and a Deep Survey/Spectrometer telescope performing a deep survey in the 80-250Å spectral region. Guest Observers will remotely access the EUVE spectrometer database located at the SOC.

The EUVE database will consist of $\sim 2 \times 10^{10}$ bytes of information in a very compact form, very similar to the raw telemetry data. A history file will be built concurrently, giving telescope parameters, command history, attitude summaries, engineering summaries, anomalous events, ephemeris summaries, etc. The history file and the database are accessed concurrently when source detection algorithms are run. Sources are expected to be rare, $\sim 10^{-(4-5)}$ of the database, so special procedures are established and a separate database is maintained for data near expected, cataloged sources and those detected during the survey. The connection to Guest Observers is expected to make use of established astronomy data networks and standardized data analysis packages.

1. Introduction

The Extreme Ultraviolet Explorer (EUVE) is designed to study the astronomical sources in the extreme ultraviolet (EUV: 80 to 800Å). Scheduled for launch in late 1990, it will carry out the first all-sky survey at these wavelengths. After six months, the EUVE will be operated entirely on a Guest Observer basis for spectroscopy. A short description follows. (For more details, see Malina *et al.*[1], and Bowyer[2]).

1.1. Instrumentation

EUVE will carry three scanning telescopes for use during the all-sky survey phase of the mission (see Figure 1). Two are used at short wavelengths (80Å $< \lambda <$ 300Å), and are of Wolter Schwarzschild Type I design giving a 5° diameter field of view, while the third is designed for longer wavelengths (300Å $< \lambda <$ 800Å) and is a Wolter Schwarzschild Type II design. A fourth telescope is oriented along the spacecraft spin axis. It has four detectors; one for the direct image and three for the different channels of the spectrometer. The mirror is a Wolter Schwarzschild Type II design.

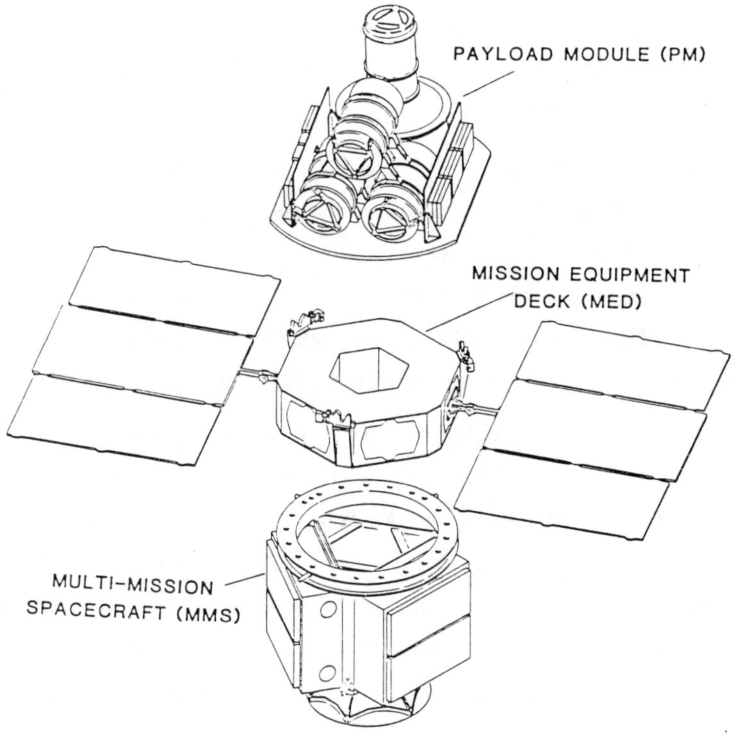

Figure 1. A schematic of the EUVE flight design. The three stacked telescopes are the scanners and the fourth is the Deep Survey/Spectrometer. The latter is oriented along the spacecraft spin axis.

All seven detectors are identical, consisting of a microchannel plate stack and a wedge and strip anode (see Siegmund *et al.*[3] for more details). Basically, the detector design gives 80μ resolution elements, resulting in 3′ angular resolution for the Type I scanners and 1′ for the Type II scanner. The rest of this paper will be primarily concerned with the Type I scanner design and the data these scanners would produce. There are ''stimulation pins'' attached to the back of the anode that are used while in orbit to calibrate the microchannel plate gains and to detect electronic drifts. There are four bandpasses planned for the all-sky survey, centered at 100, 200, 550 and 650Å. The detail of the bandpasses and their effective response when combined with the telescope is not important for the design of the software system as described here.

1.2. Mission design

During the first six months of the mission, the satellite will spin at about three revolutions per orbit, with the spin axis oriented toward the sun. The survey thus achieved will eventually cover the entire sky. A deep survey is performed simultaneously by the Deep Survey/Spectrometer telescope, pointed away from the sun and toward the earth's shadow at all times in this phase.

Data are taken primarily during nighttime in order to minimize background due to geocoronal emission lines of He. The spin rate allows a full survey of the sky in the nighttime portion of the EUVE orbit. Because the spin axis is locked on the sun, the survey detector passes through the ecliptic pole in every orbit. Sources in the ecliptic plane are scanned for five days or less, getting $\lesssim 25$ s exposure per orbit. The average exposure for a source in the all-sky survey will be ~1000 s, while the sources in the ecliptic get less exposure.

2. The End-To-End System

The End-To-End System (EES) is the software system that is used on the ground to reduce EUVE data during flight operations. It is developed in parallel to the instrument development so that data may be simulated for the data analysis system using highly accurate instrument models. The instrument models are often good enough to aid the process of instrument design and development, checkout, and calibration. Because of the high degree of overlap between the different phases, many software (S/W) modules are utilized in more than one phase of the EUVE project, thereby reducing the coding workload. Furthermore, requiring that the same S/W modules be used in development and during flight insures that this common S/W will perform as required. Using the EES, we employ source data (magnitudes, positions, spectra, etc.) to generate photons and add background, trace the photons through the optics, apply a model of the detector and associated electronics, produce and strip telemetry data, and analyze for sources, fluxes, positions, and spectral features.

Prior to launch, most of the EES software is used to generate data to test various elements of the science payload hardware, the other EES modules, and the portion of the EUVE ground system that resides at the Space Sciences Laboratory. Figure 2 shows the structure of the EES during development. Test data are generated by the Environment Model element, folded through the Satellite Model, and then analyzed using the S/W which performs the Science Operations and the Science Data Analysis functions. Tests of each S/W block are performed using these data.

Also shown in Figure 2 is a sample test. The skymaps, catalogs, and spectra produced by the Science Data Analysis block are compared to the original information which was used to generate the model sky. Algorithms in this S/W block are modified by the software development team until optimum performance is achieved. Tests of the EUVE Science Payload instrument software, and of certain key scientific features of the S/P hardware design, are also performed at this time. See Marshall *et al.*[4] for examples of EES usage and Chakrabarti *et al.*[5] for discussion of the handling of spectroscopic data and the guest observer interface.

3. Database Design

The instruments were designed to accommodate a telemetry rate of 16,000 bits per second. Because photon counting detectors are employed and the sky is scanned, each photon needs to be time-tagged. Thus the maximum expected count rate is about 450 photons per second, giving 5×10^9 photons per year during night-time data acquisition. It is impractical to store more than a few bytes of data per photon in the database, so it is expected that the archived data will be very nearly in the same form as the original telemetry, giving about 20 Gbyte for the raw data. It is expected, however that EUV sources will be rare (totalling 100 to 10,000) so that the detailed data may be stored for each individual source. Assuming that there will be about 2000 expected sources and 200 photons per source (background should be quite low) in each of 4 bandpasses, then a file may be created allocating 30 bytes per photon which may then be kept on line, will never exceed 50 Mbytes, and can be continuously updated. The general plan for this system is outlined below.

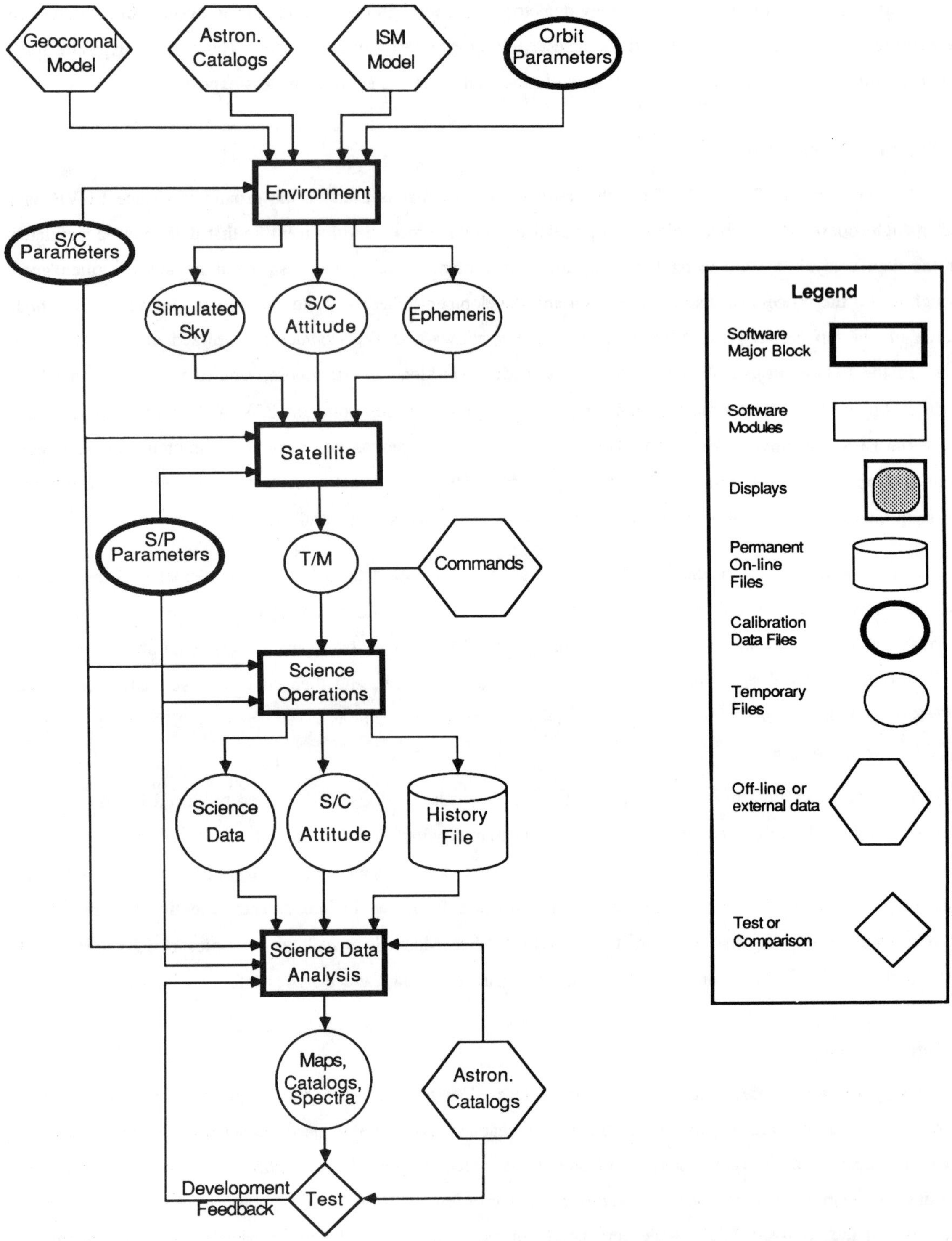

Figure 2. Overview of the End-To-End System (EES). The major software blocks consist of a number of software modules to perform functions as described in the text.

3.1. Contents

There are seven important files in this system. The archive file records all the data in a very compact form and should not exceed 20 Gbyte per year. A history file is generated from the original data which summarizes the status of the instrument each orbit and provides an index which may be perused for detailed interactive data selection. This file contains general orientation information, instrument status words such as temperatures, voltages, dead time counters and pulse height distributions, and other information that changes slowly, such as the sun angle, positions of the moon and planets, etc. Each record is less than 1500 bytes, giving a file of 5 Mbyte each year. Intermediate results can also be kept on-line if organized properly. There will be four such files consisting of detailed photon data for each cataloged source, detailed exposure information for each source, an accumulation of photons in 6′bins on the sky, and a similar accumulation of effective exposure in 6′ bins on the sky. Each of these files will be approximately 50 Mbyte in size. Data in the photon file will consist of detector (x,y), right ascension and declination, time, detector number, and a quality flag for each photon from a fixed region about the (known or preliminary) source position. The photon data for each source may be stored in separate files which we call ''pigeon holes.'' Data in the source exposure file will be used to generate detailed detector crossings for each source so that exposures, profiles, light curves, and fluxes may be calculated. Finally, the results are stored in a catalog which should be no larger than 0.2 Mbyte.

3.2. Data Base Access

Figures 3 and 4 show a flow chart for the subsection of the End-to-End (EES) System dealing with the analysis of scanner and deep survey data. Figure 3 shows the ''production'' portion of the analysis. The raw telemetry is archived and split in the Science Operations software block which also produces the history file. Data from an individual orbit (or satellite playback) are expanded into a usable form so that the permanent data files may be updated. At this point, any subsequent analyses may take place each orbit or each week, as desired.

An important aspect of this analysis design is that the data for each source are collected with the guidance of a preliminary catalog. Objects that are expected to be EUV sources on the basis of data from other instruments and catalogs are placed in the preliminary catalog. The new source positions resulting from the detection analysis are merged with this list for extraction of pigeon hole data. Thus, the list will expand as the survey results are analyzed. The extraction region is made large enough so that only rough positions are required.

The pigeon hole database is used for the accurate measurement of position, fluxes, extents and other source parameters. Sophisticated, computationally expensive techniques may be used here because of the small number of expected sources and photons. The techniques applied must account for irregular exposure caused by missing data (the detectors are turned off during passage through the South Atlantic Anomaly), varying image shape and centroid as a function of detector coordinates, varying background, asymmetric point spread functions, and the complicated dead-time corrections due to an unusual electronic ''round-robin'' algorithm employed for allocating data to the telemetry stream. It is expected that the maximum likelihood method (see Cruddace, Hasinger and Schmitt, these proceedings) will be used for determining source strengths and positions but that a separate test will be required to check the agreement between the data and the model. The latter will give an indication of source variability and/or extent.

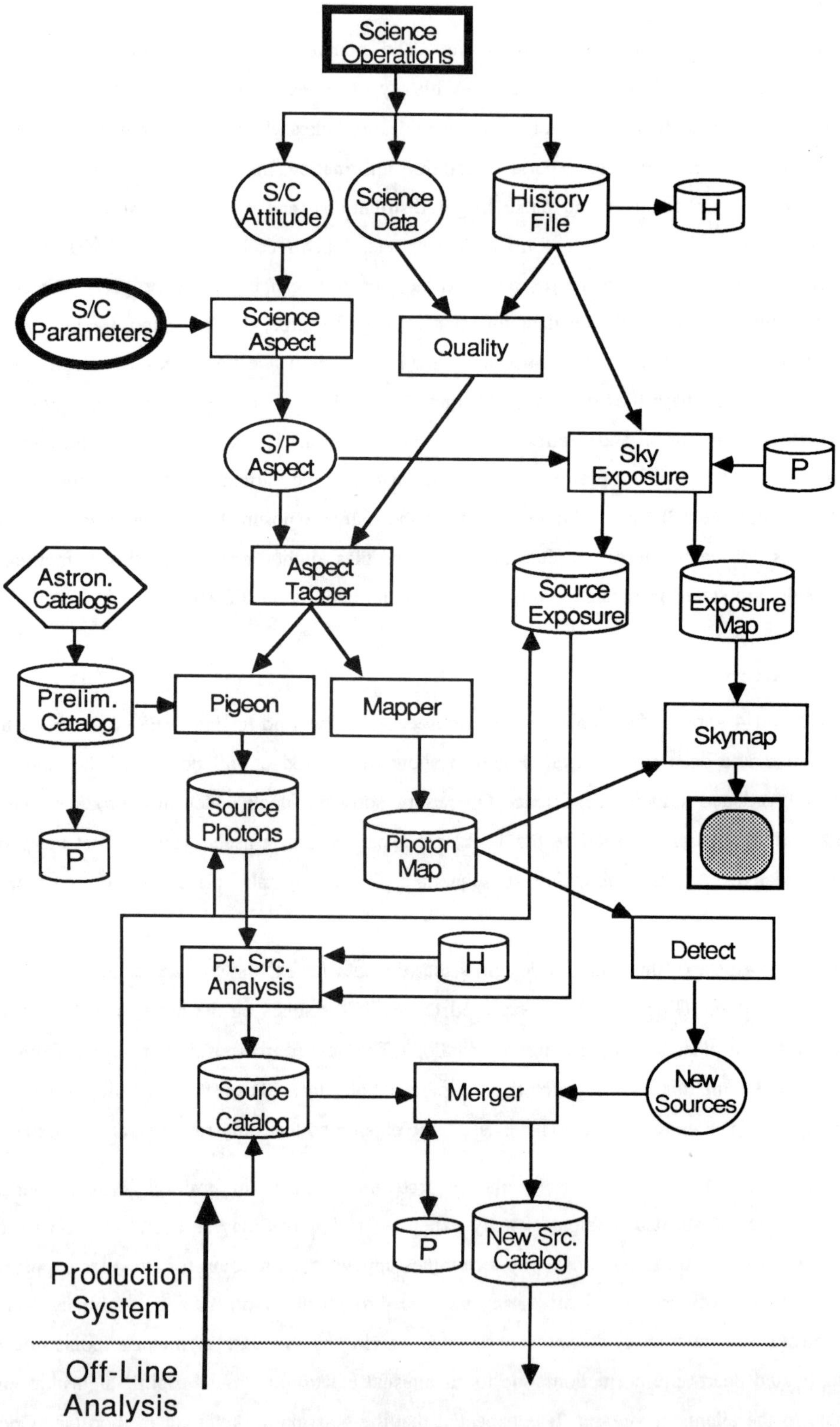

Figure 3. Flow of the Science Data Analysis section of the EES. Only the flow of the scanner data is shown. Treatment of the deep survey and spectrometer data is essentially similar. The symbols are as in Figure 2.

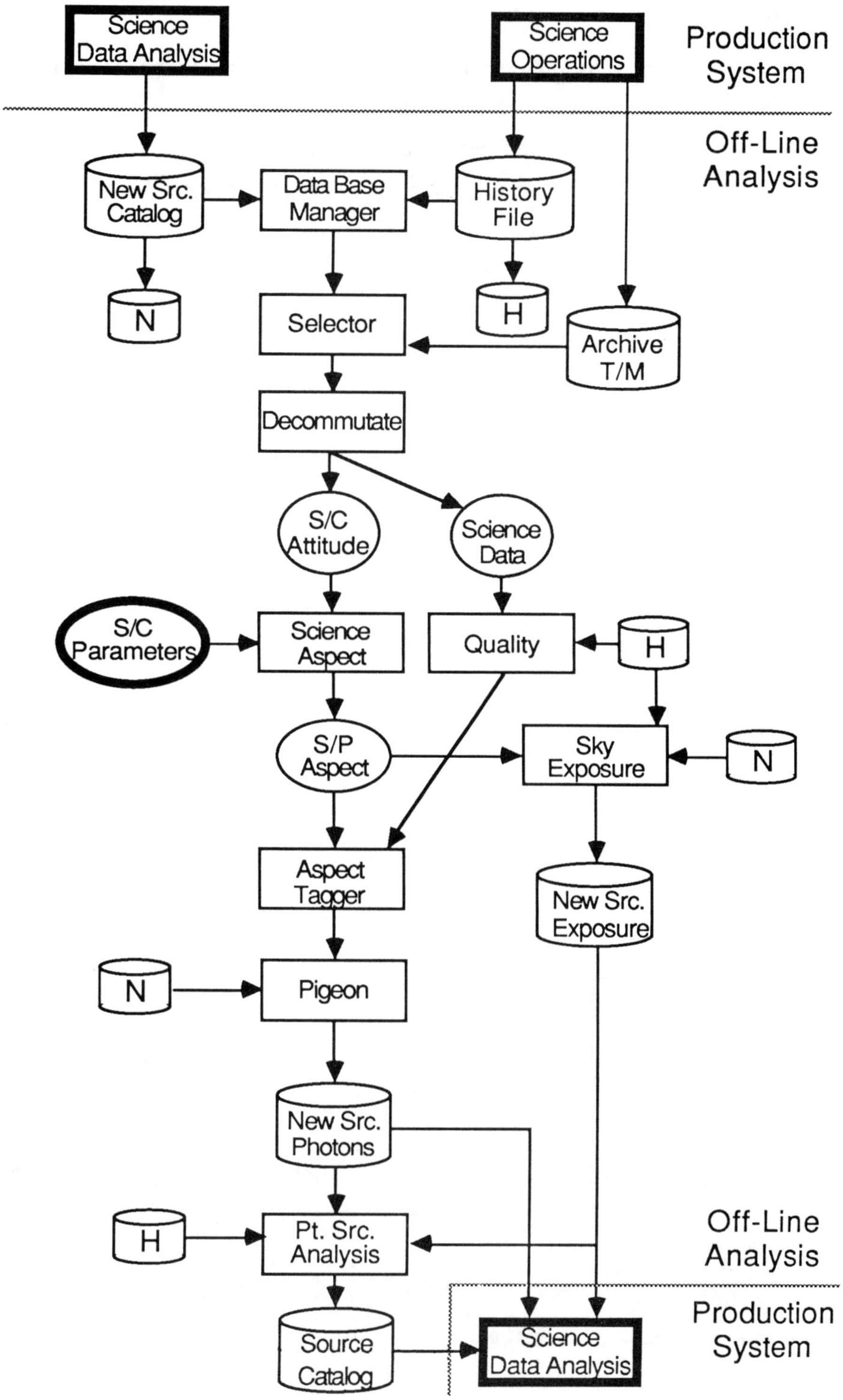

Figure 4. Flow of the off-line data analysis. Data are supplied by EES major software blocks and interactively examined for guiding the remaining processing. It is assumed that an operator controls the behavior of the data base manager software. Results are supplied to a specialized program, the data selector, and then piped to the remaining analysis which relies on software already developed for the rest of the EES.

3.3. Pipeline Activity *vs.* Off-line Analysis

Although the new sources are added to the list of objects for future extraction by the production processing, the detailed photon data for these sources from prior orbits (such as those that were used to discover the source) must be added to the permanent on-line files so that future analyses have all available data. Figure 4 shows the handling of the archived data is performed so that this goal may be accomplished.

An interactive database manager is used to read the history file so that the appropriate portions of the archived telemetry may be selected. To improve the interactive aspects of the database manager, this program does not actually cull the archived data. The program that decommutates data in the Science Operations software block is then used to split the data so that standard production programs may then be used to create files compatible with the production system. The behavior of these programs is governed by the catalog of new sources in this case and the results are then made available to the production system so that future analysis can account for these older data.

ACKNOWLEDGEMENTS

We would like to thank Garrett Jernigan and Dan Quinlan for discussions and Aileen Corelli for help in preparing the manuscript. This work was supported under NASA contracts NAS5-301xx.

REFERENCES

1. R.F. Malina, S. Bowyer, M. Lampton, D. Finley, F. Paresce, G. Penegor, and H. Heetderks, "The Extreme Ultraviolet Explorer", *Opt. Eng.*, **21**, 764, 1982.

2. S. Bowyer, "The Extreme Ultraviolet Explorer", *Adv. Space Res*, **2**, 157, 1983.

3. O.H.W. Siegmund, R.F. Malina, K. Coburn, and D. Werthimer, "Microchannel Plate EUV Detectors for the Extreme Ultraviolet Explorer", *IEEE Trans. on Nuc. Sci.*, **NS-31**, 776, 1984.

4. H.L. Marshall, C.A. Dobson, R.F. Malina, and S. Bowyer, "Analysis of Simulated Images from the Extreme Ultraviolet Explorer", *Proceedings ANRT Conference on Image Detection and Quality*, 275.

5. Chakrabarti, C. Dobson, G. Kaplan, H. Marshall, M. Lampton, R. Malina, S. Bowyer, and J. Star, "Astronomical Data Analysis from Remote Sites", these proceedings.

THE GODDARD INFRARED ASTRONOMICAL DATA BASE

Jaylee M. Mead and Daniel Y. Gezari
Goddard Space Flight Center
Greenbelt, Maryland USA
and
Marion Schmitz
Computer Sciences Corporation
Beltsville, Maryland USA

ABSTRACT

The computerized Infrared Astronomical Data Base at the NASA Goddard Space Flight Center (GSFC) contains infrared (1-1000 micron) observations of astronomical sources (outside the solar system) published in the major scientific journals from 1965 to the present, as well as the published infrared observational surveys and relevant specialized infrared catalogs. The current data base contains more than 140,000 individual observations of at least 30,000 different infrared sources.

The principal product of this data base is the Catalog of Infrared Observations (CIO) (Gezari et al., 1984), which is available in both printed and machine-readable forms. The second edition of the CIO (Gezari et al., 1987) is complete for 1965-86. It contains all Infrared Astronomical Satellite (IRAS) observations for the approximately 11,500 CIO sources reported in the IRAS Point Source Catalog (PSC) (1985).

A by-product of this data base is the Infrared Source Cross-Index (Schmitz et al., 1987), which correlates source names and positions with alternate name identifications made in the literature.

1. INTRODUCTION

The growth of infrared astronomy has been extremely rapid since the 1960's, as evidenced by the increasing number of observational articles published in scientific journals. Figure 1 shows a graph of this data for the years 1960-86. Frustration was sometimes experienced by researchers who tried to locate all the infrared observational data for a given object. The task was often complicated further by the multiple identifiers used for a single object or by the lack of uniformity of nomenclature in earlier papers. For example, workers often assigned "pet names," such as "So and So's star" or the "Red Rectangle" to newly announced objects. In subsequent papers such objects were sometimes referenced by these names without even giving coordinates--a cardinal sin in astronomy! Later, some of these objects were found to have identifications by which they were already known through observations at other wavelengths.

2. THE GODDARD INFRARED ASTRONOMICAL DATA BASE

A new generation of infrared astronomical observatory facilities, including the Infrared Astronomical Satellite (IRAS), the NASA Kuiper Airborne Observatory (KAO), the NASA Infrared Telescope Facility (ITF), and the Cosmic Background Explorer (COBE) satellite, further intensified the requirement for a comprehensive, machine-readable data base and catalog of infrared astronomical observations. Recognizing the need to collect all of this data into a single machine-readable file, the authors started in the late 1970's to establish the Goddard Infrared Astronomical Data Base by searching the most active scientific journals, infrared surveys, and catalogs. Journal articles were screened manually and cross-checked with the NASA/GSFC library RECON computer search system and the Astronomy and Astrophysics Abstracts under applicable keywords. From this data base we have produced two editions of the Catalog of Infrared Observations (CIO), and the Infrared Source Cross-Index.

The data base is processed with the Goddard NASA Space and Earth Sciences Computing Center (NSESCC) IBM 3081 computer. A magnetic tape library contains all of the observational data, bibliographic reference information, object name aliases, and stellar catalogs (for supplementary position determinations). A library of FORTRAN language programs (used to access and process the data) and a file of journal article photocopies are maintained as part of the data base.

The current extent of the literature coverage is summarized in Table 1. To date, about 3,170 journal articles and 10 major survey catalogs have been included in this data base, which contains over 140,000 individual observations.

The First Edition of the Catalog of Infrared Observations (CIO) was published in 1984 as two NASA Reference Publications (RP-1118 is the complete catalog; RP-1119 is the Far-Infrared Supplement). The Second Edition of the CIO is now available. Figure 2 shows a sample page from the new edition.

The Second Edition of the CIO differs from the First Edition in three significant ways: IRAS Point Source Catalog (PSC) identifications and fluxes have been added in the main catalog listings for all of the CIO sources detected by IRAS; the CIO data base literature search has been updated so that the data base and Catalog are now complete for 1965 through 1986; and the latest version of the Revised Air Force Geophysics Laboratory (AFGL) Survey Catalog has replaced the AFGL and AFGL Supplement data which appeared in the previous edition.

Because of its size (a 55% increase over the First Edition), the Second

Edition is divided into two parts. Part I contains the main catalog listings, including two new columns of coded IRAS data applicable to each source. Part II contains five appendices: three familiar from the First Edition (the Index of Infrared Source Positions, and two versions of the Bibliography of Infrared Astronomy, organized chronologically and alphabetically by first author) as well as two new reference documents (Atlas of Infrared Spectral Ranges, and the IRAS PSC Data for CIO Sources).

3. IRAS DATA IN THE SECOND EDITION

While it would be reasonable to include as much IRAS data as possible in the new data base, the large number of sources (245,000) in the IRAS PSC alone would clearly overwhelm the CIO. Thus, criteria had to be established to include IRAS PSC data without changing the basic nature of the CIO (i.e., an accessible data base of all published infrared astronomical observations). The final basis by which IRAS data was incorporated into the CIO was to include the IRAS PSC names, positions, and fluxes for those CIO sources detected by IRAS in the PSC Version 1.0. This provides the CIO user with accurate and easily accessible PSC data for all of the infrared sources published in the astronomical literature. The PSC name and four digit code summarizing the four IRAS band fluxes are given in two new columns in the main catalog listings (Part I), and complete PSC flux and position data are provided in an appendix (Part II). When IRAS sources were subsequently observed by other means and the results published in the literature, the new observations appear under the IRAS name in the CIO. The identifications of CIO and PSC sources were based on source identifications made in the IRAS PSC and correlated infrared source names and aliases in the Infrared Source Cross-Index data base.

About 11,500 of the individual infrared sources represented in this edition of the CIO were listed in the IRAS PSC. The PSC 12, 25, 60, and 100 micron fluxes for these CIO sources are given in an appendix to the CIO.

4. APPENDICES TO THE CIO

The following appendices serve as tools for using this data base:

The Index of Infrared Source Positions lists the CIO source positions alphabetically by source name;

The Bibliography of Infrared Astronomy, ordered first chronologically and then alphabetically by first author, links observations in the Catalog with the original references from the astronomical literature;

The Atlas of Infrared Spectral Ranges is a new reference tool, which summarizes the wavelength range over which spectra have been published for individual sources, since plotted spectra cannot be readily included in the automated data base; and

The listing of IRAS PSC Data for CIO Sources gives the full IRAS PSC name, position, and four-band flux data for all CIO sources contained in the PSC Version 1.0.

5. INFRARED SOURCE CROSS-INDEX

A new data product, the Infrared Source Cross-Index (Schmitz et al., 1987), was recently published to facilitate use of the CIO. The Cross-Index lists all infrared source names and identifications (or aliases) by which infrared sources have been identified in the astronomical literature (including the relevant IRAS Survey identifications). Also, identifications were obtained by correlating source names from the available machine-readable astronomical data bases (including visible, radio, etc.). There are some 80,000 different source names in the Cross-Index, corresponding to over 27,000 unique infrared sources. This document was distributed to regular recipients of the CIO and is also available by request to the authors.

6. SUMMARY

The Goddard Infrared Astronomical Data Base is a machine-readable compilation of data obtained by searching the astronomical literature both in scientific journals and in infrared survey catalogs. The principal products of the data base, the Catalog of Infrared Observations and its associated appendices and the Infrared Source Cross-Index, aid researchers to quickly locate additional infrared observations of the objects of interest and to provide alternative names by which a given object is known. Since the data products are available in both printed and machine-readable form, the user can adapt these tools to his specific needs.

REFERENCES

Gezari, D.Y., Schmitz, M., and Mead, J.M., 1984, Catalog of Infrared Observations - First Edition, NASA Reference Publication (NASA RP-1118).

Gezari, D.Y., Schmitz, M., and Mead, J.M., 1984, Far Infrared Supplement: Catalog of Infrared Observations (NASA RP-1119).

Gezari, D.Y., Schmitz, M., and Mead, J.M., 1987, Catalog of Infrared Observations - Second Edition (NASA RP-XXXX).

Joint IRAS Science Working Group, 1985, IRAS Point Source Catalog, U.S. GPO, Washington, D.C.

Schmitz, M., Mead, J.M., and Gezari, D.Y., 1987, Infrared Source Cross-Index (NASA RP-1182).

TABLE 1: LITERATURE INCLUDED IN THE DATA BASE

The Catalog contains observational data obtained from a search of the following infrared catalogs and scientific journals for the years 1965-86, inclusive. The number of articles in each journal containing infrared astronomical data and the journal abbreviations used in the bibliography are indicated.

Scientific Journals Searched (1965-86 complete):

220	Astronomical Journal (A.J.)
375	Astronomy and Astrophysics (Astr. & Ap.)
28	Astronomy and Astrophysics Supplement (Astr. & Ap. Suppl.)
904	Astrophysical Journal (Ap.J.)
457	Astrophysical Journal Letters (Ap. J. Letters)
41	Astrophysical Journal Supplement Series (Ap. J. Suppl.)
21	Astrophysical Letters (Ap. Letters)
13	Astrofizica
15	Communications of the Lunar and Planetary Laboratory (Comm. L.P.L.)
1	Earth and Extraterrestrial Sciences (Earth and Ext. Sci.)
56	I.A.U. Circulars (I.A.U. Circ.)
396	Monthly Notices of the Royal Astronomical Society (M.N.R.A.S.)
5	Monthly Notices of the Astronomical Society of South Africa (M.N.A.S.S.A.)
96	Nature and Nature Physical Sciences
8	Observatory
5	Proceedings of the Astronomical Society of Australia (Proc. A.S.A.)
34	Publications of the Astronomical Society of Japan (P.A.S.J.)
138	Publications of the Astronomical Society of the Pacific (P.A.S.P.)
36	Soviet Astronomy (Sov. Ast.)
25	Soviet Astronomy Letters (Sov. Ast. Letters)

Infrared Catalogs:

Caltech Two-micron Sky Survey (1969)
Equatorial Infrared Catalog (1978)
Far Infrared Sky Survey Experiment (1983)
Revised AFGL Four-Color Infrared Sky Survey Catalog (1983)
Infrared Astronomical Satellite (IRAS) Point Source Catalog (1984)
IRAS Small Scale Structure Catalog (1985)

Other Journals Searched (all years not complete):

Annales d'Astrophysique (Ann. d'Ast.)
Astrophysics and Space Sciences (Ap. and Sp. Sci.)
Chinese Astronomy (Chi. Ast.)
Comments on Astrophysics (Comm. on Ap.)
Memoirs of the Royal Astronomical Society (Mem. R.A.S.)
Science
Tokyo Astronomical Bulletin (Tokyo Ast. Bul.)
Zeitschrift fur Astrophysik (Zeit. fur Ap.)

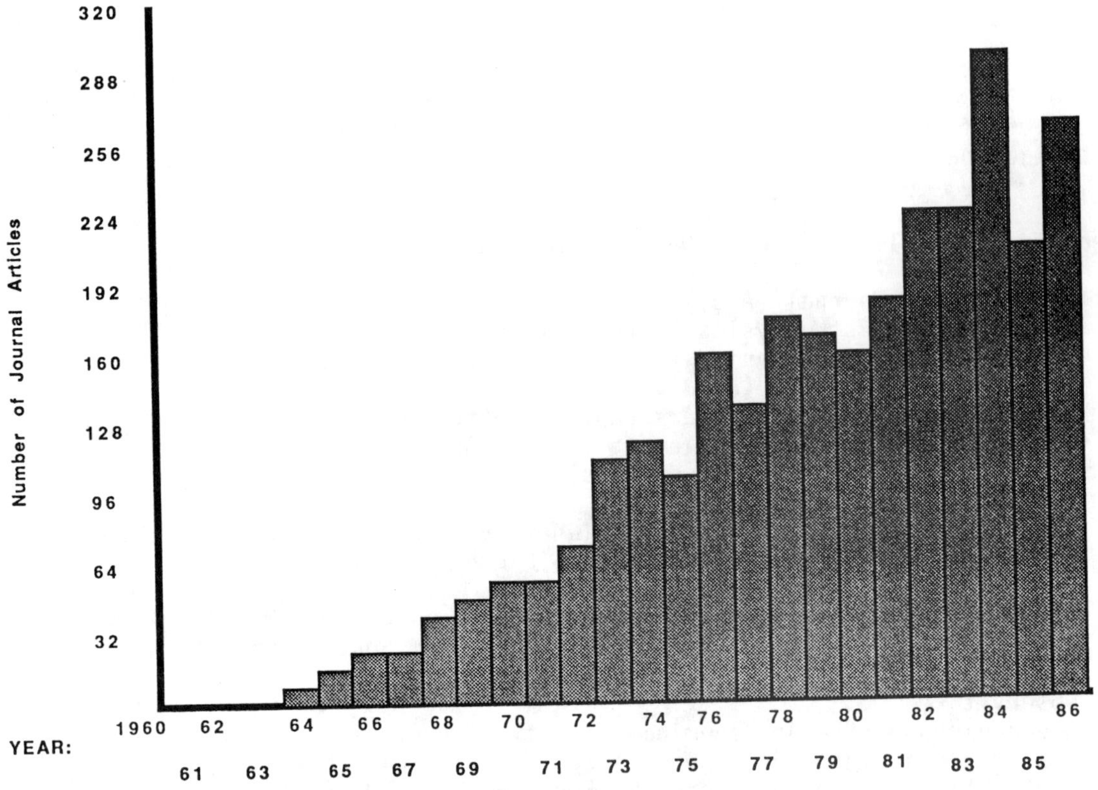

Figure 1 - Number of Infrared Observational Articles Published in Scientific Journals 1960-1986

NAME	RA (1950)	DEC	λ(μm)	FLUX	BEAM	BIBLIO	IRAS	
''	h m s ''	° '' ' "	10.6	.0014J	5.5"	''		
IRC+80001	0 11 10	+75 44 54	2.2	2.12M	10'	690001	00112+7544	1000
RAFGL 4017S	0 11 10.1	-26 17 57	4.2	1.1M	10'	830610	00111-2618	1000
IRC-30004	0 11 11	-26 17 54	2.2	2.00M	10'	690001	''	''
RAFGL 6023S	0 11 11.2	-23 42 29	20	-2.4M	10'	830610		
RAFGL 4018S	0 11 13.9	+75 44 57	4.2	1.6M	10'	''	00112+7544	1000
IRC-30005	0 11 14	-26 33 42	2.2	2.60M	10'	690001	00112-2633	0000
RAFGL 6024S	0 11 19.8	+13 45 35	20	-1.8M	10'	830610		
RAFGL 35S	0 11 19.9	+73 06 47	4.2	1.7M	10'	''	00113+7306	0000
IRC+70005	0 11 21	+73 06 54	2.2	2.91M	10'	690001	''	''
HD 1032	0 11 28.0	-85 16 19	1.2	2.45MV	-	790004	00114-8516	1000
''	''	''	1.6	1.58MV	-	''	''	''
''	''	''	2.2	1.35MV	-	''	''	''
''	''	''	3.4	1.18MV	-	''	''	''
RAFGL 6025S	0 11 28.5	+02 32 13	11	-0.4M	10'	830610		
NGC 45	0 11 32.0	-23 27 34	1.6	10.50M	81"	821013	00115-2327	0000
''	''	''	1.6	10.09M	110"	''	''	''
''	''	''	1.6	9.97M	111"	''	''	''
''	''	''	1.6	9.41M	165"	''	''	''
RAFGL 6026S	0 11 38.2	+00 52 39	20	-2.8M	10'	830610		
RAFGL 6027S	0 11 38.7	+06 36 41	11	-0.7M	10'	''		
RAFGL 5009	0 11 39.8	+00 06 16	11	-0.4M	10'	''		
''	''	''	20	-2.5M	10'	''		
RAFGL 6028S	0 11 50.3	-23 45 48	27	-2.3M	10'	''		
BS 46	0 11 54.1	-08 03 29	1.02	2.23M	-	670901	00119-0803	1100
''	''	''	1.02	2.37M	-	691002	''	''
''	''	''	1.04	2.07M	-	720002	''	''
''	''	''	1.05	2.069C	-	''	''	''
EIC 3	''	''	2.7	173F	-	780604	''	''
RAFGL 37	0 11 54.2	-08 03 31	4.2	0.2M	10'	830610	''	''
''	''	''	11	0.4M	10'	''	''	''
IRC-10005	0 11 55	-08 03 30	2.2	0.55M	10'	690001	''	''
RAFGL 6029S	0 11 59.5	-23 55 06	20	-2.5M	10'	830610		
CHI PEG	0 12 00.6	+19 55 42	1.04	2.10M	-	720002	00120+1955	1100

Figure 2 - Sample Page From Catalog of Infrared Observations (CIO) - Second Edition (1987)

TABU

A fast interactive program for processing databases

Jorge Melnick
European Southern Observatory

1. PHILOSOPHY

The final result of a wide range of experimental investigations is a table of results extracted from the data reduction. In fact the research really begins once the relevant information has been extracted from the data. However, many data processing systems, do not implement facilities for processing data tables. A notable exception is MIDAS which has comprehensive table processing capabilities. However, there are several important differences in the conceptions of TABU and the MIDAS tables. The basic design requirements of TABU were,

- Speed
- Fully interactive
- Work on format-free alphanumerical data bases created with standard editors and/or formated FORTRAN write statements
- Provide publication-quality plots and tables

In order to profit from network communication facilities it was considered essential to work on normal alphanumerical files. In addition, to increase the flexibility of TABU for working with editor-created small data files it was essential to *avoid* structuring the data in any way. Thus, TABU works with characters. The requirements of speed and format-free alphanumerical data-bases restricts the maximum size of the data sets which can be analyzed with TABU to 98 columns and 4096 rows. A new version which will extend this limit to $\sim$ 60000 rows without increasing the already substantial memory requirements (about 12 megabytes) is under development. The commands `MERGE` and `READ/ADD` allow appending and concatenating several different data sets to assemble large data-bases from "editable" files.

The present implementation of TABU on VAX computers makes use of the JPL PGPLOT graphics package and cannot be easily ported to other machines. However, TABU is a self-contained package which requires less than 1000 disk blocks and which can easily be installed in any VAX computer. PGPLOT allows plotting on a large selection of devices and therefore TABU may work with almost any graphics terminal.

2. DATA-BASE STRUCTURE

The TABU data-bases, which I will call TABLES, may have up to 98 columns separated by blanks or tabs. A special command is provided to fill-in blanks which are not column separators (as for example in NGC 1275). This is the only formatting which may be required. Internally TABU assigns a real and a character value to each entry in the TABLE and keeps a record of whether that entry is a number or a name. A column in the TABLE, which I will call a VECTOR, may contain numbers or names and the VECTOR arithmetic commands assign a large negative numerical value to names (the NULL value).

A VECTOR may be called by its COLUMN number (an integer between 0 and 98) or by a name assigned using the `DICTIONARY`. Thus, integers between 0 and 98 are **always** treated by TABU as VECTORS and the decimal point must be used to distinguish VECTORS from scalars in this range. Number 99 is used to designate the *last* COLUMN in the TABLE.

2.1 VECTOR arithmetic

Arithmetic is handled by a two-level reverse polish code. The `COMPUTE` command allows arithmetic between VECTORS, between VECTORS and scalars, and between scalars only. Every operation involving at least one VECTOR produces a new column in the TABLE which, however, may be forced to replace a previously existing one.

Examples:

```
COMPUTE 1 LOG -2.5 * 25.5 +
```

transforms COLUMN #1 to magnitudes and stores the result in the next available COLUMN.

```
DICTIONARY FLUX,1
COMPUTE/01 FLUX LOG -2.5 * 25.5 +
```

does the same after assigning the name FLUX to COLUMN #1 and stores the result back in COLUMN #1. The VECTOR previously stored in COLUMN #1 is overwritten.

Illegal operations such as arithmetic with names or operations like logarithms of negative values are given the NULL value.

2.2 Scalar arithmetic

TABU has 999 memory locations (the MEMORY) where the results of arithmetic involving only scalars are stored. The values of the MEMORY may be used in all TABU commands which involve scalars.

Examples:

```
COMPUTE 1.5 LOG 23.123 * SQR INV
```

computes $1./\sqrt{23.123 * \log(1.5)}$ and stores it in the next available MEMORY position or in the position specified in COMPUTE. The result of the computation and its MEMORY position are given.

```
COMPUTE/07 @07 1.  +
```
incremets the value stored in MEMORY #7 (@07) by one.

2.3 Logical operations

TABU allows logical operations acting on individual values of a VECTOR (SELECT) and global conditions which act on full VECTORS (FORCE).

Examples:

```
COMPUTE 1 LOG -2.5 *  DICT MAG,99
SELECT MAG GT 12.  AND MAG LT 20.
```
converts COLUMN #1 to magnitudes, assigns the name MAG to the result and restricts all further analysis to MAG values between 12 and 20.

```
SELECT IDENT EQ NGC*
LIST IDENT
```
lists all the NGC objects in the data-base.

```
FORCE MAG LT 12.  12.
DICT CORRMAG,99
```
creates a new COLUMN which is identical to MAG but where all values lower than 12. are set to 12.

```
FORCE MAG LT COMPARE REPLACE
```
replace all entries in column MAG which are less than the corresponding values in column COMPARE by the values of column REPLACE.

The commands COMPUTE/SELECT, and COMPUTE/EXCLUDE allow mixing logical and arithmetical operations. COMPUTE also implements arithmetic on FAMILIES identified by a COLUMN.

3. COMMAND LANGUAGE

TABU has its own command language, called the PLAY language, which is not too different from that of other image processing systems and which allows implementation of DO loops, conditional branching (IF), etc. Any number of PLAY files may be nested. Examples of PLAY files are given below.

4. GRAPHICS

TABU uses a full implementation of PGPLOT which allows defining fonts, viewing-windows, line types widths and colours, margins and the graphics cursor. The implementation of the CURSOR allows re-scaling the plot, eliminating or adding points, pointing to families of objects defined by COLUMNS or parts thereof etc. A few illustrations together with listings of the PLAY files used to create them are presented below.

5. OTHER TABU COMMANDS

TABU has commands for statistical analysis including principal components (`FIT/PCA`), maximum likelihood fits (`FIT/ERR`), calculation of means and moments (`STATISTICS`), commands for handling spectra (`ALICE`), commands for drawing lines and curves (`DRAW`), for smoothing (`MODEL`) etc. TABU also allows users to define their own commands using the `USER` command and to add new routines by means of a "standard" TABU interface.

The PLAY files listed below illustrate some of the TABU commands. A copy of the TABU users manual is available upon request.

EXAMPLES

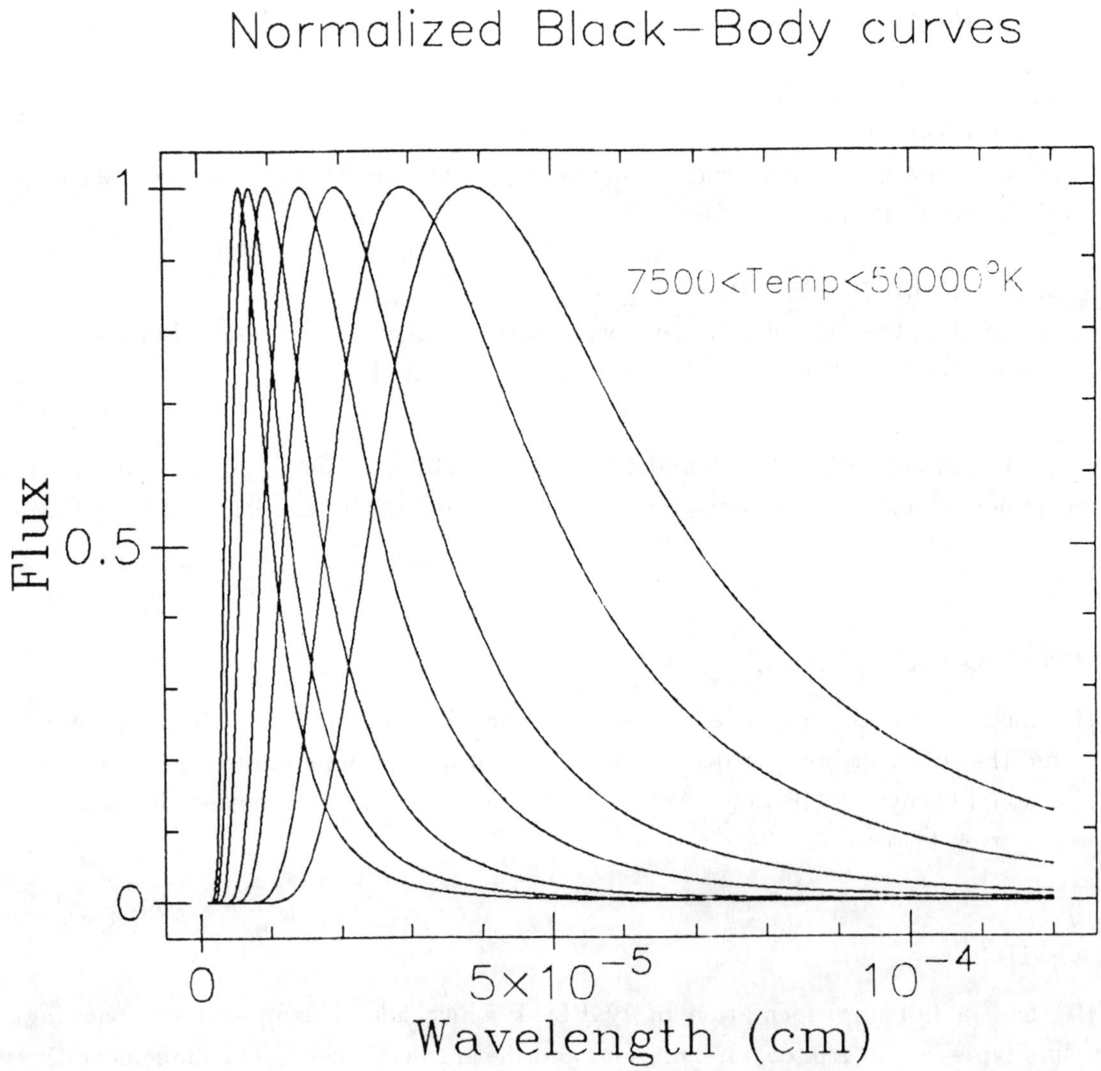

```
! EXAMPLE #1: Generate a series of Black-Body spectra
!
! Command                                        Brief description
! -------                                        -----------------
GRAF/VX .2 .8                               Smaller view-window (full page is
GRAF/VY .2 .8                               (0,1).
GRAF/WIDTH 3                                Triple width of lines in plot.
!
READ DUMMY                                  Read a dummy file to initialize
COMPUTE/77/NO 0.                            Set MEMORY #77 to 0. No echo.
ALICE/MAKE 100. 12000 10.                   Make a spectral file from 100 to
!                                           12000 angstroms in steps of 10.
COM 99 1.E+08 /                             Convert angstroms to centimeters.
DIC LAMDA,99                                Assign the name LAMDA to the result
!
:TEMP                                       LOOPING flag.
TYPE Enter temperature: $\7                 Get the Black-Body temperature from
GET 2                                       the terminal. The $ inhibits
!                                           returns and the \7 (esc 7) is BELL.
IF @02 EQ 0. GOTO LABEL                     Break the LOOP and jump to :LABEL
!
COMP/01 @02 5. ** 1.28648E-04 *             Compute the Black-Body spectrum for
COMP LAMDA -5. ** 3.74185E-05 *             that temperature.  Echo the numbers
!                                           of the new COLUMNS created.
!
COMP 1.43883 LAMDA / @02 / EXN 1. - 99 / INV @01 /
!
DICT/REPL BLAK,99                           Replace the entry BLAK in the
!                                           DICTIONARY by the new COLUMN.
IF @77 GT 0. GOTO OVER
SHOW/AXIS/LINE LAMDA BLAK                   Plot a continuous line with new
\fr Wavelength (cm)                         labels using the ROMAN font.
\fr Flux (ergs sec\u-1\d)
\fr Normalized Black-Body curves
COMPUTE/77/NO 1.                            Set MEMORY #77 to 1.  No echo.
GOTO :TEMP                                  Loop
:OVER                                       Looping FLAG
OVER/LINE LAMDA BLAK                        Overplot
GOTO :TEMP                                  Loop.
!
:LABEL                                      Label the frame. Put
LABEL/FRAME                                 blanks in X and Y labels and
                                            the temperature range
                                            in the TITLE.
7500<Temp<50000\uo\dK
READ/KEY/02 YMAX                            Store the XY limits of the plot
READ/KEY/01 YMIN                            in MEMORIES #1-#4.
READ/KEY/04 XMAX
READ/KEY/03 XMIN
COMP/05/NO @02 @01 - 5. /                   Offset the label by one fifth of
COMP/06/NO @04 @03 - 2. /                   the range in Y and one half of
COMP/10/NO @02 @05 -                        the range in X. Store the results
COMP/11/NO @04 @06 -                        MEMORIES #10 and #11.
LABEL @011 @010 0. %LAB(TITLE)              Plot the LABEL at an angle=0.
```

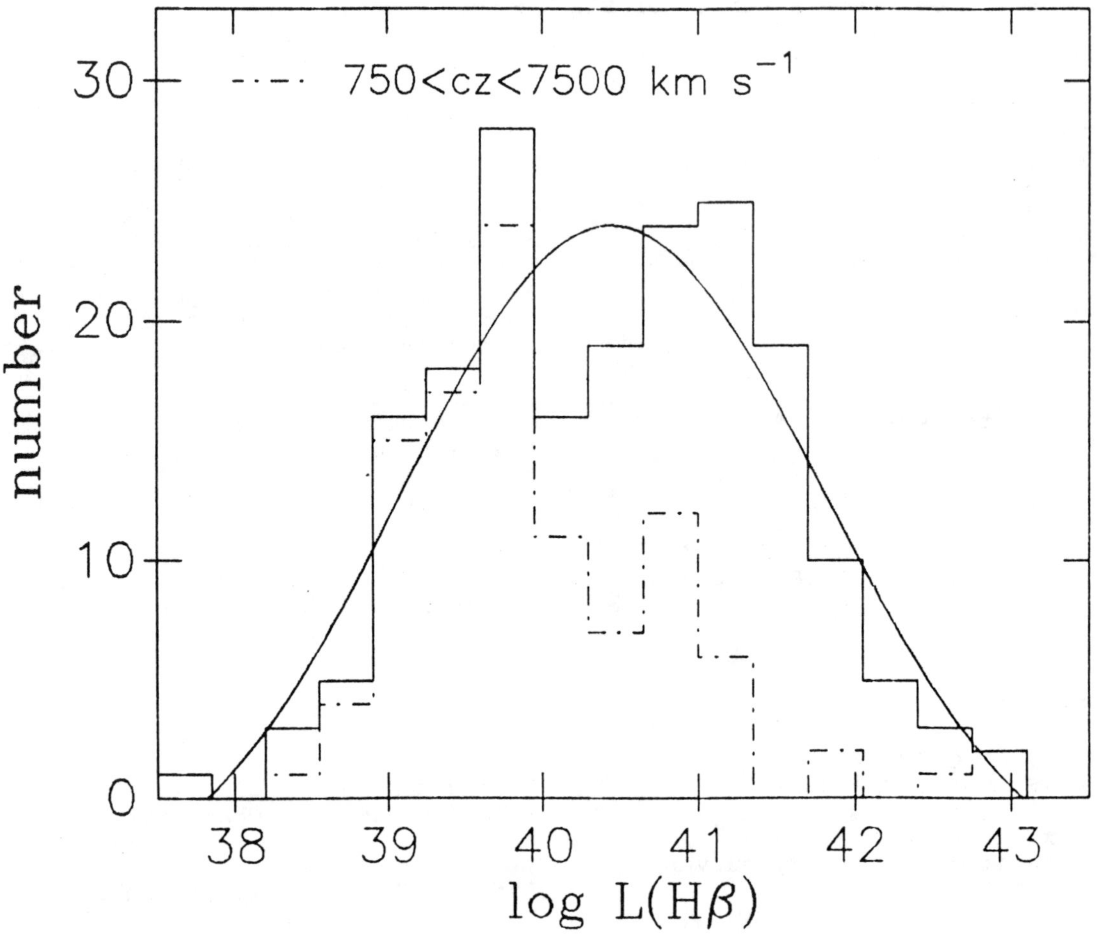


```
! EXAMPLE #2:  Plot a luminosity function from a large data-base
! ===========
!
!  Command                                    Brief description
!  -------                                    -----------------
GRAF/VX .2 .8                              Define a smaller viewing window
GRAF/VY .2 .8                              (the full page is 0,1).
GRAF/WIDTH 3                               Set the width to 3 lines
GRAF/TY 10 5                               X-tiks every 10 units 5 intervals
GRAF/TX 1 5                                Y-tiks every one unit 5 intervals
!
READ CATALOG                               Read DATA-BASE
SEL/CLE                                    Clear SELECT flags
!
TYPE Enter lower velocity limit $\7        Get data from terminal. In this case
GET 1                                      radial velocity limits.  The $ inhib
TYPE Enter upper velocity limit $\7        carriage returns and the \7 (ESC 7)
GET 2                                      rings the BELL of the terminal.
!
SH/HIST/GAUS 2 -0.35                       Plot HISTOGRAM in steps of 0.35
0 33 37.5 43.5                             Define the plot limits in Y and X.
\fr log L(H\gb)                            Axis labels using the ROMAN font.
\fr number

SELECT 3 GT @01 AND 3 LE @02               Apply velocity limits. In this examp
!                                          the limits are 750<V<7500 km/sec.
DRAW/TYPE 3                                Change line-style to dot-dashed
OVER/HIST/GAUS 2 -0.35                     Plot over old plot.
STATISTICS 2                               Type statistics on COLUMN #2
```

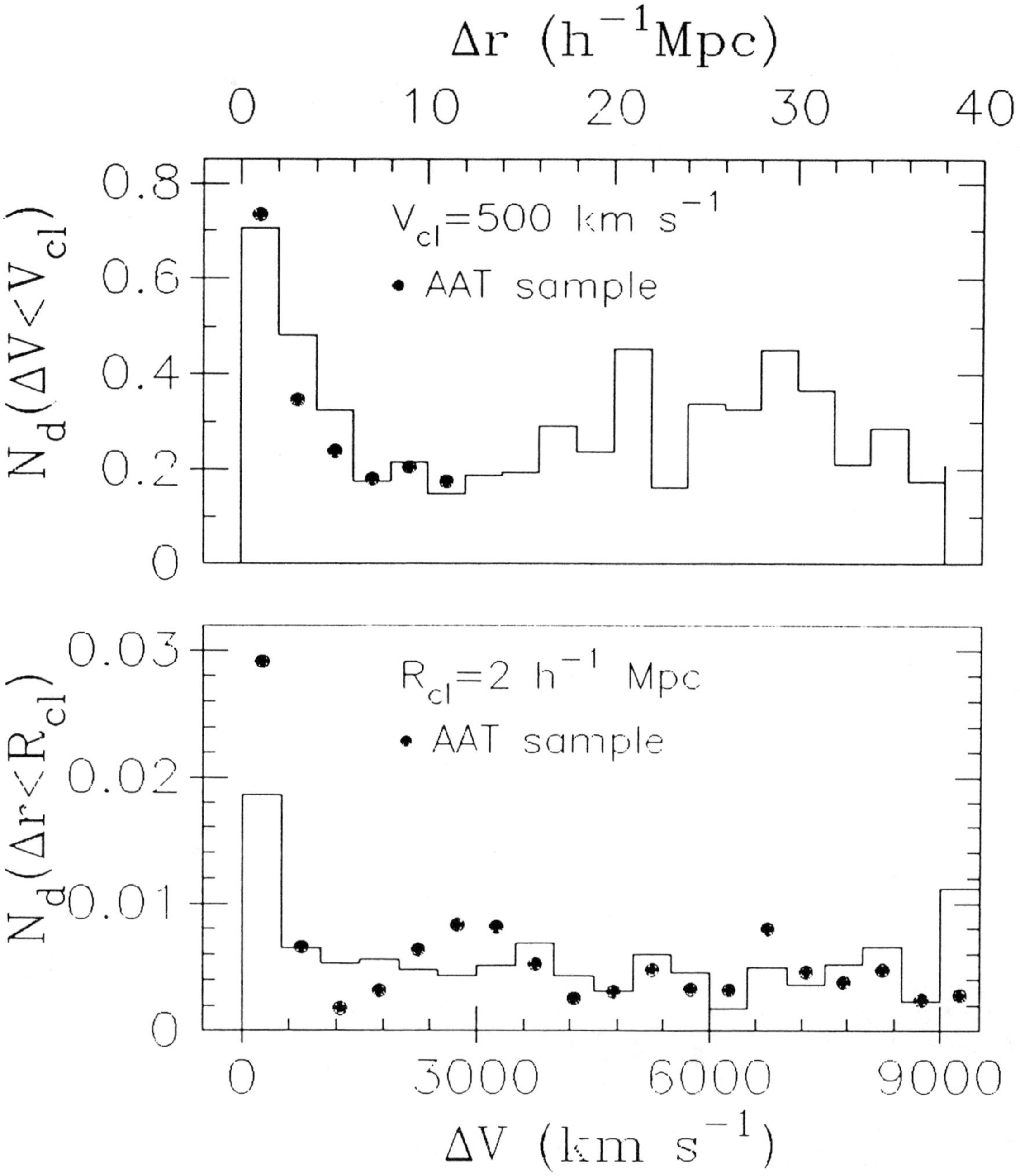
Δr (h⁻¹Mpc)
0 10 20 30 40
0.8
0.6
0.4
0.2
0
N_d(ΔV<V_cl)
V_cl=500 km s⁻¹
• AAT sample
0.03
0.02
0.01
0
N_d(Δr<R_cl)
R_cl=2 h⁻¹ Mpc
• AAT sample
0 3000 6000 9000
ΔV (km s⁻¹)

```
! EXAMPLE #3: Compare two data-bases
! ===========
!
! Command                                          Brief description
! -------                                          -----------------
GRAF/WPLOT 3                                      Define the width of the lines in
GRAF/WIDTH 3                                      the terminal and plotter.
LABEL/CLEAR                                       Reset the LABEL memory.
LABEL/WIDTH 3                                     Set the width of the LABELS
GRAF/MAR 0.03 .03                                 Plot margins in relative units
PAINT/SIZE 1.70                                   Size of the plot symbols.
PAINT -17                                         Symbol -17 = filled circle
!
READ HIIGAL                                       Read data-base HII galaxies
SEL/CLE                                           Clear al SELECT flags
!
:LOWER                                            Plot the lower panel
LABEL/FRAME                                       Label the axis.
\fr \GDV (km s\u-1\d)                             Roman font plus Greek (\g)
\fr N\dd\u(\GDr<R\dcl\u)
                                                  Blanks for no title.
GRAF/VY .18 .48                                   Define the viewing window
GRAF/TX 3000 5                                    Big tik marks every 3000 in X and
GRAF/TY 0.01 5                                    0.01 in Y with 5 intervals.
SHOW/STEP/XTOP 2 4 SCALE -500 9500 0 0.032
!                                                 Plot lines with ''ladder'' form and
!                                                 top labeling and fixed scale
READ AAT                                          Read AAT data-base
COMP//NO 2 250. +                                 Add one half bin to the velocities
DIC XX,99                                         Assign the name XX to the new COLUMN
COMP/OO/NO 17.                                    Set MEMORY #0 to 17.
COMP//NO 4 @OO /                                  Divide COLUMN #4 by MEMORY #0 (scaling
DIC YY,99                                         Assign the name YY to the ne COLUMN
OVER XX YY                                        Plot over previous plot
!                                                 Label the axis.
LABEL 2000. 0.027 0. {R\dcl\u=2 h\u-1\d Mpc}
LABEL 2020. 0.022 0. {\(0850 AAT sample}
!
:UPPER                                            Repeat for upper panel
READ HIIGAL                                       Read HIIGAL data-base again
LABEL/FRAME                                       (double reading can be avoided
\fr \gDr (h\u-1\dMpc)                             if necessary using READ/ADD or
\fr N\dd\u(\gDV<V\dcl\u)                          MERGE or optimizing the procedure
                                                  to do all plotting with one and
GRAF/VY .56 .86                                   then reading the second data-base).
GRAF/TX 10 5
GRAF/TY 0.2 2
MULT/STEP/XTOP 5 7 SCALE -2 40 0 0.85
READ AAT                                          Read AAT data base again
COMP//NO 5 1. +
DIC/REPL XX,99
COMP/OO/NO 1.5
COMP//NO 7 @OO *
DIC/REPL YY,99
SEL XX LT 12.
OVER XX YY
LABEL 8. 0.7  0. {V\dcl\u=500 km s\u-1\d}
LABEL 8.1 0.56 0. {\(0850 AAT sample}   \(0850 is a filled circle.
```

DEVELOPMENT OF A DATABASE FOR STARS IN OPEN CLUSTERS

J.-C. Mermilliod
Institut d'Astronomie de l'Université de Lausanne
CH-1290 Chavannes-des-Bois, Switzerland

ABSTRACT

Data for stars in 395 open clusters, including photometry in several systems, spectral classification, radial and rotational velocities, membership probabilities, remarks, references, bibliography, positions and cross-identifications have been organised in a database on a Sun 3/50 workstation. UNIX-style commands have been developed to handle the data and interrogate the database. Software is being written to analyse the data, compute mean values of photometric parameters and also to plot on the display various photometric diagrams, as well as cluster maps.

Key words: open clusters - data base

1. INTRODUCTION

In past years, a big effort has been devoted to the compilation of astrophysical data for stars in open clusters, great attention being paid to the standardisation of star numbering systems in each cluster. The status of the compilations was presented by Mermilliod (1985), the collections being kept up to date since that time. Until recently, the compiled data were not readily accessible since they were kept in magnetic tape files. The situation is now quite different since the Institute of Astronomy of the University of Lausanne acquired a Sun 3/50 workstation equipped with two disks of 70 Megabytes, which affords the possibility of permanent data storage on the disk. The present project was started in March 1987, and within two months most of the data were stored on the disk and the main commands written and tested.

2. CONCEPTION OF THE DATABASE

a) Definition of the objects: Open star clusters are basically collections of stars located in a limited region of the sky. The limiting magnitude reached by the photometric studies is normally well beyond the limit of astronomical catalogues and therefore the cluster stars are generally designated by a running number and identified on photographic charts. Astrometric and astrophysical data are available for the best studied clusters, often with several sources of data for each star, together with cross-identifications between the various numbering systems used in the literature. This definition implies that one or several kinds of data are generally handled for a sample of (member) stars in a given cluster rather than many data for a single star.

b) Main requirements: The database should not only be an efficient tool to store and analyse the data, but should also provide a versatile environment to develop the study of open clusters considered as distinct astrophysical entities. At the present stage of development, the way the data will be handled and the main steps of the work on star clusters provide the main requirements on the organisation of the database and the possibility if should offer:

(1) Data collection: The multiplicity of star identification represents the main problem usually encountered in the compilation of data for stars in open clusters. The database should therefore provide an easy procedure to solve it, in addition to the normal facilities of modification and addition.

(2) Data analysis: Experience shows that it is safer to make a systematic intercomparison of the data before any use. This analysis is mostly done for one kind of data for one cluster at a time and concerns, for example, the analysis of different sets of UBV photographic data and the reduction to a common system. It leads to the computation of mean values, which form the basic data set for subsequent scientific analysis, and should preferably be done with

the codes already written.

(3) Scientific analysis: The first major application foreseen is a systematic analysis of each cluster, using all the data at present available to define the most complete sample of member stars and determine reliable basic parameters, with a view to several applications in keeping with work already published.

(4) Graphic facilities: Although the study of the cluster may be automatised, it still requires a lot of graphic actions. We frequently need to visualise not only the results of the data analysis, but also the many photometric diagrams used in the determination of star membership or cluster parameters. The graphic facilities should also include the fitting of reference sequences and isochrones and the plotting of cluster maps.

3. REALISATION

a) File organisation: The first version (August 1987) of the database is a collection of some 4000 files concerning 395 open clusters, arranged within the directory hierarchy supported by the UNIX operating system run by the Sun workstation. Each cluster defines an independent working directory identified by the cluster name; the available data are distributed in distinct files. The clusters are collected in parent directories according to the source catalogues: "ngc", "ic" or "anon". The main directory is the database itself, which contains also other sub-directories: the description of the database and help information, references, bibliography, commands and programmes. When working on a cluster, access has first to be made to its directory, which is simply reached by entering its path: "amas ngc/2287". This structure allows easy inclusion of any kind of new data, provided the stars are identified with the appropriate designation.

b) File and record structure The files are sequential, not compacted and sorted by star numbers and references. The latter statement is not required by the search processes, but by the application programmes. Whenever possible, the records have the same structure: star identification, reference, data.

c) Content: The data already collected (Mermilliod, 1985) have been installed in the database and newly published data are entered regularly. This includes the photometric data extracted from the Lausanne compilations (UBV, uvby, Geneva, DDO, Walraven), spectral classification in the MK system (Mermilliod, 1986b), individual radial velocities, rotational velocities (unpublished) and membership probabilities (unpublished), positions and cross-identifications (Mermilliod, 1986a). All data are referenced. New kinds of data are also entered according to the possibilities. The bibliographic information is based on chapter 153 of the Astronomy and Astrophysics Abstracts.

d) Access to the information: Thanks to the work on stellar identification invested in the preparation of the catalogues used for building the database, the identification problem is greatly simplified: the various data for each star within a cluster are collected under a unique designation, according to the adopted numbering system. This identification is used as the main key for access to the data in all files. However, there are other possibilities for interrogating the database:

- it is possible to recover the adopted number (main key) starting with any existing identification thanks to the transit tables;

- The data may be accessed directly with the HD, DM or any other identification (Mermilliod, 1986b), or by coordinates;

- Data can be selected according to a reference number;
- Samples can be formed according to astrophysical criteria.

e) Correction and updating: Correction and updating are done by editing the files when only a few data are to be added or corrected by the author. Better solutions can be developed, but this does not represent a problem at present. If a larger amount of data has been processed by another person or received by electronic mail, the original file and the new data can be merged easily by using UNIX commands grouped in small shell programmes.

4. STRUCTURE OF THE COMMANDS

The file organisation and small size, a few hundred records, make their consultation very rapid with the tools provided by the UNIX system. The programming of the shell, together with specific fortran or C codes offer interesting possibilities for working with the database. UNIX-style commands have been developed which are named after the kind of data they handle: "ubv" is used for the UBV data, while "mk" refers to the spectral classification and "vsini" or "vsn" operates on the rotational velocities. Most commands are active in the "current" cluster and are organised in the same way:

a) Without any parameter, the commands provide a listing of the whole file, page by page;

b) Given a star number as argument, the same command returns the desired data for the chosen star;

c) With an option and a parameter, the command forms a sample according to the specification given, which may be a limiting value referring to the data or a specific reference.

The main commands often call specific functions written in the shell language. The result or sample is displayed on the standard output, but is also written in a temporary file. One straightforward application is to look for the stars having a membership probability above a given limit, say 90 percent ("prob -g 90") and then search for UBV photometric data ("aud ubv") and draw an HR diagram. The references can be consulted either in block or individually for any matter, and the bibliography either by giving a cluster identification or a topic, or both.

Using the UNIX facilities, new commands can easily be created or the output sent through a pipe for further treatment.

5. DISCUSSION AND CONCLUSION

This system has been conceived for an application specific to open clusters and is based on the facilities offered by the UNIX operating system. Its realisation has been developed by starting from the existing compilations realised by the author. The disk space occupied by the data is around 13 Mb and can be easily supported by the system. However, there will probably come a time when the data have to be compacted and the database conception completely modified. Before investing too much time in the development of voluminous software, several applications will be realised to benefit from the first effort.

Future projects include an extension of data collections and bibliography, development of a versatile procedure for computing mean values and the writing of software to realise new tasks that cannot be performed easily by programming the shell.

Under the present circumstances, the database is not transferable to non-UNIX machines. Furthermore, it is not intended to be generally accessible, due to the host computer size. However, selected data or even entire files can be printed or sent by electronic mail.

REFERENCES

Mermilliod J.-C. 1985, Mem S.A. It. 56, 549
Mermilliod J.-C. 1986a, Astron. Astrophys. Suppl. 63, 293
Mermilliod J.-C. 1987b, Bull. Inform. CDS 31, 175

DATA TRACEABILITY IN THE CONSTRUCTION OF ASTRONOMICAL ARCHIVES

Glenn Miller

Astronomy Programs, Computer Sciences Corporation [1]

Patricia Monger

Mark Johnston

Space Telescope Science Institute [2]

Abstract

A large and increasing fraction of astronomical data is taken by persons other than the astronomer: this has been the case for some time with satellite observatories and is becoming more common for ground based observatories as well. The requirements of constructing generally accessible data archives which are related to other archives and to astronomical catalogs make the problem of data traceability an acute one. It is necessary to relate exposures actually taken to those planned, and to incorporate any changes made during data collection as well as post facto information about the data. Because of the volumes of data involved, this process must be largely automated. We have addressed this problem for Hubble Space Telescope (HST) by analyzing the means to uniquely label, describe and trace proposed observations through the science operations ground systems. While some of the issues are specific to HST, many are of general concern for the design of future astronomical archive systems. We treat HST as a case study for the problem of archive construction and discuss a number of recommendations for consideration in the development of future systems.

1. Introduction

The process of devising and implementing an astronomical observing program involves several steps starting with planning the observing program, to choosing particular telescope and instrument parameters, to scheduling the telescope, and finally executing the observations. At each step, the observing program is specified in finer detail. Modifications and replanning are often necessary. Traceability is defined as the ability to track the thread of an observation from proposal through data reduction, including all the changes made during the process. When a single astronomer is responsible for all phases of an observing program, maintaining traceability is a straightforward (if sometimes tedious) task. With the increased complexity and capability of both space and ground based telescopes, it is common for many persons to be involved in the observing process. These factors, coupled with the large amount of data involved, require that traceability analysis be automated.

Traceability is also important in the routine operations of an observatory, e.g. to verify that planned observations have been executed or to determine the status of an observing program. It is essential in the construction of a scientifically useful data archive - understanding of the nature and quality of the data

---

[1] Staff Member of the Space Telescope Science Institute

[2] Operated by the Association of Universities for Research in Astronomy for the National Aeronautics and Space Administration

(including instrumental and systematic effects) will depend on an understanding of why it was gathered and how it was implemented. The task of catalog construction only strengthens the need for traceability. The problems of maintaining traceability in operations and in constructing the archive are the same. If traceability is not maintained through the operational era, then construction of the archive or catalog will be, at best, needlessly complicated, and will be, at worst, impossible.

This paper discusses the Hubble Space Telescope (HST) as a case study for the problems of data traceability. Section 2 details the HST operations, while section 3 presents a general model and recommendations. Conclusions are given in section 4.

2. Data Traceability for the Hubble Space Telescope

The starting point for traceability is the pool of accepted proposals which are entered into the Proposal Entry Processor (Pep) entry database during "Phase II" of the proposal cycle. These proposals (a few hundred per observing cycle) will received as formatted files or paper forms from the proposer. The Pep system allows the proposer to describe the observations at a high level of abstraction without having to specify detailed telescope and instrument configuration information. The Pep software automatically converts this high level description into the more detailed description used by the planning system.

Proposals in the Pep system progress through three representations. The first is a database that looks like the proposal form itself. These data are converted to a Pep internal representation (including data syntax and consistency checks). Using a rule-based expert system, Pep Transformation maps the Pep internal representation into the data structures for the planning software, the Science Planning and Scheduling System (SPSS). In SPSS, the data are organized into a hierarchical structure: exposures are grouped into alignments (nominally a single pointing of the spacecraft), which are in turn grouped into observation sets (which use a single pair of guide stars). The results of Pep Transformation are written to the SPSS database.

The HST planners construct a list of observing candidates and a calendar. First a long range calendar is constructed with only the most severely constrained observations. This calendar is then broken into successively smaller calendars, which are fully populated with observations. A week long calendar is used to construct the Science Mission Specification (SMS), which is sent to the Payload Operations Control Center (POCC) for HST computer command generation. This procedure may involve several iterations because the POCC must arrange the communications satellite (TDRSS) links and detect problems with the SMS. The final SMS must be received by the POCC at least three days before the first observations are to begin.

The HST returns science and engineering telemetry streams back to the POCC via the TDRSS. The science and astrometry data, which are the data of interest for the astronomical archive, are reformatted and sent to the Post Observation Data Processing System (PODPS), for basic processing (calibration, flat fielding, etc.) and archiving to tape. The Data Management Facility (DMF) provides convenient user access to the HST archive, including traceability information from the Pep, SPSS and PODPS systems. The quality of the archive depends on the data traceability through the various systems.

From the above sketch of HST, it can be seen that an astronomer's observing program is represented in different ways by different components of the ground systems. The fact that some components in the ground system were designed separately has certainly contributed to the number of different representations. However, it is important to note that some changes in representation are inherent in any observatory of this scope, as the specification supplied by the proposer (e.g. target location, instrument configuration, exposure, time, relationship to other exposures) is at a much higher level than the specification required by the telescope to execute the observations (e.g commands for slewing, detailed commands for setting up the observation and the instrument configuration, communications commands). The need to spare the proposer from determining this detailed information was a motivation for the Pep system. Changes in data representation introduce a need for a tracking system. An example taken from HST will illustrate this point. The HST proposal exposure logsheet allows the proposer to define a skeleton set of exposures as an exposure *sequence*. Consider a *sequence* definition of *N* lines. Each USE *<sequence>* line is expanded by Pep to *N* lines, with variables such as target name and exposure time filled in from the values in the USE *<sequence>* line. The next step in Pep processing takes each of these *N* lines and expands the number of exposures field to give *M* lines. These *M* exposures are put on the calendar by the planners. Each of these exposures is implemented by the science commanding staff as a set of telescope commands which appear on the SMS. The number of commands depends on the exact nature of the observation, as well as the what will be observed just prior to and just after the observation.

As observations are executed, the science data as well as telescope engineering information are formatted by the HST in "packets" to be sent to a TDRSS satellite and thence to the ground system. A packet includes header information about an observation and the status of the telescope and may include more than one observation. PODPS breaks the packets into separate files for the header information, the science data, and logging and error message files. When calibrations are applied to the data, the header files are edited to show this. Finally the header and log files, and calibrated and uncalibrated science data are archived to tape.

The scenario presented here represents the best case situation: e.g. no data are lost en route (by failures of the communications link or by machine crashes), the transformations are straightforward, the data in one system are not modified after transformation to a succeeding system. In a real system, any one of these conditions may occur. Other examples from the HST system illustrate this. The HST system allows for an observer to propose for so-called Targets of Opportunity. These are events which cannot be predicted (e.g. supernovae). When a Target of Opportunity presents itself, the observations, which have been outlined in skeleton form by the proposer (e.g. supplying instrument configuration and mode, but not exposure time and target position), will be patched into the SMS, replacing whatever other observations had already been scheduled for that time. The displaced observations must be replanned, thereby rendering invalid the previous transformation from calendar to SMS, and from planning database to calendar. HST also has a capability for limited real time choices in observations (e.g. choice of filters, or choice between preplanned branches). The results of real time decisions must be traceable to the original program and final data.

Another case arises if modifications must be made after planning has begun (e.g. the proposer discovers an error in a proposed observation or new information about the target is discovered). If this known sufficiently far in advance of the execution of the observation, it may be possible to make the correction in the planning database, even if the proposal has been transformed by Pep. This could leave the Pep version of the observation in an inconsistent state as compared to the planning version. In the case of multi-year proposals, modifications are practically inevitable.

3. Model for Data Traceability and a User Interface

Given this background on the HST, we now examine data traceability in terms of a generalized model of the systems and a traceability user interface. Given a collection of N coupled *systems*, the goal is to track an observation through the systems. *Observation* is defined as the atom of scientific information as seen by the astronomer, and may include target acquisitions, calibrations, different filters, etc. In general, an observation corresponds to multiple instrument operations, data readouts and observation files.

Each system deals with a different *representation* of an observation and uses a distinct set of *identifiers* for units within the system. For example, the Pep system operates on an astronomer-oriented description of observations using proposal id and exposure logsheet line number to identify data, SPSS manipulates data structures related to scheduling and spacecraft operations using proposal id, observation set id, alignment id and exposure id, while PODPS uses observation id as a key field.

The process of casting an observation representation (and its identifiers) from one system to another is called *transformation*. A transformation may be as simple as a change in database keys, or may be a change in data abstraction involving a complex, many-to-many mapping between representations in two systems. The transformation from Pep exposure to SPSS observation set is an example of a change in the level of abstraction, while the transformation of data in PODPS is a simple repackaging of data received from the HST.

As noted in section 2, there are a number of complications possible with any transformation: Although the process of observing is often described as a predetermined progression of events, changes to a program and replanning will occur (e.g. patching an SMS for a target of opportunity requires that already scheduled observations be sent back for replanning). At each system interface data may be garbled or lost (e.g. loss of data during readout of an instrument).

In order to provide tracking, each individual system must maintain two kinds of knowledge: 1) history and status of observation representations within that system and 2) history and status of transformations into and out of the system, including a mapping between the representations of each system (e.g. both Pep and SPSS retain the mapping between Pep logsheet lines and SPSS observation sets). At a minimum, one of the two systems must maintain the history and mapping, and having both systems maintain this information will allow a better ability to recover from inconsistent or incomplete information.

The systems must be designed so that software procedures automatically record history, status, and mapping information without special user action. This is especially true for "last minute" or exceptional conditions where normal operational procedures must be circumvented; in this situation data traceability can easily be lost.

The telescope and instrumentation must be considered as distinct systems in this model, that is, they should be designed to supply observation identifiers, status and history information, especially when capable of autonomous actions (e.g. the HST High Resolution Spectrograph can terminate an observation which will be underexposed and can cycle through a series of observations).

From the point of view of traceability, an observatory should be designed to minimize the number of systems and transformations, but this must be carefully balanced against the need for well-designed, modular systems with representations tailored to a particular need.

Routine checks of all systems for traceability inconsistencies must be made, otherwise problems will only be discovered when one is fortunate enough to inquire about certain observations.

An observatory following the above model provides the basic data for traceability, but in order to answer queries from users, there must be a *traceability user interface.* This user interface must be logically (if not physically) distinct from the individual systems, since the answer to a query may require information from several different systems. Note that this user interface is distinct from the interface used to access the data archive (e.g. the DMF). The traceability user interface supports observatory operations and supplies the archive with an accounting of the observation.

An essential requirement for the user interface is *reasoning with incomplete or inconsistent information from the systems and diagnosing the causes for this situation* (several examples were given earlier how tracking information could be lost or scrambled). It is this point which distinguishes the traceability user interface from a simple database query system or procedural accounting system. We have identified several techniques which will be useful in building a traceability user interface: rule based expert systems, hypothetical reasoning, and frame-based data representation. These techniques are available in a variety of commercial software systems.

Hypothetical reasoning allows the creation of an alternate "world view" which differs from an existing set of facts in one or more ways. In this hypothetical world, rules can be applied. If a conclusion which violates some constraint is drawn, then the original hypothesis is invalid. If no constraints are violated, then this is a possible explanation for an inconsistency in the traceability.

For example, an observation scheduled for execution on an SMS may not have a corresponding data file in PODPS. If there is a data file in PODPS that does not match a scheduled observation on the SMS, one could hypothesize that the PODPS header had been garbled and that in fact the data file corresponded to SMS observation. To test this hypothesis, one could query the planning database for information about the scheduled observation and compare the database information with the PODPS file (e.g. target name, instrument configuration). If these do not match, the hypothesis is invalid. If they do match, the user is informed that a possible explanation for both the "missing" observation from the SMS and the "extra" PODPS data file is that the identification numbers were altered in transmission. The user could examine the science data to verify this conclusion.

In the traceability interface there may be more than one instance of a given observation representation (e.g. the exposure time for a particular exposure id is changed) and a given representation may be transformed many times. The relationships between the observation representations will in general be many-to-many. Frame-based representation is useful in expressing these kinds of relations.

For some queries there will be additional information that should be displayed. For example, the shorthand allowed in the HST proposal syntax means that simply displaying the logsheet line to which an observation file maps may not provide a complete view of an observation should the logsheet line be part of a group of lines (sequence, repeat, group within, etc.). Since the relation between objects is, in general, many-to-many and transformations may be repeated, a graphical display is essential to the user interface.

A brief sketch of the processing in the user interface is as follows: A user poses a query to the interface in terms of the observation representation identifiers of one of the systems (e.g. a Pep logsheet line or a PODPS file name). The user interface queries that system about these identifiers and from the response determines if the user query is satisfied. If not, then the user interface must issue another query (to the same or a different system). This query will be constructed using information gained in the first query, e.g. information on the mapping of identifiers from one system to another. The user interface repeats this process until the users original query is satisfied or the interface finds it cannot satisfy the query (e.g. no more queries possible or an inconsistent situation). The user interface is only allowed to execute queries of a system; it is not allowed to change data within any system. The user interface is of course subject to any data security restrictions.

Currently we are developing a prototype traceability user interface to demonstrate the utility of these techniques for reasoning with incomplete or inconsistent information.

4. Conclusions

The construction of scientifically useful astronomical archives requires that traceability be maintained throughout the observing process. Using HST as a case study we have presented a general model of traceability in an observatory and identified a number of requirements for the components of the observatory. We have discussed a traceability user interface which can respond to queries about observations, even in the presence of incomplete or inconsistent tracking information.

Our thanks go to R. Doxsey, J. Isaacs, R. Jackson, S. Lubow, D. McElroy, S. Parsons and F. Schiffer for useful discussions on traceability in the HST ground system.

Connecting an Archive with Large Astronomical Catalogues

François Ochsenbein

European Southern Observatory
Karl-Schwarzschild-Straße 2, D-8046 Garching bei München (FRG)

Abstract:

This paper summarises how the observations made with the EFOSC intrument are archived at ESO. The SIMBAD database is especially used to improve the reliability of the main astronomical parameters of the observed targets: positions, designations, spectral or morphological classifications. A software allowing the remote querying was developed for this purpose.

1 Introduction

A test archiving of the EFOSC (ESO Faint Object Spectrograph and Camera) data started in April 1986, to be used as a test before general Archiving at ESO. Over a one-year period (95 nights at the 3.6m telescope), 6574 data files were created, including 2865 calibration files and 3709 astronomical observations.

The Archive is organized in two parts:

1. The observations, with a complete description of each image called the "header", are stored on slow access media (presently tapes, and later optical disks). This part does not need to be organised as a data-base, each observation being stored as a sequential file designated by a *filename*.

2. The catalogue of observations, comparable to the log-book, is organised as a data-base, and is accessible at ESO through STARCAT (Space Telescope Archive and Catalogue, see *e.g.* Richmond *et al.*., 1987). The contents of such a catalogue should be such that "The information given should be sufficient to allow the astronomer to locate the data and to judge whether they will be suitable and of sufficiently high quality to justify the effort involved in obtaining them" (Wilkins, 1982). The contents of the catalogue may be divided in the following sections:

   - astronomical specifications: instrument, detector, purpose of the observation, telescope position, object designation ...
   - instrument setting parameters, like filters, grating ...
   - exposure specifications: exposure time, *filename*, location of calibration data , observer's name ...
   - secondary data extracted from other sources added to facilitate retrieval purposes, like a classification of the observed object.
   - evolving data: specifications about protection and availability. and various comments *e.g.* about quality.

2 Overview of the Archiving Process

A copy of the result of the observations is provided by the La Silla station, on magnetic tape. The data are originally stored with IHAP (an ESO-specific format) on magnetic tapes. The data are locally transformed to FITS format (Flexible Image Transport System, Wells *et al.*, 1982), to ensure better portability and flexibility. The Tape archive is physically organised by instrument, and for each instrument the *astronomical* and the *calibration* data are separated.

The present archiving processing involves the following steps:

1. *File conversion* from IHAP to FITS. Bad files or files specified by the observer as *not to be archived* are marked as "to be deleted". About 10% of the files are ignored, generally being data processed from other instruments.
   As many automatic corrections as possible are made in this step, mainly date and time errors.

2. *Supply the observer's name* and program identification associated to each file, since this information does not exist in the original files. The automatic procedure simply uses the telescope and the date as an index to derive these items.

3. *Instrument setting* checking : mainly the missing information is added for averaged calibration data (instrument setting specifications are removed from the IHAP directory for averaged files). This is done interactively, from a visual inspection of the contents of the neighbouring files. It is however frequent, especially for calibration data, that this information cannot be retrieved.

4. *Astronomical parameters* checking: mainly the telescope position versus the designation provided by the observer, using existing astronomical catalogues. This step may lead to reclassify some files, *e.g.* as calibration data or irrelevant data. The J2000 positions, and a "standard" target designation are derived if the comparison is successful.

5. *Generate the catalogue* entries from the corrected headers, with specification of the release date (the observations performed with the ESO telescopes are property of the observers). The Britton-Lee IDM 500/1 relational database machine is presently used to manage the Catalogue.

6. *Astronomical parameters* additions, derived from other sources (catalogues) to facilitate later queries.

7. *Add Files into the Archive :* this step involves copying the original data, with the information added from the processing, onto the Archive tapes, and keeping track of the position of each file.

The fourth step is especially useful for the usage of the Archive: it ensures that observations can be located in the Archive from commonly used designations, provides a correlation with already existing data, and allows to check *e.g.* the reliability of the telescope position. SIMBAD (see *e.g.* Egret, 1986; Egret and Wenger, 1987) is presently the largest source of astronomical data (stellar and non-stellar), and its bibliographical part is kept up to date.

3 Usage of SIMBAD

The main problem in using SIMBAD from ESO is the networking, since this database is physically located near Paris, in France, implemented on a different hardware, and uses a non-standard query language.

The software which was developed for the Archiving purpose uses the standard X.25 connection protocol, and dialogues with SIMBAD without a direct intervention of the operator: the remote login is done if necessary, the SIMBAD program is started, the questions are reformulated in the SIMBAD-specific way, and the results which are of interest for the Archive are stored. The program makes use of Proteus (Richmond, 1987) for the user interface. The usage of this program is made as simple as possible: Fig. 1 summarizes the commands, consisting of single letters, available to classify each observation.

Phase 2 of Archiving mainly consists in retrieving the TARGET name and position in J2000 of the observed objects. The **SIMBAD** data-base is used for this purpose.

The following menu is available:

- **C**: Question SIMBAD from coordinates
- **D**: Delete this object (transform to Calibration data)
- **I**: Question SIMBAD from a user-supplied identification
- **H**: Help on this phase
- **L**: Insert a comment in the FITS header
- =: The target is the same as the previous one
- **N**: Give a name to the target (without SIMBAD)
- **Q**: Quit (can't solve the problem ...)
- **R**: Read a file
- **$**: Spawn another process
- **S**: Setup to menu mode

Figure 1: Help Screen *Available facilities for Archiving*

For each observation, either the problem can be solved locally (the observation was *e.g.* misclassified, or the target is the same as the previous one), or SIMBAD is to be queried for more information. The C command of Fig. 1, for instance, remotely submits the query of *"Retrieve all known objects within a 5′ target circle around the telescope position, and present them by increasing distance from the supplied position"*. The existing objects, if any, are displayed on the screen, and the choice is generally easy by a visual comparison of the name provided by the observer and the set of designations existing in SIMBAD. The astronomical parameters which are of interest for the Archive Catalogue (2000 position, designations, classifications) are added as complementary data to the observation.

4 Results

About 60% of the targets observed with the EFOSC instrument were identified in SIMBAD without ambiguity; it should be noticed that, for faint sources, SIMBAD only deals with objects already quoted in the astronomical literature.

For point-like sources (stars), the recorded telescope position was closer than $1'$ to the target, except in the $0°$ to $-1°$ zone where the recorded declination has the wrong sign. For direct imaging, the recorded positions are therefore reliable enough to derive a "standard" identification of the target.

```
                       ESO Archive - Astronomical observations
+--------Setting---------+ +-------------Position--------------+ +--Pixels--+
|Tel.     3.6m  La Silla | |          R.A.     Dec.    Air.M/Ep| | x     337|
|Instrum.        EFOSC   | +-----------------------------------| | y     520|
|Aperture FREE           | |Tel.     219.712   -26.4825    1.02| |min    167|
|Filter   V BESSEL  # 420| |Target   219.5500  -26.4667 1986.27| |Max  16383|
|Grism    FREE           | |Target   219.7500  -26.5333  J2000 | |mean   704|
+------------------------+ +-----------------------------------+ +----------+
+---When, Who, Where-----+ +--------Target Designation & Classification------+
|date 05:14:27 1986/04/10| |Original    NGC5496 V 2MIN    Type                |
|Release date:           | |Names  NGC 5694                                  |
|Exposure time        120| |        C 1436 -263                              |
|Obs. Fusi Pecci         | |                                                 |
|Tape AD033.030  seq# 4401| |                                                |
+------------------------+ +-------------------------------------------------+
 Comments


+--------------The Program: Identification, Authors, Title---------------------+
| 037.05-020  by Fusi Pecci/Buonanno/Corsi/Renzini/ at Bologna/Roma/Berkeley  |
| White dwarfs and lower main sequence stars in nearby globular clusters.     |
|                                                                             |
+-----------------------------------------------------------------------------+
```

Figure 2: Example of a Display from ESO Archive

The layout of the resulting Archive Catalogue, as seen by the astronomer for querying, is shown in Fig. 2; any combination of the fields displayed in Fig. 2 may be used for queries, *e.g.* a combination of requirements on position, filter and exposure time. This figure incidentally shows also typical clerical errors (inversion of digits) that are easily corrected with the used procedure.

The used procedure, however, is not quite fast: if the result from the query on SIMBAD by position is generally displayed after a few seconds, the delay may be as high as 30 seconds during peak hours, due likely to both computer and network loads. It also happens, more frequently than wished, that the connection cannot be established at all.

If such a procedure cannot be claimed to be quite stable — modifications in the SIMBAD dialogue must be reflected in SIMBAD interface of the Archiving process — it however demonstrates the feasibility of remote querying with a minimal human intervention.

5 Future developements

STARCAT allows an access to several (nearly 40) astronomical catalogues; it is planned to develop the same kind of interface as used for Archiving to remotely get more information about the objects selected from any catalogue accessible in STARCAT.

The facility of loading the located files from a query in the Archive Catalogue will be implemented in the near future; the File Handler, a piece of software designed to optimize file transfers in Archives, is described by Richmond et al. (1987) and will be implemented in STARCAT for further image processing.

References

Egret D., 1986: *Bull. Inf. Centre de Données Stellaires* **30**, 25

Egret D., Wenger M., 1987: *SIMBAD – The present status and the future*, Astronomy from Large Databases (October 1987)

Richmond A., 1987: *The Proteus Software Handbook*, ST-ECF O-02 Science Data Archive vol. VIII (September 1987), version 1.3

Richmond A., McGlynn T., Ochsenbein F., Russo G, 1987: *The Design of a Large Astronomical Database System*, Astronomy from Large Databases (October 1987)

Wells D.C., Greisen E.W., Harten R.H, 1981: *Astron. Astrophys., Suppl. Ser.* **44**, 363

Wilkins G.A., 1982: in *Automated Data Retrieval in Astronomy, C. Jaschek and W. Heintz (eds), D. Reidel*, 193

THE EXTRAGALACTIC DATABASE OF THE LYON-MEUDON OBSERVATORIES

G. Paturel (1), L. Bottinelli (2), P. Fouque (2), L. Gouguenheim (2)
(1) Observatoire de Lyon, 69230 Saint-Genis Laval, France
(2) Observatoire de Meudon, DERAD'N, 92195 Meudon Principal, France

ABSTRACT:

In 1983, a general extragalactic database was created at the Observatoire de Lyon. Since 1985, this database is managed by both the Observatoire de Lyon and the DERAD'N of the Observatoire de Meudon. The data are now accessible at Meudon and Lyon Observatories.The database contains 59747 galaxies with their positions and cross identifications (104848 identificators as given in 23 catalogues). Raw data covering different domains of astrophysics are continuously collected for each galaxy. The aim of this work is to produce a permanently updated catalog of mean homogeneous data, similar to the BGC or RC2 catalogues made by G. and A. de Vaucouleurs and their collaborators. The mean data of this preliminary catalogue and also the raw data from the database are used in various astrophysical researches including: the statistical analysis of raw data, preparation of future observations, distance scale and Malmquist bias analysis, Large scale distribution of galaxies, etc... We report some results recently obtained. Apart from updating the database, we started the creation of an image database allowing the management of 2D informations.

A portion of this database will be available to the community, together with a simple database management software system. A version for microcomputers will be available too.

INTRODUCTION

It was an old dream to create an extragalactic general catalogue permanently updated. In 1930, the famous swedish astronomer K. Lundmark had already devised such a project, but this dream was never realized because of practical difficulties. The use of computers has made more easy the management of large sets of data by means of Data Base programs.

We will first describe what was the aim of our work when creating our extragalactic database (EDB). Then, we will describe what we did to reach such a target and what were the preliminary astronomical results obtained from this work. We will conclude with a brief description of projects we started recently to improve the EDB.

I - PHILOSOPHY OF THE DATA MANAGEMENT

The method of compilation is the one initiated by G. and A. de Vaucouleurs when they produced the RC1 (also known as the BGC) and RC2 catalogues. That means that we want to obtain mean homogeneous measurements in various fields of extragalactic research.

Figure 1: Scheme of our philosophy for data management

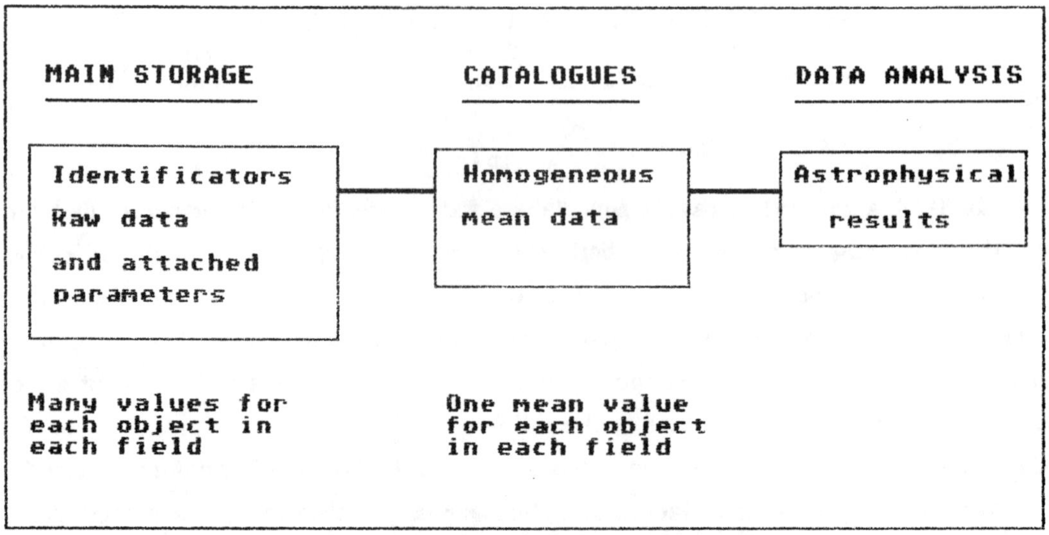

Starting from raw data, we apply corrections to put them in a homogeneous system, which then is used to calculate mean values. The consequence of such a method is that we only have to collect and to save raw measurements with parameters describing how the measurements have been obtained The scheme of such a method is illustrated in Figure 1

Such a method has several advantages. It is possible to produce a new catalogue at any time. The Request Language program (program for communicating with the computer when we have to use the data) is written to work just with the final catalogue, not with all the raw data. Thus, it is more simple and faster. One can say that this method imposes a given pre-treatment of data which is not necessarily the one wished by everybody. Fortunately, the treatment can easily be changed by modifying the program of reduction, on special request. However, the treatment used at a given time is generally the best known at that time.

II- WHAT HAS BEEN DONE

First we wrote the FORTRAN program to create the structure of the data base. The FORTRAN language has been chosen because it is used on many scientific computers. We used the most simple and most widespread instructions in order to insure the compatibility. The structure of the EDB is given on Figure 2.

The master file contains the 1950 Right Ascension and Declination, and ten pointers associated to ten data files. The description of the content of these data files is given in Table 1.

The inverse identification file permits the retrieval of a galaxy from any identificators (but of course a galaxy can be accessed by position or directly by its record number). Each record in each data file contains a pointer to address the next record, the data itself and some attached parameters, including the reference code. This reference code allows the retrieval the bibliographic reference. Today the data base contains 59747 objects with their most important data. The content of each data file is given in Table 1.

Figure 2. The structure of our EDB

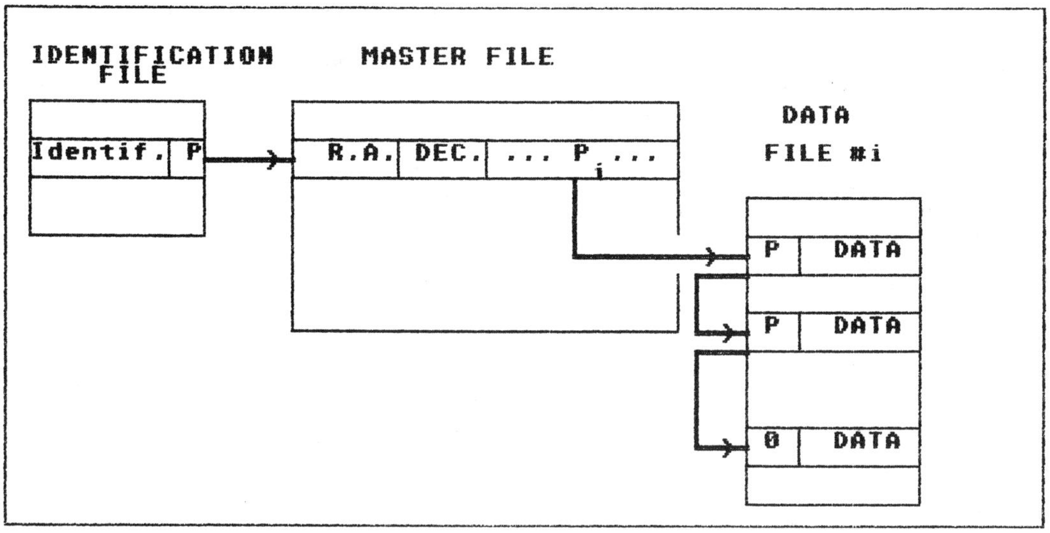

Table 1: Content of the data files

#	type of data	number	#	type of data	number
1	Identificators	104848	6	Dimensions	99506
2	morphology	57863	7	flux	111727
3	position angle	22112	8	vel. rotation	15792
4	neighbourhood	4945	9	vel. dispersion	878
5	brightness	30479	10	radial velocity	20303

Then, starting from large catalogues (MCG, UGC, ESO, CGCG, etc... see Paturel et al. 1987) we created the frame on which we built the data base by adding the data, collected in the litterature by ourselves or available on tape.

To test the completeness of our EDB we plotted logN vs. m, where N is the number of galaxies being brighter than magnitude m. To do that, we have estimated the apparent blue magnitude for 98% of the total sample using either measurements of magnitudes or blue diameters. The result appears on Figure 3, where we can see that the EDB is complete up to the m=15.0 magnitude (i.e. about 28000 galaxies). It is interesting to note that from the theoretical curve we can expect about 1,800,000 galaxies in a data base which would be complete up to m=18.0 mag.

III - WHAT RESULTS COME FROM OUR DATABASE

The EDB has been used in a large variety of extragalactic fields. It is possible to divide these studies in two categories: i) studies based on raw data ii) studies based on mean homogeneous data.

III-1 Results obtained from raw data.

Figure 3. Cumulative function logN=f(m). The line corresponds to the expected one for a complete catalogue, assuming an homogeneous Universe. The EDB is complete up the m=15.0 magnitude.

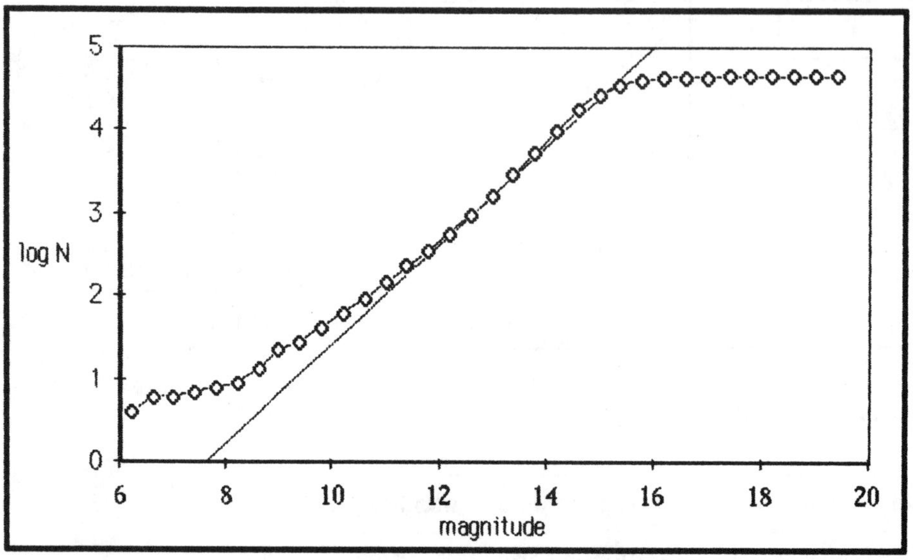

1) **Reduction of diameters to the standard D_{25} system**. The visual diameters from large catalogues (UGC, MCG, ESO, and SGC) have been compared with photometric ones leading to new relations to reduce them to the standard system D_{25}, defined by an isophotal limiting surface brightness of 25 mag.arsec^{-2}. (Fouque and Paturel 1985, Paturel et al. 1987).

2) **Homogeneization of HI-data**. All the 21-cm line widths have been converted to measurements at both 20% and 50% of the peak. HI fluxes and heliocentric HI-radial velocities have also been converted to a mean system.

3) **morphological types**. A preliminary solution has been reached. Some significant discrepancies have been found between the different scales. In addition, two new relations connected to morphological types have been found: the ratio of blue and red visual diameters from Nilson's catalogue and also the difference of inner and outer brightness codes as given by Vorontsov- Velyaminov can be used to roughly estimate the morphological type

4) **Magnitude**. Combining D25-diameters and mean surface brightness or morphological type, it is possible to estimate magnitudes for some statistical studies (Paturel et al. 1987). These relations have been used for the study of the completeness of the EDB (see for example Fruscione, 1987)

5) **Large Scale Structure of the Universe**. From a Flamsteed's projection in supergalactic coordinates (Figure 5), we suggested that some clusters could be connected to form a huge flat

system. The pole of this "hypergalactic" plane has been determined (Paturel et al. 1987). This plane has an S-shape in supergalactic Flamsteed's projection.

Figure 5. Flamsteed's equal area projection showing all the galaxies of the EDB in supergalactic coordinates. A S-shaped concentration appears which could be a huge "hypergalactic" system.

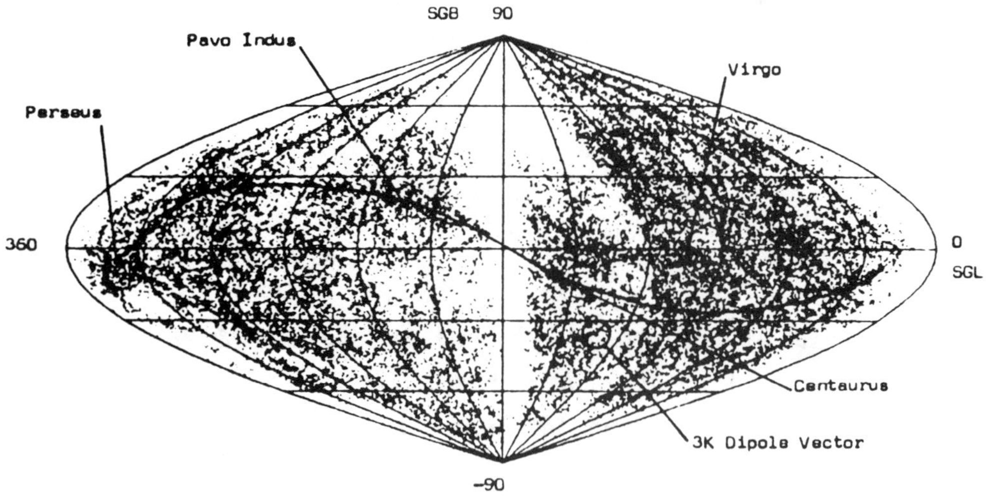

III-2 Results obtained from mean homogeneized data

1) **Groups of galaxies.** Using the large sample of galaxies with known radial velocities we re-analyzed the groups of galaxies with a method similar to the one introduced by Huchra and Geller. The preliminary results are reported by Fruscione (1987).

2) Preparation of observations. The data base has been used to prepare new observations, mainly for radio astronomy.

Hidden galaxies. Using IRAS data, we selected objects near the galactic plane and having IRAS-colors typical of galaxies. Among these objects, 7 were demonstrated to be galaxies when they were detected with the radiotelescope of Nancay at 21-cm (Bottinelli et al., in preparation)

Megamasers. By selecting IRAS objects having high infrared surface brightness $m(IRAS)+5\log D_{25}$, 10 Megamasers were discovered in Nancay at 18 cm (Bottinelli et al. 1985a, 1985b, 1986).

Elliptical IRAS galaxies. Eleven E or L IRAS galaxies selected from the EDB have been detected in Nancay (Boisson et al., in preparation)

3) Another use of the EDB is, of course, the preparation of catalogues. A general catalogue of cross-identifications is ready and a part of the EDB will be used in collaboration with G. de Vaucouleurs and his co-workers at the University of Texas to prepare the Third Reference Catalogue (RC3).

IV - WHAT PROJECTS ?

First of all, we want to maintain the EDB by adding new published data and new objects (e.g. faint galaxy companions, and faint cluster galaxies).

The second and more ambitious project we just started is the creation of an image data base allowing us to obtain an image of a galaxy and of its environment in the same way we obtain any other information. That will be interesting for many reasons: Identifications will be more secure, interaction and morphology will be well described and such a data base will allow us to add more easily some faint cluster galaxies. In addition, it would be possible to save some 2-D informations (line profiles, isophotal maps, etc...).

REFERENCES

Bottinelli, L., Fraix-Burnet, D., Gouguenheim, L., Kazes, I., Le Squeren, A.M., Patey, I., Rickard, L.J., Turner, B.E., 1985a, Astronomy and Astrophysics, *151*, L7

Bottinelli, L., Gouguenheim, L., Le Squeren, A.M., Dennefeld, M., Paturel, G., 1985b, IAU Circ. No 4106

Bottinelli, L., Gouguenheim, L., Le Squeren, A.M., Dennefeld, M., Paturel, G., 1986, IAU Circ. 4231

Fruscione, 1987, D. E.A., Observatoire de Meudon, DERAD'N

Fouque, P., Paturel, G., 1985, Astronomy and Astrophysics, *150* 192

Paturel, G., Bottinelli, L., Gouguenheim, L., Fouque, P., 1987, Astronomy and Astrophysics, (in press)

Paturel, G., Fouque, P., Lauberts, A., Valentijn, E.A., Corwin, H.G., Vaucouleurs, G.de, 1987, Astronomy and Astrophysics, (in press)

Vaucouleurs, G.de, Vaucouleurs, A., 1964, Reference Catalogue of Bright Galaxies, Texas University Press, Austin (BGC or RC1)

Vaucouleurs, G.de, Vaucouleurs, A., Corwin H.G. Jr, 1976, Second Reference Catalogue of Bright Galaxies, Texas University Press, Austin (RC2)

MIDAS TABLES: Present Status and Future Evolution

D. Ponz* and F. Murtagh*†

* European Southern Observatory, Karl-Schwarzschild-Straße 2, D-8046 Garching.

†Space Telescope — European Coordinating Facility. Affiliated to Astrophysics Div., Space Science Dept., European Space Agency.

1 Introduction

This paper describes the current state and development of the MIDAS table system in the context of scientific and statistical database management (SSDBMS). This latter — a burgeoning field in many application domains — has much to offer in terms of recommendable and achievable goals.

Astronomical catalogues of objects and small numbers of unstructured attributes — the major type of astronomical data stored and archived to date — give rise to problems which are more typical of the information retrieval area than of the DBMS field. Efficiency of access may be a problem, but data formatting and query language were, in the past, fairly straightforward. This is changing, with the increasing storage of data of growing complexity. The evolution of the MIDAS table system is indicative of this trend.

We begin with a survey of the field of SSDBMS (— issues relating to astronomical databases are overviewed in Ochsenbein, 1986). We then describe the evolution of MIDAS tables within this context, concluding with some of the more immediate future directions which will be followed.

2 Scientific and Statistical Databases

Extensive database capabilities have grown out of commercial statistical packages, which are widely used from the business to the scientific arenas. Unlike scientific database problems, the statistical database area could be said to have a solid tradition. Hext (1982) discusses the data storage and manipulation features of many statistical systems. Among those covered in this article, with database properties tabulated and cross-referenced, are: SAM (a microcomputer system), SAS ("Statistical Analysis System"), SPSS ("Statistical Package for the Social Sciences"), BMDP ("Biomedical Package"), Genstat, Minitab, and P-Stat. All of the foregoing are well-known statistical systems. Also covered are: SIR (a DBMS system specialising in scientific data); RAPPORT (a relational DBMS, from Logica, London); IDMS (a network DBMS, from Cullinet Software Corp.); and RAPID (a relational system developed by Statistics Canada).

The mainstay of most traditional DBMSs lies in business applications. Scientific and statistical data often present different characteristics. Härder and Reuter (1985) include the following areas among those where "non-standard" DBMSs are required:

- image and speech processing systems (where image frames and spectra are to be stored);
- text and information retrieval;
- geographic information systems;
- office automation, involving forms systems (hierarchically structured user interface units of data);
- expert system knowledge bases.

together with scientific and engineering (CAD/CAE, — computer aided design and engineering) databases.

Some of the characteristics of scientific databases (Shoshani and Wong, 1985) are:

1. In traditional commercial applications, capabilities for frequent updating must be supported in the form of transactions, whereas in the scientific domain, the experimental (or observational or statistical) results are rarely updated but rather are "reduced" as they are given.

2. Ancillary data (calibration data, instrument modes, experiment and environment details) are often stored for later use; while inextricably part of the data, ancillary data nonetheless give rise to added complexity.

3. While "entities" are often the basic building blocks of traditional systems, "cases" (with many values) may be more relevant for scientific and statistical data. Examples include vectors of time–varying values, astronomical or geographical coordinates, spectra, or closely associated textual information.

4. Special data types (temporal data, spectral data, survey data, irregularly sampled measurements) may need to be supported. (For temporal databases, see McKenzie, 1986). Special operations may need support (e.g. for statistical sampling and aggregation; interpolation and estimation; for handling various types of missing data values; or for data reliability estimates).

5. Scientific and statistical data are often available is very large quantities. Efficient compression techniques may be required for voluminous, sparse data. Transposed data matrices may be required for efficiency in addition to the usual "horizontal" record structure. Multidimensional data structures (e.g. grid files, multidimensional binary search trees, etc.) and search methods (e.g. best match or nearest neighbour searching) may be needed.

6. Finally, more "intelligence" may be needed for checking on data integrity, producing summary tables, and so on. This need for more semantic knowledge in databases is especially required in well–focussed application areas such as astronomy.

With extra programming, such requirements can be handled on top of vendor–supplied DBMSs. But this is inconvenient for the individual scientist or statistician, and effectiveness is compromised by less than suitable storage structures, query optimization methods and buffer management strategies. A consequence of the requirements looked at is that DBMSs and query languages for scientific and statistical databases may need to go beyond what is desirable in more traditional (business) fields. For further references in this area, the following may be recommended: Eastman (1985), EUROSTAT (1986), Gebhardt (1987), and Murtagh (1987).

Not all the characteristics indicated above are equally relevant for astronomical applications. In Section 4 below we look at some of the more important issues in more detail.

3 The Present Table Concept

It is obvious that the availability of modern detectors and sophisticated instrumentation has changed the methodology of the experimental astronomer during the last decades. The change is due not only to new technology, but also to the quantitative increase, by several orders of magnitude, in the volume of observational material.

The astronomical community has dedicated a substantial effort to the development of image processing systems to solve the critical points indicated above (see Di Gesù *et al.*, 1985 and 1986). Several image processing systems are now well established in the community, and are widely available for data analysis; we will not enter here into a detailed description of the different systems. Our interest will be focussed instead on an aspect of MIDAS (Banse *et al.*, 1987) that is not found in any of the systems metioned above, and it is especially relevant in the context of this conference. This aspect is the concept of tabular data, the MIDAS Table System (Grosbøl and Ponz, 1985).

The data structures in MIDAS were designed for astronomical applications. There are three high level data structures available to the user: images, tables and functions.

- Images are raster type data, used to store uniformly sampled information. Pixels in an image can be referred to by pixel number or by users' coordinates.

- Tables are collections of n–tuples which represent the properties of objects in a structured form. They are used to store the results of the analysis of images, and to keep catalogues and small databases inside the image processing environment.

- Functions are linear combinations of a basic set of system-defined functions. They are used to analyse images and tabels and to model phenomena.

At present, MIDAS provides an extensive set of algorithms operating on these structures, and allowing information to be transfered between them.

The original idea of the tabular data structure grew out of the developments of two-dimensional photometry in the IHAP system, the image processing system which was a precursor of MIDAS and is still in use today (Middelburg, 1983): the analysis of images led to lists of coordinates, associated flux values and morphological parameters. In this sense, most image processing systems use tables in one form or another, whether visible to the user or not.

The next step taken in MIDAS was to implement this idea in an efficient form, generalizing the concept of a simple list of numbers. Given the mixed type of information to be stored, the first requirement was to support different data types, including character strings, integer numbers and floating point representation in either single or double precision. A simple implementation, using a standard text file, would have lead to a very restrictive design, not conforming to the required accuracy and flexibility in the representation of the data.

The next problem in the implementation was to decide on the physical organization of the files. The traditional approach is to organize the data horizontally, i.e. all the variables corresponding to an object are stored together in a record. This organization optimizes the process of creating the table, but access to a few of the variables requires access to the full data set. The alternative is to organize the data vertically, i.e. the values of a variable are stored together for all the objects. This organization, called a transposed file, optimizes access to the variables. Given that statistical analysis and graphics typically involves only a few of the variables at each step, this last organization was selected. This implementation provides also more flexibility in the definition of new variables generated during the analysis of the data, which can be dynamically appended to the original table.

Additional requirements were a flexible user interaction with the data via on-line editors and graphics and an easy-to-use interface with the image processing algorithms.

The final design was, therefore, a binary transposed file, corresponding to the base table in the relational model. Rows in the table can be accessed by sequence number, or by a key value in a previously defined reference column. Columns can be accessed by sequence number or by label.

A table contains additionally a selection column, which allows the user to define subsets in the table by means of logical expressions involving the data values in columns of the table. Missing values are implemented system-wide, the internal manipulation being transparent to the user.

An extensive set of functions can be performed on the tables, for input/output, restructuring, mathematical operations and analysis. Tables can be created from, or output to, standard text files. Restructuring includes sorting, copying, deleting, and concatenating variables or objects in the table. New variables in a table can be derived from previous variables using standard mathematical expressions. Information in the table can be analysed using deterministic models, or multivariate statistical methods

4 The Future Outlook

During several years of use. the concept described in the previous section has proven to be very efficient, not only in relation to data analysis, but also for implementation of system-related data structures required by the image processing system. This includes the storage of look-up-tables, data catalogues and implementation of "virtual cursors" for emulating interaction in batch processing.

However, some questions, not considered in the original design, become relevant if the amount of information exceeds a certain threshold. This is particularly important in the case of distributed architectures and in connection with the archiving of large data sets in specialized devices (e.g. database machines or optical storage media). We summarize here some of these questions which will point to future developments of the present concept of table.

1. **Security**: The present approach to the problem of security is based completely on the services provided by the underlying operating system. This provides enough functionallity if tables are private files in the users' workspace. However, the increase of information, and the maintenance of centralized information, requires some new considerations about data security. As a very illustrative example, we can cite the large astronomical and instrumental databases used by the Hubble Space Telescope (HST) modelling package (Rosa and Baade, 1986); the requirement is not to restrict access to the information, but to avoid unauthorized modifications of the database. Some level of security can be achieved by defining a suitable policy for the management of the data by the system administrator, but this policy has to be complemented by system functions, protecting the data at the level of a variable in the table.

   An extension of the present history mechanism, in order to record the modifications of the table in more detail and, where necessary, to generate old versions of the file, is a more ambitious plan for the security of the data.

2. **Data integrity**: This aspect refers to the accuracy and correctness of the information in the table. The issue is especially relevant in the context of SSDBMS, although it is normally handled with *ad-hoc* solutions. Currently, most of the integrity checking is done by user-written code in a given application. Some extra description, associated with the table variables, will substantially contribute to the integrity of the information. In particular, the processing of the units associated with each variable and the definition of the variable domain, either implicitly by the associated units, or explicitly (e.g., data has to be positive), is a first step in this direction. Future plans include also the definition of high level data types, like *measurement*, that will have an error associated.

3. **Semantics**: At present, the only information about a variable that the system "understands" is its internal format. Implicit conversion between this internal format and the external representation, defined by the user, is done automatically. In this way, one can use a single precision algorithm with a double precision variable in a table, without explicit data conversion. Therefore, there is some basic knowledge in the current design. The next step is to introduce high level knowledge — astronomical or image processing related — associated with the variables. This knowledge will be introduced by a systematic treatment of the units associated with the variables and, more importantly, by introducing high level data types (*measurement*, associating a value with its error) and table types (time series, table containing a variable with increasing values, etc.).

4. **Extending the relational model**: As indicated in the previous section, the present implementation in MIDAS corresponds to the concept of *base table*, with operators providing some of the functions of the relational model. Recent years have seen the relational model coming to the fore, with the feeling that previously widely-used models (network and hierarchical) were less useful in practice and were not as theoretically well-founded. Much current research aims at pushing the relational model further. The need for semantic enhancements is a widely-felt need (see Gebhardt, 1987).

   Approaches to extending current relational DBMS capabilities can involve one of the following strategies.

   (a) Extra features may be added to standard DBMSs to cater for special requirments. Alternatively, interface systems can connect together DBMSs with statistical packages and graphics software.

   (b) Formal extensions to the relational model may be defined. Nonfirst normal form relations use both atoms and also arbitrarily nested sets of these as tuple components. Extended nonfirst normal form models allow the handling of various data types (e.g. ordered sets or matrices).

   The nonfirst normal model (referred to as the NF^2 model) allows the attribute values of relations to possibly be themselves relations. Fig. 1 (similar to a figure from Pistor and

Traunmueller, 1986) illustrates a possible NF^2 table in the scientific domain. A theoretical treatment of the NF^2 model is available in Schek and Scholl (1986). In some astronomical applications, involving collections of spectra, of images, or of other data, the user ought to be provided with at least some of the functionality which results from such extensions of the relational model.

observations					
object	*data*				
	date	*exposure*	*spectra*		
			start	*step*	*counts*
SN 1988a	Jan 20	300	3400.	1.5	402. 420. 510. ...
	Jan 28	300	3400.	1.5	402. 420. 510. ...
	Feb 2	300	3400.	1.5	402. 420. 510. ...
SN 1988b	Apr 1	300	3400.	1.5	402. 420. 510. ...
	Apr 8	300	3400.	1.5	402. 420. 510. ...
	Apr 16	300	3400.	1.5	402. 420. 510. ...

Figure 1. Example of nonfirst normal form where attribute values of relations are themselves relations.

5. **Query Languages**: As a result of the special needs in the scientific area, suitable query languages are a subject of much interest. A survey of recent work involving query languages for statistical databases is available in Ozsoyoglu and Ozsoyoglu (1985). As discussed by Ghosh (1986), deletion and updating are not often practised in the statistical — nor also the scientific — database context: outlier handling or the use of robust procedures are made use of instead. This indicates that query languages in the scientific domain differ in required functionality from traditional (i.e. business) domains.

   Extensions of the basic relational model which would suitably support collections of images, including spectra, were mentioned above. The query interface is part of this area. Although many new query languages in the area of SSDBS are still at a prototype stage, commercial implementations are becoming available (e.g. PISA/DB from INFODAS which claims to support NF^2 relations).

References

1. K. Banse, D. Ponz, C. Ounnas, P. Grosbøl and R. Warmels, "The MIDAS Image Processing System", in L.B. Robinson (ed.), *Instrumentation for Ground-Based Optical Astronomy.* Springer Verlag, 1987 (in press).
2. V. Di Gesù, L. Scarsi, P. Crane, J.H. Friedman and S. Levialdi (eds.). *Data Analysis in Astronomy I.* Plenum Press, New York, 1985.
3. V. Di Gesù, L. Scarsi, P. Crane, J.H. Friedman and S. Levialdi (eds.). *Data Analysis in Astronomy II.* Plenum Press, New York, 1986.
4. P. Grosbøl and D. Ponz, "The MIDAS Table System", *Mem. S. A. It.*, **56**, 429–437, 1985.
5. C.M. Eastman, "Database management systems", in M.E. Williams (ed.), *Annual Review of Information Science and Technology*, Vol. 20, Knowledge Industry Publications Inc., White Plains, New York, 1985, pp. 91–115.

6. EUROSTAT, *Proceedings of the Third International Workshop on Statistical and Scientific Database Management*, Statistical Office of the European Commission, Special Publication, 1986.

7. F. Gebhardt, "Semantisches Wissen in Datenbanken — Ein Literaturbericht", *Informatik-Spektrum*, **10**, 79–98. 1987.

8. S.P. Ghosh, "Statistical relational tables for statistical database management", *IEEE Transactions on Software Engineering*, **SE–12**, 1106–1116, 1986.

9. T. Härder and A. Reuter, "Architektur von Datenbanksysteme für Non–Standard–Anwendungen", in A. Blaser and P. Pistor (eds.) *Datenbank–Systeme für Büro, Technik und Wissenschaft*, Informatik–Fachberichte Nr. 94, Springer–Verlag, Heidelberg and New York, 253–286, 1985.

10. G.R. Hext, "A comparison of types of database system used in statistical work", in H. Caussinus et al. (eds.) *COMPSTAT 1982*, Physica–Verlag, Vienna, 272–277, 1982.

11. E. McKenzie, "Bibliography: temporal databases", *SIGMOD Record*, **15**, 40–52, Dec. 1986.

12. F. Middelburg, *IHAP Users Guide*, ESO Operating Manual, 1983.

13. F. Murtagh, "Scientific databases: a review of current issues", (forthcoming, 1987).

14. F. Ochsenbein, "Data storage and retrieval in astronomy", in V. Di Gesù, L. Scarsi, P. Crane, J.H. Friedman and S. Levialdi (eds.), *Data Analysis in Astronomy II*, Plenum Press, New York, 305–313, 1986.

15. G. Ozsoyoglu and Z.M. Ozsoyoglu, "Statistical database query languages", *IEEE Transactions on Software Engineering*, **SE–11**, 1071–1081, 1985.

16. P. Pistor and R. Traunmueller, "A database language for sets, lists and tables", *Information Systems*, **11**, 323–336, 1986.

17. D. Ponz and F. Murtagh, "Image processing, databases and statistical software: the common interface in MIDAS", *Statistical Software Newsletter*, **12**, 129–132, 1986.

18. M. Rosa and D. Baade, "Modelling Space Telescope Observations", *The Messenger*, **45**, 22–27, 1986.

19. H.–J. Schek and M.H. Scholl, "The relational model with relation–valued attributes", *Information Systems*, **11**, 137–147, 1986.

20. A. Shoshani and H.K.T. Wong, "Statistical and scientific database issues", *IEEE Transactions on Software Engineering*, **SE–11**, 1040–1047, 1985.

THE XUV WIDE FIELD CAMERA ON ROSAT: PLANS FOR THE DATA HANDLING AND ANALYSIS

J.P.Pye and C.G.Page
X-Ray Astronomy Group, Physics Department, Leicester University, Leicester, UK.
G.K.Rochester, T.J.Sumner, S.D.Sidher and J.J.Quenby
Astrophysics Group, Blackett Laboratory, Imperial College, London, UK.

ABSTRACT. In addition to its large X-ray telescope, the W.German ROSAT satellite will carry an XUV telescope (the Wide Field Camera, WFC) provided by the United Kingdom, and covering the wavelength range from 60 to about 600 Å. We describe the plans for the handling and analysis of ROSAT data in the UK, and discuss some of the results expected from the all-sky XUV survey.

1 Introduction

The W.German ROSAT satellite (see e.g. Trümper 1984) will carry an X-ray telescope (XRT) and a UK XUV telescope (the Wide Field Camera, WFC[1]; see e.g. Courtier 1986, Pye 1986, Wells 1986). The primary scientific objective is to perform all-sky surveys, over a 6 month period, in the X-ray (6–80 Å) and XUV (60–200 Å) wavebands. Both telescopes view in the same direction.

The WFC telescope consists of three nested, gold-coated aluminium mirrors with a microchannel plate (MCP) detector at their common focus. Any one of 8 filters, mounted on an aperture wheel in front of the detectors, can be selected to define the wavelength passbands and suppress geocoronal background radiation which would otherwise saturate the detector count rate. During the survey, 2 filters will be used, covering the waveband from about 60 up to 200 Å. For pointed observations, additional filters will be available to extend the spectral response to longer wavelengths. The sensitivity of the instrument with each filter is given in Table 1.

The spatial resolution of the WFC is about 1 arcmin FWHM within about 1 degree of the optical axis, and for the survey is about 2 arcmin FWHM averaged over the field-of-view. The location accuracy (error circle 90% confidence radius) for point sources will vary from ~ 1 arcmin for the weakest ones (set by photon counting statistics) to $\lesssim 25$ arcsec for strong sources (limited by the WFC attitude determination and systematic errors).

During the survey, the viewing axis will scan in ecliptic latitude at a rate of ≈ 4 degrees/minute and in ecliptic longitude at ≈ 1 degree/day. Hence, a source will come within the WFC field-of-view (FOV) for about $5/\cos(\textit{ecliptic latitude of source})$ days for durations of up to about a minute every 1.5 hours.

ROSAT is expected to be launched in early 1990. The Flight Model WFC is currently (October 1987) undergoing integration and test before delivery to the spacecraft in the Spring of 1988.

The aims of this paper are to describe the plans for the handling and analysis of ROSAT data in the UK. The primary tasks are: (i) to produce from the WFC survey data, a catalogue

[1]The WFC project is supported by the UK Science and Engineering Research Council. The WFC instrumentation and the Ground System for WFC mission operations and data processing are being developed by a consortium of five UK institutes: University of Leicester, SERC's Rutherford Appleton Laboratory (RAL), the Mullard Space Science Laboratory—University College London, University of Birmingham and Imperial College of Science and Technology—London. The WFC design and development is the responsibility of the many WFC personnel at the Consortium institutes.

of fluxes and celestial positions of 'point-like' XUV sources; and (ii) to provide a flexible system for interactive analysis of WFC data from pointed observations and the survey, and of the UK share of XRT pointed observation data. We also discuss some of the results expected from the all-sky XUV survey.

2 The Ground System

The WFC Ground System comprises the necessary hardware, software, manpower, operational procedures and documentation for handling of all WFC observations and for the UK share of XRT pointed observations. The Ground System is responsible for:

- Generation of WFC command procedures.
- Monitoring WFC status, health and performance.
- Analysis of WFC all-sky survey data to produce a catalogue of XUV sources.
- Pointed observations:
  - Administration of UK observation requests within UK.
  - Processing, distribution and analysis of data from WFC observations.
  - Distribution and analysis of data from UK XRT pointed observations.

2.1 WFC Ground System Sites

Control of ROSAT and reception of telemetry data will be handled by the German Space Operations Centre (GSOC) at Oberpfaffenhofen, using their ground station at Weilheim. The main elements of the WFC Ground System are :

- **WFC Quick Look Facility (QLF),** at the ROSAT Scientific Data Centre (RSDC), MPE, Garching — for WFC command procedure generation, WFC health monitoring, QL analysis of WFC data, and acting as the WFCC interface with GSOC and MPE for mission operations.
- **ROSAT UK Data Centre (UKDC)** in UK, at RAL — for routine processing and distribution of WFC survey and pointed observation data.
- **ROSAT UK Survey Centre (UKSC)** in UK, at Leicester University — for full and long-term health and performance monitoring of the WFC, WFC calibration, final survey analysis, and large-scale background analysis.
- **WFC Consortium Institutes (WFCCI)** — for analysis of survey and pointed observation data.

The data flow between Ground System sites is shown in Figure 1.

2.2 Software Development

The data reduction and analysis software is being developed jointly by the five groups in the WFC consortium. At an early stage in the software development programme the tasks were divided into modular packages and each group given responsibility for one set of modules. The interfaces between these modules were defined very carefully and subsequently put under a change control procedure. Software portability has not been a significant problem since all those involved in ROSAT are using VAX computers running under VMS. The VAX computers of the WFC groups are all inter-connected via the UK Joint Academic Network (JANET) which is used very heavily for file and mail transfers.

Program coding standards were also established at the outset: programs are written in ANSI Fortran-77 with certain restrictions but with a few extensions allowed. All of these extensions appear in a similar form in the current draft of Fortran-8x.

Software for the initial reduction stages, up to and including the conversion of photon events into images, is specific to the WFC and the data files use ordinary VMS file structures. The scientific analysis software will be designed to form part of the Starlink *Asterix* package which was developed to handle data from *Exosat* (Peden 1987). This is now being converted to run under the new Starlink Software environment called ADAM. All files produced by the *Asterix* package will use the Hierarchical Data System (HDS) which is the most widely used part of the Starlink software environment. HDS structures are open-ended and self describing; this makes them particularly suitable for calibration files. If large arrays (such as images) are stored in HDS files they may also be accessed via the VMS mapping mechanism.

3 Processing of the Science Data

The coordinates and arrival times of the events registered by the focal plane detector will form the primary scientific data from the WFC. Besides photon events, there are 'background' events caused by charged particles etc. The raw telemetry data from the WFC will occupy $\sim$ 50 Mbyte/day, i.e. $\sim$ 10 Gbyte for the 6-month all-sky survey and $\sim$ 20 Gbyte per year of pointed observations.

3.1 All-Sky Survey

During the scanning phase the WFC is expected to produce an average of $\sim$ 100 events/second for 6 months. Effectively, these have to be sorted into a fixed frame and imaged with a resolution of better than 1 arcminute. In total this means that some 1.6×10^9 events have to be sorted into a set of $\sim 3 \times 10^8$ pixels. This will be handled by dividing the celestial sphere into a coarse grid along lines of ecliptic longitude and latitude with a spacing of about about 3°. The attitude of the instrument, computed from star-tracker data, will be used to sort each incoming photon into the appropriate celestial cell. The original (instrument) coordinates of each photon event are preserved; a set of linked lists is used, one for each celestial cell, the whole structure being stored on a random-access file. One such file will be produced at the UK Data Centre for each day of operations; this will be distributed to the WFCC institutes for analysis.

As the survey observations proceed, the data will be analysed in the UK with a time lag of about 2 weeks. Each WFCC institute will receive all the WFC data but will have the prime responsibility for the analysis of a different region of the celestial sphere. The first operation will be to merge data from the current day with the data base which has accumulated events collected earlier in the survey. Each linked list will be merged with the linked list for the corresponding celestial cell in the data base. The scan path advances in ecliptic longitude by 1° per day, so that a new strip of sky will be available for analysis every few days. The data for this strip will be contained in a small number of cells, i.e. linked lists, on the data base. After analysis these lists can be archived and the disc space released for re-use by a garbage collection phase. The analysis of each strip of sky will proceed by forming images and searching these for sources. The exposure time of each detected source will be computed, and the flux derived. A time-series may be extracted from the data for the stronger sources. A source on the ecliptic will be scanned some 80 times over a five-day period; those at higher ecliptic latitude will be in the field of view for rather longer. Small regions around each ecliptic pole are likely to be scanned for the whole of the survey period and will be treated specially in the analysis.

Since it is unlikely that a uniform survey analysis at maximum sensitivity will be possible while the survey data are still being accumulated, it is planned to perform a re-analysis of the complete survey database after the end of the survey phase of the mission and following review of the experience gained during the previous analysis.

In order to monitor the general progress of the survey, the QLF will analyse a small fraction ($\sim 20\%$) of the data within about a day of the observations being made.

3.2 Pointed Observations

The pointed phase processing will be somewhat simpler: the incoming events will be sorted into data-sets structured with a 'small-map' scheme along the lines of those used by *Einstein* and *Exosat*; these pointed observation data-sets will be compatible with those described above for the survey data. A software package will be provided so that users can form images and extract temporal and spectral information in the usual way.

3.3 Generation of Test Data

Pre-launch, data to test the Ground System software are available from two sources: (i) the WFC instrument, (ii) a software scientific data simulator. The latter is also used to predict background count rates and exposure times.

4 Handling of Calibration Data

Instrument calibration data, derived both from pre-launch laboratory measurements and in-orbit observations, have to be available in suitable form to the scientific analysis programs. In-orbit calibrations will use both observations of selected celestial targets and an on-board calibration system. The latter projects, from a mounting rigidly locked to the mirror assembly, a rectangular array of about 80 fiducial spots on to the focal plane detector, to allow linearisation of the electrical readout of the detector.

Any method of handling the derived calibration data must allow both for parameters which vary with time and for improved knowledge of parameters as the mission progresses. These requirements are being met by use of HDS, which provides a flexible system for storing, updating, enlarging, deleting, retrieving, or altering the form of any particular subset of the data. Two structured 'container' files are used, one of which holds all of the spatial linearisation matrices (since these may vary with time) while the other stores the latest versions of all of the remaining calibration data.

5 Predicted Numbers of Sources

Due to the very limited observations so far made in the XUV band, any predictions of numbers of sources observable by the WFC will have large uncertainties. However, it is useful for mission planning purposes at least, to have some rough estimates. The two major contributors to the XUV source detections are expected to be cool stars and hot white dwarfs, with each yielding $\gtrsim$ 1000 sources in the all-sky survey, giving the potential for detailed statistical studies.

Pye and McHardy (1987) have discussed cool star observations with the WFC. The maximum distance at which quiescent emission from main sequence stars is expected to be detected is about 25 pc. At this distance the absorbing column of ISM is still low ($n_H \approx 5 \times 10^{18}$ cm^{-2}, transmission at λ100 Å $\approx$ 0.85) and the limiting factor is simply the low intrinsic luminosity of these objects. RS CVn systems, with their higher luminosities, are expected to be observable out to at least 100 pc, where the column density then becomes significant ($n_H \approx 2 \times 10^{19}$ cm^{-2}, transmission at λ100 Å $\approx$ 0.5). The WFC is expected to detect $\gtrsim$ 1000 cool main sequence stars. The apparent magnitudes (m_V) of main sequence stars at a distance of 25 pc are (Allen 1973): F0—4.6, F5—5.4, G0—6.4, G5—7.1, K0—7.9, K5—9.3, M0—11.0, M5—13.8. The corresponding surface densities (stars per 4π steradians) are: F—200, G—400, K—650, M—3300. Hence, for each spectral type, we would expect to detect in the WFC survey about 50%, 30%, 30%, 20% respectively of the total numbers of these stars in the volume out to 25 pc.

Barstow and Pounds (1987) have predicted the numbers of hot white dwarfs detectable by the WFC. For a white dwarf (WD) of a given temperature, the WFC is much more sensitive to a H-dominated stellar atmosphere than to one dominated by He. Hot WDs are sufficiently luminous in the XUV to be potentially observable out to distances approaching several kpc. However, it is likely that in practice, photoelectric absorption in the ISM will largely determine the viewing horizon. The WFC is expected to detect $\sim$ 2500 WDs during the all-sky survey. A total of approximately 1500 WDs is presently known, of which about 30% have temperatures to which the WFC is sensitive. Comparing this figure with the number of WDs that the WFC is expected to see, shows the value of the all-sky XUV survey for statistical analyses.

References

Allen, C.W., 1973, **Astrophysical Quantities**, Athlone Press, London.

Barstow, M.A. and Pounds, K.A., 1987, Proceedings of the NATO Advanced Study Institute *Hot Thin Plasmas in Astrophysics,* in press.

Courtier, G.M., 1986, *Journal of the British Interplanetary Society,* **39**, 217.

Peden, J.C.M., 1987, *Journal of the British Interplanetary Society,* **40**, 185.

Pye, J.P., 1986, *Adv. Space Res.,* **6**, No.2, 139.

Pye, J.P. and McHardy, I.M., 1987, Proceedings of the conference on *Activity in Cool Star Envelopes,* Tromsø, in press.

Trümper, J., 1984, *Physica Scripta,* **T7**, 209.

Wells, A., 1986, in **X-Ray Instrumentation in Astronomy,** ed. J.L.Culhane, *Proc. SPIE,* **597**, 146.

Table 1. ROSAT WFC Filters, Wavebands and Sensitivity

Filter Type [a]	Survey (S) or Pointed (P)	FOV Diam. (deg.)	'Mean' Wavelength (Å)	Bandpass (Å) (at 10% of peak efficiency)	Point-source Sensitivity (μJy) [c]
C/Lexan/B (×2)	S + P	5	100	60–140	1.0
Be/Lexan (×2)	S + P	5	140	112–200	1.4
Al/Lexan	P	2.5	180	150–220	7.3
Sn/Al [b]	P	2.5	600	530–720	220
TBD	P	2.5	TBD		

[a] An eighth filter is an interference type for use with the WFC UV calibration source.
[b] Provisional.
[c] For 5σ significance, exposure time of 2000 s (a typical value for each filter for the survey and for pointed observations) and 'typical' background.

Figure 1. ROSAT WFC Ground System: data flow between sites.

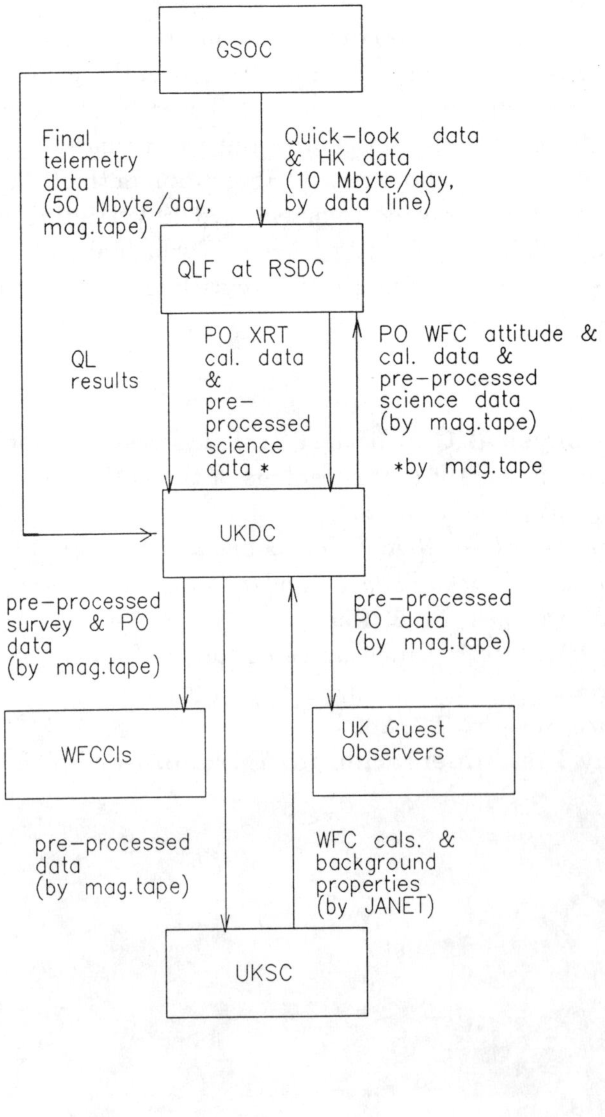

ACCESS TO ASTRONOMICAL LITERATURE THROUGH COMMERCIAL DATABASES

Joyce M. Rey-Watson
Center for Astrophysics
Cambridge, MASS., USA

Until recent years, astronomy and astrophysics have been the orphan children of the physical sciences from the point of view of online access to bibliographical information. Since such services became generally available, in the late 1960s and early 1970s, it appeared that the topic was either overlooked or ignored, at a time when the databases of physics and chemistry (e.g., *CHEMICAL ABSTRACTS* and *SCIENCE ABSTRACTS A: PHYSICS ABSTRACTS*) were charging ahead. (Medical science was the first discipline to appreciate the potential of the new technology in the mid-1960s, and a vast body of information is now available to the medical profession.)

It must be appreciated, however, that before the advent of the "space age," in 1957, the one major abstracting journal available was probably sufficient. *ASTRONOMY AND ASTROPHYSICS ABSTRACTS*, in English, made its appearance in 1969, superseding the *ASTRONOMISCHER JAHRESBERICHT*, which had run from 1899 to 1968, under different publishers. *ASTRONOMY AND ASTROPHYSICS ABSTRACTS* (compiled by the Astronomisches Rechen-Institut in Heidelberg and published by Springer-Verlag) is certainly still the first and most comprehensive source for most astronomers performing their own searches. There is a serious drawback to its usefulness, however, in that it is published only twice a year, which detracts from its timeliness in these days when the first stop for astronomers visiting the library is the latest collection of preprints. The Astronomisches Rechen-Institut has announced the intention of increasing the publication of the *ABSTRACTS* to four volumes per year.

While it would not be feasible, economically or otherwise, to make *ASTRONOMY AND ASTROPHYSICS ABSTRACTS* available for online searching retrospectively, it is to be hoped that the bulk of the current and future volumes will, eventually, be made accessible online. This undertaking will be described below, and as the process is still in its early stages, any information I may venture to give may already be out of date by the time this paper is published. (Any new developments will be discussed at the meeting.)

There are several ways in which literature searches can be undertaken, with good results, through databases marketed by vendors, in the United States, in Europe, and in Japan. Let us make it clear from the outset that we are discussing SUBJECT searching here, as searching by object is a frustrating and painful undertaking through these sources, giving poor, if any, results. We are now blessed with access to *SIMBAD*, about which you will hear much at this meeting from experts, and I will only sing its praises, and extol it as the only source for searching by OBJECT. Thus one can utilize the two approaches in combination – a search for Supernovae Remnants as a subject will be fruitful in the following sources, to be followed by a *SIMBAD* search to give information on the specific SN, defined as an object.

The database vendors mentioned here are primarily Dialog Information Systems, Inc., a wholly owned subsidiary of the Lockheed Company, and STN International, marketed by Fachsinformationzentrum (FIZ) of Karlsruhe, FDR, ESA-IRS and BRS. There are others, but for our purposes, and in the interests of brevity, we will confine ourselves primarily to these. There is much duplication of databases among vendors, and specific questions concerning availability should be referred to their Customer Service Departments. Contracts are required, and in return one receives a password and identification number, which enable one to access the database of choice. Charges are on a connect-time basis, and include telecommunications costs. Printouts of the results of the search can be obtained, either on or off line, in a choice of formats, with or without abstracts. In addition, some databases, though not all, provide a document delivery service, whereby copies of citations can also be ordered online, and are mailed promptly. There are also many document delivery services that are not specific to individual databases, and photocopies of articles may be obtained from them.

Searches may be performed by specifying author, title, subject, class code, or corporate source, in most databases. These elements can either be defined with a prefix or suffix, or searched "free text," which covers all occurrences of the terms searched, both in the title and the abstract. Keyword terms are assigned to items, and thesauri are usually available. Complex searches are conducted by the use of Boolean operators, "and," "or," and "not," and publication years, or ranges of years, can be specified. It should be mentioned that databases such as those described here are based upon printed equivalents, and that only terms included in the title, abstract, and assigned keywords can be searched. Using *SIMBAD*, no Boolean operators are needed, and objects are indexed wherever they appear in the text of an article. Thus, "free text" searching in these databases has a different meaning from the literal one that characterizes *SIMBAD*.

The databases described below were selected for their coverage of astronomy and astrophysics only. There are many others where other aspects of space science may profitably be searched for related fields of chemistry, physics, engineering, and instrumentation, for example. No reference is made to the costs involved. These change (rise!) frequently, and include charges for different telecommunications carriers. Those interested in performing online searches should contact vendors for price lists, contracting requirements, etc.

INSPEC

The online version of *SCIENCE ABSTRACTS, SECTION A: PHYSICS ABSTRACTS*, which also includes *SECTION B: ELECTRICAL AND ELECTRONICS ABSTRACTS, SECTION C: COMPUTER AND CONTROL ABSTRACTS.* (Institution of Electrical Engineers, Ltd., London, England. Distributed internationally by Information Services in Physics, Electrotechnology, Computers and Control.) Coverage is from 1969 to date, and it is possible to specify which section should be searched. It is updated every two weeks, corresponding to the issues of the printed version. This database is divided into two files on Dialog, 1969-1976, and 1977 to date, and is available as one file on STN International.

While *INSPEC* covers the whole field of physics, it does include the core journals of astronomy and astrophysics, and good results can usually be obtained. It is particularly valuable in searching peripheral topics, such as atomic and molecular physics, gravitation, cosmology, etc.

INSPEC does not provide a document delivery service.

PHYS

PHYS is the online version of the printed publication *PHYSIKALISCHE BERICHTE*, or *PHYSICS BRIEFS*, published by FIZ Karlsruhe, FDR, in cooperation with the American Institute of Physics. This file is available on STN International only, both from Karlsruhe and, in the United States, from Columbus, OH. It contains abstracts for journals, books, patents, reports, theses, and conference papers published since 1979 in all major languages, and all aspects of physics.

During the last year or more, a concerted effort has been made to increase the coverage of astronomy and astrophysics in *PHYS.* Anyone searching astronomy-related topics would do well to become proficient in the STN *PHYS* file. Not only has the coverage already been increased, but negotiations are in process to incorporate the *ASTRONOMY AND ASTROPHYSICS ABSTRACTS* tapes on the near future. When these are mounted, *PHYS* will undoubtedly become the database of choice in this field. In recent months, listings of objects have been added in the title field, thus adding to its flexibility. PHYS also, of course, provides good coverage in related areas of physics, though it is necessary to revert to INSPEC for citations preceeding 1979. Updating is very prompt.

An online document delivery service is available.

NASA/RECON and AEROSPACE DATABASE

The *NASA/RECON* database is the oldest, and most comprehensive, of databases that encompasses the entire field of space and space-related science, including astronomy, and dates back to 1962. This represents the printed versions of the *NASA SCIENTIFIC AND TECHNICAL AEROSPACE REPORTS (STAR)* and the *INTERNATIONAL AEROSPACE ABSTRACTS.* (*STAR*: U.S. National Aeronautics and Space Administration, Washington, D.C., and *IAA*: American Institute of Aeronautics and Astronautics, Inc., New York, N.Y.) The printed versions of these volumes are complementary, as both cover government-funded research. *STAR* covers unpublished report literature, including research funded not only by NASA but by other government agencies, foreign material, translations, contributed papers emanating from NASA-sponsored meetings, and other miscellaneous material. *IAA* similarly covers the open literature, monographs, journal articles, published conference proceedings, including individual papers presented, etc.

Access to the *NASA/RECON* database was originally limited to NASA-related facilities only, but was opened to NASA contractors, and academic and other interested institutions in the early 1980s, and is also available via ESA-IRS in Europe. In 1985, a somewhat limited version was made available as the *AEROSPACE DATABASE* through Dialog in the United States only. This does include all material from *STAR* and *IAA*, but omits miscellaneous material, such as work in progress, which is indexed in *NASA/RECON.*

Document delivery is available for all cited material: NASA reports (N numbers) may be obtained by online ordering either from the NASA/SCIENTIFIC AND TECHNICAL INFORMATION FACILITY, or from the NATIONAL TECHNICAL INFORMATION SERVICE (NTIS), and *IAA* items (A numbers) from the AMERICAN INSTITUTE OF AERONAUTICS AND ASTRONAUTICS/TECHNICAL INFORMATION SERVICE.

SCISEARCH

This is the online version of the *SCIENCE CITATION INDEX*, published by the Institute for Scientific Information, Philadelphia, PA. It is a multidisciplinary index for all fields of science and technology, and contains well in excess of six million entries, dating back to 1974.

The uses of *SCISEARCH* for our purposes vary from other databases described here. No abstracts are included, so subject searches must be performed on title entries only. It covers a wide selection of astronomical journals, and searching by author produces good results. However, the exclusive and most valuable feature of *SCISEARCH* lies in the indexing of citations. All references to previous work cited by an author in footnotes or bibliography of an article are indexed. Thus it is possible to input a reference, say, from

1975, and see a list of all subsequent papers which cited that reference. It is therefore possible to track the direction that research on a topic has taken from the original work reported, and which scientists and institutions are involved. Nor is there a limitation to references from 1974, as earlier works are included as citations appear.

Other useful applications of this feature are obvious. A quick check can be made of the publications of a scientist, and an assessment made of the quality of his or her research by the number of times the works were cited, and by whom. Thus bibliographies can be generated, and the citations to each item obtained.

Document delivery is available from ISI for any reference cited.

CURRENT CONTENTS: PHYSICAL, CHEMICAL AND EARTH SCIENCES

This very useful, weekly publication of the Institute for Scientific Information, has recently become available online through the BRS system. This is literally a collection of reproductions of tables of contents of journals, which are often published here before the journal issue itself is received. The coverage of astronomical and related journals is good, and the database is also an excellent source of items published in obscure or unusual journals that might otherwise be overlooked.

A document delivery service is available from ISI, as described in *SCISEARCH* above.

ASTRONOMY AND ASTROPHYSICS MONTHLY INDEX

This is a privately-produced index covering commercially available journals, observatory publications, preprints, and conference proceedings. Requests should be addressed to Olivetree Associates, Sierra Madre, California. This is a quick announcement service which attempts to bridge the gap between the semi-annual issues of *ASTRONOMY AND ASTROPHYSICS ABSTRACTS*. It is available in a printed version, and as monthly tapes distributed for mounting on inhouse computers for local searching. There are no abstracts and no annual index, as the accent is on timeliness, and the average monthly issue contains in excess of 1000 new articles.

The following databases are also useful in their coverage of certain aspects of the subject area, though not subject-specific as those described above:

CONF: A database available on STN International, which includes coverage of space science, astronomy and astrophysics. Schedules are included as soon as a conference is announced, sometimes years in advance. Coverage is from 1976 to date.

NTIS: (National Technical Information Service). This is based on *GOVERNMENT*

REPORTS ANNOUNCEMENTS, and is a very large database accessible from most vendors (including all those mentioned above). Its principle use lies in the coverage of the report literature on research supported by the Department of Defense and other government agencies, and possibly omitted from the NASA databases.

SPIN: (Searchable Physics Information Notes). This is the database of the American Institute of Physics, and has no printed equivalent. Basically this is confined to AIP and member societies' journals, including their conference series, and cover-to-cover translations of Russian titles, although other material is included.

DISSERTATION ABSTRACTS ONLINE

This is a listing of all theses accepted by universities in the United States from the year 1861. It is based on the printed source *Dissertation Abstracts International*, and prepared and published by University Microfilms International Inc., of Ann Arbor, Michigan. This includes theses on all astronomical topics, with lengthy abstracts given.

Copies of all theses are available from the publisher.

PASCAL

This is the French database which includes *PASCAL EXPLORE, E48: TERRESTRIAL COSMIC ENVIRONMENT, ASTRONOMY AND EXTRATERRESTRIAL GEOLOGY*, published by the Centre National de la Recherche Scientifique (CNRS). This is a very large, multidisciplinary database, which is particularly useful for searching peripheral topics. Abstracts both in the printed version and online are in French. This is also useful for its coverage of original Russian literature, and is the only available source of information on theses accepted by French universities.

A LIBRARY COLLECTION OF SOFTWARE DOCUMENTATION SPECIFIC TO ASTRONOMICAL DATA REDUCTION

Carol Rhodes, Michael Kurtz and Joyce Rey-Watson
Harvard-Smithsonian Center for Astrophysics
60 Garden Street
Cambridge, Massachusetts 02138 USA

ABSTRACT

We are creating a library collection of software documentation used in astronomical data reduction. In this paper, we discuss our objectives in establishing this collection. We present a list of acquired documentation with brief descriptions of the software and the hardware required. The collection will be catalogued and available for interlibrary loan through the Smithsonian Institution Library. Instructions on accessing this collection are included. The project is summarized with special consideration given to future directions of this collection.

OBJECTIVES AND FUTURE DIRECTIONS

Early in 1987 we began the uncertain task of establishing a library collection of documentation for data analysis systems used in astronomy. By uncertain we mean we had no prior model to follow, so that we could only be sure of our intentions to create a collection that would benefit the astronomical community.

Data analysis systems are much like a living organism in that they undergo frequent revisions. Scientists in return must strive to keep up to date while attending to their primary responsibilities. The task of updating the collection could be handled by library staff, freeing the scientists of the task and thus simplifying the scientists job of data reduction and analysis.

It is also urgent that we preserve the documentation for systems which are no longer in use, or are being phased out (e.g. systems that reduced data from now extinct satellites). Systems such as these are of significant historical interest. In order to establish an archive, we will be directing more effort towards acquiring the documentation for these systems.

The major software systems who distribute their documentation free and without restriction are frequently burdened with a high demand for documentation from scientists within a single large institution. A publicized and central location could help reduce unnecessary requests.

SOFTWARE DESCRIPTION

The following list describes the documentation we have thus far acquired. The descriptions have been taken from the documentation and edited to fit the format of this list. All documentation is in English unless otherwise specified. Dots indicate uncatalogued documentation and circled dots, catalogued.

- AIPS. The Astronomical Image Processing System is developed and maintained by the National Radio Astronomy Observatory (NRAO). A system of hardware/software for: processing *uv* data to make, improve the quality of, and display images in a variety of ways and for analyzing and combining images to obtain useful parameters. AIPS is based on POPS. FORTRAN.
  - AIPS COOKBOOK. Rev. and ed. by A. Bridle and E. Greisen. Provides a recipe approach to basic AIPS operations to assist beginning users. 5th ed. 15 Oct 1986.
  - ⊙ AIPS MEMO. A memoranda series.
  - AIPSLETTER. Ed. by E. Greisen, A. Bridle and D. Wells. A quarterly newsletter announcing new releases, revisions and updates.
  - GOING AIPS: A Programmers Guide to the NRAO Astronomical Image Processing System. Two volumes by W. Cotton et al. VOLUME 1 A reference manual for applications programmers. VOLUME 2 For those developing AIPS for new peripheral devices or computers. Version 15APR87.
- American National Standards & ISO International Standards for Information Technology. Published by the American National Standards Institute (ANSI). Agreed-upon solutions to a wide range of information technology problems. May 1987.
- ASTRONET. Maintained by the Osservatorio Astronomico di Trieste (OAT), this Italian Astronomical Network for the handling, processing and analysis of astronomical information, supports software developed by or contributed to and accepted by it. DEC VAX with DEC VMS operating system.
  - AGL Reference Manual. Developed by the Astronet Graphics Working Group (AGWG), the main author of the Astronet Graphic Library is L. Fini. For graphic applications of an intermediate level of complexity. FORTRAN. Version 2.1-15 Feb 1986.
  - DIRA manual. Developed by the Astronet Working Group on Database and Documentation, the main author of The Distributed Information Retrieval from Astronomical files is M. Nanni. Strictly VAX/VMS. FORTRAN and DCL. Version 1.4
  - HAGL User's Guide. Developed by the AGWG the main author of the High-Level Astronet Graphic Library is F. Pasian. For graphics production. FORTRAN Version 2.0 Dec 4 1985.
  - Structure and Utilities of the Astronet Documentation System User's Guide (Local User Note (TRIESTE) no. 7). By F. Pasian. Describes the structure of the Astronet Documentation System and the related utilities. 25 May 1985.
- ASURV: A Software Package for Statistical Analysis of Astronomical Data with Upper

Limits. By T. Isobe and E. Feigelson of Pennsylvsania State University. ASURV consists of 50 subroutines of FORTRAN totalling about 1/4 MBy. Installation requires FORTRAN Compilation and linking, correction of any compiler-specific bugs and executing. Runs on UNIX-based supermicro and on Mackintosh PC. Code written consistent with FORTRAN 77 conventions. rev. 0.0 test version. Fall 1987.

⊙ Beginner's Manual for the Interactive Image Processing System CASSANDRA. The foundation written by R. Deverill. D. Schneider later merged P. Young's KAREN (direct imaging) and LYNNE (spectroscopy), and included Deverill's updated system routines to produce CASSANDRA. Accomodates direct image and spectroscopic data. Developed on the Space Telescope Wide Field Planetary Instrument Definition Team VAX 11/780 in Pasadena. FORTRAN. Version 1.6-Jun 1986.

- CCD Analysis Program. Photometry package by S. Kent of the Center for Astrophysics (CfA), designed for advanced processing of CCD images, beyond the NOVA computers' simple flat-fielding. Written for VAX, in FORTRAN, some procedures mainly in C and Macro. Nov 13 1982.

- CDF Implementer's Guide. By M. Gough of the National Space Science Data Center (NSSDC), CDF is a data structure used to store and retrieve data on disk. VAX/VMS. FORTRAN. Jan 1 1987.

⊙ Computer Programs for Robust Locally Weighted Regression. By W. Cleveland of Bell Labs. For smoothing scatterplots by locally weighted regression. FORTRAN.

⊙ COSMIC. Produced by NASA's Computer Software Management and Information Center, COSMIC is NASA's software catalog.

- DAOPHOT User's Manual. Developed by P. Stetson, E. Olszewski and L. Stryker at the Dominion Astrophysical Observatory (DAO) through modification of J. Mould's and Keith Shortridge's POORMAN. Digital stellar photometry reduction program. Jun 19 1987.

- ELSPEC/11 Interactive Spectrogram Processing Package Applications. By F. Pasian, L. Rusconi and G. Sedmak. General purpose software package from OAT for processing long format photographic spectra. Two volumes: I. Applications (OAT publication N. 806 1982) and and II. Software Sources (OAT N. 807 1982). Running on DEC PDP11 under DEC RSX11M. FORTRAN IV.

⊙ EVE. Developed by M. Caillat and D. Pelat of Observatoire de Meudon and A. Bijaoui of Observatoire de Nice. Image processing package. Uses the MIDAS files (Image and Tables). Call to subroutines corresponding to the algorithms. Runs on VAX under VAX/VMS.

  ⊙ Manuel des commandes. In English and French.

  ⊙ Systeme EVE Guide du Programmeur. In French by A. Bijaoui. 5 Jan 1987.

- ⊙ EVE Commands, Commands for Image Processing. In English by M. Caillat, D. Pelat and A. Bijaoui. 2 Mar 1987.
- ⊙ LIBRARY SAI Algorithms for Image Processing. In English by A. Bijaoui, mainly using the CDCA and STII programs. 2 Mar 1987.

- EXOSAT. From European Space Agency/European Space Research and Technology Centre (ESA/ESTEC). Four independent reduction packages, one for each EXOSAT experiment.
  - The LE Interactive Analysis System User's Guide. More than 50 programs that support all the basic operations needed for the detailed analysis of CMA data.
  - The ME Interactive System User's Guide. Version 1.0-23 Apr 1985.
  - The GS Interactive Analysis System User's Guide.
  - EXOSAT Timing Analysis System. For detailed analysis of temporal data. Version 2A.
- FOCAS. By F. Valdes of the National Optical Astronomy Observatory (NOAO). The Faint Object Classification and Analysis System was initially developed by J. Jarvis and J. Tyson at Bell Labs. Documentation includes: The FOCAS User's Manual and a paper, "Faint Object Classification and Analysis System," both by F. Valdes. Strictly UNIX 4.1 BSD. Only tested on NOAO's VAX/750. Version 3.
- ⊙ Groth documentation. By E. Groth of Princeton University. A collection of unsupported VAX/VMS specific programs. Includes: Groth Programs, Groth Calc, Groth Routines, Groth G-Plot and Groth Tapes.
- IHAP. see MIDAS
- ⊙ IRAF. Originated and designed by D. Tody, the Image Reduction Analysis Facility system is produced and maintained at the NOAO. Runs on the VOS. UNIX and VMS versions available. Written in Command Language (CL). Version 2.3.
  - ⊙ IRAF User Handbook. Volume 1 IRAF System. Volume 2A NOAO Packages. Volume 2B NOAO Cookbooks.
  - ⊙ IRAF System Handbook. Volume 3A covers System Information and CL Information. Volume 3B covers FORTRAN and SPP/VOS programming and includes Historical Documents.
  - IRAF Newsletter. Announces new releases, revisions, system updates and status reports.
- IUE Data Reduction. By G. Nassiopoulos of the CfA. Summarizes IUE Data Reduction Programs available on SUN/UNIX computers.
- MIDAS. The Munich Image Data Analysis System is developed and maintained by The Image Processing Group of The European Southern Observatory (ESO). Runs on VAX.

FORTRAN. Version 4.4 strictly DEC/VMS versions 4 or later. Jul 15 1987.

  - MIDAS User's Guide, ESO Operating Manual No. 1.
  - MIDAS Environment. Version 15-Jul 1987
  - MIDAS proto-type IDI-routines.
  - IHAP Manual. Hewlett packard based. Mar 1985.

- MONGO: An Interactive Graphics Program. By J. Tonry. Runs on VAX computers running the VAX/VMS or the 4.2 BSD UNIX system. FORTRAN 77. Revision 1987.2
- NRAO Packages:
  - ⊙ CONDAR. By L. Morgan, H. Payne, C. Salter, B. Stobie and B. Vance of NRAO. Describes the Continuum Analysis System for the NRAO Single Dish Telescope. Based on POPS language. May 1984.
  - NOD2 Commands and Programs. Consists of a subroutine library of astronomical routines, a set of DCL commands using these routines and a large collection of programs. Mar 1986.
  - Continuum Mapping With The 12-M Telescope. By C. Salter. Describes the continuum mapping system at NRAO's 12-M telescope. Use in conjunction with the Observer's Reference Manual. Jul 1985.
  - Spectral Analysis System for the 12-M Telescope. By E. Stobie. Describes the spectral line data reduction program running in the analysis PDP 11/40 computer at the 36-foot telescope at NRAO. Apr 1981.
- PDM Period Determination Package. By R. Stellingwerf of Mission Research Corporation. In text format on 360K IBM-PC format floppy disk. Jul 1986.
- RETICENT. Adapted from CFHT/HP RETICENT and modified for the VAX by Christopher Pritchet. Developed at University of British Columbia and at DAO. A FORTRAN-coded command language for manipulating spectrophotometric data. Nov 1983.
- Software Library Definition. By Space Telescope-European Coordinating Facility (ST-ECF) at ESO. Descibes the focus of the software in the ST-ECF Software Library. Explains how to access materials from and contribute software to the Library. Describes dissemination of software information and structure of relevant on-line directories.
- Starlink. The computing facility provided for UK astronomers by the Science and Engineering Research Council at Rutherford Appleton Lab. Hardware based on UK-wide network of DEC VAX computers.
  - Starlink User Notes. For general system users. Typically, contain description of program or package use. Volume 1 and Volume 2.
  - Starlink General Papers. Contains a wide variety of material concerning development and management of Starlink.

  - Starlink Systems Notes. Describes system software and gives implementation and installation details about application software.

⊙ STII. VAX version of the CDCA (running with a PDP 11 and RT11).

  - ⊙ Notice Preliminaire d'Utilisation du Logiciel STII. In French.
  - ⊙ Systeme STII= Guide du Programmeur. In French by A. Bijaoui and G. Mars. 2 Mar 1987.
  - ⊙ Documentation du Systeme STII/CPAI. In French by A. Bijaoui. 2 Mar 1987.

- ST Model User's Guide. Performs realistic simulations of observations which might be made with the Hubble Space Telescope. Refers to MIDAS Manual. Release 87APR30.
- SYNTH: A Procedure to Compute A Stellar Synthetic Spectrum. By F. Castelli. Runs on the DEC VAX 11/750. FORTRAN IV. This is a revised version (Jul 14 1987) of Local User Note (TRIESTE) no. 6. Apr 22 1985.
- TAE news. Transportable Applications Executive newsletter published by NASA.
- VICAR Image Processing System. The Video Image Communication And Retrieval system produced by the Jet Propulsion Laboratory's (JPL) Multimission Image Processing Laboratory (MIPL) is a general purpose image processing system. Runs on VAX under VAX/VMS.
  - Volume 1. Introduction, release notes, installation guide etc.
  - Volume 2. Describes libraries. Jun 15 1987.
  - Volume 3 VICAR User's Guide. By S. LaVoie, C. Avis, H. Mortensen, C. Stanley, and L. Wainio. Covers through VICAR version 3A/10A. May 15 1987.
- XANADU. Developed at the Institute of Astronomy by T. McGlynn, R. Schafer, A. Tennant, and K. Arnaud for the analysis of EXOSAT (x-ray satellite) data, XANADU has since been ported to the CfA and extended to allow some analysis of Einstein IPC, HRI and MPC data. Originally written under VMS on a VAX 11/780, this is a first attempt at a SUN/UNIX distribution.

To acquire information about this collection, write to

Software Documentation Collection,
Smithsonian Institution Library
60 Garden St., MS 56
Cambridge, Massachusetts 02138

THE DESIGN OF A LARGE ASTRONOMICAL DATABASE SYSTEM

Alan Richmond[1], Thomas McGlynn[2], François Ochsenbein[3], Fred Romelfanger[2], Guido Russo[1,4]

[1]Space Telescope–European Coordinating Facility, ESO

[2]Space Telescope Science Institute, 3700 San Martin Drive, Baltimore, Maryland, 21218, USA.

[3]European Southern Observatory (ESO), Karl Schwarzschild-Str. 2, D-8046 Garching bei München, FRG.

[4]On leave from Dipartimento di Fisica, Università di Napoli, Italy.

Abstract.

The user interface / relational database / optical disk archive catalog system that we have developed has been used successfully for an astronomical database containing over 200 gigabytes and is planned to be used for a number of other applications. The Space Telescope Archive is the most demanding of the applications for which this database system was designed. It stores the most data, has by far the most complex catalog, and will probably face the greatest demand by users. Astronomical data is stored in datasets which comprise FITS files of the observational data itself, and associated header files which contain parameters relating to how the observation was scheduled, taken or processed.

1 Introduction.

We describe the implementation of a system for large astronomical databases, developed as a cooperative project of the Space Telescope Science Institute (STScI) in Baltimore, Maryland, and the Space Telescope–European Coordinating Facility (ST–ECF), located at the European Southern Observatory (ESO) outside Munich, Germany; and of ESO itself.

In this section we shall discuss the general requirements that these database systems were required to meet; the driving requirement was the need to accomodate the Hubble Space Telescope archive. The second section discusses the implementation of the database system; in particular we shall concentrate on the user interface. Each of these databases has been or will be implemented using basically the same software systems over a variety of hardware configurations. The next section describes the first operational databases using the system – for Guide Star Selection; and access to nearly 40 astronomical catalogues with STARCAT. The final section discusses some databases we expect to use with these system.

Before we began the design of the system, we attempted to identify and isolate the requirements which we had to meet. This included not only our desires for the capabilities of the final system, but our recognition of practical constraints in time and money that we faced. Below the most critical of these requirements are itemized.

1.1 Astronomical Requirements

1. **Total capacity of at least 15 Terabytes.** This was set by the need to accommodate the Hubble Space Telescope archive [1]. The size of the archive implies that a low–cost storage media must be used.

2. **Archive lifetime of several decades.** Astronomical data have an essentially infinite useful lifetime. The ST mission lifetime is defined as 17 years, but many space astronomy probes have exceeded their rated lifetimes by a factor or two or more.

3. **Data portability.** While the internal archive media and format need not be the same as that used for distribution, it is clearly more efficient if conversions are kept to a minimum. More importantly, any ties to commercial software or hardware must be evaluated in terms of the extended life-time of the archive.

1.2 Usage Requirements

1. **User friendliness.** This much abused term is a critical item in a number of our archives. Both the Space Telescope organizations and the ESO expect to provide a general archive service to the astronomical community.

2. **Uniformity.** Astronomers may use this system at more than one site, but should not have to relearn the system at each institution. A careful layering of the software hides the hardware configurations present at each site, allowing the user interface to be quite uniform.

3. **Network access.** Any modern archive must be able to link into existing computer networks, and where data rates are adequately high, should permit direct transfer of data from the archive to the remote site.

1.3 Software Requirements

1. **Modularity.** Using the standard 3-year doubling time for computer technology, it is clear that many portions of the database system hardware and software will be replaced.

2. **Software portability.** For the same reasons that the data should not depend upon specific vendor formats, the software design for the archive has been made maximally vendor independent.

3. **Heterogeneous support.** It is clear that given a decades-long life, the archive will eventually consist of an agglomeration of different computers, storage devices, networks, etc. The system must be able to properly coordinate and use these diverse elements.

2 Implementation

In designing the database system it was clear that the problem could be almost entirely decoupled into two separate areas: The first area is the physical management of the archive which may contain many Terabytes of information in millions of separate files. The system must reliably store this information and be able to quickly retrieve information from the archive. This File Handler simply appears as a vast reservoir of data. The storage media, and internal organization are invisible to the user [12].

The second critical element is the human interface to the system. E.g. no astronomer can be expected to know the detailled scientific requirements of Space Telescope observations, critical engineering information that reflects on the quality of particular images or spectra, the organization of the archive, and the list of observations that have been made. A user interface had to be provided that guides an astronomer into an intelligent choice of data files to be retrieved and analyzed. Without such guidance the utility of the archive would be drastically reduced. This user interface uses a catalog which contains pointers to data in the archive and indexes that information and cross-references it to other relevant data.

This environment is called the Data Management Facility (DMF). See Fig. 1. It comprises:

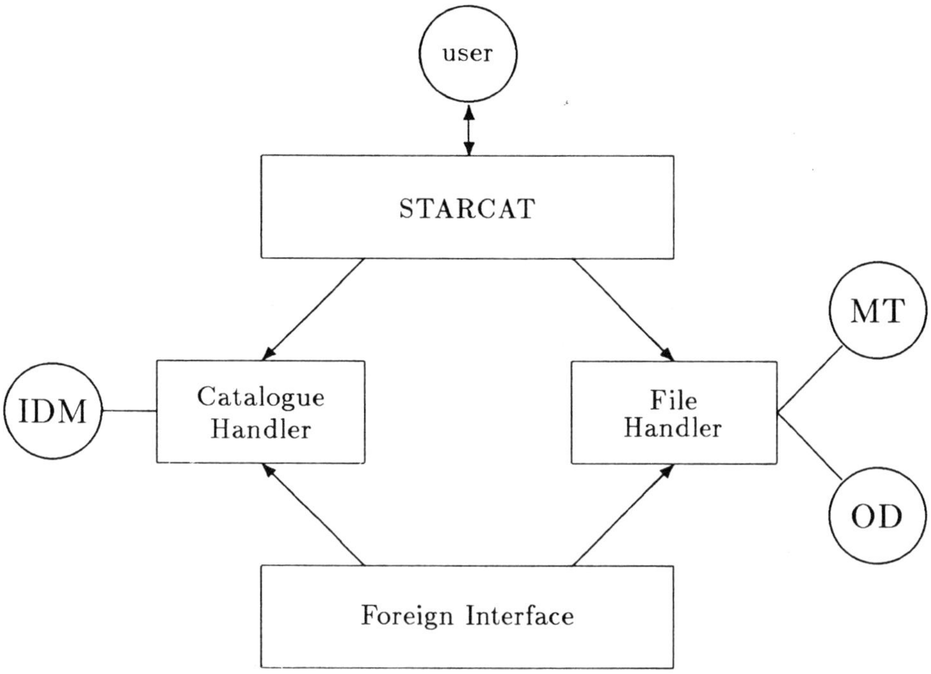

Figure 1: Data Management Facility

1. **STARCAT User Interface.** STARCAT (**S**pace **T**elescope **Ar**chive and **Cat**alogue) is the user interface to the Data Management Facility (DMF) of the Hubble Space Telescope and to the ESO and Software Catalogues [13]. It provides the following major capabilities:

   (a) **Queries on catalogs** can be entered via the `catalogs` command, with an option to output files for MIDAS input [2].

   (b) **Retrieval of files** from the optical disk archive. Some catalogues will contain pointers to data in the archive; when a choice of entries has been made, a request is sent to the File Handler asking it to retrieve the corresponding file (or files).

   (c) **Access to SIMBAD**, the Stellar Data Center data-base containing roughly 600,000 stars and 70,000 galaxies with regularly updated bibliography [3].

2. **Catalogue Handler.** The catalogues are maintained on a relational database machine (IDM500/1, Britton Lee) which is accessed via Signal Technology's OMNIBASE.

3. **Archive Handler.** The images are to be stored on 1 Gbyte optical disks, and accessed via the File Handler developed by T. McGlynn and colleagues at ST ScI.

4. **Foreign Interface.** The set of procedures for creating the catalogues and archives.

2.1 The User Interface.

STARCAT is supported by both application-specific and *generic* software. The latter kind of 'user interface software' is becoming increasingly recognised as a good investment area for research and development efforts, and even has an acronym: UIMS (User Interface Management System). A UIMS should do for user interfaces what a DBMS does for data — decouple them from the applications, so that they can be managed independently [8].

The UIMS used by STARCAT is Proteus [9], which provides a *toolkit* for user interface construction (written in C), and is intended to be adaptable (hence the name). It uses, and is supplied with, a window management system written by François Ochsenbein of ESO [7], called TermWindows, which not only allows use of most terminal types through the use of a terminal capability database, but also allows a User Guide to be obtained directly from the on-line help, using LaTeX. The software architecture is layered [10]:

STARCAT	Primary interface between users and the DMF; it is concerned only with the structure of the dialogue between user and the DMF.
Proteus	Manages user interface entities, e.g. menus, forms, prompts, help, etc. Some of these entities may be altered by the dialogue.
TermWindows	Maps objects to rectangular areas of the screen, and takes input from the active window.
Support	Various utilities and services used throughout the above levels, e.g. database management system.

Special attention was paid during the design of Proteus, to *reusability*, to reduce duplication of software development effort. Reusability applies at several levels, e.g. source and object code, and documentation. A good example of the latter is that the User Guide and the on-line help both derive directly from the same files.

Proteus was designed with the following criteria in mind:

1. **Improved service to the user**, e.g. intuitive interaction, meaningful error messages, on-line help, and a selection of dialogue modes, e.g. prompting, menus, forms.

2. **Separation of application tasks from user interface details.** A generalised user interface relieves application programs from all details of communications with the user. The application programmer is relieved of the burden of writing his own user interface management software, and his code is simplified. All the 'frills' of a comfortable, friendly user interface can be encoded separately from the applications program, e.g. prompts, menus, tables, screen layout, input validation, and so on. The user interface can then be distributed towards the end user, unloading centralised resources.

3. **Modularity.** Allowing the designer to use a 'building block' approach, and to have control over the end product style. The user interface should provide a very high-level interface for easy, fast construction (e.g. prototypes), with defaults, e.g. screen layout, and lower levels allowing the system designer more control.

4. **Evolution pathway** to windows and other workstation / PC concepts, without requiring or prohibiting use of modern hardware. Proteus provides an easy pathway for gentle evolution from traditional 'prompt and read' programs, to menus, forms, windows, etc.

5. **Portability.** The user interface should be transportable to various host environments and terminal types. It must also be allowed that existing interaction packages, e.g. window management systems (WMS), may be made use of.

2.2 The File Handler

The archive subsystem is seen from the outside world effectively as a single device. Messages may be passed to the system indicating the existence of a file to be archived or a request to retrieve a file (or files). Thus the hardware implementation of the system is invisible and can (and does) include a mixture of both optical and magnetic media. Currently all of the archives use 12" optical disk drives as their primary storage media, with magnetic disk being used for the superdirectory. However should some other technology, e.g. magneto-optical storage, eventually become more desirable, it can easily be incorporated by designing an appropriate device management process.

The file handler is implemented as a number of separate processes. A single *communications process* receives all requests from, and relays all replies to, the outside world; it also decodes a given user request and assigns the sub-elements to various other processes in the system. A unique *device management process* is assigned to perform I/O to each of the storage devices in the system. A third class of processes manages the I/O to the *robotics controllers* of jukeboxes. Finally a single *superdirectory process* keeps track of the location of all of the files in the archive.

Consider a request to retrieve a single file. The request is received by the communications process which initiates a request of the superdirectory process to find where the file resides. It might then ask a robotics process to dismount a platter from one of the optical disk drives in a particular optical disk jukebox and then mount the appropriate disk. Finally it sends a request to the optical disk process controlling the optical disk drive to read the file and send it back to the user. When this process is complete a message is sent back to the user indicating that the request has been satisfied. If any of the required tasks is already engaged in satisfying a previous request, the current request is queued and sent automatically when the previous request (or requests) is satisfied.

The division of all of these activities into separate processes has a number of desirable effects. Modularity is an axiom of modern computer programming practice, and the standard benefits of easier coding and testing apply. Since communications may be over a network interface, specialized hardware can be used for any process that requires it, e.g. the superdirectory process can (though in the present implementation does not) use a relational database machine. In UNIX asynchronous I/O is not generally supported for a single task, so using multiple tasks for the different I/O devices ensures that all storage devices are being used where possible. Additional drives can be added or deleted from the system without affecting the rest of the system. Tasks which talk to different devices can be readily optimized. A crash in any one task need not seriously degrade system performance. The cost of these benefits is the requirement to pass messages among the many processes. With the requisite optical disk I/O for a typical request requiring at least several seconds, the message passing overhead is not significant.

2.3 Network Interfaces.

The network interfaces of the database system were critical to both the design and implementation; most usage of the ST catalog is expected to be by remote users of these sites, with data transmitted electronically or by mail as appropriate.

1. **Local Area:** The overall system was broken into a number of cooperating sub-processes. Messages are passed between the processes using a meta-protocol which is built on top of existing network protocols and which is capable of linking together processes which are in separated protocol domains; e.g. one process might run on a computer in a DECnet

domain and communicate with another expecting TCP/IP where the connection between the two computers involves a link where X.25 protocols must be used. All of this happens transparently to the processes involved, although once several links are needed the speed of the transmissions can be compromised.

2. **Wide Area:** STScI has direct links to *ARPAnet and SPAN*, and will participate in the NASA PSCN high-speed network. *X.25* links are available to ESO, NRAO in Charlottesville and NOAO in Tucson. It is a *Telnet node*, which provides virtually complete coverage of the United States and through gateways much of Europe, though at somewhat lower data rates. STScI will also be a major node in the *Astronomy Internet* which will link major observatories and astronomical research institutions throughout the United States.

   ST ScI and ST-ECF / ESO are directly linked via SPAN and X.25 links, and through gateways in SPAN and ARPAnet, which permit access to both BITnet and EARN and provide extensive linkages with European networks.

   ST-ECF / ESO is a *SPAN node* with connections to EARN, ASTROnet in Italy and STARLINK in the British Isles. *X.25* connections are available not only to STScI but also to ESOC in Darmstadt, FRG, ESTEC in Nordwijk, Netherlands, the CDS in Strasbourg, France, etc. As European national and international networks mature, the ST-ECF and ESO will ensure connectivity with their archives.

3 Current Applications.

The software system described above has been used to support a number of archives in both Europe and the United States. Here we discuss some of these projects.

3.1 The Guide Star Selection System Archive Support Program

One of the major projects undertaken by STScI has been digitization of the entire Palomar Sky Survey and ESO Southern Survey. The primary motivation for this effort has been determination of the Guide Star catalog [4] for the pointing of the Space Telescope. Approximately 1500 plates were scanned, generating more than 4600 tapes and 600 GB of data. Each plate generated a single 13999 $\times$14000 pixel FITS file [14], about 400 MB. 522 plates have been archived to optical disk so far (September 1987), comprising more than 200 gigabytes of information, starting with those fields which are most frequently requested or which are needed to support early ST observations. 2 hours are required to archive each tape, from where a particular 512$\times$512 arc-second region can be extracted in about 1 minute. Procurement of a jukebox to fully automate access to the optical disks has been delayed, but disks can be mounted manually. When the human intervention to locate and mount the disk is included, the total time to retrieve an image increases to about 10 minutes. Approximately 200 requests for fields have been satisfied.

The catalog contains the coordinates of each of the plates. The catalog for this archive contains information on the coordinates of each of the plates, and the user need merely indicate the coordinates desired, and determines which plates the region is located on, and whether this region has yet been archived to optical disk. Several different types of projects have been supported. A number of observers have used retrieved regions and used them along with the preliminary Guide Star Catalog to get accurate astrometry for proposed Space Telescope or ground based observations. Others have used the images for studies of galaxy morphology and photometry. A major continuing project is determination of photometric sequences for the new Palomar Sky Survey.

3.2 STARCAT access to Astronomical Catalogues

As a preliminary to developing the Hubble Telescope Archive, the ECF loaded a number of astronomical catalogs onto its database machine. While originally intended for prototyping studies, the facility to rapidly access astronomical catalogs on-line has become a major resource and is available at both the ECF and the Science Institute. There are nearly 40 catalogues [11] accessible on-line with STARCAT, e.g: DIXON (Master List of Radio and Non-Stellar Sources), IRAS (InfraRed Astronomical Satellite Point Source Catalogue), IUE (International Ultraviolet Explorer Log), and RC2 (Second Reference Catalogue of Bright Galaxies).

4 The Future

All of the astronomical projects discussed above are merely preliminary to the main projects that we expect to support with this archive system. In both Europe and America, the primary impetus behind this project was the support of the Space Telescope Archive. Another goal of the European facility is to support the archiving of data taken with digital detectors (e.g. CCD's) at the ESO observatory at La Silla, Chile.

ST data will normally be proprietary for one year after it is taken, and during the proprietary period the data will be kept at STScI. Updates of all non-proprietary data will be sent every month to the ST-ECF, the Canadian Space Astronomy Data Center (in Victoria, BC) and possibly other sites around the world. However all sites will have up-to-date catalogs indicating all observations that have been taken, proprietary or not.

Given the complexity of the instrumentation in ST, the catalog is quite complex, with tables describing scheduling, engineering parameters, and the observations themselves. To assist astronomers not versed in these arcana, the user interface provides relatively simple views of the database, permitting queries on standard astronomical fields, but also indicating where the peculiarities of ST observations must be taken into account.

Other projects include:

1. **ESO archive** of data taken with digital detectors at La Silla [6].

2. **The IUE Line-by-line image archive:** The International Ultraviolet Explorer (IUE) is a NASA/ESA scientific satellite launched in 1978. About 60,000 spectra have been acquired so far, two thirds in low-resolution mode. The expected lifetime of three years has been exceeded by a factor of three now, and the instrument is still operational. Archival request has therefore become more important than it was foreseen at the beginning, but both the VILSPA (ESA) station and the GSFC (NASA) station provide – or will provide soon – on-line access only to the extracted spectra, while images are stored on tape. ST–ECF started, at end of 1986, a project to build an optical disk archive of the subset of the raw images that contain low-resolution spectra. Since the entire IUE log of observations was already on-line in the ST–ECF database, a software for searching and then requesting the desired images was relatively easy to achieve. The original raw images are stored in the NASA SPEC format. We convert this to a standard FITS format before copying to optical disks. The File Handler described in the previous sections takes care of the user requests.

3. **The ESO–LV project** (Lauberts–Valentijn) will provide automated parameter extraction for the 16000 galaxies in the ESO/Uppsala catalogue [5]. The scans will be copied to optical disk, from where they can be extracted for processing in MIDAS, using components of the large database system we have described.

Acknowledgements

We are grateful to P. Shames (ST ScI), R. Albrecht (ST–ECF), and P. Benvenuti (ST–ECF), for discussions and guidance in the development of the Data Management Facility.

References

[1] Albrecht, R., Russo, G., *Hubble Space Telescope data analysis in Europe*, SPIE Conference Proceedings Vol. 628, Advanced Technology Optical Telescopes III (1986)

[2] Banse, K., Ponz, D., Ounnas, C., Grosbøl, P., Warmels, R., *The MIDAS Image Processing System*, in: Instrumentation for Ground–Based Astronomy, ed. L. B. Robinson, Springer–Verlag (1987, in press)

[3] Egret, D., Wenger, M., *SIMBAD — The present status and the future*, ALDB[1] (October 1987)

[4] Jenkner, H., *The Guide Star catalog — structure, publication, and future plans*, ALDB (October 1987)

[5] Lauberts, A., Valentijn, E., *The ESO–LV project — automated parameter extraction for the 16000 galaxies in the ESO/Uppsala catalogue*, ALDB (October 1987)

[6] Ochsenbein, F., *Connecting an Archive with Large Astronomical Catalogues*, ALDB (October 1987)

[7] Ochsenbein, F., *TermWindows: A Terminal–Independent Screen Formatting and Window Package*, Image Processing Group, European Southern Observatory (July 1987)

[8] Pfaff, G. *ed.*, *User Interface Management Systems*, Springer–Verlag (1985)

[9] Richmond, A., *The Proteus Software Handbook*, ST–ECF O–02 ST Science Data Archive vol. VIII (September 1987) version 1.3

[10] Richmond, A., *Software Design by Object–oriented Functional Layering*, Computer Physics Communications 41, North–Holland, Amsterdam (1986) pp. 377–384

[11] Russo, G., *et al.*, *STARCAT Documentation for on-line catalogues*, ST–ECF O–02 ST Science Data Archive vol. VIII (October 1987) version 1.9

[12] Russo, G., Albrecht, R., Ochsenbein, F., *Optical mass storage for a 10 Terabyte archive: system integration problems and user interface*, SPIE's 30th Annual International Technical Symposium on "Optical and Optoelectronic Applied Sciences and Engineering", Session on "Optical Mass Data Storage", San Diego, California (August 1986)

[13] Russo, G., Richmond, A., and Albrecht, R., *The European Scientific Data Archive for the Hubble Space Telescope*, Proceedings of the International Workshop on Data Analysis in Astronomy, Erice, Sicily, in: Data Analysis in Astronomy II, ed. V. Di Gesù, L. Scarsi, P. Crane, J. H. Friedman, S. Levialdi. Plenum Press (1986) pp. 193–200

[14] Wells, D. C., Greisen, E. W., Harten, R. H., *FITS : A Flexible Image Transport System*, Astron. Astrophys. Suppl. Ser. 44 (1981) pp. 363–370

[1] ALDB = Astronomy from Large Databases: Scientific Objectives and Methodological Approaches. Garching, Munich, FRG

ASTRONOMICAL ARCHIVES AND DATABASES: CURRENT TRENDS

G. Russo *
Space Telescope European Coordinating Facility
European Southern Observatory
D-8046 Garching bei München, F.R.G.

Abstract

The large amount of data which is now obtainable in astronomy with modern detectors and telescopes has raised the need for data archiving systems. Several projects in this area have started in the past few years, and more are planned for the near future. New hardware technologies are being experienced, which allow for fast access to data, while browsing through catalogues has now become a matter of seconds. However, even if the user interface has in general become very user-friendly, the software techniques required to handle properly systems which are supposed to live for decades have become more and more sophisticated. The present problems in handling catalogues and archives are summarized here, with examples of existing implementations. The trends in hardware and software techniques are also presented.

1. Introduction

Astronomy is probably the physical science that makes the most usage of data for long times after they have been taken: a typical high energy physics experiment may produce thousands of high-density tapes, but after the scientific information has been fully extracted, they are no longer useful, and are discarded after a few years (< 10). On the contrary, astronomical photographic plates taken at the end of the XIX century contain information that is still to be extracted for scientific purposes, and in some cases (Rousseau and Guilbert, 1984) this is done. Astronomers are quite accustomed to this concept of re-use of data, and therefore in almost any observatory there are a lot of "stored" data, from photographic plates, to punched cards, to tapes, to simple lists of photometric data. The space astronomy era has only given further impetus to this need for data storage, without changing its characteristics. But when can we consider a collection of stored data as an archive? To give an answer, we must recall that the purpose of an archive is the re-use of data; to accomplish this,

* On leave from Dipartimento di Fisica, Università di Napoli, Italy

not only do the data have to be stored in a retrievable way, but the description of the data has to be stored also. This description, often called "header" (Albrecht, 1982) must contain information such as:

- location
- time
- instrument and receptor characteristics
- observer
- engineering data

among other things.

Therefore, an archive is not a simple collection, but "an ordered collection of a large number of observations of a certain kind made for a certain purpose and carried out at a certain place and time". This description, taken from Jaschek (1985) does not make any distinction regarding the type of data. We will adopt the following: we will use the term "Archive" if the data are 1-D or 2-D images, and "Catalogue" for the other cases. We will consider the two cases separately in the next two sections.

2. Archives

We defined the archives on the basis of the type of data that they store, i.e. 1-D or 2-D images. There are several kinds of images, however:

- raw data
- edited
- calibrated data.

As the main purpose of an archive is the re-use of its data, it is clear that raw data are not of general interest, except for the observation site, where they have to be maintained for a while. Edited data may be needed for a longer time, since the calibration is never the best one, in the first years of an instrument: re-calibration is to be considered a normal event. But of course the astronomer is interested in the calibrated data only, from which he can extract scientific information directly. These data have in principle to be maintained for a very long period, that is several decades, at least, and in an easily accessible way.

What the astronomer needs, in fact, is a system that allows retrieval of the required images from the archive, to a place where he can work on them, i.e. into an Image Processing System.

The present situation is, in general, the following: the data are maintained in a central place on tapes, in a homogeneous (not standard) format.

The user requests the data, which are mailed to him after a period of time which spans from two to six weeks, in the best cases. When the data are received, sometimes in FITS format but often in an ad-hoc format, they have to be read with a special program, and converted to the local internal format. This situation is obviously unsatisfactory. Astronomers want to get the data into the work area of the system they use for reduction directly from the archive system, without the need for an intermediate storage like tapes. This is possible, in principle, but requires, among other things, major improvements in two areas: data storage and telecommunications. We are progressing in both areas, and the time when we will be able to realize the above mentioned scenario is not far from now. On a reduced scale, some systems will allow it very soon, like the IUE ULDA project (Uniform Low Dispersion Archive) which will enter the final stage in February 1988: a copy of all IUE low dispersion spectra, uniformly processed and calibrated, will be maintained in each ESA country, with a software package (USSP) which allows a selection and downloading of spectra from any observatory equipped with minimal network capabilities (Driessen, 1987).

3. Catalogues and Databases

As described by Jaschek (1985), catalogues can be classified as follows:

- Observational Catalogues, which contain a list of observations made by the author;

- Bibliographic Compilation Catalogues, which collect observations from many sources;

- Critical Compilation Catalogues, which contain a list of critically evaluated data;

- General Compilation Catalogues, which contain a list of average values for many different parameters.

We will add to these catalogues two special cases: the instrument logs, which contain a list of all observations made with a certain instrument and which are stored in an archive, and the bibliography databases, such as the Astronomy and Astrophysics Abstracts, prepared by the Astronomisches Recheninstitut in Heidelberg in machine-readable form.

We are not interested here in discussing the catalogues themselves, but their organization. It is clear that, in order to be easily accessible, the catalogues must be organized in a database. There are three basic types of databases: network, hierarchical and relational, but the relational model is now considered as the best approach for catalogue organization.

Data are organized in "databases" which contain "relations" (= catalogues or tables). Each relation has some "attributes" (= fields, or columns) of defined type, and the data are stored in "tuples" (= records, or rows). Therefore, data have a tabular form, and this is of course a natural form for astronomical catalogues, or at least it should be. But the situation is not so simple. A catalogue, which generally appears first in printed form, is made of three parts: i) a description, ii) the data, iii) the notes and references. The first and last part are essential to make proper use of the catalogue; however, only the data part is in general loaded in a database; instead, the notes and references could and should be loaded too, in a separate table, linked to the data table. But there are also problems with the data themselves. A typical problem with the data is that if a field is missing (e.g. an observation not available) an entry must still be supplied in the database table. Consider the case of a colour index: how can you distinguish between (B-V) = 0 and a missing entry for the (B-V)? Also in this case, the need for notes is evident, but the database manager has to supply its own note for that record, as the corresponding printed version will not have this problem.

Another important issue concerns the "errata". Regular publication of errata to catalogues is only done, at present, by the CDS in their bulletin. How many people use this information to correct their databases?

The conclusion is that having a copy of a catalogue is not sufficient to provide full support for it. This must still be done in specialized centers like the CDS (Jaschek, 1982); or the NSSDC (Warren, 1985). However, having the major astronomical catalogues loaded in the local database means providing astronomers with an easy tool for their work, and this approach should be encouraged, not forgetting that the "master" copy is the one in the principal center, and from time to time a reload is necessary.

Let us now discuss the two special cases of instrument logs and bibliographic databases. For the first, it is clear that the content should be selected to provide the user with a quick idea of which kind of observations have been done of a certain object, and where to find them (Wilkins, 1982). This latter item, which is often a pointer to a file, e.g. an image number, is the primary field of the log; all the other information is given to support the user search, except for the object identifier. This is often inserted in the original log by the observer, who follows his own criteria for it: a typical example is the IUE log, where there is a complete inhomogeneity in the object identifier, and a lot of work has been needed to convert them to a standard, so that the same object has always the same name. Also this work has been done at the CDS, which seems to be one of the few places where these concepts are fully applied.

Concerning bibliographic databases, I will mention that the Astronomy and Astrophysics Abstracts will shortly be available on-line, and that the ESIS project (Albrecht et al., 1987) is devoting special attention to these kind of data.

Having discussed data organization, we must look to the methods available to integrate the database. The situation is much clearer now than it was a few years ago, when several query languages were available, and effectively used in different centers. The SQL, developed in the early 1970's at IBM San Jose by Codd et al., has now evolved to a very stable status which has been accepted as ANSI standard. It is a non-procedural query language, very close to a relational algebra (e.g. supports union and intersection operations on tables). In SQL, a typical query looks like this:

```
select e.ra, e.decl, e.map, e.name
from catalog e
where e.decl > 45.0
group by e.obj
having min(e.period) > 10.
```

There are two issues involved here:

First, we cannot assume that every institute has SQL. There are money and manpower investments to be saved, in all the places which started with different languages (e.g. IQL, or QUEL) or even with different database models (e.g. the ADABAS system used at VILSPA, or the DIRA system of ASTRONET). A solution to this is to adopt a programming interface that allows a program to query a database in a relatively simple, vendor independent way. The first implementation, to my knowledge, is the STDB package by Lubow (1987).

The second issue concerns the use of SQL. Although not difficult, it cannot be learned immediately by non-specialists. Most astronomers don't want to use it. There is therefore the need for an "astronomical" query language, devised ad-hoc for our purpose. Work is in progress to (at least) define it, within the ST-ECF Database Working Group, which periodically meets to coordinate the work in this area.

But probably the best approach is that of devoting more functionalities to the software system that handles the dialogue with the astronomer, i.e. the user interface. This will be treated in the next section.

4. Accessing the Databases and Archives: the User Interface

In order to allow astronomers easy access to archives and databases, particular emphasis should be given to the User Interface, that is to the software that handles the interaction with the user. The traditional prompt-and-reply technique can now be considered as one of the available options, but much more powerful mechanisms are available, which provide intuitive interaction, clean syntax, meaningful error messages, on-line help etc.

In general, a User Interface Management System (UIMS) should allow a selection of modes, from menus to forms to prompting, in order to accommodate the user preferences, and should allow usage of different devices, from simple ASCII terminals with keyboard input to workstation-style windowing systems with mouse input. The Proteus software (Richmond et al., 1987) is a very good implementation of such specifications, and the first general one used in astronomical applications. If the user interface is properly layered, it should generally be possible to decouple it from the application specific part. This would allow not only the re-usability of the user interface, but also the insertion, between it and the application, of a layer that acts as a dispatcher over a network, so that remote databases can be accessed in a way transparent to the user. In particular, for Local Area Networks with mixed vendor host computers (e.g. VAX/VMS and Unix machines), which is a configuration very common in astronomical environments, this allows the use of a dedicated database machine, which acts as a server for queries coming from different hosts, where these are formulated via the local user interface. More specific User Interface aspects can be found in the paper by Richmond et al. in this volume.

5. Technology Trends in Data Storage

Photographic plates have the unique characteristic of being both the detector and the storage device. The problem of archiving a plate is therefore easy to solve, at least in principle, and the lifetime is very high (~ 100 years). Plates can in fact be considered as WORM devices (Write Once, Read Many times) of analogue type.

The advent of re-usable detectors led to the differentiation of the two functionalities (detection and storage), and for storage the astronomers started using magnetic recording devices: tapes, floppy disks, hard disks, etc. While this was probably the only possible choice, it is also clear that this was only a temporary choice, because of the low reliability with time (tapes are unreadable after a few years) and the low limit for the surface density obtainable ($\sim 10^6$ bit/cm^2 on normal magnetic disks). But in the last few years a new technology has become available, that of digital optical disks, and this has immediately received the acceptance of the astronomical world. Their principal characteristics are: they are WORM devices, like plates, but of digital type, with a long lifetime (decades, but not yet tested), and the surface density obtainable is very high ($\sim 10^9$ bit/cm^2).

The basic concept of optical disk technology is the following[1]: a thin film of a special material (tellurium alloy) is deposited between two adjacent surfaces of glass (or plastic), of circular form: a laser beam, which can be set at two different levels (low/high) moves radially over the surface, while the disk rotates, and can thus reach every area of the disk surface, with high positioning precision. When the light beam is pulsed at high power, a small area is burned, thus creating a hole, i.e. a positive bit of information: the sequences of burned-unburned bits are translated into the usual sequence of 0's and 1's used in digital computers. These sequences are organized in sectors, usually of 1024 or 2048 bytes, which is also the minimum unit read or written by the host computer with a single input/output operation. Once a sector is written, its information cannot be altered: another write operation would destroy the sector. It can instead be read an (almost) infinite number of times by using the laser beam at low power, and detecting the light as it is reflected by the substrate: a hole will not reflect light. Once the entire disk is written, it can be put in read-only mode, thus avoiding accidental erasure. The limit surface density is not yet available on the market, but this is only a matter of time, probably 2-3 years. Concerning the capacity, many manufacturers offer 1 Gigabyte per side on 12" media (e.g. OSI Laserdrive 1200, Alcatel-Thomson Gidadisc and Optimem 1000), and recently also 5"1/4 disks have appeared (Maxtor RXT-800S) with a capacity of 400 Mbyte per side and a very low cost per Megabyte. It is safe to say that these capacities will be doubled within two years.

As already said, this new technology has received a positive acceptance in astronomy, but under what conditions can one use optical disks as a storage medium for an astronomical archive? Again, to give the answer one should

---

1) This is only the very basic principle, and the actual implementation may be slightly different.

consider that an archive exists for re-use of the data. Therefore any implementation which is heavily dependent on specific computer dependencies should be avoided. For the hardware part, the interface is always the SCSI one, thus assuring standardization, but media interchangeability is not yet assured. This will hopefully be solved in the near future, but builders of astronomical archives cannot do much about that, except perhaps by adopting a common device. For the software part, on the contrary, there is something we can do now, and this is to avoid host dependencies in the way data are stored. With the wish to use the new technology immediately, several people are copying data from their tape storage to optical disk using for the latter the organization provided by the host operating system (often, the ODS-2 of VAX/VMS systems). This approach is a step backwards for astronomical archives: only recently we agreed on the FITS standard (Harten et al., 1985) for tapes, and we all benefit from the data interchangeability provided by this IAU standard. STScI, ST-ECF and ESO have designed (Russo et al., 1986, 1987) and implemented a simple data organization for optical disks which is host-independent and accommodates the FITS data format. Although still susceptible to modifications, this approach is to be preferred to the other option for obvious reasons. Other organizations, among them ASTRONET (Benacchio, 1987) are considering this format.

It is also important to note that astronomy needs long term storage, of the order of a century, not of a decade, and the WORM devices we described are only a first step. This point has recently been raised by Jaschek, but does not seem to be fully accepted by astronomers.

We will just mention here that other technologies are being experimented with, like the magneto-optical recording, which will add the possibility of re-writing to the characteristics of optical disks; for short term storage, instead, the magnetic recording on new media, like the IBM cartridge tapes, will allow high volume storage (> 5 Gbytes per cartridge) and seems to many a good way of transporting the observational data from the telescope site to the data reduction site, as a replacement of the present half inch, 6250 bpi tapes. VHS recorders are also being used for the same purpose, but we can probably consider these as a temporary solution.

6. Conclusions

The general archiving system which emerges from the previous considerations turns out to be layered as follows:

- High-level user interface, decoupled from application via network-independent interfaces and able to access distributed archives

- Archive server, accessible via LAN (Local Area Networks) and/or WAN (Wide Area Networks), with read-only devices and high volume (~ 100 GB)

- Catalogue server, accessible via LAN and WAN with ~ 1 GB of data on line (all major astronomical catalogues)

- Utility package, e.g. EasyMail (Loodts and Defert, 1987) to allow telecommunications and file transfer.

For the s/w and h/w trends, we have the following scenario:

- FITS standard for all types of data exchange, with extensions to accommodate new needs;

- Relational database model and SQL query language for catalogue organization and retrieval, respectively, but used through a generalized STDB-like interface layer;

- WORM optical disks for permanent data storage; cartridge tapes (IBM-like) for temporary one.

Acknowledgements

I wish to thank F. Ochsenbein and A. Richmond for stimulating discussions during the preparation of this talk.

References

Albrecht M., Richmond A., Russo G., Hepgood M.: 1987, ESIS User Requirement Document, Oct. 1987, ESRIN internal document.

Albrecht R.: 1982, in Automated Data Retrieval in Astronomy, C. Jaschek and W. Heintz (eds.), D. Reidel, p. 87.

Benacchio L.: 1987, private communication.

Driessen C.: 1987, USSP Architectural Design Document, ESA internal document (available from VILSPA).

Harten H.R., Grosbøl P., Tritton K.P., Greisen E.W., Wells D.C.: 1985, Mem. Soc. Astron. It. **56**, 437.

Jaschek C.: 1982, in Automated Data Retrieval in Astronomy, C. Jaschek and W. Heintz (eds.), D. Reidel, p. 3.

Loodts, B., Defert, P.: 1987, Easy-Mail Specifications, ESO internal report.

Lubow, S.: 1987, Space Telescope Science Institute internal document.

Ochsenbein F.: 1986, in Data Analysis in Astronomy II, V. di Gesù et al. (eds.), Plenum Press, p. 305.

Richmond, A., Russo, G., Ochsenbein, F., McGlynn, T., Romelfanger, F.: 1987, The Designing of a Large Astronomical Database, in this volume.

Rousseau M., Guibert J.: 1984, Bull. Inf. Centre de Données Stellaires **28**, 133.

Russo G., Albrecht R., Ochsenbein F.: 1986, Proceedings of SPIE Conference No. 695, San Diego, Aug. 86, p. 366.

Russo G., Ochsenbein F., Richmond A., McGlynn T.: 1987, Proceedings of Optica '87 Conference, Amsterdam, Apr. 87, p. 331.

Vette J.I.: 1985, Mem. Soc. Astron. It. **56**, 269.

Warren W.H. Jr.: 1985, Mem. Soc. Astron. It. **56**, 285.

Wilkins G.A.: 1982, in Automated Data Retrieval in Astronomy, C. Jaschek and W. Heintz (eds.), D. Reidel, p. 193.

THE *EINSTEIN* OBSERVATORY STELLAR X-RAY DATABASE: AN OVERVIEW

S. Sciortino[1], F.R. Harnden, Jr.[2],
A. Maggio[1], G. Micela[1], G.S. Vaiana[1,2],
J. Schmitt[3], R. Rosner[2,†]

[1] Osservatorio Astronomico di Palermo
[2] Harvard Smithsonian Center for Astrophysics
[3] Max Plank für Extrarrestrische Physik

ABSTRACT

We present the motivations for and the methodology followed in building the "*Einstein* Observatory Stellar X-ray Database" based on the uniform analysis of all *Einstein* Observatory Imaging Proportional Counter fields obtained during the life of the HEAO-2 mission. The database has been implemented using the *INGRES*^{TM} database system, so that statistical analyses of the properties of the full detection catalog are relatively easily and flexibly accomplished. Some illustrative examples will furnish a general view both of the kind and the amount of the archived information, and of the statistical approach used in analyzing the global properties of the data.

1. INTRODUCTION

The *Einstein* Observatory was the first imaging x-ray telescope to study the x-ray properties of stars. In consequence, the early results reported from the Observatory were largely of a discovery nature, as x-ray emission was observed from virtually every type of star in the H-R diagram (Harnden *et al.* 1979, Ku and Chanan 1979, Seward *et al.* 1979, Long and White 1980, Feigelson and Kriss 1981, Feigelson and De Campli 1981, Vaiana *et al.* 1981, Helfand and Caillault 1982). Subsequent efforts focussed upon the characterization of the x-ray properties of stars, based on then-current data reduction software; the results showed (for recent reviews cf. Rosner, Golub and Vaiana 1985, Vaiana and Sciortino 1987) that, in the main, x-ray emission is indeed characteristic of most stars, with the principal exceptions of the A dwarf stars (Schmitt *et al.* 1985) and late-spectral type giants and supergiants (Ayres *et al.* 1981, Haisch, Maggio and Vaiana 1987).

The purpose of this paper is to describe the database we are using to study stellar x-ray emission as observed by the Imaging Proportional Counter (IPC) onboard the *Einstein* Observatory, and to describe how the database was constructed.

2. THE DATABASE

The data used to construct the database are derived from the latest reprocessing of all of the *Einstein* Observatory IPC fields, using the so-called 'Rev 1' data reduction and analysis software (Harnden *et al.* 1984). The entire database contains more than 16,000 distinct detections, i.e., localized regions with an x-ray flux above a specified detection threshold which depends upon the determination method and the intensity of the background. Some 5000 of these detections have been tentatively identified with counterparts from 64 individual catalogs (mainly radio- and optically

† Presently at Department of Astronomy and E. Fermi Institute, Chicago University
INGRES is a trademark of Relational Technology

identified QSOs, galaxies and stars) that make up the *Einstein* Master List Catalogue. Of this tentative ientifications, approximately 2500 are identified with stars.

(a) Database Organization

Our database was established by porting the results of the standard, *Einstein* IPC, on-line processing, and those of stellar specially-designed on-line processing of all of the *Einstein* IPC fields to a *INGRES*^{TM} database system based on *SUN* and *DEC VAX*^{TM} computers. The standard analysis consists of locating all x-ray detected positions against those of all the objects in the set of optical and other catalogs, either to establish tentative identifications or to determine x-ray upper limits (of which there are over 30.000, the majority of which are from stars from the SAO Catalog). Use of the *INGRES*^{TM} relational database has allowed us to approach the formidable task of analyzing the properties of this vast quantity of detections and upper bounds.

In organizing the *Einstein* Observatory stellar database within the *INGRES*^{TM} system, we have divided the relevant information in two logically-separated sets of tables (or relations, to use the database terminology): (i) a book-keeping set consisting of 9 main tables‡ where general processing information are maintained (this includes tables which track the various updates and correction required in porting the 'Rev 1' data to *INGRES*^{TM} system; (ii) a scientific set consisting of 13 main tables that contains the basic x-ray data or x-ray related data. These scientific tables are complemented by other tables that archive various stellar catalog, such as the Yale Bright Stars Catalog (Hoffleit and Jascheck 1982, Hoffleit, Saladyga and Wlasuk 1984) the Woolley Catalog (Woolley *et al.* 1970), the Garmany O star catalog (Garmany, Conti and Chiosi 1982), etc.

The scientific tables are organized in four groups: (i) one containing general information on x-ray images, such as exposure time, number of detections in the fields, intensity of background, etc. (ii) one containing information on x-ray detections, such as position, flux, spectral hardness ratio, raw and fitted spectra (for more intense sources), etc., (iii) one containing information on possible identifications for x-ray detections, (iii) one containing the x-ray upper limits flux for all catalogued stellar objects.

INGRES^{TM} is a *relational* database management system (DBMS), in which data are stored in tables, organized in rows and columns. A significant characteristic of a relational DBMS is the capability of processing data in sets, i.e. in groups of rows, as opposed to the row-by-row scheme. *INGRES*^{TM} provides a QUEry Language (QUEL) to manipulate data, define basic data structures and their relationship to each other. A basic characteristic of QUEL is that of being a high level non-procedural language, i.e., a language in which the end users has only to specify what to do, not how to accomplish the desired task. These (and other) features makes the *INGRES*^{TM} DBMS well suited for building a scientific database. Particularly useful capabilities, typical of any relational DBMS, are the ability to sort and query on any of the fields (columns) of a given table, to cross-correlate many independent tables using any given field(s) as cross-reference index(es), and to index on multiple columns, which can significantly increase the query speed. In our database, for example, a unique number identifying the IPC image (sequence number) and a unique number identifying the x-ray detection inside a given IPC image (detection number) are the two main cross-reference fields implemented. While the *INGRES*^{TM} DBMS allows a graphical display of data query results, we have preferred the adoption of the *S*^{TM} interactive statistical package as main system to perform the scientific data analysis and display, due to investigatory nature of our research.

VAX is a trademark of Digital Corporation.

‡ "Main table" in this context indicates an *INGRES*^{TM} relation storing fixed, reference information that can be modified only by the database administrator but can be accessed by any authorized user to retrieve needed information in their own working relations.

S is a trademark of AT&T Company.

(b) Some General Data Characterizations

The full *Einstein* data set involves about 4000 IPC fields, covering approximately 3,500 square degrees, the sky coverage attained by the *Einstein* Observatory IPC images is shown in Figure 1; the distribution of the effective on-time of all images is shown in Figure 2.

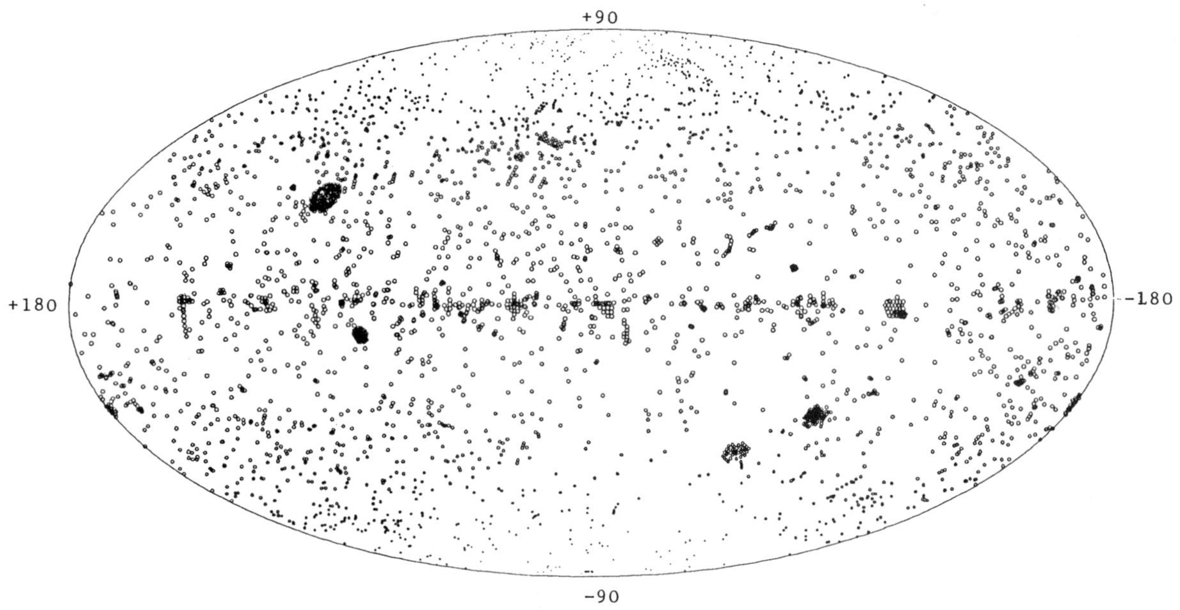

Fig. 1 - Galactic coordinates sky projection of the coverage attained by the *Einstein* Observatory IPC images, each image is represented by one circle, whose area is equal to the projected image area. Note that the actual IPC field of view is a (rotated) square.

(c) Some Scientific Applications

The first application of this database has been the extensive (63 distinct images were analyzed) survey of the stars in the Hyades-Cluster Region (Micela et al. 1988). This survey involved about 150 stars, about 300 x-ray detection and more than 200 x-ray upper bounds. One of the scientific result of this survey is the conclusion that the level of stellar x-ray luminosity L_X is not only a function of stellar age τ, but is also a function of color, i.e., $L_X = L_X(\tau, \text{B-V})$, (cf. Figure 3).

As a further scientific application we can quote results of the ongoing survey of about 400 late-type giants and super-giants observed by the *Einstein* Observatory and included in the Bright Stars Catalog and its supplement. One of the conclusion of this survey is the statistical significant decrease of x-ray luminosity level beyond the spectral type K0, cf. Figure 4 (Maggio 1987).

Acknowledgments: We would like to thank R. Giacconi for having made this effort possible in the first place. We acknowledge partial support of the Consiglio Nazionale delle Ricerche (PSN) and the Ministero Pubblica Istruzione (Italy), and support from NASA contract NAS 8-30751.

BIBLIOGRAPHY

Ayres, T. R., Linsky, J.L., Vaiana, G.S., Golub, L., Rosner, R. 1981, *Ap. J.*, **250,** 293.

Feigelson, E.D., and Kriss, J.A. 1981, *Ap. J. (Letters)*, **248,** L35.

Feigelson, E.D., and DeCampli, W.M. 1981, *Ap. J. (Letters)*, **243,** L89.

Garmany, C.D., Conti, P.S., and Chiosi, C. 1982, *Ap. J.*, **263,** 777.

Harnden, Jr., F.R., *et al.*, 1979, *Ap. J. (Letters)*, **234,** L51.

Harnden, Jr., F.R., Fabricant, D.G., Harris, D.E., and Schwartz, J. 1984, *SAO Special Report*, **393**.

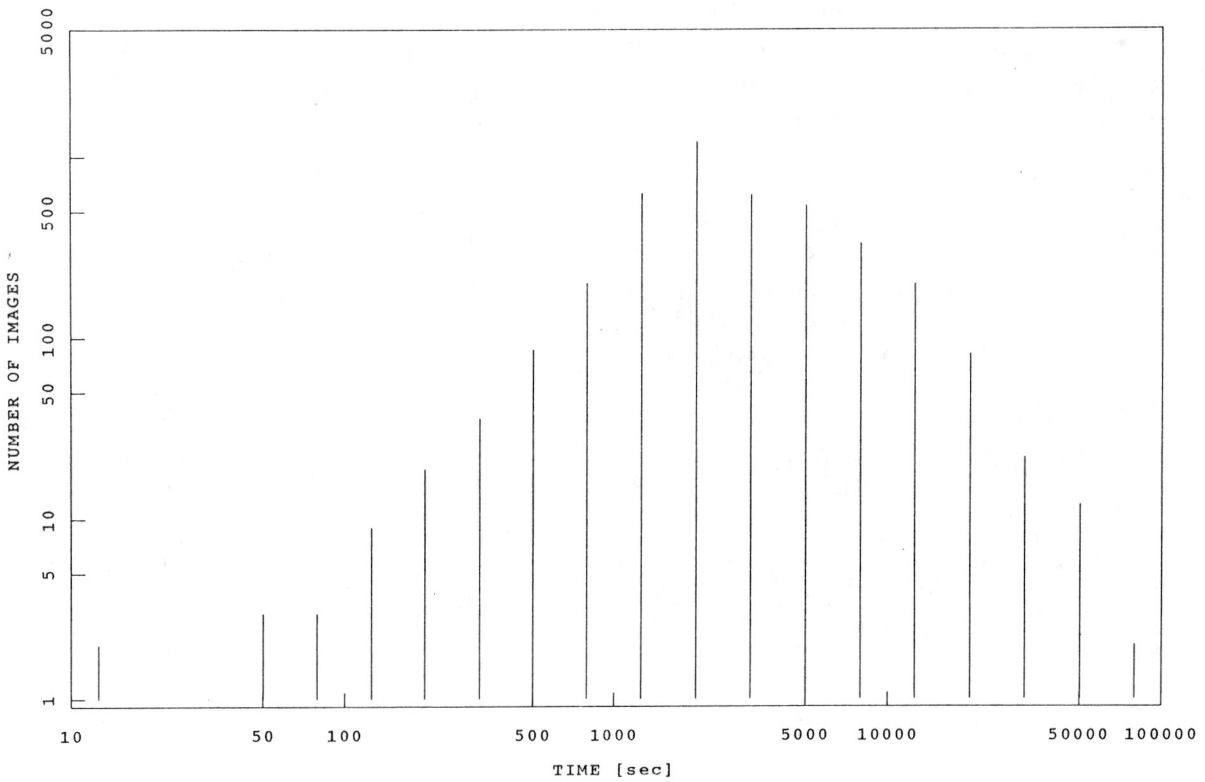

Fig. 2 - Effective on-time distribution of the entire IPC data set, the majority of the images have a typical exposure time of few times in 10^3 sec.

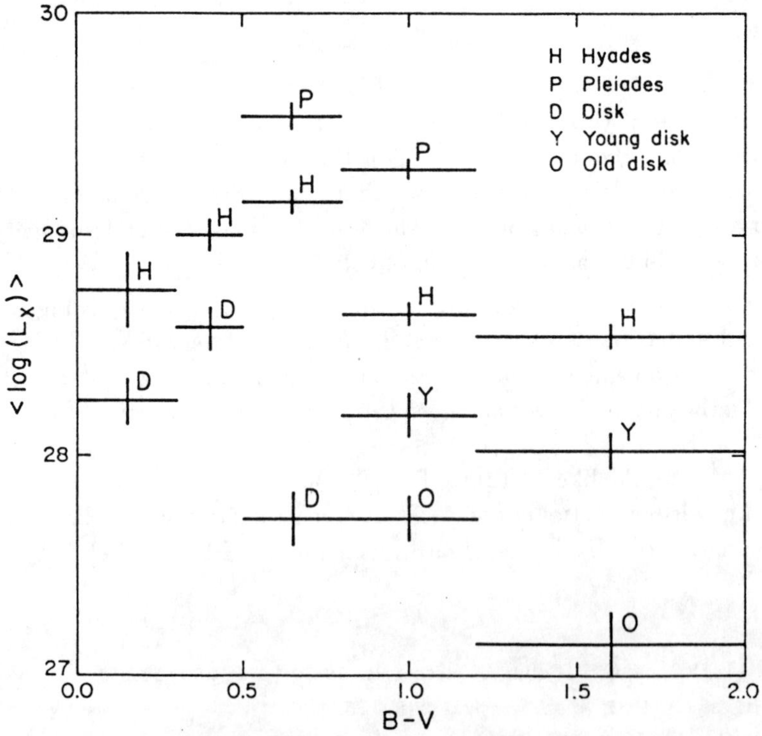

Fig. 3 - Summary of the behavior of <Log(L_x)> as function of spectral type and stellar age, as deduced from volume limited sample of nearby stars, and analysis of samples of coeval stars.

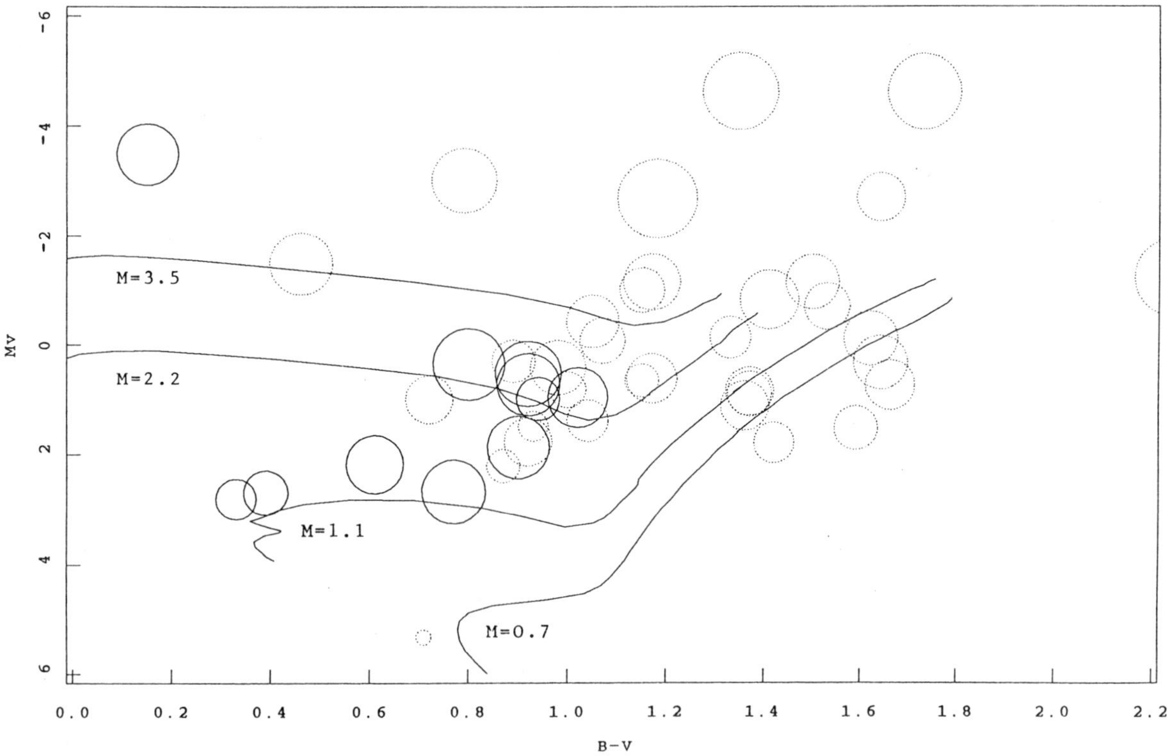

Fig. 4 - HR diagram of the single late-type giants and supergiants observed by the *Einstein* Observatory. Bubble dimension is proportional to Log(L_X), solid bubble indicates x-ray detection, while dotted bubble indicates x-ray upper bound. The solid line are the evolutionary tracks (adapted from Mengel and Sweigart 1979, and Sweigart & Gross 1978) for stars of indicated masses (in unit of solar mass). It is noteworthy the drastic reduction of the number of x-ray detections vs. that of x-ray upper bounds at B-V $\sim$ 1.0.

Haisch, B., Maggio, A. and Vaiana, G.S. 1987, in *Hot Loops, Cool Loops and the Coronal Dividing Line,* in press.

Helfand, D. J., and Caillault, J. P., 1982, *Ap. J.,* **253,** 766.

Hoffleit, D., and Jaschek, C. 1982, *The Bright Stars Catalog (4th revised edition),* (New Haven: Yale University Observatory).

Hoffleit, D., Saladyga, M., and Wlasuk, P. 1984, *A Supplement to the Bright Stars Catalog,* (New Haven: Yale University Observatory).

Ku, W., H.-M., Chanan, G. A. 1979, *Ap. J. (Letters),* **234,** L59.

Long, K., White, R. 1980, *Ap. J. (Letters),* **239,** L27.

Maggio, A., 1987, private communication

Mengel, J.G., and Sweigart, A.V. 1979, *Ap. J.,* **229,** 624.

Micela, G., Sciortino, S., Vaiana, G.S., Schmitt, J.H.M.M., Stern, R. A., Harnden, Jr., F.R., and Rosner, R. 1988, *Ap. J.,* in press.

Pallavicini, R.P., Golub, L., Rosner, R., Vaiana, G.S., Ayres, T., and Linsky, J. 1981, *Ap. J.,* **248,** 279.

Rosner, R., Golub, L., Vaiana, G.S. 1985, *Ann. Rev. Astron. Ap.,* **23,** 413.

Schmitt, J.H.M.M., Golub, L., Harnden, Jr., F.R., Maxson, C.W., Rosner, R., and Vaiana, G.S. 1985, *Ap. J.,* **290,** 307.

Seward, F., *et al.* 1979, *Ap. J. (Letters),* **234,** L55.

Sweigart, A.V., and Gross P.G. 1978, *Ap. J. (Suppl.),* **36,** 405.

Vaiana, G.S., Sciortino, S. 1987, in *Circumstellar Matter,* I. Appenzeller and C. Jordan eds., Reidel Publishing Company, Dordrecht, Holland, p. 333

Vaiana, G.S., *et al.* 1981, *Ap. J.,* **245,** 163.

Woolley, R., Epps., E.A., Penston, M.J., and Pocock, S.B. 1970, *Royal Observatory Annals,* **3**.

NASA ASTROPHYSICS DATA SYSTEM (ADS) STUDY

G. F. Squibb, California Institute of Technology
Cynthia Y. Cheung, BDM Corporation
ADS Workshop Participants

INTRODUCTION AND BACKGROUND

The forthcoming launches of the Great Observatories: the Hubble Space Telescope (HST), the Gamma Ray Observatory (GRO), the Advanced X-ray Astrophysics Facility (AXAF), and the Space Infrared Telescope Facility (SIRTF), create an opportunity for multi-spectral research in astronomy that will extend into the next century. The simultaneous operation of several large, long-lived astronomical observatories presents a number of operational as well as budgetary problems. Recognizing that both the operational and the budgetary aspects of the Great Observatories must be addressed in a timely manner, the Director of the NASA Astrophysics Division, in 1985, initiated a thorough study of the operation and data management of astrophysics missions.

The study was divided into two phases. The first phase dealt mainly with the project management practices of NASA and was completed in April, 1986. The current study, which is the second phase of the activity, was initiated in early 1987. Specifically, this study has the goal to define the functional concepts for an astrophysics data system that will serve the astrophysics community into and through the 1990s. The data system should be flexible enough to accommodate the evolution of data processing technologies, yet remain accessible to both the scientist with modest processing and analysis capabilities and to the scientist with access to major computational facilities. It also should be capable for accommodating data from future missions while providing continued access to existing data sets.

The first step in the study was to survey user requirements and existing data center facilities. These surveys were carried out through a data user's questionnaire distributed at the joint meeting of the American and Canadian Astronomical Societies in Vancouver, British Columbia, and through a data systems survey of the NASA mission data centers, respectively. The surveys were followed by the workshop held at Annapolis, Maryland, August 18-20, 1987. Members of the scientific community and data systems experts were brought together to review the existing and planned capabilities of NASA data facilities, and to participate in panel discussions on five different scientific and technical aspects of the data system: data accessiblity, data format standards and quality, data analysis and reduction software, user scenarios, and observation planning and operations.

This paper is the executive summary of the major conclusions and recommendations of the workshop. The detailed recommendations have been published in the Astrophysics Data System Workshop Report in September, 1987.

The architecture of the NASA astrophysics data system is currently being defined. It will rely heavily on the recommendations of the Annapolis workshop and on basic services and capabilities being developed by NASA to support their entire scientific community. This new data system architecture will be presented to the community at a second workshop to be held on November 16 and 17, 1987. Comments from the second workshop will be integrated into a report on the final architecture, to be published by January 1988.

NASA ASTROPHYSICS DATA SYSTEM WORKSHOP

SUMMARY AND RECOMMENDATIONS

As NASA astronomers increasingly make use of existing archives of astronomical data, undertake multi-spectral analysis of astronomical objects, and prepare for the era of the Great Observatories, the need for interactive data analysis capabilities and for ready access to distributed data sets becomes nearly universal. Thus far, most NASA-sponsored data analysis support has been specifically associated with individual NASA flight programs and their guest observer programs. Emphasis has, therefore, been placed on the establishment of centralized computing facilities or dedicated mission-specific systems. Because data analysis requirements are becoming more unified from project to project and discipline to discipline, and because astronomy research itself is becoming more dependent on the use of data from a variety of missions, this emphasis on mission-specific systems must change. NASA should recognize that certain capabilities are needed by all astronomers who perform NASA-sponsored research, regardless of their mission or discipline orientation, and should reflect this need in its mission operations and data analysis plans and projections. The Agency should also recognize that funding continuity is essential for sustained scientific productivity and understanding.

In order to prepare for this new era of astronomy data analysis, the participants at the Astrophysics Data System Workshop identified six major functional areas in which specific actions should be taken by NASA. These areas, which include availability of data, observation planning and operations, inter- and intra-agency interfaces, research support and distibuted system attributes, are discussed in more detail below. In addition, NASA is encouraged to provide mechanisms for the support of multi-spectral analysis and to be receptive to scientific proposals where this analysis is the thrust of the proposal. Such proposals might include the establishment of short lived topical centers for focussed studies, e.g. Supernova 1987a activities. Finally, the workshop endorses and supports the Space Astrophysics Data Analysis Program (SADAP) and gives some suggestions on how to improve the process and scope of the program.

A. Availability of Data

To ensure that data from NASA-sponsored space missions are fully exploited by the astronomical community, the data must be of high quality, in standard formats, easily locatable, and should be provided along with the associated standard analysis software in a well-documented and readily usable form. The participants also emphasized the need for access to data from non-NASA space missions and from ground-based observatories. The following paragraphs summarized the recommendations for improving data availability.

1. The workshop endorses the FITS format as the standard for exchange of information between astrophysics projects and urges NASA to adopt this standard within all of its Astrophysics Data System activities. Several suggestions are made in Section III. A. to assist flight projects in implementing this recommendation.

2. The need for standard analysis software (S/W) such as the Astronomical Image Processing System (AIPS) and the Image Reduction and Analysis Facility (IRAF) is supported across the panels. In addition to delivering data products, such as the Infrared Astronomical Satellite (IRAS) Point Source Catalog, projects in the future are encouraged to develop standard analysis software, delivered with their data products, to enable the immediate and rapid use of the data. An example of such software would be a simple catalog access program for the IRAS point source catalog which would satisfy a majority of the users. The need to support large S/W programs with a support staff responsible for maintenance, enhancement and distribution of the S/W, to answer user questions and assist in bringing the S/W up in the small university environmnet, was also identified.

During the development of scientific software, the close involvement of the user/scientist is strongly recommended. Experience on past and current missions has shown that the development of scientific processing software, both analysis and pipeline, is more efficient and produces better results if the software is written in-house in a NASA center or a science operations center, or a science operations center, with the scientists participating in the development effort. NASA should factor this into its program and S/W development policies. The model used in the past by projects of writing requirements, and then having an outside contractor implement a system to those requirements with little interaction with the science users, is not the recommended mode of operation.

3. There is an urgent need for a master directory of all NASA space-based observations. It is recommended that the directory should include all past observations and currently planned observations from NASA observatories, and the past and planned observations from ground-based observatories where possible. NASA and NSF should enter into discussions regarding how this can be accomplished. It is clear that in performing analysis on observations, and in planning future observations, the observational scientist has a close and necessary tie to ground-based data. To ignore this is to essentially develop half a system for the community.

Further recommendations regarding directories, which range from ways that current data catalogs can be made available in the near term (i.e., on-line via networks or on CD-ROM) to the need for archives of ground-based data and data from non-U.S. sources, may be found in section III. C.

4. The astronomical data base SIMBAD at the Observatoire Astronomique, Strasbourg, France with its on-line capability is recognized and endorsed as a required capability for the U.S. astronomy researcher. This system may be accessed by all astronomers through standard communication lines. NASA is encouraged to find a more economical means of paying for such communications costs than to have each astronomer pay the individual commercial line charges.

5. The concept of having a scientific users' group, separate from the flight project, to review the project data management plan and the project deliverables from the point of view of the archive user, is recommended.

Six separate areas are identified where current and planned activities would be significantly enhanced by the active participation of a scientific users group. These areas are:

(1) Astronomical activities of the NSSDC;

(2) Astrophysics Data System (this study effort and beyond);

(3) Definition of Project Data Deliverables;

(4) The proposed FITS Format Office at NSSDC;

(5) Standard S/W Analysis Package; and

(6) Data Archival Systems - Cataloging, etc.

B. Observation Planning and Operations

A major problem with previous missions, and a potential problem for future missions, has been that the ground system, spacecraft and instruments have not been treated as an integrated data system. There is, thus, little continuity in the equipment, computer programs, computer languages and data products used during the different stages of project development and mission operations. There is also, typically, little scientist/user involvement in the development of these data systems until the late stages of project development. In light of these facts, the Operations Panel (Panel 5) concluded that it is imperative, that the scientists responsible for science and data operations (in the case of the Great Observatories, the Science Operations Centers) should be identified and participate in a project from its inception.

The workshop report discusses in detail the concept of expert centers where the data are acquired, processed and archived for either large scale projects or by discipline or topical area. The issues discussed herein are constrained to those capabilities required to support the observation and planning aspects of a project as well as the pipeline processing of the data at these centers.

1. On-line catalogs of completed and planned observations for all NASA missions are recommended to facilitate the observation planning process. The previously recommended master directory of data sets corresponding to these observations is also important during the observation/proposal generation phase. Finally, on-line access to the HST Guide Star Catalog and to the digitized sky survey plates completes the set of NASA controlled information required.

2. The submission via networks of electronic rather than paper

proposals should also be supported by NASA. (This will also involve the development of new and up-to-date procedures for the legal process of proposal submission to NASA from the proposing institutions.) Peer review via the networks should also be a goal of the revised proposal generation and approval process.

3. The conduct of observations simultaneously from space observatories and ground observatories needs to be addressed and solutions proposed. The similar problem of simultaneous observations from two or more future NASA observatories also needs to be addressed and the solutions implemented within the next several years. A small workshop including experts in the observation and/or sequence planning field and a scientific users group might be a suitable means to start the process of identifying requirements and opportunities for such simultaneous observations.

4. Methods of providing models of instrument response to assist in the prediction of observational results should be studied. Such modeling could be used to aid the proposer in developing the best observation proposals and to assist selected investigators in the defining of their observations.

5. The responsibility for developing the prototype pipeline processing software for each science instrument should lie with the instrument builders since they must develop such software as an integral part of their instrument development and calibration activities.

C. Interface Recommendations

In planning for and conducting his or her research, the astronomer needs access to NASA-acquired space data, ground-acquired data and both ground and space data acquired from other countries. NASA is requested to pursue interfaces with both NSF, non-government research centers, and foreign research centers. These interfaces would also aid the coorindation of scientific data analysis systems. In some cases, where some seed money would assist in establishing a capability that does not now exist, it might be appropriate for NASA to assist in developing capabilities in non-NASA areas (i.e., NSF ground archives).

The ongoing and planned work of the Earth Science and Applications Data System (ESADS) activities, the Science and Applications Information System (SAIS) activity and the work supported by the Office of Space Science and Applications (OSSA) Information System Office was recognized. Where solutions to problems identified in this report are currently being studied or appropriate solutions implemented by these other groups, it is recommended that the Astrophysics Division joins with them for a common solution.

D. Support of Research

Several areas are identified where the participants feel that research is needed to establish the necessary capabilities. The areas most in need of such research is that of data bases and directories. This capability is essential to the distributed astrophysics data system concept and the remaining problems must be solved as soon as possible. It is recommended that NASA takes the necessary steps to ensure that sufficient research is carried out in the area of data bases and distributed catalog access to ensure the rapid solution of the remaining problem.

Longer term research topics that should receive research attention are those of observation planning and the possible use of artificial intelligence techniques for helping to solve this problem area. It is also recommended that NASA considers applying some resources in the area of algorithm development for maximizing the retrieval of information gathered by our space-based systems.

E. Distributed System Attributes

The workshop participants envision a distributed data system for astronomy, consisting of three components:

(1) Portable data analysis systems readily available to astronomers.
(2) Distributed data centers, containing catalogued and easily accessible mission-specific or discipline-specific astronomical data in the curatorship of active astronomers who understand and use the data and who assist outside astronomers in accessing and using the data.
(3) Networks which join the astronomy community with these data

centers and other resources, such as computational or observational facilities.

The distributed Astrophysics data system centers around the NASA Science Internet (NSI). Astronomers believe that this net has the capability to support the Astrophysics needs and that their needs are not a driver on the capabilities of the NSI. It is important that the NSI link not only the major universities with the archive sites and science operation centers, but also that the small universities where individual and small group research is performed have access to such facilities. Access to this net should be of low cost to the user and yet should not significantly impact the science research funding currently planned by the Astrophysics Division.

Standardized interface protocols are required to make efficient use of distributed processing concepts and the workshop participants endorse the TCP/IP protocol as the standard to be used, with the eventual migration to the ISO/OSI protocol when practical. This implies solving inter-operability problems between DECNET and TCP/IP in the near term and phasing the SPAN network over to TCP/IP.

The network should not only provide astronomers with easy access to international astronomical data bases and capabilities but should also link them to NSF facilities and capabilities. Thus the recommended network implies interfaces, both technical and managerial, with other government agencies (NSF) and international agencies.

Workstations in every astronomer's office will become standard during the next several years. Workshop participants urge NASA to recognize this and start taking steps to make this transition easy for the individual astronomers by evaluating new workstations and assisting the small universities with the installation of the workstations. NASA should support the porting of widely used analysis packages into these environments.

Although not used extensively in the past, the participants believe that astronomers will begin to use supercomputers in the future as an effective tool for such tasks as the deconvolution of diffraction limited images from HST. The network should provide access to such supercomputers, and the astronomy community should be involved in defining some discipline-specific requirements for the new supercomputer to be procured by NASA/GSFC.

THE PREPARATION OF HIGH QUALITY MACHINE-READABLE CLASSICAL CATALOGS FOR INCORPORATION INTO ASTRONOMICAL DATABASES

Wayne H. Warren Jr.
Astronomical Data Center
National Space Science Data Center
NASA Goddard Space Flight Center
Greenbelt, Maryland USA

ABSTRACT

The merging of data from various catalogs and their incorporation into existing databases such as SIMBAD is a demanding task, mainly because of the diverse and inconsistent astronomical nomenclature that has developed over the years. Although the cross identification of objects by computer is in principle an extremely simple procedure, numerous pitfalls have been encountered in practice because of errors and mutually exclusive identifications in existing catalogs. The incorporation of classical and new data into SIMBAD is clearly simplified if machine versions of catalogs are prepared with a view toward this problem. New versions of a number of important catalogs have been prepared and fully documented at the Astronomical Data Center, NASA Goddard Space Flight Center with consideration of both independent use and data integration. Recently completed catalogs and current projects will be reviewed in this presentation.

I. INTRODUCTION

The SIMBAD data bank is clearly a marvelous development for observational and statistical astronomy. Its design as an object oriented system has resulted in a uniqueness and versatility never before accomplished in computerized data handling in astronomy, particularly in the area of bibliographical reference retrieval, which had developed to a sophisicated level for author and subject searches, but which had never been possible in the object oriented mode generally required by astronomers.

It is quite clear from recent meetings and SIMBAD usage statistics that there is strong support for the further development of the data bank and for improvement in its software system capabilities. The most pressing need is for the incorporation of additional data into SIMBAD, particularly observations made from space vehicles and data for extragalactic objects. While the procedure of merging new data into SIMBAD would seem to be simple and straightforward once the data become available in machine-readable form, past experience has shown this not to be the case. The difficulties encountered in the cross identification of objects are responsible for most of the merging problems and the time consuming task of pre-preparing each catalog for incorpo ration has often resulted in a time requirement of months to check a catalog for errors and inconsistencies and to reformat the identifications to allow the matching of objects with those already in the data bank.

These problems can be largely overcome by the careful preparation of machine versions of data to be incorporated into SIMBAD. The treatment of object identifications will be discussed in this paper, as illustrated by new versions of classical catalogs recently produced at the ADC. It will also be emphasized that, even though the identifiers and data for many of these catalogs are already in or will be incorporated into SIMBAD, efforts should still be made to prepare high quality complete versions of all catalogs for independent archiving and distribution to the scientific community.

II. OBJECT IDENTIFICATIONS

The incorporation of new data into SIMBAD from existing machine-readable files requires the correct cross identification of objects contained within each file to their counterparts already in the data bank. For objects new to SIMBAD, standard nomenclature must be used for identification and cross indexing must be provided to other designations by which an object is known. This procedure can necessitate time consuming programming and cross indexing requirements if the objects in the new file are not identified with standard designations structured according to the formats used in the data bank. Adherence to the following guidelines will help to ensure that objects can be correctly identified with their data bank equivalents or can be entered as new objects with standard designations:

1. If a special numbering system is employed, cross indexing to well-known designations should be provided. Standard designations can be found either by using SIMBAD to retrieve some of the objects and/or by consulting the *First Dictionary of the Nomenclature of Celestial Objects* (Fernandez, Lortet, and Spite 1983) and its *First Supplement* (Lortet and Spite 1986).

2. Designations should be formatted in a standard way and should be uniform in a computer sense. For example:

   (*a*) Numerical files should be aligned and spaced according to standard SIMBAD usage:

   ```
   HD 140573
   SAO 121157
   BD +06 3088
   IRC +10..294
   ```

   (*b*) Character prefixes should be used to identify catalog designations and there should be case agreement for the characters used:

   ```
   CD
   CP
   Ci 18
   GC
   GMb
   ```

   (*c*) Signs should be included where appropriate, even though most computers do not specifically insert a plus sign if a quantity is positive. All signs should be aligned and occupy the same column in every record:

   ```
   15 41 48.151+06 34 53.93 +0.75 -1.247
   00 01 02.472-00 26 31.23 -0.62 +0.004
   09 00 00.026+00 16 42.87  0.00 +2.744
   ```

3. An established hierarchy of designations should be followed; *e.g.*, if a star has an HD number, it should be present. If special numbering systems are used, *e.g.*, for cluster stars, HD or Durchmusterung (DM) designations should be included for objects that have them. Cross indexing can be realized with SIMBAD or by making use of such catalogs as those contained in category IV of the CDS list of available catalogs (category 4 of the *ADC Status Report on Machine-Readable Catalogs*).

4. Cross indexing should be as complete as possible for the designations included, *i.e.*, partial cross indexing should be avoided because a user can be misled about the existence of a particular major designation for an object.

5. Each major designation should occupy its own field if possible. For example, stars may be designated by HR, HD, and/or DM numbers if these exist, or by miscellaneous designations if they do not; however, there should be at least one or two data fields reserved for specific designations and an additional field for miscellaneous identifiers. On the other hand, if there is insufficient space for multiple identifiers, an established hierarchy should be employed and prefix identifers should be used. This is illustrated by the two examples following:

(*a*) Uniform table of identifications:

HD	DM	Name	Data
28	BD-06 6357	33 Psc	
10697	BD+19 282	109 Psc	
177716	CD-2713564	Tau Sgr	
197635	CP-69 3138	Sig Pav	

(*b*) Single designation with identifier prefix and suffix:

Identification	Data
HD 28	
BD+04 3561a	
CD-27 13564	
CP-69 3138B	

(*c*) Uniform designation field with separate parts for alphanumeric (prefix), numeric, and suffix identifiers:

Identification				Data
HD			46264A	
BD		-06	2124	
BD		+04	3561a	
CYG	OB 2		9	
LS	III	+62	31	

III. THE PREPARATION OF EXISTING CATALOGS FOR ARCHIVING AND DATABASES

Most of the classical catalogs, such as the *Durchmusterungen*, *The Henry Draper Catalogue*, the *Astronomische Gesellschaft Katalog 3* (AGK3), the *Smithsonian Astrophysical Observatory Star Catalog* (SAO), and even parts of the *Astrographic Catalogue*, are available in machine-readable form. Some of these catalogs have been improved by the correction of known errors and the addition of cross identifications to other well-known catalogs. Although many classical catalogs have been incorporated into SIMBAD and are available in their original form for distribution to the scientific community, most have not yet become completely stable because errors are constantly being discovered. These catalogs should continue to be improved independently by the periodic incorporation of corrections. The original data should not be replaced by more modern values because the newer data are available in SIMBAD and the availability of the original catalogs is still of value. All changes and corrections to the original catalogs should be documented and cumulative lists of changes to the original published catalogs should be available in the documentation for each new edition.

Many existing catalogs contain duplicate entries because the same objects were cataloged more than once during a long project. Duplicate objects should be effectively deleted from newer machine versions, but the original records should be retained and flagged. The retention of deleted records allows their reactivation if new evidence indicates that the wrong entries were deleted. Although it is not always clear which of a set of duplicates should be deleted, judgments can usually be made on the basis of source data and agreement with more modern determinations.

The above precepts can be illustrated by recent work on several classical catalogs redone at the ADC. The following summaries are provided in more detail in full documentation that has been prepared and is distributed with every machine-readable copy of a catalog.

1. *Durchmusterung* (DM) *Catalogs*

There are four DM catalogs that have been produced in machine-readable form through an international collaborative effort. These include the *Bonner Durchmusterung* and its *Southern Durchmusterung* extension, the *Córdoba Durchmusterung*, and the *Cape Photographic Durchmusterung*. The format has been kept uniform thoughout, even though two extra bytes had to be allocated to all catalogs to accommodate greater precision declination data in the southern zones of the CPD. Catalog prefixes have been used throughout so that the data can be merged if desired and individual catalogs still maintain their identities. All errata and corrigenda in the published catalogs have been incorporated along with corrections determined during the course of the work. Published and unpublished DM-cataloged stars found to be missing from the sky have been flagged so that they need not be plotted on charts, but the identifications and data are left intact.

2. *Smithsonian Astrophysical Observatory Star Catalog* (SAO)

The cross indexing of SAO stars in the original machine version was complicated by the fact that there were multiple SAO stars with the same DM numbers. These were analyzed individually and component identifications from the Lick *Index Catalogue of Visual Double Stars* were assigned. Component identifications were also assigned to southern stars for which different DM numbers (CD and CPD) had been assigned to components of double and multiple stars. Supplemental DM identifiers (lower case letters) were attached where appropriate and DM numbers were determined for stars not having them (FK3 double stars omitted from the FK4). DM numbers were also determined for stars in the AG zones +50° to +59° where only AG numbers are given in *The Henry Draper Extension*; HDE numbers were then determined for these stars and GC numbers were subsequently assigned.

All known errata were corrected and duplicate entries were flagged after analyzing each redundancy and selecting an entry to be retained, usually based on the estimated accuracy of source positions.

3. *The Henry Draper Catalogue* (HD)

Individual known errata were corrected and duplicate entries were flagged but not deleted.

DM supplemental identifiers were added and all DM identifiers were expanded to full representation with prefixes in order to remove special coding that had been present. DM numbers from southern zones were added where they were missing because of the HD convention used for DM zone catalogs.

Rectangular coordinates for AG zones +50° to +59° and for the Magellanic Clouds were converted to equatorial coordinates for stars in HDE 100 and the format for the entire catalog was made uniform.

4. *Morphological Catalogue of Galaxies* (MCG)

The entire machine-readable catalog was broken down into the individual regions of the published catalog after numerous errors, omissions, and duplicate entries were discovered. Region-by-region counts were made in the published and machine versions and it was found that entire pages in the published version had been missed and large numbers of objects in some positive zones had been placed into their corresponding negative zones because of sign errors.

Component galaxy designations were added and changed to match the published catalog. The format was then modified for uniformity and storage efficiency and the entire catalog was resorted by MCG number. All individual changes are listed in the catalog documentation.

IV. FUTURE ENHANCEMENTS TO CLASSICAL CATALOGS

1. DM Catalogs

An analysis of the southern catalogs (CD, CPD) will be made in an attempt to determine missing objects and erroneous positions, particularly in the CD. A full cross index between the catalogs has been prepared in collaboration with B. N. Rappaport, who did the machine cross indexing. Many individual problem cases have yet to be resolved.

2. SAO

Equatorial coordinates and proper motions on J2000 and the FK5 reference frame will be computed and added to the machine version of the catalog after publication of the FK5. Errata discovered since the last version will be incorporated into the new version.

3. HD and HDE

The HD spectral types for stars in the second extension have been entered into a file at the ADC and positions have been determined by workers at L'Observatoire de Strasbourg. A collaborative project is underway to combine these data, prepare a uniformly formatted version of HDE **112**, and incorporate the data into the present machine version to produce a complete catalog in machine-readable form.

4. MCG

The present machine version contains only the first four volumes of the published catalog; however, the fifth volume has recently been computerized by our colleagues at the Abastumani Astrophysical Observatory under the direction of Dr. N. G. Kogoshvili and at the Soviet Center for Astronomical Data in Moscow under the direction of Dr. O. B. Dluzhnevskaya. This volume will be incorporated into the existing machine catalog to produce a complete MCG.

REFERENCES

Fernandez, A., Lortet, M.-C., and Spite, F. 1983, *Astron. Astrophys. Suppl.* **52**, No. 4.

Lortet, M.-C. and Spite, F. 1986, *Astron. Astrophys. Suppl.* **64**, 329.

Author Index

LIST OF PARTICIPANTS

M. ALBRECHT
ESRIN
via Galileo Galilei
C.P. 64
I-00044 Frascati
Italy

F. ARENOU
Observatoire de Meudon - DASGAL
Place J. Janssen
F-92195 Meudon Principal Cedex
France

J. BANFIELD
Department of Statistics, GN-22
University of Washington
Seattle, WA 98195
USA

M. BARYLAK
ESA Villafranca
Apartado 54065
E-28080 Madrid
Spain

R.K. BHATIA
International Centre for Theoretical Physics
P.O. Box 586
I-34100 Trieste
Italy

A. BIJAOUI
Observatoire de Nice
B.P. 139
F-06003 Nice Cedex
France

F. BONNAREL
Max-Plack-Institut für Astronomie
Königstuhl
D-6900 Heidelberg
FRG

M.L. BOUGEARD
Observatoire de Paris
UA 1125 CNRS
61 Av. de l'Observatoire
F-75014 Paris
and
Université Paris X
Dep. Statistiques, I.U.T.
1 Chemin Desvallières
F-92410 Ville d'Avray
France

E.W. BRUGEL
University of Colorado
CASA Campus Box 391
Boulder, Co. 80309
USA

M. BRUYNOOGHE
99, rue Frederic Mistral
F-84120 Pertuis
France

R. BURG
Space Telescope Science Institute
3700 San Martin Drive
Baltimore, MD 21218
USA

E.K. BUSCH
Information Access Systems, Inc.
3340 Mitchell L. #200
Boulder, CO 80301
USA

L.M. BUSON
Osservatorio Astrofisico di Asiago
I-36012 Asiago (VI)
Italy

F. CASTELLI
Osservatorio Astronomico
Via G.B. Tiepolo 11
I-34131 Trieste
Italy

S. CHAKRABARTI
Space Sciences Laboratory
University of California
Berkeley, CA 94720
USA

L. CHIAPPETTI
Istituto di Fisica Cosmica del CNR
via Bassini 15
I-20133 Milano
Italy

R.G. CLOWES
Royal Observatory
Blackford Hill
Edinburgh, EH9 3HJ
United Kingdom

D.R. CRABTREE
Dominion Astrophysical Observatory
Canadian Space Astronomy Data Center
5071 W. Saanich Rd.
Victoria, B.C.
Canada V8X 4M6

R.G. CRUDDACE
Naval Research Laboratory
Washington D.C. 20375
USA

E. DAVOUST
Observatoire de Toulouse
14, Avenue E. Belin
F-31400 Toulouse
France

V. DI GESÙ
University of Palermo
Institute of Cosmic Physics and Informatics CNR
Via Mariano Stabile
I-90130 Palermo
Italy

P. DUBOIS
Observatoire de Strasbourg
11, rue de l'Université
F-67000 Strasbourg
France

D. DURAND
Canadian Space Astronomy Data Center
5071 W. Saanich Rd.
Victoria, B.C.
Canada V8X 4M6

D. EGRET
Observatoire de Strasbourg
11, rue de l'Université
F-67000 Strasbourg
France

V. GERICKE
Astronomisches Institut Münster
Wilhelm-Klemm-Str. 10
D-4400 Münster
FRG

P. GIOMMI
EXOSAT Observatory
ESTEC / SAE
Postbus 299
NL-2200 AG Noordwijk
The Netherlands

G. GUERRERO
Osservatorio Astronomico di Milano
Via E. Bianchi 46
I-222055 Merate (Como)
Italy

J. GUIBERT
I.N.S.U.
77, avenue Denfert-Rochereau
F-75014 Paris
France

G.R. HASINGER
Max-Planck-Institut für Extraterrestrische Physik
Karl-Schwarzschild-Str. 2
D-8046 Garching
FRG

S.R. HEAP
Code 681
NASA, Goddard Space Flight Center
Greenbelt, MD 20771
USA

A. HECK
Observatoire de Strasbourg
11, rue de l'Université
F-67000 Strasbourg
France

H.M. HECKATHORN
U.S. Naval Research Laboratory
Ultraviolet Astronomy Section
Space Science Division, Code 4109
Washington, D.C. 20375
USA

G. HELOU
IPAC
California Institute of Technology
M.S. 100-22
Pasadena, CA 91125
USA

L.A. HIGGS
Dominion Radio Astrophysical Observatory
P.O. Box 248
Penticton, B.C.
Canada V2A 6K3

H. JAHREISS
Astronomisches Rechen-Institut
Mönchhofstr. 12-14
D-6900 Heidelberg
FRG

C. JASCHEK
Observatoire de Strasbourg
11, rue de l'Univeristé
F-67000 Strasbourg
France

M. JASCHEK
Observatoire de Strasbourg
11, rue de l'Univeristé
F-67000 Strasbourg
France

H. JENKNER
Space Telescope Science Institute
3700 San Martin Drive
Baltimore, MD 21218
USA

M. JOHNSTON
Space Telescope Science Institute
3700 San Martin Drive
Baltimore, MD 21218
USA

J.R. KENNEDY
NOAO / NSO
950 N. Cherry Avenue
Tucson, AZ 85718
USA

G. KIERS
D. Reidel Publishing Co.
P.O. Box 17
NL-3300 AA Dordrecht
The Netherlands

D.G. KOCH
Smithsonian Astrophysical Observatory
Mail Stop 65
60 Garden Street
Cambridge, MA 02138
USA

J. KRAUTTER
Landessternwarte
Königstuhl
D-6900 Heidelberg
FRG

M.J. KURTZ
Image Processing Laboratory
Center for Astrophysics
60 Garden Street
Cambridge, Massachusetts 02138
USA

J. LASALA
Physics Department
University of Southern Maine
96 Falmouth Street
Portland, ME 04103
USA

H. LAURILA
Helsinki University of Technology
Metsähovi Radio Research Station
SF-02540 Kylmälä
Finland

A. LLEBARIA
Laboratoire d'Astronomie Spatiale
Traverse du Syphon (Les 3 Lucs)
F-13012 Marseille
France

S. LUBOW
Space Telescope Science Institute
3700 San Martin Drive
Baltimore, MD 21218
USA

G. LYNGÅ
Lund Observatory
Box 43
S-221 00 Lund
Sweden

M.C. MACCARONE
Ist. Fisica Cosmica e Applicazioni dell'Informatica,
C.N.R.
Via Marino Stabile 172
I-90139 Palermo
Italy

C. MACERONI
Osservatorio Astronomico di Roma
Viale del Parco Mellini 84
I-00136 Roma
Italy

H.T. MACGILLIVRAY
Royal Observatory
Blackford Hill
Edinburgh, EH9 3HJ
United Kingdom

M.L. MALAGNINI
Università degli Studi di Trieste
Dipartimento di Astronomia
Via G.B. Tiepolo, 11
I-34131 Trieste
Italy

H. MARSHALL
Space Science Lab.
University of California
Berkeley, CA 94720
USA

B. MCLEAN
Space Telescope Science Institute
3700 San Martin Drive
Baltimore, MD 21218
USA

J.L. MEAD
NASA Goddard Space Flight Center
Code 630
Greenbelt, MD 20771
USA

J.-C. MERMILLIOD
Institut d'Astronomie
Université de Lausanne
CH-1290 Chavannes-des-Bois
Switzerland

G. MILLER
Space Telescope Science Institute
3700 San Martin Drive
Baltimore, MD 21218
USA

D. MORIN
Observatoire de Meudon - DASGAL
Place Janssen
F-92195 Meudon Principal Cedex
France

M.B. NEGRI
c/o CNR/PSN (National Space Plan)
Viale Regina Margherita 202
I-00198 Roma
Italy

C.G. PAGE
Physics Department
Leicester University
Leicester LE1 7RH
United Kingdom

R. PAPOULAR
Service d'Astrophysique
CEN SACLAY
F-91191 Gif-sur-Yvette Cedex
France

A. PARMAR
EXOSAT Observatory
ESTEC / SAE
Postbus 299
NL-2200 AG Noordwijk
The Netherlands

F. PASIAN
Osservatorio Astronomico di Trieste
via G.B. Tiepolo 11
I-34131 Trieste
Italy

L.E. PASINETTI FRACASSINI
Dipartimento di Fisica
Università di Milano
Via G. Celoria 16
I-20133 Milano
Italy

L. PASTORI
Osservatorio di Milano
via E. Bianchi 46
I-22055 Merate (Co)
Italy

G. PATUREL
Observatoire de Lyon
F-69230 Saint-Genis-Laval
France

J. PINTAR
National Solar Observatory
Post Office Box 26732
Tucson, AZ 85726
USA

I. PORCEDDU
Università di Cagliari
Osservatorio Astronomico
Via Ospedale 72
I-09100 Cagliari
Italy

M. PUCILLO
Osservatorio Astronomico
Cas. Post. Succ. 5
I-34131 Trieste
Italy

J.P. PYE
Physics Department
Leicester University
Leicester LE1 7RH
United Kingdom

E. RAIMOND
Netherlands Foundation for Radio Astronomy
(NFRA)
Postbus 2
NL-7990 AA Dwingeloo
The Netherlands

H. RALAMBONDRAINY
INRIA
Domaine de Voluceau
Rocquencourt
B.P. 105
F-78153 Le Chesnay Cedex
France

B.N. RAPPAPORT
New Mexico State University
Department of Astronomy
Box 4500
Las Cruces, NM 88003
USA

J.M. REY-WATSON
Smithsonian Astrophysical Observatory
60 Garden Street
Cambridge, MA 02138
USA

P.J. RICHARDS
Rutherford Appleton Laboratory
Chilton, Didcot
Oxon OX11 0QX
United Kingdom

O.G. RICHTER
Space Telescope Science Institute
3700 San Martin Drive
Baltimore, MD 21218
USA

N.R. ROBBA
Ist. Fisica
Università di Palermo
Via Archirafi N. 36
I-90123 Palermo
Italy

D. ROSENTHAL
NASA Ames Research Center
M/S 244-17
Moffett Field, CA 94035
USA

L. RUSCONI
Università degli Studi -
Dipartimento di Astronomia
via G.B. Tiepolo 11
I-34131 Trieste
Italy

P. SANTIN
Osservatorio Astronomico di Trieste
via G.B. Tiepolo 11
I-34131 Trieste
Italy

D.R. SCALES
Astronomisches Institut der Universität Tübingen
Waldhäuserstr. 64
D-7400 Tübingen
FRG

S. SCIORTINO
Osservatorio Astronomico di Palermo
Palazzo dei Normanni
I-90134 Palermo
Italy

G. SEDMAK
Osservatorio Astronomico di Trieste
via G.B. Tiepolo 11
I-34131 Trieste
Italy

G. SHAW
Information Access Systems, Inc.
3340 Mitchell L. #200
Boulder, CO 80301
USA

J. SPICKER
Ruhr-Universität Bochum
Astronomisches Institut
Postfach 10 21 48
D-4630 Bochum 1
FRG

G.F. SQUIBB
California Institute of Technology
Mail Code - IPAC 100-22
Pasadena, CA 91125
USA

J.R. STERNBERG
ESOC
Robert-Bosch-Str. 5
6100 Darmstadt
FRG

J.A. STÜWE
Ruhr-Universität Bochum
Astronomisches Institut
Postfach 10 21 48
D-4630 Bochum
FRG

D. TEUBER
Astronomisches Institut
Universität Münster
Wilhelm-Klemm-Str. 10
D-4400 Münster
FRG

D. TREVESE
Istituto Astronomico
Università di Roma
Via Lancisi 29
I-00161 Roma
Italy

M. TRIFOGLIO
CNR Istituto TESRE
Via de'Castagnoli, 1
Bologna
Italy

C. TURON
Observatoire de Paris - Section de Meudon
F-92195 Meudon Cedex
France

E.A. VALENTIJN
Kapteyn Laboratorium
P.O. Box 800
NL-9700 AV Groningen
The Netherlands

G. VIRAGHALMY
Konkoly Observatory
Box 67
1525 Budapest XII
Hungary

W. VOGES
Max-Planck-Institut für Extraterrestrische Physik
Karl-Schwarzschild-Str. 2
D-8046 Garching
FRG

S. WAGNER
Landessternwarte
Königstuhl 1012
6900 Heidelberg
FRG

R. WALTER
Observatoire de Genève
CH-1290 Sauverny
Switzerland

W.H. WARREN Jr.
NASA Goddard Space Flight Center
Code 633
Greenbelt, MD 20771
USA

M. WENGER
Observatoire de Strasbourg
11, rue de l'Université
F-67000 Strasbourg
France

P.R. WESSELIUS
Space Research
P.O. Box 800
NL-9700 AV Groningen
The Netherlands

G. WESTERHOUT
U.S. Naval Observatory
Washington, D.C. 20392
USA

N. WHITE
EXOSAT Observatory
ESTEC / SAE
Postbus 299
NL-2200 AG Noordwijk
The Netherlands

W. WISNIEWSKI
University of Arizona
Space Science Bld.
Tucson, AZ 85721
USA

U. ZIMMERMANN
Max-Planck-Institut für Extraterrestrische Physik
Karl-Schwarzschild-Str. 2
D-8046 Garching
FRG

PARTICIPANTS FROM:

- ESO GARCHING

K. BANSE

S. D'ODORICO

D. ENARD

E. GIRAUD

P. GROSBØL

L. GUZZO

A. LAUBERTS

J. MELNICK

E. MEURS

A. MOORWOOD

F. OCHSENBEIN

D. PONZ

M. TAPIA

M.-H. ULRICH

R. WARMELS

- ST-ECF GARCHING

H.-M. ADORF

R. ALBRECHT

P. BENVENUTI

R.A.E. FOSBURY

F. MURTAGH

R. RAMPAZZO

A. RICHMOND

G. RUSSO

S. DI SEREGO ALIGHIERI